T0178153

Food Science Text Series

The *Food Science Text Series* provides faculty with the leading teaching tools. The Editorial Board has outlined the most appropriate and complete content for each food science course in a typical food science program and has identified textbooks of the highest quality, written by the leading food science educators.

More information about this series at http://www.springer.com/series/5999

Norman G. Marriott · M. Wes Schilling
Robert B. Gravani

Principles of Food Sanitation

Sixth Edition

 Springer

Norman G. Marriott
Virginia Polytechnic Institute
State University
Blacksburg, Virginia, USA

Robert B. Gravani
Department of Food Science
Cornell University
Ithaca, New York, USA

M. Wes Schilling
Department of Food Science
Mississippi State University
Mississippi State, Mississippi, USA

ISSN 1572-0330   ISSN 2214-7799   (electronic)
Food Science Text Series
ISBN 978-3-030-09792-9   ISBN 978-3-319-67166-6   (eBook)
https://doi.org/10.1007/978-3-319-67166-6

Printed on acid-free paper

This Springer imprint is published by the registered company Springer International Publishing
AG part of Springer Nature.
The registered company address is: Gewerbestrasse 11, 6330 Cham, Switzerland

Preface

Effective sanitation is mandatory for the attainment of a safe food supply. The continued interest in and consumer demand for food safety and security and high-volume food processing and preparation operations have increased the need for improved sanitary practices from processing to consumption. This trend presents a challenge for the food processing and food preparation industry to adopt rigid sanitation practices.

Sanitation is an applied science that involves the attainment of hygienic conditions. Because of the emphasis on food safety, sanitation has increased in importance to the food industry. In the past, sanitation workers, including sanitation program managers, were inexperienced employees with limited skills who have received little or no training and have had only limited exposure to this important function. However, it is crucial that sanitation employees have knowledge about the attainment of hygienic conditions. Technical information has been limited primarily to training manuals provided by regulatory agencies, industry and association manuals, and recommendations from equipment and cleaning compound organizations. A large percentage of this material lacks specific information about the selection of appropriate cleaning methods, cleaning equipment, cleaning compounds, sanitizers, pest control, and waste disposal for maintaining hygienic conditions in food processing and preparation facilities.

The purpose of this book, as with previous editions, is to provide sanitation information needed to ensure hygienic practices and a safe food supply. Sanitation is a broad and somewhat complex subject; thus, this text addresses principles related to contamination, cleaning compounds, sanitizers, cleaning equipment, allergen control, and pest control, as well as specific directions for applying these concepts to attain hygienic conditions in food processing and food preparation operations.

The discussion of this treatise begins with the importance of sanitation with information about regulations, including the Food Safety Modernization Act (FSMA). Increased concerns about biosecurity necessitated the need for an update and expansion of Chap. 2, which addresses this subject. To enable the reader to understand more fully the fundamentals of food sanitation, an updated Chap. 3 is devoted to microorganisms and their effects on food products. This chapter contains additional information about pathogenic microorganisms and rapid microbial determination methods. The ubiquity of allergens and concern of those affected suggest the need to update and expand Chap. 4. A discussion of contamination sources and hygiene has been rewritten and updated (Chaps. 5 and 6), including how management can encourage improved sanitation. Chapter 7 provides an updated discussion on Hazard Analysis Critical Control Points (HACCP) including additional information about hazard analysis and electronic HACCP.

Chapter 8 is about quality assurance (QA) and sanitation. The updated information given here presents specific details on how to organize, implement, and monitor an effective sanitation program.

Chapter 9 discusses cleaning compounds and contains updated information on this subject. It examines characteristics of soil deposits and identifies the appropriate generic cleaning compounds for the removal of various soils. Furthermore, it relates how cleaning compounds function, identifies their chemical and physical properties, and offers information on their appropriate handling. Because of the importance of sanitizing, Chap. 10 discusses updated information about sanitizers and their characteristics. This chapter discusses specific generic compounds for various equipment and areas, as well as updated information on such compounds.

Chapter 11 provides rewritten and updated information on cleaning and sanitizing equipment most effective for various applications in the food industry. It provides detailed descriptions, including new illustrations of most cleaning equipment found in food processing and food preparation facilities.

Chapter 12 discusses waste product handling, which remains a major challenge for the food industry. This chapter contains updated information about the treatment and monitoring of liquid and solid wastes. Pest control is another problem for the food industry. Chapter 13 provides an updated discussion about common pests found in the food industry; their prevention, including chemical poisoning; integrated pest management (IPM) and biological control; and the potential advantages and limitations of each method. Chapter 14 contains a discussion of additional hygienic design and construction information for food establishments.

Because sanitation is so important in dairy, meat and poultry, seafood, fruit and vegetable, and beverage plants, a chapter is devoted to each of these areas. Chapters 15, 16, 17, 18, 19, and 20, which were rewritten, present updated information about plant construction, cleaning compounds, sanitizers, and cleaning equipment that applies to those segments of the industry. These chapters provide the food industry with valuable guidelines for sanitation operations and specific cleaning procedures.

Chapter 21 addresses existing and updated sanitation information for the foodservice industry. It provides instructions on how to clean specific areas and major equipment found in a foodservice operation.

Effective management practices can promote improved sanitation, a topic addressed in Chap. 22. The intent is not to provide an extensive discussion of management principles but to suggest how effective management practices can improve sanitation.

This book provides an updated and concise discussion about sanitation for low-, intermediate-, and high-moisture foods. It offers value as a text for college students and continuing education courses about sanitation. It will serve as a reference for food processing courses, industry-sponsored courses, and the food industry itself. The authors acknowledge and sincerely thank all of the companies, organizations, and agencies that have provided photos, illustrations, and information that has been used in this book.

The authors wish to express their appreciation to those organizations that provided figures that gave further insight to the information discussed. Furthermore, we acknowledge the support and patience of our families during the preparation of this revised edition.

Blacksburg, Virginia, USA   Norman G. Marriott
Mississippi, MS, USA   M. Wes Schilling
Ithaca, NY, USA   Robert B. Gravani

Contents

List of Figures

List of Tables

Sanitation and the Food Industry

1

Abstract

Large-volume food processing, retail, and preparation operations have increased the need for sanitary practices and hygienic conditions in the food industry. Even in hygienically designed plants, foods can be contaminated with spoilage microorganisms or those causing foodborne illness if proper sanitary practices are not properly and routinely followed.

Sanitation is the creation and maintenance of hygienic and healthful conditions. It is an applied science that incorporates principles regarding the design, development, implementation, and maintenance of hygienic practices and conditions. Sanitation is also considered to be a foundation for food safety assurance systems.

The Food, Drug, and Cosmetic Act covers food commodities except meat and poultry products from harvest through processing and distribution channels. Meat and poultry products are under the jurisdiction of the U.S. Department of Agriculture (USDA). Good Manufacturing Practices (GMPs) regulations are specific requirements developed to establish minimum criteria for sanitation practices. A number of statutes related to pollution control of the air, water, and other resources are enforced through the Environmental Protection Agency (EPA).

The progressive company, including food processors, food retailers, and foodservice operators, should take responsibility for establishing and maintaining sanitary practices. An effective sanitation program is the foundation of a food safety assurance system that is essential in reducing the risk of biological, chemical, and physical hazards, meeting regulatory requirements; protecting brand, image, and product reputation; and ensuring product safety and quality.

Keywords

Food processing • Food system • HACCP • Laws • Pathogens • Regulations

© Springer International Publishing AG, part of Springer Nature 2018
N.G. Marriott et al., *Principles of Food Sanitation*, Food Science Text Series,
https://doi.org/10.1007/978-3-319-67166-6_1

Introduction

The food system is the technical, social, and economic structure that supplies food to the population. It is a complex, dynamic, and international chain of activities that begins with production and harvesting of raw agricultural commodities on farms, ranches and in fishing operations and moves to value-added processed and preserved products and then to retail food stores and foodservice establishments (restaurants and institutions) where these foods are prepared, merchandised and sold to consumers. In addition, food is also sold in a variety of places including farm stands, farmers' markets, convenience stores, gas stations, pharmacies, and food trucks and even delivered to homes. This system is influenced by a myriad of public policy concerns including political and economic issues, sociocultural issues, environmental issues, transportation issues, as well as food sanitation and safety issues. To fully comprehend the role of sanitation and food safety in the food industry, it is important to understand the size, scope, and uniqueness of each sector of the food system.

Production Agriculture

Agricultural production is the beginning of the food chain and provides fruits and vegetables for direct consumption and the raw materials for use in further processed and preserved products. Today, there are about 2.1 million farms in the United States (USDA 2012) employing more than 2.6 million workers (USDA 2017a). Today's farmer produces 262% more food with 2% fewer inputs than in 1950, and average farmer produces enough food each year to feed 168 people worldwide (American Farm Bureau Federation 2015). Even though the number of farms is decreasing, increased productivity, arising from innovation and changes in science and technology, has resulted in a wide variety of foods being made available to US consumers. Proportionately, less is spent on food in the United States (9.8% of disposable income in 2014) than by consumers in other countries around the world (USDA 2017b).

Although the structure of production agriculture and farming practices has changed dramatically over the years, the results have been a larger, less expensive, more diverse, and safer food supply.

Food Processing and Manufacturing

Food and beverage processing facilities transform raw agricultural materials into intermediate food stuffs, ingredients, and/or edible products. In the United States, there are about 31,000 food processing plants owned by 22,000 companies. These plants employ 1.8 million workers with meat and poultry plants employing the largest percentage of workers, followed by bakeries and fruit and vegetable processing plants (USDA 2017c). Food manufacturing accounts for 14% of all US manufacturing employees (USDA 2017c). In recent years, the food processing industry has become more consolidated than ever before with mergers and acquisitions. From 2010–2015, there were 2,238 mergers and acquisitions in the food industry (The Food Institute 2016). To continue attracting customers and increase sales, profits, and market share, food processors are restructuring and expanding opportunities, reducing costs, and developing new, unique, value-added products. In 2015, there were about 17,143 new food and beverage products developed in the United States (Mintel GNPD 2017). The major focus of this new product development was on nutritious, refrigerated, and fresh foods, gourmet products, and convenient foods, and this trend appears to continue with food manufacturers appealing to health-conscious, environmentally concerned, and on-the-go consumers. The sanitation, safety, quality and labeling of foods continue to be an important consumer concern.

Food Retailing

In the last two decades, the US retail food industry has experienced significant changes as consolidation and structural changes through mergers, acquisitions, internal growth, and new competitors (drugstores and nontraditional food stores) have occurred. In 2015, there were 76 mergers among food retailers and 36 mergers of convenience stores (The Food Institute 2016). According to a Progressive Grocer Report (2016), there are over 228,000 food stores in the United States with different formats including conventional supermarkets, supercenters, warehouse club stores, convenience stores, gas stations, and other nontraditional places where food is sold. Retail food stores employ over 3.4 million workers throughout the United States and stock about 40,000 food items that provides consumers with a wide variety of products (Food Marketing Institute 2017).

To meet demands and increase customer satisfaction, food retailers have expanded their offerings of fresh prepared and convenience foods and provided other unique food items for sale. Expanded fresh produce and seafood departments, ethnic foods, international foods, salad and hot bars, in-store cafes, and restaurants have attracted health conscious consumers who have an increased need for convenience. Consumers select their primary food store for many reasons, but top among them are fresh and healthy foods, store cleanliness, convenience, and store maintenance. Store cleanliness and food sanitation clearly play a very important role in supermarket choice.

Foodservice (Restaurants and Institutions)

In 2017, there were more than one million restaurants in the United States that provide employment for approximately 14.7 million people (about one in ten working Americans) (NRA 2017a). The restaurant industry is the nation's second largest private sector employer (NRA 2017b). According to the USDA, in 2014, food purchases away from home accounted for about 50.1% of the total amount that people spent on food (USDA 2017b). Since the 1970s, a number of factors have caused this trend of greater eating away from home, including more women working outside of the home, more two income-earner households, and higher-income and smaller households (USDA 2017d). Most eating and

drinking establishments are small businesses, with approximately 90% of restaurants employing fewer than 50 employees, while seven in ten are single-unit operations (NRA 2017a). In 2015, foodservice sales worldwide grew 5.7% more than the previous year (Statistica 2017). Prepared, home-delivered meals, as well as meal kits (meals in a box), where food and ingredients are precisely proportioned, packaged, and delivered to the home for preparation, are becoming increasingly popular with young professionals. Food trucks that often offer quick, convenient, and unique foods are also becoming popular. The restaurant industry will face some challenges in the years ahead including legislative and regulatory pressures, increased labor costs, and cybersecurity issues. In addition, the retention and recruitment of employees will be a top priority, as a tighter labor market means greater competition with other industries for employees. Changes in the workforce indicate a shift to older workers, as the pool of young labor shrinks (NRA 2016). All of these factors have implications for food safety, sanitation, and employee training and retention.

Consumers

Demographic changes have resulted in an unprecedented shift in the size and structure of the US population. Today, there are about 325 million people in the United States, with approximately 2.2 million people added since 2016 (US Census Bureau 2016 and 2017a). The population is also aging. As baby boomers (those born between 1946 and 1964) reach retirement age, the proportion of the elderly population (65 years old and older) is projected to more than double from 2012 to 2050, when 21% of the population will be 65 or older (Orman et al. 2014). The US population is becoming more diverse with Hispanics being the nation's largest minority (US Census Bureau 2017b). More women are working and postponing marriage and childbirth or not having children at all. Today, women comprise approximately 47% of the total US workforce (US Department of Labor 2010). In 2014,

US consumers spent almost $1.5 trillion on food, and 50.1% of this was spent on food away from home (USDA 2017b). As mentioned previously, Americans spent 9.8% of their 2014 disposable income on food. This is the smallest percentage of disposable income spent on food anywhere in the world. With the increasing diverse population and more adventurous eating patterns of Americans, consumers are seeking a wider variety of high-quality ethnic and international foods. Today, foods imported into the United States make up about 19% of food eaten by Americans (USDA 2016).

These dynamic and significant changes in all sectors of the food system highlight the importance of food safety and sanitation in ensuring a safe and wholesome food supply. Each sector needs to work in partnership and collaboration to assure a seamless food safety system.

As the food industry has become larger, more concentrated and diversified, and international in scope and as new hazards have emerged to cause concern, food safety and sanitary practices have taken on a new importance in protecting public health. Many companies and organizations are aggressively addressing food safety issues in their facilities to prevent biological, chemical, and physical hazards from causing illness and injury to consumers. These issues have increased the need for food workers to understand the critical importance of food safety and sanitary practices and how to attain and maintain hygienic practices in all food facilities from the packing house on the farm and in fishing boats to supermarkets, restaurants, and institutions, as well as households. Those who truly understand the biological basis behind these practices, the reasons why they need to be performed properly and regularly, will be more effective in assuring the safety of products that they grow, catch, manufacture, prepare, and sell.

What Is Sanitation?

The word *sanitation* is derived from the Latin word *sanitas*, which means "health." When applied to the food industry, sanitation is "the creation and maintenance of hygienic and healthful conditions." It is the application of science to:

- Provide wholesome food that is stored, processed, transported, prepared, merchandised, and sold in a clean environment by healthy workers.
- Prevent contamination by biological, chemical, or physical hazards that cause foodborne illness or injury.
- Minimize the proliferation of food spoilage microorganisms.

Effective sanitation refers to all procedures and protocols that help accomplish these important goals.

Sanitation: An Applied Science

Sanitation is an applied science that incorporates the principles of design, development, implementation, maintenance, restoration, and/or improvement of hygienic practices and conditions. The application of sanitation refers to practices designed to maintain a clean and wholesome environment for food production, processing, preparation, and storage. Sanitation is more than just cleanliness and, when done properly, can improve the aesthetic qualities and conditions of commercial operations, public facilities, and homes. Sanitation can also reduce waste and improve waste disposal (Chap. 12), which results in less pollution and improved sustainability. When properly applied, food sanitation and general sanitary practices have a beneficial impact on our health, well-being, and environment.

Sanitation is considered to be an applied science because of its importance in protecting human health and its relationship to environmental factors that relate to health. It relates to the control of biological, chemical, and physical hazards in a food environment. Those who practice sanitation (sanitarians) must be familiar with all of these hazards and thoroughly understand basic food microbiology and the bacteria, viruses, parasites, and molds that are most likely to affect human health. By identifying, evaluating, and controlling hazards and through effective application of sanitary practices, a safe and wholesome food supply can be assured.

Sanitary Design and Sanitation

When building a new food facility or upgrading or expanding an existing facility whether it is a packing house on a farm, a food processing plant, a retail food store, or a foodservice operation, it is important to incorporate the principles of sanitary design into that facility and the equipment contained in it (Stout 2003). The use of sanitary design can prevent development of microbial niches and harborage sites, facilitate cleaning and sanitation, maintain or increase product shelf life, and improve product safety by reducing potential of foodborne illness and injury and very costly food product recalls (Stout 2003). Sanitation is one of the foundation blocks that food safety management systems are built upon and is vital to the production of safe food.

Gravani (1997) stated that never in recent history have Americans been more concerned about the quality and safety of the food supply than now. Through the media, the Internet, and social media, consumers have more information about food (factual as well as opinion) than ever before. In the 2016 International Food Information Council (IFIC), Food and Health Survey of over 1,000 Americans ages 18–80, 66% of the respondents indicated that they were confident in the safety of the food supply, while one third were not confident or not sure about the safety of the food supply. Foodborne illness from bacteria was the most important food safety concern of 57% of those surveyed (International Food Information Council 2016).

The Centers for Disease Control and Prevention (CDC) estimates that approximately 48 million people will become ill from foodborne illness, 328,000 will be hospitalized, and about 3,000 will die in the United States each year (Centers for Disease Control and Prevention 2017). The national impact of these illnesses is estimated to be $15.6 billion (in 2013 dollars)

(Hoffman et al. 2015). Gone are the days when some food processing, retail food store, and foodservice operators can offer excuses for poor sanitation in their establishment(s). Yet, the reasons for establishing and maintaining such programs are more compelling, because they relate to public health, to customer satisfaction, and to the bottom line of a profit and loss statement. Many consumer surveys indicate that shoppers and diners are interested in patronizing establishments that are clean and sanitary. A sanitation program is "a planned way of practicing sanitation" and is one of the key foundation blocks of a food safety management system. Food sanitation failures can result in consumer complaints, lost customers, adverse publicity, as well as local, state, and/or federal regulatory actions. The old adage, "Sanitation does not cost, it pays," is still true today.

Most owners and operators of food facilities want a clean and sanitary operation. However, unsanitary operations frequently result from a lack of understanding of the principles of sanitation and the benefits that an effective sanitation program will provide. The following brief discussion of these benefits show that sanitation is not a "dirty" word:

1. Inspection is more stringent because inspectors are using the Hazard Analysis Critical Control Points (HACCP) concept and beginning to address preventive controls to establish compliance. HACCP-based inspections focus primarily on the items critical to the safety of foods and less on aesthetics. Thus, an effective sanitation program is essential.
2. Foodborne illness can be controlled when sanitation is properly implemented in all food operations. Common problems caused by poor sanitation are food spoilage through off-odor and flavor. Spoiled foods are objectionable to consumers and cause reduced sales, increased consumer complaints, and increased claims. Off-condition products convey the lack of an effective sanitation program. Poor sanitation can often lead to microbial niches, biofilm formation, and food contaminated with pathogenic microorganisms that can

cause foodborne illness. When consumers think that they have become ill from food, they often notify regulatory authorities and contact attorneys to seek compensation for their illness and inconvenience.

3. An effective sanitation program can improve product quality and shelf life because the microbial population can be reduced. Increased labor, trim loss, packaging costs, and reduced product value due to poor sanitation can cause a decrease of 5–10% of profit of meat operations in a supermarket. A well-developed and maintained sanitation program can increase the shelf life of food, providing a product quality dividend.

4. An effective sanitation program includes regular cleaning and sanitizing of all equipment and utensils in a facility, including heating, air conditioning, and refrigeration equipment. Dirty, clogged coils harbor microorganisms, and blowers and fans can spread microbial flora throughout the facility. Clean and sanitized coils lower the risk of airborne contamination, are more effective heat exchangers, and can reduce energy and maintenance costs by up to 20%. Insurance carriers may reduce rates for clean establishments as a result of improved working conditions as well as fewer customer complaint claims.

5. Various, less tangible benefits of an effective sanitation program include (a) improved product acceptability, (b) increased product shelf life, (c) satisfied and perhaps even delighted customers, (d) reduced public health risks, (e) increased trust of regulatory agencies and their inspectors, (f) decreased product waste and removal, and (g) improved employee morale.

Sanitation: A Foundation for Food Safety Assurance

Proper sanitation practices provide the foundation that food safety management systems are built upon. Poor hygienic and sanitary practices can contribute to outbreaks of foodborne illnesses and cause injury. In the last several years,

there have been some major food safety incidents that have made headlines and focused attention on poor sanitary practices and safety controls in all sectors of the food system. Products that caused foodborne outbreaks and were recalled from the market place included contaminated cookie dough, peanut butter paste, ice cream, flour, cantaloupes, and many other food items. Some of these incidents are shown in Table 1.1 and explained below.

A large *Salmonella enteritidis* outbreak in ice cream was caused by the cross-contamination of pasteurized ice cream mix. The pasteurized mix was transported from premix plants to a freezing operation in contract tanker trucks that had previously been used to haul raw, liquid eggs. The eggs were contaminated with *Salmonella enteritidis*. The hauler was supposed to wash and sanitize the trucks before the ice cream mix was loaded, but this procedure was often bypassed. Investigators found egg residue in one tanker truck after cleaning and noted soiled gaskets, inadequate records, and the lack of inspection and documentation of cleaning and sanitization procedures. There was a nationwide recall of almost 140 million pounds (6.3 million kg) of ice cream products before the incident was resolved. It was estimated that approximately 224,000 people became ill in this outbreak. The proper cleaning and sanitization of the tanker trucks could have prevented this incident (Hennessey et.al. 1996).

Table 1.1 Major food safety incidents

Agent	Food	Effect
S. enteritidis	Ice cream	~224,000 ill
E. coli O157:H7	Hamburgers	732 ill, 4 deaths
Benzene	Mineral water	Worldwide recall of 160 million bottles
Listeria monocytogenes	Hot dogs	101 ill, 21 deaths
Big "8" allergens	Many foods	4–5% of US population have food allergies. 150–200 people die each year
Glass	Bottled beer	15.4 million bottles were recalled, destroyed, and replaced

In another large outbreak, *E. coli* O157:H7 in contaminated and undercooked ground beef patties caused 732 illnesses and 4 deaths in four states (CDC 1993). Ground beef contaminated at the meat processing plant was undercooked in the fast food restaurant resulting in this outbreak. Over 225,000 ground beef patties were recalled from the chains' restaurants. This was the largest *E. coli* O157:H7 outbreak in US history and was estimated to cost between $229 and $610 million. The company took bold, innovative steps to develop a state-of-the-art food safety program and improve their reputation and brand image. Today, this company enjoys the reputation of being one of the most stringent food safety programs in the foodservice industry.

During the past, a popular brand of imported, bottled water was contaminated with benzene. The natural gas present in the spring water source contained a number of impurities. The carbon filters that were used to remove these impurities became clogged. A faulty warning light on the process control panel went undetected by employees for 6 months, allowing the filters to become clogged. When the benzene-contaminated water was discovered, the company recalled 160 million bottles of water from 120 countries. This incident was estimated to cost the bottler about $263 million.

An outbreak of *Listeria monocytogenes* in frankfurters resulted in 101 cases of illness and 21 deaths in 22 states. Although the frankfurters were processed, they were contaminated after processing and before packaging. It was reported that major renovations were being made in the processing plant when the contamination occurred. A nationwide recall of frankfurters made in this plant was undertaken to prevent additional cases of illness.

Today, almost 5% of the US population or about 15 million Americans have food allergies (including approximately 5.9 million children) (Food Allery Research and Education [FARE] 2017), and approximately 200 people die each year from food allergic reactions (Cianferoni and Spergel 2009). The prevalence of food allergies has increased in the last decade, and this trend will continue in the years ahead. The "Big 8"

food allergens including milk, eggs, fish, crustacean shellfish, peanuts, tree nuts, wheat, and soybean account for about 90% of all food allergic reaction in the United States. These proteins must be clearly noted (in "plain English") on the labels of any processed food. Since trace amounts of the offending food will trigger reactions, people with food allergies depend on accurate labels on processed foods, as well as knowledgeable chefs, wait staff, and food workers in foodservice operations and retail food stores.

In the early 1990s, a European beer maker inadvertently used defective glass to make beer bottles. When transported or opened, glass splinters could fall into the beer and cause injury. No one was injured as a result of the glass splinters, but the beer manufacturer recalled, destroyed, and replaced 15.4 million bottles. At the time, the company estimated the loss to be between $10 and $50 million.

Major food safety incidents have common characteristics and include biological, chemical, or physical hazards. They occur throughout the food system and have occurred globally and often result from one or a combination of factors including:

- Contaminated raw materials
- Errors in transportation, processing, preparation, handling, or storage
- Packaging problems
- Food tampering/malicious contamination
- Mishandling
- Changes in formulation or processing
- Inadequate maintenance of equipment or facilities
- Addition of incorrect ingredient(s)

These are examples of the importance of sanitation during food processing and preparation, as well as proper cleaning and sanitizing of food manufacturing and foodservice equipment and facilities. The consequences of improper sanitation are severe and include loss of sales, reduced profits, damaged product acceptability, loss of trust and consumer confidence, adverse publicity, erosion of brand image, loss of market share, and, often, legal action. Sanitary practices coupled with an effective food safety management system can prevent these problems. Consumers certainly have the right to expect and receive wholesome and safe food products.

Foodborne illnesses are a real concern to public health professionals, food scientists, microbiologists, and sanitarians. Today there are more than 200 known diseases transmitted through foods, and many of the pathogens of greatest concern were not recognized as causes of foodborne illness 20 years ago. Thirty-one major pathogens are responsible for the burden of foodborne illness in the United States (Scallan, et al. 2011). Most cases of foodborne illness involve gastrointestinal symptoms (nausea, vomiting, and diarrhea) and are usually acute, self-limiting, and of short duration and can range from mild to severe. Deaths from acute foodborne illnesses typically occur in the very young, the elderly, or in persons with compromised immune systems. The US Food and Drug Administration (FDA) estimates that 2–3% of all acute foodborne illnesses develop secondary long-term complications often referred to as chronic sequelae. These sequelae can occur in any part of the body such as the heart, kidney, nervous system, or joints and can be quite debilitating and, in severe cases, can cause death.

There are many factors associated with the emergence of "new" foodborne pathogens and outbreaks of foodborne illnesses. Some of these factors include:

Demographics
There is an increase in the number of elderly and chronically ill persons in the United States. The population aged 65 and older was 43 million in 2012 and is expected to double by 2050 making up 21% of the US population. Significant portions of older Americans suffer from chronic health conditions, including heart disease, cancer, diabetes, as well as serious intestinal disorders, decreased gastric activity, and HIV/AIDS, and that makes them more susceptible to foodborne illness. For example, persons with AIDS or late-stage HIV infections have a 20 times higher possibility of developing salmonellosis than healthy people. These individuals are also at a 200–300 times higher risk to develop listeriosis.

Some people have had transplants and are taking immunosuppressive drugs, making them vulnerable to foodborne illnesses. As a person ages, their immune system function decreases, so people have a decreased resistance to pathogens as they get older.

Changes in Consumer Practices

US consumers have varied levels of awareness of specific microbial hazards, risk factors for foodborne illness, and the importance of good personal hygiene during the preparation and serving of foods. Consumers have a relatively poor knowledge of safe food preparation practices in their homes. Overall, some changes in behavior have occurred, but consumer habits are still frequently less than ideal. An observational study on handwashing habits sponsored by the American Society for Microbiology revealed that 85% of adults washed their hands in public restrooms, compared with 77% in 2007. The 85% total was actually the highest observed since these studies began in 1996 (ASM 2010), but according to the survey, Americans said that they are likely to wash hands:

- After using the bathroom at home (89%)
- Before handling or eating foods (77%)
- After changing a diaper (82%)
- After petting an animal (42%)
- After coughing or sneezing (39%)

Clearly there are handwashing challenges that need to be addressed, and more effective information about the importance of handwashing is needed.

Changes in Food Preferences and Eating Habits

In 2014, US consumers spent a majority of their food dollar away from home. The sheer volume of meals prepared each day stresses the need for knowledgeable, well-trained foodservice and retail food store employees who understand and practice the principles of safe food preparation every day. Food preferences have also changed with many people now eating raw foods of animal origin or lightly cooked foods that can increase the risk of foodborne illnesses.

Complexity of the Food System

As explained earlier, the food system is a complex, concentrated, and dynamic chain of activities that moves food from farm to table. Greater food system complexity provides more interfaces with food and increased chance of errors. When errors occur, major food incidents can result. Multiple handling of foods (or ingredients) increases the chances for contamination and subsequent temperature abuses. The key is to develop close working relationships and strong partnerships between and among the different sectors of the system to assure a safe and wholesome food supply.

Globalization of the Food Supply

The international sourcing of food and food ingredients has enabled US consumers to enjoy a consistent supply of a wide variety of products from around the world. The main concern is that the sanitary standards and safety assurance systems in some countries may not be as stringent as those in the United States. Strategic partnerships with suppliers and an enhanced ingredient traceability system are key components in a company's food safety management plan.

Today, with increasing international travel, a pathogenic microorganism that causes a problem in one part of the world can be easily transported to another country very quickly. Rapid detection, early intervention, and vigilance are important in preventing the spread of foodborne illness from country to country.

Changes in Food Processing Technologies

As the food industry strives for fresher and longer shelf-life products, product developers must be aware of how food composition, processing parameters, packaging systems, and storage conditions influence the microorganisms that are present. Food safety must be built into the product while it is being developed or reformulated to identify, evaluate, and control food safety hazards that may be present. New and novel processing technologies that are used must be validated to insure the safety of products. There has been a greater awareness of the environmental conditions in processing plants, retail food stores, and foodservice establishments and the need to ensure that biofilms and microbial niches do not develop.

Diagnostic Techniques

In the last decade, there have been significant improvements in foodborne disease surveillance and responses to outbreaks, improved diagnostic techniques, and better medical interventions when illnesses occur. More rapid microbial tests have been developed, and electronic databases such as FoodNet, PulseNet, and eLEXNet have been developed to provide better surveillance of foodborne illnesses, improved information sharing, and more rapid responses when outbreaks occur. A new laboratory technique called whole-genome sequencing is being used to determine the complete DNA makeup of an organism, enabling government agencies like the FDA to quickly identify pathogens during a foodborne outbreak. This powerful tool can be used to identify the source of the contamination, focus and speed up investigations resulting in the removal of the contaminated food or ingredient from the marketplace.

Changes in Foodborne Pathogens

There have been many changes in the microorganisms that cause foodborne illnesses. Scientists have observed more virulent strains of organisms, where a few cells can cause severe illness. An example is *Salmonella enteritidis* and *E. coli* O157:H7. Adaptive stress responses have also been observed where organisms have adapted to environmental conditions to survive and grow such as psychrotrophic pathogens that grow (slowly) at refrigerated temperatures. Organisms such as *Yersinia enterocolitica*, *Listeria monocytogenes*, and *Clostridium botulinum* type E are examples of bacteria capable of growing at refrigerator temperatures. In recent years, increased resistance to antibiotics has been observed in *Salmonella typhimurium* DT104. A number of outbreaks in produce and unpasteurized apple cider have been caused by the protozoan parasites, *Cyclospora cayetanensis* and *Cryptosporidium parvum*.

All of these factors have played and continue to play a role in the emergence of foodborne pathogens and foodborne illnesses. In a discussion of food safety issues, a CEO of a retail food chain made the following comment: "Today, we're facing a new enemy; it is not business as usual." This statement clearly describes the fact that we live in a changing world and must be proactive in assuring food safety.

Sanitation Laws and Regulations and Guidelines

Since thousands of laws, regulations, and guidelines are currently in effect to control the production, processing, and preparation of food in the United States, it would be impossible to address all of these rules in this book. Therefore, it is not the intent of this chapter or this book to emphasize the specific details of the regulations governing food processing or preparation. Only the major agencies involved with food safety and their primary responsibilities will be discussed Curtis (2013). The reader should consult regulations available from various jurisdictions to determine specific requirements for the food operation and area where it is located. It is inappropriate to discuss regulatory requirements for cities and counties because they have designated governmental entities with their own food safety criteria that often differ from one area to another and can change periodically.

Sanitation requirements developed by legislative bodies and regulatory agencies in response to public demands are detailed in laws and regulations. They are not static but change in response to sanitation; public health; new scientific and technical information regarding biological, chemical, and physical hazards; and other important issues brought to public attention.

Laws are passed by legislators and must be signed by the chief executive. After a law has been passed, the agency responsible for its enforcement prepares regulations designed to implement the intention of the law or act. Regulations are developed to cover a wide range of requirements and are more specific and detailed than are laws. Regulations for food provide standards for facility design, equipment design, commodities, tolerances for chemical or other food additives, sanitary practices and qualifications, labeling requirements, and training for positions that require certification.

Regulation development is a multistep process. For example, in the federal process, the relevant agency prepares the proposed regulation, which is then published as a proposed rule in the Federal Register. The Federal Register is the official daily publication for rules, proposed rules, and notices of federal agencies and organizations, as well as executive orders and other presidential documents. Accompanying the proposal is information related to background. Any comments, suggestions, or recommendations are to be directed to the agency, usually within 60 days after proposal publication, although time extensions are frequently provided. Regulatory agencies often hold "listening sessions" with the food industry, nongovernment organizations, consumers, and consumer groups, so they can get input and feedback on their proposed regulation. The regulation is published in final form after all of the comments on the proposal have been reviewed, with another statement of how the comments were handled and specifying effective dates for compliance. This statement suggests that comments on matters not previously considered in the regulations may be submitted for further review. Amendments may be initiated by any individual, organization, other government office, or the agency itself. A petition is necessary, with appropriate documents that justify the request.

There are two types of regulations: substantive and advisory. Substantive regulations are the more important because they have the power of law. Advisory regulations are intended to serve as guidelines. Sanitation regulations are substantive because food must be made safe for the public. In regulations, the use of the word shall means a requirement, whereas should implies a recommendation. Several regulations important to sanitation by various governmental agencies will now be addressed.

Food and Drug Administration Regulations

The Food and Drug Administration (FDA), responsible for enforcing the Food, Drug, and Cosmetic Act, as well as other statutes, has wide-ranging authority. It is under the jurisdiction of the US Department of Health and Human Services. This agency has had a profound impact on the food industry, especially in the control of adulterated foods. Under the Food, Drug, and Cosmetic Act, food is considered as adulterated if it contains any filth or putrid and/or decomposed material or if it is otherwise unfit as food. This act states that food prepared, packed, or held under unsanitary conditions that may cause contamination from filth or that is injurious to health is adulterated. The act gives the FDA inspector authority, after proper identification and presentation of a written notice to the person in charge, to enter and inspect any establishment where food is processed, packaged, or held for shipment in interstate commerce or after shipment. Also, the inspector has the authority to enter and inspect vehicles used to transport or hold food in interstate commerce. This official can check all pertinent equipment, finished products, containers, and labeling.

Adulterated or misbranded products that are in interstate commerce are subject to seizure. Although the FDA initiates action through the federal district courts, seizure is performed by the US Marshals office.

Legal action can also be taken against an organization through an injunction. This form of legal action is usually taken when serious violations occur. However, the FDA can prevent interstate shipments of adulterated or misbranded products by requesting a court injunction or restraining order against the involved firm or individual. This order is effective until the FDA is assured that the violations have been corrected. To correct flagrant violations, the FDA has taken legal steps against finished products made from interstate raw materials, even though they were never shipped outside the state. The FDA can also seek criminal prosecution by the Justice Department for violations of the FD&C Act. This type of legal action depends on a number of factors, including the severity and scope of the public health threat and whether the violations are part of a pattern of criminal behavior (FDA 2017). In a 2008/2009 highly publicized foodborne illness case involving *Salmonella*-contaminated peanut butter that sickened 714 people in 46 states and caused the death of 9 people, the president of the company was criminally prosecuted for shipping product that he knew was contaminated with *Salmonella typhimurium*. He was sentenced to 28 years in federal prison. The contaminated peanut butter was used in over 3, 913 other food products that had to be recalled from the marketplace.

The FDA does not approve cleaning compounds and sanitizers for food plants by their trade names. However, the FDA regulations indicate approved sanitizing compounds by their chemical names. For example, sodium hypochlorite is approved for "bleach-type" sanitizers, sodium or potassium salts of isocyanuric acid for "organic chlorine" sanitizers, n-alkyl dimethyl benzyl ammonium chloride for quaternary ammonium products, sodium dodecylbenzenesulfonate as an acid anionic sanitizer component, and oxypolyethoxy-ethanol-iodine complex for iodophor sanitizers. A statement of maximum allowable use concentrations for these compounds without a potable water rinse on product contact surfaces after use is also provided.

Good Manufacturing Practices

On April 26, 1969, the FDA published the first Good Manufacturing Practices (GMPs) regulations, commonly referred to as the umbrella GMPs. These regulations deal primarily with sanitation in manufacturing, processing, packing, or holding food.

The sanitary operation section establishes basic, minimum rules for sanitation in a food establishment. General requirements are provided for the maintenance of physical facilities, cleaning and sanitizing of equipment and utensils, storage and handling of clean equipment and utensils, pest control, and the proper use and storage of cleaning compounds, sanitizers, and pesticides. Minimum demands for sanitary facilities are included through requirements for water, plumbing design, sewage disposal, toilet and handwashing facilities and supplies, and solid waste disposal. There is also a short section on education and training of employees. Specific GMPs supplement the umbrella GMPs and emphasize wholesomeness and safety of several manufactured products.

Each regulation covers a specific industry or a closely related class of foods. The critical steps in the processing operations are addressed in specific detail, including time-and-temperature relationships, storage conditions, use of additives, cleaning and sanitizing, testing procedures, and specialized employee training.

According to Marriott et al. (1991), inspections are used by regulatory agencies to assure compliance with food safety regulations. However, this approach has limitations because laws that are supposed to be enforced by inspectors are frequently not clearly written and what constitutes compliance is questionable. Furthermore, it is sometimes difficult to distinguish between requirements critical to safety and those related to aesthetics. In recent years, regulatory agencies have recognized these problems and revised their inspection procedures and forms. Now, many agencies have two major categories to differentiate food safety items and aesthetic issues. There are critical deficiencies that address items, when left unattended, could lead to foodborne illness and general deficiencies related to aesthetic items.

In 1995, the FDA issued the procedures for the Safe and Sanitary Processing and Import of Fish and Fishery Products—Final Rule which is the Seafood HACCP regulation. This first HACCP regulation in the United States requires processors of fish and fishery products to develop and implement HACCP systems for their operations.

As a consequence of several large foodborne outbreaks related to raw juices processed in commercial facilities, the FDA published a final rule in 2001 mandating that all juices processed for inter- or intrastate sale be produced under a HACCP plan. This rule was designed to improve the safety of fruit and vegetable juice and juice products and is known as the Juice HACCP regulation.

On January 4, 2011, President Obama signed the FDA Food Safety Modernization Act (FSMA) into law. FSMA was created to strengthen the food safety system and is the most sweeping reform of food safety laws in the United States in over 70 years. The focus of FSMA is on the prevention of food safety problems rather than relying on reacting to incidents once they have occurred (FDA 2017). FSMA contains several sections that address the following areas:

Section 103: Hazard Analysis and Risk-Based Preventive Controls (animal and human food)
Section 105: Standards for Produce Safety
Section 106: Protection Against Intentional Adulteration (Food Defense)
Section 206: Mandatory Recall Authority
Section 111: Sanitary Transportation of Human and Animal Food
Section 204: Enhancing Tracking and Tracing of Human Food and Recordkeeping
Section 206: Mandatory Recall Authority
Section 301: Foreign Supplier Verification Program
Section 302: Voluntary Qualified Importer Program
Section 307: Accreditation of Third-Party Auditors

Some areas of the law will go into effect quickly, while other will require FDA to prepare and issue regulations and guidance documents. The scope and details of each section of FSMA are broad and complex and beyond the scope of this book. The reader is referred to the specific sections of FSMA that pertain to their food operation.

US Department of Agriculture Regulations

The US Department of Agriculture (USDA) has jurisdiction over three areas of food processing, based on the following laws: the Federal Meat Inspection Act, the Poultry Products Inspection Act, and the Egg Products Inspection Act. The agency that administers the area of inspection is the Food Safety and Inspection Service (FSIS), established in 1981.

By design, federal jurisdiction usually involves only interstate commerce. However, the three statutes on meat, poultry, and eggs have extended USDA jurisdiction to the intrastate level if state inspection programs are unable to provide proper enforcement as required by federal law. Products shipped from official USDA-inspected plants into distribution channels and subsequently identified as adulterated or misbranded come under the jurisdiction of the Food, Drug, and Cosmetic Act. The FDA can take legal steps to remove this product from the market. Normally, the product is referred back to the USDA for disposition.

In 1994, the FSIS began an evaluation, review, and revision of existing food safety regulations for meat and poultry. This review led to the 1996 publication of the Pathogen Reduction/Hazard Analysis and Critical Control Points (PR/HACCP) Final Rule (USDA-FSIS 1996). The objective of this regulation was to reduce foodborne illnesses associated with meat and poultry products. The meat and poultry HACCP regulation requires all meat and poultry slaughter and processing establishments to design and implement a HACCP system for their operations.

Environmental Regulations

The Environmental Protection Agency (EPA) enforces provisions for numerous statutes related to the environment, many of which affect food establishments. Environmental regulations that affect sanitation of the food facility include the Federal Water Pollution Control Act; Clean Air Act; Federal Insecticide, Fungicide, and Rodenticide Act (FIFRA); and the Resource Conservation and Recovery Act.

The EPA is involved in the registration of sanitizers by both their trade and chemical names. Sanitizing compounds are recognized through federal regulators as pesticides; thus, their uses are derived from the FIFRA. The EPA requires environmental impact, antimicrobial efficacy, and toxicological profiles. Furthermore, specific label information and technical literature that detail recommended use of applications and specific directions for use are required. Disinfectants must be identified by the phrase "It is a violation of federal law to use this product in a manner inconsistent with its labeling."

Federal Water Pollution Control Act

This act is important to the food industry because it provides for an administrative permit procedure for controlling water pollution. The National Pollutant Discharge Elimination System (NPDES), which is under this permit system, requires that industrial, municipal, and other point source dischargers obtain permits that establish specific limitations on the discharge of pollutants into navigable waters. The purpose of this permit is to effect the gradual reduction of pollutants discharged into streams and lakes. Effluent guidelines and standards have been developed specific to industry groups or product groups. Regulations for meat products and selected seafood products, grain and cereal products, dairy products, selected fruit and vegetable products, and beet and cane sugar refining are published by the EPA.

Clean Air Act

This act, devised to reduce air pollution, gives the EPA direct control over polluting sources in the industry, such as emission controls on automobiles. Generally, state and local agencies set pollution standards based on EPA recommendations and are responsible for their enforcement. This statute is of concern to food operations that may discharge air pollutants through odors, smoke-stacks, incineration, or other methods.

Federal Insecticide, Fungicide, and Rodenticide Act

The Federal Insecticide, Fungicide, and Rodenticide Act (FIFRA) authorized EPA control of the manufacture, composition, labeling, classification, and application of pesticides. Through the registration provisions of the act, the EPA must classify each pesticide either for restricted use or for common use, with periodic reclassification and registration as necessary. A pesticide classified for restricted use must be applied only by or under the direct supervision and guidance of a certified applicator. Those who are certified, either by the EPA or by a state, to use or supervise the use of restricted pesticides must meet certain standards, demonstrated through written examination and/or performance testing. Commercial applicators are required to have certain standards of competence in the specific category in which they are certified.

Current EPA regulations permit the use of certain residual insecticides for crack and crevice treatment in food areas of food establishments. The EPA lists residual pesticides that are permitted in crack and crevice treatment during an interim period of 6 months, while registrants apply for label modification.

Resource Conservation and Recovery Act

Through the Resource Conservation and Recovery Act, a national program was designed to control solid waste disposal. The act authorizes the EPA to recommend guidelines in cooperation with federal, state, and local agencies for solid waste management. It also authorizes funds for research, construction, disposal, and utilization projects in solid waste management at all regulatory levels.

Hazard Analysis Critical Control Points (HACCP) Concept

Although other voluntary programs have been developed in the United States and throughout the world, the Hazard Analysis Critical Control Points (HACCP) concept is the approach that is being emphasized. After this concept was developed jointly by the Pillsbury Company, the National Aeronautics and Space Administration (NASA) and the US Army Natick Laboratories in the late 1960s, HACCP was adopted for use in the space program. Recognizing its application in other areas, the HACCP concept was shared with the food industry at the 1970 Conference for Food Protection. Since then it has been adopted as a voluntary or mandatory program to assure food safety through the identification, evaluation, and control of biological, chemical, and physical hazards in a food facility. A large number of these hazards are clearly affected by the effectiveness of sanitary measures adopted. Although HACCP was initially voluntary, several regulations that have been previously mentioned were developed by the FDA and USDA that require HACCP plan development, implementation, and maintenance in specific sectors of the food industry and have changed the status of this program from voluntary to mandatory (seafood and fishery products, juice, and meat and poultry). With the passage of FSMA, the Hazard Analysis and Risk-Based Preventive Controls not only focuses on the traditional risk-based HACCP approach that includes process-related hazards and critical control points (CCPs) but also controls for hazards related to food allergens, sanitation, suppliers, and others requiring preventive controls (FSPCA 2015).

Because of the importance of HACCP and Hazard Analysis and Risk-Based Preventive Controls, this subject will be discussed in detail in Chap. 7.

Establishment of Sanitary Practices

Sanitation, good manufacturing practices, and other environmental and operating conditions necessary for the production of safe, wholesome food are known as prerequisite programs. These prerequisite programs provide the foundation for HACCP and are a vital component in a company food safety assurance system. So, the design and development of this entire system in a food facility begins with the establishment of basic sanitary practices.

The employer is responsible for establishing and maintaining sanitary practices to protect the public health and maintain a positive image. The problem of establishing, implementing, and maintaining sanitary practices within the food industry is certainly a challenge! The person in charge of this important area must assure that the sanitary practices keep low-risk, potential hazards from becoming serious hazards that could cause illness or injury. The sanitarian is both the guardian of public health and the counselor to company management on quality and safety issues which are influenced by sanitary practices.

A large food processing company should have a separate food safety department on the same organizational level, as production or research, that is responsible for food safety at all operating plants. A sanitation department or team should exist in a plant on a level with other departments. In a large organization, sanitation should be separated from production and mechanical maintenance, an arrangement that will enable the sanitation department team to exercise company-wide surveillance of sanitary practices and maintain a high level of activity. Production practices, quality control, and sanitary practices are not always compatible when administered by a single department or individual; but all of these functions are complementary and are best performed when properly coordinated and synchronized.

Ideally, an organization should have a full-time sanitarian with assistants, but this is not always practical. Instead, a trained individual who was originally employed as a quality control technician, a production foreman, a superintendent, or some other individual experienced in production can be charged with the responsibility of the sanitation operation. This situation is common and usually effective. However, unless the sanitarian has an assistant to take care of some of the routine tasks and is given sufficient time for

proper attention to sanitary details, the program may not succeed.

A one-person safety assurance department with a full schedule of control work will be generally inadequate to assume the tasks of a sanitarian. However, with proper assistance, quality assurance and sanitation supervision can be successfully conducted through a qualified individual that can divide his or her effort between sanitation and quality assurance. It is beneficial for this person to have the advice and service of an outside agency, such as a university, trade association, or private consultant, to avoid becoming submerged in the conflicting interests of different departments. The extra expense can be a worthwhile investment.

A planned sanitation maintenance program is essential to meet legal requirements and protect brand and product reputation, product safety, quality, and freedom from contamination. All phases of food production and plant sanitation should be included in the program to supplement the cleaning and sanitizing procedures for equipment in the facility. A safety assurance program should start with compliance inspection and audit of the entire facility.

The inspection and audit should be comprehensive and critical. As each item is considered, the ideal solution should be noted, irrespective of cost. When the audit is completed, all items should be reevaluated and more practical and/or economic solutions determined. All items that need attention should be prioritized, and an action plan for completion should be established. Attention should be clearly focused on critical deficiencies throughout the facility. Aesthetic sanitary practices should not be adopted without clear evidence of their ability to pay dividends in increased sales or because they are necessary to meet competitive sales pressure.

Study Questions

1. What is sanitation?
2. What is a law?
3. What is a regulation?
4. What is an advisory regulation?

5. What is a substantive regulation?
6. What is the significance of HACCP?
7. What are examples of how microorganisms can mutate?
8. Which acts affect environmental regulations in the food industry?
9. What are prerequisite programs?
10. What US agency administers the Clean Air Act?

References

American Farm Bureau Federation (2015). *Food and farm facts*. Washington, DC.

American Society for Microbiology (2010). Public handwashing takes a hike. http://www.cleaninginstitute.org/public_handwashing_takes_a_hike/

Centers for Disease Control and Prevention (CDC) (1993). Update: Multistate outbreak of Escherichia coli O157:H7 infections from Hamburgers – Western United States, 1992–1993. *MMWR Morb Mortal Wkly Rep* April 19, 42(14): 258.

Centers for Disease Control and Prevention (CDC) (2017). Estimates of foodborne illness in the U.S. https://www.cdc.gov/foodborneburden/

Cianferoni A, Spergel JM (2009). Food allergy: Review, classification and diagnosis. *Alergol Int* 58(4): 457.

Curtis PA (2013). Federal, state and local laws. In *Guide to U.S. food laws and regulations*, ed. Curtis PA. 2nd ed. Wiley Blackwell, John Wiley and Sons: Hoboken, NJ.

Food Allergy Research and Education (FARE) (2017). Facts and statistics. https://www.foodallergy.org/facts-and-stats

Food Safety Preventive Controls Alliance (2015). Preventive controls for human food. Training Curriculum. Participant Manual. 16 Chapters and 6 Appendicies.

Gravani RB (1997). Coordinated approach to food safety education is needed. *FoodTech* 51 s(7): 160.

Hennesey TW, Hedberg CW, Slutsker L, White KE, Besser-Wiek JM, Moen ME, Feldman J, Coleman WW, Edmonson LM, MacDonald KL, Osterholm MT (1996). A National outbreak of Salmonella enteritidis infection from Ice Cream. The investigation team. *N Engl J Med* May 16, 334(20): 1281.

Hoffman S, Maculloch B, Batz M (2015). Economic burden of major foodborne illnesses acquired in the United States. EIB-140. U.S. Department of Agriculture, Economic Research Service. May 2015. https://www.ers.usda.gov/webdocs/publications/eib140/52807_eib140.pdf

International Food Information Council (IFIC) (2016). *2016 Food and health survey*. International Food Information Council: Washington, DC. http://www.foodinsight.org/sites/default/files/2016-Food-and-Health-Survey-Report_FINAL1.pdf

Marriott NG et al. (1991). Quality assurance manual for the food industry. Virginia Cooperative Extension, Virginia Polytechnic Institute and State University, Publication No. 458-013.

Mintel GNPD (2017). Personal communication. New food products developed in the U.S. in 2016. Mintel Group Ltd.: Chicago, IL.

National Restaurant Association (2016). News and Research. Restaurant Industry to Navigate Continued Challenges in 2016. http://www.restaurant.org/News-Research/News/Restaurant-industry-to-navigate-continued-challeng

National Restaurant Association (NRA) (2017a). Facts at a glance. http://www.restaurant.org/News-Research/Research/Facts-at-a-Glance

National Restaurant Association (NRA) (2017b). America works here. 2017. http://www.restaurant.org/Industry-Impact/Employing-America/America-work-here

Orman JM, Velkoff VM, Hogan H (2014). An aging nation: The older population in the United States. Current Population Reports. U.S. Census Bureau: Washington, DC. p. 25.

Progressive Grocer (2016). Progressive Grocer's annual report of the grocery industry. Chicago, IL.

Scallan E, Hoekstra RM, Angulo FJ, Tauxe RV, Widdowson M, Roy SL, Griffin PM (2011). Foodborne illness acquired in the United States—Major pathogens. *Emerg Infect Dis* 17(1): 7. https://doi.org/10.3201/eid1701.P11101.

Statistica (2017). Food service sales growth worldwide in 2014 and 2015. https://www.statista.com/statistics/525210/foodservice-sales-growth/

Stout J (2003). Ten principles of equipment design for ready-to-eat processing operations. *Food Saf Mag* June/July. http://www.foodsafetymagazine.com/magazine-archive1/junejuly-2003/10-principles-of-equipment-design-for-ready-to-eat-processing-operations/

The Food Institute (2016). *Food business mergers and acquisitions*. 2015 Edition. The Food Institute: Upper Saddle River, NJ.

The Food Marketing Institute (2017). *Supermarket facts, 2014*. The Food Marketing Institute: Arlington, VA. http://www.fmi.org/research-resources/supermarket-facts

U.S. Census Bureau (2016). U.S. and world population projections. https://www.census.gov/newsroom/press-releases/2016/cb16-tps158.html

U.S. Census Bureau (2017a). U.S. and world population clock. https://www.census.gov/popclock/

U.S. Census Bureau (2017b). Quick facts United States. https://www.census.gov/quickfacts/table/PST045216/00

U.S. Department of Agriculture, Economic Research Service (2016). Import share of consumption. https://www.ers.usda.gov/topics/international-markets-trade/us-agricultural-trade/import-share-of-consumption.aspx

U.S. Department of Agriculture, Economic Research Service (2017a). Ag and food sectors and the economy. https://www.ers.usda.gov/data-products/ag-and-food-statistics-charting-the-essentials/ag-and-food-sectors-and-the-economy.aspx

U.S. Department of Agriculture, Economic Research Service (2017b). Food prices and spending. https://www.ers.usda.gov/data-products/ag-and-food-statistics-charting-the-essentials/food-prices-and-spending/

U.S. Department of Agriculture, Economic Research Service (2017c). Food and beverage manufacturing. https://www.ers.usda.gov/topics/food-markets-prices/processing-marketing/manufacturing.aspx

U.S. Department of Agriculture, Economic Research Service (2017d). Food away from home. https://www.ers.usda.gov/topics/food-choices-health/food-consumption-demand/food-away-from-home.aspx

U.S. Department of Agriculture, National Agricultural Statistics Service (2012). Census of agriculture. https://www.agcensus.usda.gov/Publications/2012/Online_Resources/Highlights/Farms_and_Farmland/Highlights_Farms_and_Farmland.pdf

U.S. Department of Labor. Women's Division (2010). Women in the workforce, 2010. https://www.dol.gov/wb/factsheets/qf-laborforce-10.htm

U. S. Department of Agriculture, Food Safety Inspection Service (1996). Pathogen reduction; hazard analysis and critical control point (HACCP) systems; Final Rule 9 CFR 304. https://www.usda.gov/wps/wcm/connect/e113b115a-837c-46af-8303-73f7c11fb666/93-016f.pdf?MOD=AJPERES

US FDA (2017). Food Safety Modernization Act. https://www.fda.gov/Food/GuidanceRegulation/FSMA/default.htm

The Relationship of Biosecurity to Sanitation

2

Abstract

Knowledge of the threat of bioterrorism in food processing and preparation is essential to the maintenance of a safe food supply. The US Department of Agriculture (USDA) has provided some beneficial guidelines for the processing, storage, and protection against bioterrorism, and the Food and Drug Administration (FDA) has implemented guidelines for enforcement of the Bioterrorism Act. The development of a functional food defense plan that is developed, implemented, tested, reviewed, and maintained is the cornerstone of a food defense program.

Since pest management is an integral part of food security, the training of pest management personnel is a viable method to improve food safety through monitoring the premises for indications of bioterrorism. Biosecurity and pest management personnel should collaborate to create a set of common goals and training opportunities. The FDA and USDA have websites for the food industry that includes an extensive amount of information about biosecurity including food defense plans, risk mitigation tools, and a food defense plan builder.

Keywords

Biosecurity • Bioterrorism • Cybersecurity • Food defense plan • FDA • Records • RFID • Security • Traceability

Introduction

The food industry is susceptible to cybersecurity threats, like all other industries. Therefore, a comprehensive cybersecurity plan that complies with company objectives and federal and state government regulations is essential for protection.

Cybersecurity best practices should include a security assessment to establish security gaps and determine potential risks to deal with business operations (Straka 2014). Security gaps may include weak security configurations, outdated firewalls, insecure remote access, operating system flaws, lack of staff training, flawed security policies, and negligence. Some basic practices to address cybersecurity are establishing a plan to eliminate significant vulnerabilities, developing systems to identify and prevent potential attacks, updating of authorized application software, and creating an incident response plan prior to the occurrence of any incidents.

Agroterrorism, the intentional contamination of the food supply with a goal of terrorizing the population and causing harm, is an increasing risk. The food industry has become aware of the importance of addressing threats to food safety from foodborne disease outbreaks and inadvertent contaminations to isolated occurrences of product extortion and tampering. However, the food industry must now guard against the intentional, widespread contamination of the food supply. Food biosecurity is no longer addressed in hypothetical terms as the potential for the food supply being a target or tool of terrorism. Furthermore, optimism and complacency are no longer a viable option.

Food biosecurity training and a food defense plan are essential components for a food plant to help maintain a safe food supply. Food sanitarians and other employees involved with sanitation must be knowledgeable about food contaminants including microorganisms, allergens, physical hazards, pests, and contamination through bioterrorism. The food industry is vulnerable to threats and possible damage to food, which makes it important for food plants to have a functional food defense plan that includes sanitation.

In 2003, the US Homeland Security Secretary indicated the possibility that terrorists may select popular food products to deliver chemical or biological warfare. Thus, it is essential to protect consumers from bioterrorism in addition to accidental infestations or contamination from inadequate sanitation. Now, it is necessary for the food industry to protect against intentional interference and the possibility that food products could be used as weapons of destruction.

The food industry has previously faced biosecurity challenges. During the 1980s, a major security challenge increased emphasis on maintaining a drug-free workplace. In the last decade of the twentieth century, there was an increased emphasis on preventing workplace violence. During this time, the threat of biological and chemical weapons was intensified. After the terrorism events of 2001 in the United States, bioterrorism became a key security issue and necessitated that the food industry take this issue very seriously. Since 2003, food plants have been encouraged by the USDA and the Food and Drug Administration to implement food defense plans in order to minimize the potential risks of foodborne bioterrorism. The FDA and USDA have provided resources to make this feasible for food production and processing facilities. In 2016, a final rule was instituted under the Food Safety Modernization Act (FSMA) that requires all food plants registered with FDA to develop a food defense plan.

Potential Risks of Foodborne Bioterrorism

After terrorist attacks in the United States during 2001, a scenario pondered by individuals was reminiscent of the anthrax letters scare during 2001 and the Tylenol-laced cyanide of the early 1980s. DeSorbo (2004) reported that less than a month after being hired, four employees mysteriously disappeared from a dairy plant in California and became wanted in connection with an Al-Qaeda-backed attack and subsequent botulism outbreak that killed 800 and caused more than 16,000 to become ill. The scenario was continued 3 weeks after the attack. Recalls of dairy products manufactured by the California firm reduced the impact of the botulism outbreak with subsequent dairy shortages being reported throughout Southern California. Other possible threat agents are hemorrhagic fever viruses, ricin toxin, and botulinum toxin.

According to Applebaum (2004), the food industry has focused on three areas that are referred to as the "three Ps" of protection:

Personnel: Food companies have increased employee screening and supervision.
Product: Food companies have established additional controls for ingredients and products during receiving, production, and distribution, to ensure a high level of food safety.

Property: Food companies have established additional controls to ensure that they have the highest barriers in place to guard against possible intruders.

Applebaum (2004) further stated that the criteria for accurate risk assessment are to evaluate a firm's assets and determine the type of potential threat that exists and the establishment's vulnerabilities. This author further stated that where a company's assets and vulnerabilities overlap with potential threats, the risk of bioterrorism is increased. Although risk cannot be eliminated totally, it is essential to apply risk management to ensure deterrence and prevention and to apply the "prevent to protect" policy. Since food companies cannot completely prevent bioterrorism before it occurs, they must have the knowledge and tools to detect and mitigate any possible biosecurity breaches. Thus, the goal is to detect problems before it is necessary to mitigate their potential impact.

Bioterrorism Protection Measures

The US food industry has the responsibility of ensuring that approximately 400,000 domestic and foreign facilities that manufacture, process, package, or store food for human or animal consumption are properly registered with the FDA and that all companies that export food products or ingredients to the United States are meeting the prior notice requirements established by the Bioterrorism Act. The Bioterrorism Act directed the FDA to implement regulations for the registration of food facilities; prior notice of imported food shipments; the establishment, maintenance, and availability of records; and the administrative detention of food for human or animal consumption.

The food industry has been especially active in the review of existing food security programs and the implementation of preventive measures and effective controls—especially after the US terrorist attacks of 2001. Progressive companies in the United States and other countries have increased their commitment and vigilance to ensure that preventive measures are in place to

minimize and, if possible, eliminate the threat of international contamination of the food supply.

To ensure successful security efforts, food companies should establish a "security mentality" through increased knowledge of security, security needs, and the establishment of security priorities. They should review their current security practices and procedures, crisis management, and security program (if such programs exist) to determine what revisions or additions are needed. Applebaum (2004) has suggested that "food security" and "food safety" are not the same. Food safety addresses accidents such as cross-contamination and process failure during production, whereas food security is a broader issue which can include intentional manipulating of the food supply to damage it or make it too hazardous for consumption. Thus, food security addresses hazards are induced deliberately and intentionally and food safety addresses hazards that may occur unplanned and accidentally. Both of these activities have a common goal, which is to prevent problems that could undermine the safety of food products. Although the food industry must accept the responsibility of providing consumers a secure food supply, biosecurity should not impede food production, distribution, and consumption. Thus, changes either to food industry security activities or to the regulations governing food security should be realistic and workable.

Radio-Frequency Identification

An important component of compliance with industry regulations such as the FSMA is data collection to access and utilize information to trace products throughout the food supply chain. Radio-frequency identification (RFID) is a technology that provides an opportunity to get as close as possible to real-time traceability in the food supply chain. A large retailer has mandated that the larger vendors provide products tagged with RFID for products at the case and pallet levels. The utility of this technique is that RFID recordkeeping builds long-term data records that benchmark supply deficiencies and provide traceability.

RFID provides records for supply chain deviation and necessary corrective actions. Through low-power radio waves, information is transmitted instantly from specialty RFID tags to the reader. Passive, ultrahigh-frequency tags are the most commonly utilized (O'Boyle 2016). These tags are capable of tracking large volumes such as bins or pallets with a continuous flow. Active or Wi-Fi-based tags draw from their own internal power supply to transmit signals to standard wireless access points. This concept provides real-time location information for tracking high-value, mobile assets. Active tags are more expensive than passive tags but have a greater read range of up to 90 m (300 ft). Bluetooth low-energy tags are less expensive than active tags and are easier to deploy since all they require is a connection to a Bluetooth-enabled device such as a smartphone or mobile computer. This device permits the manufacturer to obtain the same real-time location information as with a Wi-Fi/active tag, but without the need for new infrastructure or multiple access points. Hybrid RFID systems that combine active and passive technology offer the potential for food manufacturers to track both high-volume and low-cost assets. According to O'Boyle (2016), new hybrid systems provide a unified visibility solution for tracking all types of assets and offer more flexibility and affordability.

Traceability

Traceability is the ability to verify the identity, history, or location of an item through documented information as it moves through the supply chain. A traceability study conducted by the US Department of Health and Human Services revealed that only 5 of 40 food items purchased for the investigation could actually have all of their individual ingredients traced back through the supply chain to their origin.

Fernandez (2015) suggested that the food industry needs improved collaboration and a more holistic (or "whole-chain") approach to the food supply chain to better track and trace food. Whole-chain traceability is achieved when a firm's internal data and processes used within its own operations are integrated into a larger system of external data exchange and business processes that occur between trading partners. Enabling whole-chain traceability involves linking internal proprietary traceability systems with external systems through the use of one global language of business—the GS1 system of standards—across the entire supply chain. GS1 standards enable trading partners in the supply chain to communicate with each other through the identification encoded in the various bar codes.

According to Fernandez (2015), whole-chain traceability can have the following positive impacts on the food supply chain:

1. Precise location of potentially harmful products through supply chain visibility—the most critical piece of traceability.
2. Ensuring trustworthy product information and data quality—by the involvement of industry leaders to identify challenges and develop potential solutions for more efficiencies and enhanced risk management.
3. Reduced food waste-adopting standard-based traceability procedures will enhance more precise inventory planning.
4. Operational efficiency enhancement—better collaboration with external trading partners and improved internal gains.

Traceability is an essential component of a proactive retrieval of lots implicated in a potential or confirmed pathogen detection event. Tracking lot-associated data over time can also lead to the identification of patterns that affect quality, profitability, and safety. Data-rich internal traceability can capture the specific timing of key transfers and movements from production through shipment.

The FDA traceback methods have become more refined (Karas 2014). The agency has offered training for its staff and members of state and local agencies to provide more insight on the techniques and methods that have been developed and standardized. Interim final rules for the registration of food facilities and prior notice of imported food shipments were issued by the FDA and became effective on December 12, 2003.

One of the purposes of FSMA's development was to improve food product tracing. The identification of a common food as the vehicle for a foodborne illness can prevent the specific food or ingredient from entering the food supply. Thus, traceback can be used to identify the sources of ingredients as the contamination source.

There are links in the food supply chain that differ in requirements for data capture, recording, and retrieval. An example is core transactional business (CTE) processes that include receipt of bulk materials, ingredients, and packaging. Technology systems are available that include enterprise resource planning (ERP) software. However, not every available ERP includes an industry-specific functionality. Angus-Lee (2014–2015) suggested that to reduce the cost, risk, and time involved in implementation, companies should seek the best-in-breed software that can effectively meet the requirements of the user.

Suppliers should assign a batch-lot number for case-level traceability. A serial number can be included in addition to the batch-lot number as a more specific product identifier within the batch-lot. The serial number will indicate what is in the container and other containers that receive its own dedicated serial number.

Suslow (2009) suggested the following key ingredients to successful quality-based traceability:

1. Integrate quality and traceability data—an integrated quality-based traceability system marries lot- and batch-level traceability information with all required quality documentation.
2. Involve suppliers with an on-line system—this access to current specifications, test procedures, non-conformances, and audits, as well as transaction-based electronic certificates of analysis (e-COA), provides the control needed for a complete system.
3. Automate COA validation—suppliers can provide shipment and e-COA data electronically where test data is immediately validated prior to shipment.
4. Eliminate manual entry of data—data can be collected from suppliers and multiple internal systems electronically without the need for manual data entry, leading to a more complete, accurate, and real-time data needed for effective traceability and root cause analysis
5. Involve multiple tiers of suppliers—this level of visibility for key ingredients is essential in the global marketplace.

Biosecurity Through Simulation

Although the food industry must accept the responsibility for the maintenance of biosecurity, the ability to test the effectiveness of preventive and reactive procedures to an act of bioterrorism remains a challenge. Role-playing and simulation can assist with the assessment of the value of biosecurity programs. Simulation has been developed by academia for such an assessment (Reckowsky 2004). The intent of this technique has been to provide companies an opportunity to test their security plans on a realistic scenario in conjunction with the pressures of time, publicity, and finances. Most decisions involved with simulation were based on information received from multiple inputs such as government releases, media relations, and communications between each other. Effective communication enhanced the traceback of contaminated products and ingredients. Participants have been optimistic about role-playing and simulation and consider this approach to be vital to the increase of industry awareness and readiness for a bioterrorism attack. It appears that simulation can be utilized to advance preparedness and strengthen decision-making abilities related to biosecurity threats. One tool that can be used to help companies use simulation to evaluate the effectiveness of their food defense plan is the Food Defense and Recall Preparedness: A Scenario-Based Exercise Tool that is provided by the USDA (2016).

Biosecurity Guidelines

Guidelines for Biosecurity and food defense plans are provided by the US Department of

Agriculture, Food Safety Inspection Service, and the US Food and Drug Administration (USDA 2016 and FDA 2016) and are listed below:

1. Organize a food defense team.
2. Develop a comprehensive transportation and storage security plan.
3. Assess and identify viable locations for contamination throughout the production and distribution process by the use of a flow diagram.
4. Identify and implement controls to prevent product adulteration or contamination during processing, storage, and transportation.
5. Provide a method to identify and track food products during storage and distribution including the use of tamper-resistant seals.
6. Verify that contract transporters and storage facilities have a security program in effect.

According to the US Department of Agriculture, security measures for purchasing and distribution include:

1. Procedures for the immediate recall of unsafe products
2. Procedures for handling biosecurity or other threats and an evacuation plan
3. Appropriate handling, separation, and disposal of unsafe products
4. Documentation method for the handling of both safe and unsafe products
5. Documented instructions for the rejection of unsafe material
6. Procedures for the handling of off-hour deliveries
7. Current list of contacts for local, state, federal, Homeland Security, and public health officials
8. Procedures for the notification of appropriate authorities if the need materializes
9. Notification of all entry and exit points available during an emergency
10. Strategy for communication of beneficial information to the news media
11. Appropriate training of biosecurity team members
12. Periodic conduct of practice drills and review of security measures

The following screening and educating measures should be considered:

1. Appropriate background and criminal checks should be conducted.
2. References should be verified for all potential employers.
3. Personnel without background checks should be under constant supervision, and their access to sensitive areas of the facility should be restricted.
4. Employees should be trained on food production practices and vigilance, specifically how to prevent, detect, and respond to threats of terrorist actions.
5. Ongoing promotion of security consciousness and the importance of security procedures should be practiced.
6. Appropriate personnel should be trained in security procedures for incoming mail, supplies, raw materials, and other deliveries.
7. Employees should be encouraged to report any suspicious activities, such as signs of possible product tampering or breaks in the food security system.
8. Ensure that employs know emergency procedures and contact information.

The following security measures are appropriate:

1. A positive ID system should be required for all employees.
2. Visitors should be escorted at all times throughout the facility.
3. When a staff member is no longer employed, company-issued IDs and keys should be collected and lock combinations changed.
4. Restricted access to facilities, transportation vehicles, locker rooms, and all storage areas is essential.
5. Specific entry and exit points for people and vehicles should be designated.
6. All access and exit doors, vent openings, windows, outside refrigeration and storage units, trailer bodies, and bulk storage tanks should be secured.
7. Access to the water supply and airflow systems should be secured and restricted.

8. Adequate light should be provided in the perimeter areas.
9. Incoming mail should be handled in an area of the facility separate from food handling.
10. Employees should be monitored for unusual behavior (e.g., staying unusually late, arriving unusually early, taking pictures of the establishment, or moving company documents from the facility).
11. All food ingredients, products, and packaging materials should be purchased only from known, reputable suppliers with accompanying letters of guaranty.
12. Advance notification from suppliers for all incoming deliveries, including shipment details, driver's name, and seal numbers, should be required.
13. Locked or sealed vehicles for delivery should be required.
14. Products known or suspected of being adulterated should be rejected.
15. Unscheduled deliveries should be retained outside of the premises pending verification of the shipper and cargo.
16. A supervisor or other agent should be required to break seals and sign off in the trucker's logbook, noting on the bill of lading any problems with product condition.
17. The broker, seal numbers, and truck or trailer number should be documented.
18. A plan should exist to ensure product integrity when a seal has to be broken prior to delivery due to multiple deliveries or for inspection by government officials.
19. Unloading of incoming products should be supervised.
20. Inbound deliveries should be verified for seal integrity, seal number, and shipping location.
21. Incoming products and their containers should be examined for evidence of tampering all or a corporation.
22. Foods should be checked for unusual color or appearance.
23. A procedural checklist for incoming and outgoing shipments should be developed.
24. All outgoing shipments should be sealed with tamper-evident numbered seals with notation on the shipping documents.
25. Employees should be aware of and report any suspicious activity to appropriate authorities.
26. Forward-shippers and backward-retailers, wholesalers, carriers, and others should be traced anterior should have systems in place for quickly and effectively locating products that had been distributed.
27. Threats or reports of suspicious activity should be investigated promptly.
28. If a food security emergency occurs, the local law enforcement agency should be contacted.

The US Department of Agriculture suggests the following precautions to address biosecurity on the outside of food plants:

1. Plant boundaries should be secured to prevent unauthorized entry.
2. "No trespassing" signs should be posted.
3. Integrity of the plant perimeter should be monitored for signs of suspicious activity or an unauthorized entry.
4. Outside lighting should be sufficient to permit detection of unusual activities.
5. Establishment entrances should be secured through guards, alarms, cameras, or other security hardware consistent with national and local fire and safety codes.
6. Emergency exits should be alarmed and have the self-locking doors that can be opened only from the inside.
7. Doors, windows, roof openings, vent openings, trader bodies, railcars, and bulk storage tanks should be secured at all times.
8. Outside storage tanks for hazardous materials and potable water supply should be protected from, and monitored for, unauthorized access.
9. A current list of plant personnel with open or restricted access to the establishment should be maintained at the security office.
10. Establishment entry should be controlled through required positive identification (e.g., picture IDs, sign-in and sign-out at security or reception, etc.).
11. Incoming or outgoing vehicles (both private and commercial) should be inspected for unusual cargo or activity.

12. Parking areas for visitors and guests should be identified and located at a safe distance from the main facility.
13. Deliveries should be verified against a scheduled roster.
14. Unscheduled deliveries should be retained outside the plant premises, if possible, pending verification of shipper and cargo.
15. Outside access to wells, potable water tanks, and ice-making equipment and storage should be secured from unauthorized entry.
16. Potable and non-potable water lines into processing areas should be inspected periodically for possible hampering.
17. The establishment should arrange for immediate notification of local health officials in the event the potability of the public water supply is compromised.
18. The establishment should determine and enforce a policy on which personal items may and may not be permitted inside the plant and within production areas.

The recommended biosecurity precautions provided by the US Department of Agriculture for the inside of food establishments include:

1. Restricted areas inside the plant should be clearly marked and secured.
2. Access to central controls for airflow, water systems, electricity, and gas should be restricted and controlled.
3. Current flat layout schematics should be available at strategic and secured locations within the plant.
4. Airflow systems should include a provision for immediate isolation of contaminated areas or rooms.
5. Emergency alert equipment should be fully operational, and the location of controls should be clearly marked.
6. Access to in-plant laboratories should be controlled.
7. Computer data processing should be protected using passwords, network firewalls, and effective and current virus detection systems.

The Food Safety Modernization Act

The FDA Food Safety Modernization Act was signed into law by President Obama on January 4, 2011, with the objective of ensuring that the US food supply is safe by shifting the focus of federal regulators from responding to contamination to preventing contamination.

FSMA specifies a final rule "Focused Mitigation Strategies to Protect Food Against Intentional Adulteration" (Agres 2014; FDA 2016). This foundational rule specifies that all domestic and foreign facilities that are registered under Section 415 of Code of Federal Regulations under the Federal Food, Drug, and Cosmetic Act will be required to review the four activities that are the most vulnerable to intentional adulteration. These four activities include (1) bulk liquid receiving and unloading, (2) liquid storage, and holding, (3) mixing and combining food ingredients together, and (4) ingredient handling. This adoption is very similar to using Hazard Analysis and Critical Control Points (HACCP) to prevent food safety issues in the meat and poultry industries including, biological, chemical, and physical contamination. FSMA states that companies must prepare a food defense plan, conduct training, take and monitor corrective actions, and maintain records of documentation.

(1) Bulk liquid receiving and loading activities include process steps where a liquid ingredient is being received and unloaded at a facility or loaded into an outbound transport vehicle. This activity type incorporates the actions of opening the transport vehicle, attaching any pumping equipment or hoses, and opening any venting hatches. **(2) Bulk liquid storage and non-bulk liquid holding and surge tanks often involve agitation and may be located in isolated areas of the facility allowing access and dispersion of a contaminant.** Access hatches may not be locked or alarmed. With regard to surge tanks in the production area, there may not be lids present or locking hatches to limit accessibility to the liquid ingredient or product. **(3) Coating, mixing, grinding, and rework activities may allow for even distribution of a contaminant.** The effect of any of these processes is that an agent added to the process could be evenly mixed throughout the product batch, contaminating the total servings produced from the contaminated batch, and includes but is not limited to mixers, blenders, homogenizers, cascade breeders, millers, grinders, pulverizers, etc. **(4) Ingredient staging, preparation, and addition activities are open process steps that may provide a point of access to introduce a contaminant into the product stream.**

The six other foundational rules implemented under FSMA include (1) Current Good Manufacturing Practice and Hazard Analysis and Risk-Based Preventive Controls for Human Food; (2) Standards for the Growing, Harvesting, Packing, and Holding of Produce for Human Consumption; (3) Current Good Manufacturing Practice and Hazard Analysis and Risk-Based Preventive Controls for Food for Animals; (4) Foreign Supplier Verification Programs (FSVP) for Importers of Food for Humans and Animals; (5) Accreditation of Third-Party Auditors/Certification Bodies to Conduct Food Safety Audits and to Issue Certifications; and (6) Sanitary Transport of Human and Animal Food (FDA 2016).

Food Defense Plan

A food defense plan is required for all food companies that are registered under Section 415 of the Code of Federal Regulations (CFR) under the Federal Food, Drug, and Cosmetic Act. The FDA has a food defense plan builder (FDA 2016) that can help plants develop their food defense plan. Many of the items listed in the Biosecurity Guidelines sections can be included in a food defense plan. A food defense plan is a written plan used to record and document the practices implemented at a facility to control/minimize the potential for intentional contamination.

The basic elements of a food defense plan include (1) assessment of the broad mitigation strategies currently implemented at the facility, (2) action items resulting from the broad mitigation assessment, (3) vulnerability assessment and the critical process steps identified, and (4) focused mitigation strategies selected to be

implemented at the facility. **Broad mitigation strategies** include procedures implemented to secure the facility, storage areas, shipping and receiving areas, utilities, and personnel. **Focused mitigation strategies** provide an additional level of security to those vulnerable areas within food processing or production steps that are inherently open to direct human contact. Food defense plans and food safety plans may overlap and can even be combined into a single plan. In some cases, food defense measures may overlap with practices in a firm's sanitation standard operating procedures (SSOPs) and HACCP. In these cases, there is no need to recreate something that is already in place when developing a food defense plan. Food defense plans will differ from facility to facility according to the food defense training 101 at FDA (2016), but all plans should include certain elements including:

1. Company information.
2. Broad mitigation strategies and action plans.
3. Vulnerability assessments and focused mitigation strategies.
4. Plan reassessment procedures.
5. Contact information for response plan.
6. Process steps that have been identified in the vulnerability assessment as critical process steps should be identified in the food defense plan. Those process steps that are determined as critical and pose a threat to intentional contamination should be identified, similar to how HACCP plans identify the critical control points for unintentional contamination in the food process.
7. Focused mitigation strategies employed at the facility to minimize or eliminate vulnerabilities at the critical process steps of the food facility should be documented in the food defense plan.

Food Defense Team

According to FDA (2016), the roles and the responsibilities of a food defense team are:

1. Conduct evaluations of the broad mitigation strategies established at the facility.

2. Develop action items to address gaps identified in the broad mitigation strategy assessment.
3. Conduct vulnerability assessments for each food (or food group) process and identify the critical steps of each process
4. Implement mitigation strategies to minimize the vulnerabilities identified at the critical process steps.
5. Document the assessments, vulnerabilities, and mitigation strategies and any food defense policies or procedures.
6. Prepare a response plan and identify emergency contacts.
7. Determine practical guidelines for managing the plan, such as testing of the plan procedures, and reassessment of the plan.

The food defense coordinator leads the food defense team and manages the development, implementation, and maintenance of the food defense plan. The food defense coordinator should have knowledge of the overall operations of the facility as well as a background and training in food defense and may be a Quality Assurance/Quality Control manager or a facility manager. Members of a food defense team may be from facility management, human resources, production, quality control, and security.

Vulnerability Assessment As individual steps of a facility's food processes are assessed for accessibility and vulnerability, those with the highest overall vulnerability to intentional contamination should be considered "critical" process steps. Vulnerability assessments are similar to conducting a hazard analysis and identifying critical control points in food safety. According to FDA (2016), one way to rate the risk of a process step is to use a scoring system, accessibility and vulnerability are each assigned a score of 1 (low vulnerability/accessibility) to 10 (high vulnerability/accessibility), and the two scores are summed for an overall vulnerability rating. The steps for conducting vulnerability assessment includes (1) identify all food products, (2) create a flowchart for each food product, (3) identify the process steps of each food product, (4) evaluate the risk of each process step, and (5) rank process steps by overall vulnerabilities.

Employee Responsibilities Employees should be aware of and trained in food defense procedures. Supervisors need to provide leadership to frontline employees to help them implement mitigation strategies and food defense procedures. In addition, employees should receive training on food defense and their responsibilities in the food defense plan. All employees need to know (1) suspicious activities that should be reported, (2) the appropriate person(s) to report suspicious activities, (3) procedure for contacting authorities, and (4) their specific responsibilities pertaining to the defense plan.

Food Defense Resources

FDA provides multiple guidance documents and tools for industry in food defense planning on their website (FDA 2016). These resources include:

1. **Preventive Measures Guidance (including self-assessment).**
2. **Vulnerability Assessment Software Tool:** A prioritization tool that can be used to assess the vulnerabilities within a system or infrastructure in the food industry.
3. **Food Defense Mitigation Strategies Database:** This database provides a range of preventive measures that a firm may choose to implement to better protect their facility, personnel, products, and operations.
4. **Food Defense Plan Builder:** A comprehensive tool that walks the user through all the steps of developing a food defense plan.

After completing the steps in the tool, it will automatically generate a food defense plan. Other resources provided by FDA include **FREE-B** (the Food-Related Emergency Exercise Bundle), which includes scenarios based on both intentional and unintentional food contamination events, and **Employees FIRST,** an FDA initiative that food companies can use to train employees on food defense.

USDA-FSIS also contains the following information to help companies develop, utilize, and update a food safety plan:

1. **CARVER + Shock Primer:** This tool can be used to assess the vulnerabilities within a system or infrastructure to an intentional attack.
2. **Developing a Food Defense Plan for Meat and Poultry Slaughter and Processing Plants:** This guide provides an easy, practical, and achievable three-step method for creating a food defense plan.
3. **Elements of a Functional Food Defense Plan.**
4. **Food Defense Risk Mitigation Tool.**

The Role of Pest Management in Biosecurity

Since pest management is an integral part of food security, the training of pest management personnel is a viable method to improve food safety through monitoring the premises for indications of bioterrorism. This is a logical approach since pest management technicians have the responsibility of investigating conditions that do not contribute to wholesome foods. A link exists between pest exclusion and food safety and security (Anon 2004) since pest management technicians monitor the interior and exterior of food facilities for abnormal conditions that may jeopardize food safety.

Biosecurity and pest management personnel should collaborate to create a set of common goals and training opportunities. The security team can mentor pest management technicians on what to observe when they conduct their daily inspections, such as unusual footprints near the perimeter or abandoned packages in the plant, and indicate the necessary actions. Pest management personnel can teach security about monitoring potential water contamination sites such as drains and sewers, identifying signs of contamination of raw materials, and choosing security solutions that minimize pest problems, such as opting for sodium vapor lights instead of mercury vapor lights, which attract pests (Anon 2004).

If a contract test management company is utilized, it should be a reputable firm with technicians that are specifically trained in food pest management. These technicians should they cleared with a security background check and possess knowledge about bioterrorism prevention strategies. These experienced technicians know how to advise the food company on the latest techniques for pest management and food security. Normally, in-house technicians did not have the access to the expertise and ongoing training that pest management vendors possess, and they cannot store chemicals off-site. This limitation creates sanitation and bioterrorism hazards within a facility. If pest management chemicals are stored on the premises, accidental contamination risk increases, and it is more convenient for disgruntled workers or terrorists to intentionally poison products and destroy a firm's reputation.

Additional Bioterrorism Information

Food Detention

This portion of the Act authorizes Health and Human Services (HHS), through the FDA, to order the retention of food if an officer or qualified employee has credible evidence or even information which suggests that a foodstuff presents a threat of serious adverse health consequences or death to humans or animals. The HHS, through the FDA, is required to issue final regulations to expedite enforcement actions on perishable foods.

Registration of Food and Animal Feed Facilities

The Bioterrorism Act requires the owner, operator, or agent in charge of a domestic or foreign facility to register with FDA by December 12, 2003. A facility is considered to be any factory, warehouse, or establishment, including importers that manufacture, process, pack, or store food for human or animal consumption in the United States. Exemptions include farms, restaurants, retail food establishments, nonprofit establishments that prepare or serve food, and fishing vessels not engaged in processing. Foreign facilities are also exempt if the food from the establishment is designated for other processing or packaging by another facility before it is exported to the

United States or if the establishment performs a minimal activity such as labeling. Such a registration roster will enable the FDA to rapidly identify and locate affected food processors and other establishments if deliberate or accidental contamination of food occurs.

Establishment and Maintenance of Records

The Secretary of Health and Human Services is required to establish requirements for the creation and maintenance of records needed to determine the immediate previous sources and the subsequent recipients of food. Such records permit the FDA to address credible threats of serious adverse health consequences or death to humans or animals. Entities that are subject to these provisions are those that manufacture, process, pack, transport, distribute, receive, store, or import food. Farms and restaurants are exempt from these requirements.

Prior Notice of Imported Food Shipments

The Bioterrorism Act requires that prior notice of imported food shipments be given to the FDA. The notice must include a description of the article, manufacturer, shipper, grower (if known), country of origin, country from which the article is shipped, and the anticipated port of entry. This regulation mandates that importers of food must give the FDA prior notice of every shipment of food before it can enter into the United States. Issued jointly with the US Bureau of Customs and Border Protection, the advance notification of what shipments contained in when they will arrive at US ports of entry is designated to assist these federal agencies to better target painting art inspections of imported foods. Currently, the FDA requires that companies provide prior notice and receive FDA confirmation no more than 5 days before anticipated arrival at a US port of entry and no fewer than 2 h before arrival by land via road, 4 h before arrival by air or by land via rail, or 8 h before arrival by water.

Study Questions

1. Why is biosecurity a major concern to the food industry?
2. What are the "three Ps" of protection against bioterrorism?
3. What is the significance of the Bioterrorism Act?
4. How does biosecurity and pest management interface?
5. How can biosecurity and pest management personnel complement each other?
6. What has the US Department of Agriculture done to promote food biosecurity?
7. What has the Food and Drug Administration done to enhance food biosecurity?
8. How have attacks by terrorists in the United States in 2001affected biosecurity among food processors?
9. What are the components of a food defense plan?
10. Which employees of a company should be involved in a food defense plan?
11. What are the responsibilities of a food defense team?
12. What is RFID?
13. What is e-COA?
14. What is agroterrorism?
15. What is traceability?

References

Agres T (2014). Food security. *Food Qual Saf* [Internet]. [cited Jul 21 2016]. Available from www.foodquality-andsafety.com/article/food-security/3/. Accessed July 21, 2016.

Angus-Lee H (2014–2015). The evolution of traceability in the meat & poultry industry. *Food Saf Mag* 20(6): 42.

Anon (2004). How your pest management technician can protect your company against bioterrorism. *Food Saf Mag* 10(1): 36.

Applebaum RS (2004). Protecting the nation's food supply from bioterrorism. *Food Saf Mag* 10(1): 30.

DeSorbo MA (2004). Security: The new component of food quality. *Food Qual* 11(4): 24.

Fernandez A (2015). A holistic approach to traceability. *Food Qual Saf* 22(3): 54.

Karas D (2014). Traceback investigations: Mapping the maze. *Food Saf Mag* 20(5): 26.

O'Boyle T (2016). RFID: A taste of traceability. *Food Qual Saf* 23(3): 44.

Reckowsky M (2004). Preparing for bioterrorism through simulation. *Food Technol* 58(8): 108.

Straka C (2014). Cybersecurity in food and beverage industry. *Food Qual Saf* 21(5): 21.

Suslow T (2009). Produce traceability and trace-back: From seed to shelf and beyond. *Food Saf Mag* 15(2): 32.

United States Department of Agriculture [Internet] (2016). [cited Jul 21 2016]. Available from http://www.fsis.usda.gov/wps/portal/fsis/topics/food-defense-defense-and-emergency-response/tools-resources-training/resources

United States Food and Drug Administration [Internet] (2016). [cited Jul 21 2016]. Available from http://www.fda.gov/Food/FoodDefense/ToolsEducationalMaterials/ucm349888.htm

The Relationship of Microorganisms to Sanitation

3

Abstract

Microorganisms cause food spoilage through degradation of appearance and flavor, and foodborne illness occurs through the ingestion of food containing microorganisms or toxins of public health concern. Control of microbial load from equipment, establishments, and foods is part of a sanitation program.

Microorganisms have a growth pattern similar to the shape of a bell curve and tend to proliferate and die at a logarithmic rate. Extrinsic factors that have the most effect on microbial growth kinetics are temperature, oxygen availability, and relative humidity. Intrinsic factors that affect growth rate most are water activity (A_w) and pH levels, oxidation-reduction potential, nutrient requirements, and presence of inhibitory substances. Chemical changes from microbial degradation occur primarily through enzymes, produced by microorganisms, which degrade proteins, lipids, carbohydrates, and other complex molecules into simpler compounds.

The most common methods of microbial destruction are heat, chemicals, and irradiation, whereas the most common methods for inhibiting microbial growth are refrigeration, dehydration, and fermentation. Microbial load and taxonomy are measurements of the effectiveness of a sanitation program by various tests and diagnoses.

Keywords

Bacteria • Diagnostic tests • Foodborne illness • Microorganisms • Molds • Pathogens • Viruses • Yeasts

Introduction

Microorganisms (also called *microbes* and *microbial flora*) exist throughout the natural environment. Effective sanitation combats their proliferation and activity.

How Microorganisms Relate to Food Sanitation

Microbiology is the science of microscopic forms of life known as *microorganisms*. Knowledge of microorganisms is important to the sanitation specialist because their control is part of a sanitation program.

What Are Microorganisms?

A microorganism is a microscopic form of life found on all non-sterilized matter that can be decomposed. The word is of Greek origin and means "small" and "living beings." These organisms metabolize in a manner similar to humans through nourishment intake, discharge of waste products, and reproduction. Most foods are highly perishable because they contain nutrients required for microbial growth. Microbial proliferation control is essential to reduce food spoilage and eliminate foodborne illness. Improper sanitation practices during food processing, preparation, and serving increase the rate and extent of deteriorative changes that lead to spoilage.

Three types of microorganisms occur in foods. They may be beneficial and pathogenic or cause spoilage. Beneficial microorganisms include those that may produce new foods or food ingredients through fermentation(s) (e.g., yeasts and lactic acid bacteria) or probiotics. Spoilage microorganisms, through their growth and ultimately enzymatic action, alter the taste of foods through flavor, texture, or color degradation. Pathogenic microorganisms can cause human illness. Two types of pathogenic microorganisms found in foods are those that cause (1) intoxication and (2) infection. Intoxication results from microorganisms growing and producing toxin (which causes the illness) in a food. An infection is an illness that results from ingestion of a disease-causing microorganism. Infectious microorganisms may cause illness by the production of enterotoxins in the gastrointestinal tract or adhesion to and/or invasion of the tissues.

Microorganisms Common to Food

A major challenge for the sanitarian is to protect the production area and other involved locations against microbes that can reduce the wholesomeness of food. Microorganisms can contaminate and affect food, with dangerous consequences to consumers. The microorganisms most common to food are *bacteria* and *fungi*. The fungi, which are less common than bacteria, consist of two major microorganisms: molds (which are multicellular) and yeasts (which are usually unicellular). Bacteria, which usually grow at the expense of fungi, are unicellular. Viruses, although transmitted more from person to person through poor employee hygiene than via food, should also be mentioned because they may contaminate food.

Molds

Molds are multicellular microorganisms (eukaryotic cells) with mycelial (filamentous) morphology. They consist of tubular cells, ranging from 30–100 μm in diameter, called h*yphae*, which form a macroscopic mass called a *mycelium*. Molds display a variety of colors with a mildewy or fuzzy, cotton-like appearance. They can develop numerous tiny spores found in the air and spread by air currents. These can produce new mold growth if they are in a location that has conditions conducive to germination. Molds generally withstand greater variations in pH than do bacteria and yeasts and frequently tolerate greater temperature variations. Although molds thrive best at or near a pH of 7.0, they tolerate a range from 2.0–8.0, although an acid-to-neutral pH is preferred. Molds are thriftier at ambient temperature than in a colder environment, even though growth can occur below 0 °C (32 °F).

Although they prefer an A_w of approximately 0.90, growth of a few osmiophilic molds can and does occur at a level as low as 0.60. At an A_w of 0.90 or higher, bacteria and yeasts grow more effectively and normally utilize available nutrients for growth at the expense of molds. When the A_w goes below 0.90, molds are more likely to grow. Foods such as pastries, cheeses, and nuts that are low in moisture content are more likely to spoil from mold growth.

Molds have been considered beneficial and troublesome, ubiquitous microorganisms. They often work in combination with yeasts and bacteria to produce numerous indigenous fermented foods and are involved in industrial processes to produce organic acids and enzymes. Molds are a major contributor to food product recalls. Most do not cause health hazards, but some produce mycotoxins that are toxic, carcinogenic, mutagenic, or teratogenic to humans and animals.

Because they may be airborne, molds spread easily. These fungi cause various degrees of visible deterioration and decomposition of foods. Their growth is identifiable through rot spots, scabs, slime, cottony mycelium, or colored sporulating mold. Molds may produce abnormal flavors and odors due to fermentative, lipolytic, and proteolytic changes caused by enzymatic reactions with carbohydrates, fats, and proteins in foods.

Mold growth has an absolute requirement for oxygen and will not occur with high levels of carbon dioxide (5–8%). Their diversity is evident through the ability to function as oxygen scavengers and to grow at very low levels of oxygen and even in vacuum packages. Some halophilic molds can tolerate a salt concentration of over 20%.

Yeasts

Yeasts are generally unicellular. They differ from bacteria in their larger cell sizes and morphology and because they produce buds during the process of reproduction by fission. The generation time of yeasts is slower than that of bacteria, with a typical time of 2–3 h in foods, leading from an original contamination of one yeast/g of food to

spoilage in approximately 40–60 h. Like molds, yeasts spread through the air or by other means and can alight on the surface of foodstuffs. Yeast colonies are generally moist or slimy in appearance and creamy white. Yeasts prefer an A_w of 0.90–0.94, but can grow below 0.90. In fact, some osmiophilic yeasts can grow at an A_w as low as 0.60. These microorganisms grow best in the intermediate acid range, a pH from 4.0–4.5. Yeasts are more likely to grow on foods with lower pH and on those that are vacuum packaged. Food that is highly contaminated with yeasts will frequently have a slightly fruity odor.

Bacteria

Bacteria are unicellular microorganisms (prokaryotic cells) that are approximately 1 μm in diameter, with morphology variation from short and elongated rods (bacilli) to spherical (cocci) or ovoid forms. Individual bacteria closely combine in various forms, according to genera. Some sphere-shaped bacteria occur in clusters similar to a bunch of grapes (e.g., staphylococci). Other bacteria (rod-shaped or sphere-shaped) are linked together to form chains (e.g., streptococci). Certain genera of sphere-shaped bacteria are formed together in pairs (diploid formation), such as pneumococci. Microorganisms, such as *Sarcina* spp., form as a group of four (tetrad formation). Other genera appear as an individual bacterium. Some bacteria possess flagella and are motile.

Bacteria produce pigments ranging from variations of yellow to dark shades, such as brown or black. Certain bacteria have pigmentation of intermediate colors—red, pink, orange, blue, green, or purple. These bacteria cause food discoloration, especially among foods with unstable color pigments, such as meat. Some bacteria also cause discoloration by slime formation.

Some species of bacteria produce spores, which may be resistant to heat, chemicals, and other environmental conditions. Some of these spore-forming bacteria are thermophilic microorganisms that produce a toxin that can cause foodborne illness.

Viruses

Viruses, a leading cause of foodborne illness in the United States, account for approximately 50% of foodborne diseases, and according to the Centers for Disease Control and Prevention (CDC), noroviruses affect approximately 20 million people each year. Norovirus and hepatitis A are the two main viruses of concern for the food industry for imported products. The norovirus is an extremely contagious virus. Thus, food contamination is likely to occur before harvest through irrigation water contaminated with feces or by infected food workers.

Viruses are infective microorganisms with dimensions that range from 20–300 nm or about 1/100–1/10 the size of a bacterium. Most viruses are visible only with an electron microscope. A virus particle consists of a single molecule of DNA or RNA, surrounded by a coat made from protein. Viruses are typically present in foods in low numbers, making their detection in traditional cell cultures difficult. However, diagnostic advances such as real-time reverse transcription-polymerase chain reaction have advanced the detection of foodborne viruses. Viruses cannot reproduce outside of another organism and are obligate parasites of all living organisms, such as bacteria, fungi, algae, protozoa, higher plants, and invertebrate and vertebrate animals. When a protein cell attaches to the surface of the appropriate host cell, either the host cell engulfs the virus particle or the nucleic acid enters from the virus particle into the host cell, as with bacteriophages active against bacteria.

In animals, some infected host cells die, but others survive infection with the virus and resume their normal function. It is not necessary for the host cells to die for the host organism—in the case of humans—to become ill (Shapton and Shapton 1991). Employees may serve as carriers and transmit viruses to food. An infected food handler can excrete the organism through the feces and respiratory tract infection. Transmission occurs through coughing, sneezing, touching a runny nose, and from not washing the hands after using the toilet. The inability of host cells to perform their normal function causes illness. After the normal function is reestablished, recovery from illness occurs. The inability of viruses to reproduce themselves outside the host and their small size complicates their isolation from foods suspected of being the cause of illness in humans. There is no evidence of the human immunodeficiency virus (HIV) (acquired immune deficiency syndrome [AIDS]) being transmitted by foods. Sanitizers such as the iodophors can destroy viruses (see Chap. 10), but they may not be inactivated by a pH as low as 3.0. A 70% ethanol and 10 mg/L free residual chlorine inactivates viruses (Caul 2000).

Foodborne viruses cause diseases through viral gastroenteritis or viral hepatitis. A virus that has caused a major increase in outbreaks in restaurants during the past 10 years is hepatitis A. Intravenous drug use is one factor that accounts for some of this rise. Infectious hepatitis A occurs in food not handled in a sanitary manner. The onset is 1–7 weeks with an average length of 30 days. Symptoms include nausea, cramps, vomiting, diarrhea, and, sometimes, jaundice, which can last from a week to several months. A major source of hepatitis is raw shellfish from polluted waters. The most likely foods to transmit viral illnesses are those handled frequently and those that receive no heating after handling, such as sandwiches, salads, and desserts. Because this disease is highly contagious, it is mandatory that employees handling food practice thorough handwashing. Viruses also cause diseases such as influenza and the common cold.

Microbial Growth Kinetics

With minor exceptions, multiplication of microbial cells by binary fission occurs in a growth pattern of various phases, according to the typical microbial growth curve illustrated in Fig. 3.1.

Lag Phase

After contamination occurs, the period of adjustment (or adaptation) to the environment, with a slight decrease in microbial load due to stress (Fig. 3.1), followed by limited growth in the

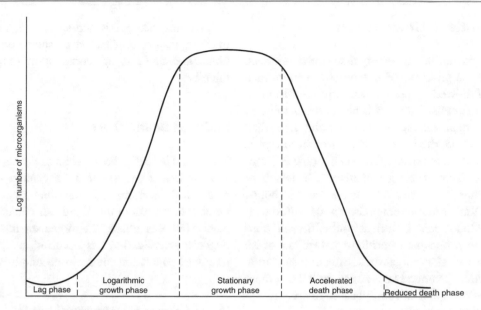

Fig. 3.1 A typical growth curve for bacteria

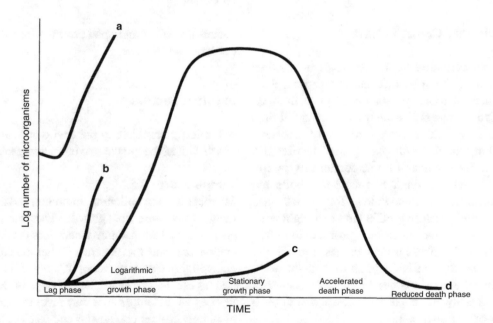

Fig. 3.2 Effect of initial contamination and lag phase on the growth curve of microorganisms: (*a*) high initial contamination and poor temperature control (short lag phase), (*b*) low initial contamination and ineffective temperature control (short lag phase), (*c*), low initial contamination and rigid temperature control (long lag phase), and (*d*) typical growth curve

number of microbes, is called the *lag phase* of microbial growth. Less microbial proliferation through reduced temperature or other preservation techniques extends the lag phase. This increases the "generation interval" of microorganisms. Decreasing the number of microbes that contaminate food, equipment, or buildings

retards microbial proliferation. Lower initial counts of microbes through improved sanitation and hygienic practices extend the lag phase and entry into the next growth phase deferred. Figure 3.2 illustrates how differences in temperature and initial contamination load can affect microbial proliferation.

Logarithmic Growth Phase

Bacteria multiply by binary fission, characterized by the duplication of components within each cell, followed by prompt separation to form two daughter cells. During this phase, the number of microorganisms increases to the point that, when cells divide, the increase in number of microbes occurs at an exponential rate until some environmental factor becomes limiting. The length of this phase may vary from two to several hours. The number of microorganisms and environmental factors, such as nutrient availability and temperature, affect the logarithmic growth rate of the number of microorganisms. Effective sanitation to reduce the microbial load can limit the number of microbes that can contribute to microbial proliferation during this growth phase.

Stationary Growth Phase

When environmental factors such as nutrient availability, temperature, and competition from another microbial population become limiting, the growth rate slows and reaches an equilibrium point. Growth becomes relatively constant, resulting in the stationary phase. During this phase, the number of microorganisms is frequently large enough that their metabolic by-products and competition for space and nourishment reduce proliferation to a slight or no decrease in the microbial proliferation. The length of this phase usually ranges from 24 h to more than 30 days but depends on both the availability of energy sources for the maintenance of cell viability and the degree of pollution in (hostility of) the environment.

Accelerated Death Phase

Lack of nutrients, metabolic waste products, and competition from other microbial populations contribute to the death of microbial cells at an exponential rate. Accelerated death rate is similar to logarithmic growth rate and ranges from 24 h to 30 days but depends on temperature, nutrient

supply, microbial genus and species, age of the microorganisms, application of sanitation techniques and sanitizers, and competition from other microbes.

Reduced Death Phase

This phase is nearly the opposite of the lag phase and is attributable to a sustained accelerated death phase with a decreased microbial population number to the extent that the death rate decelerates. After this phase, the organism has been degraded, sterilization has occurred, or another microbial population continues decomposition.

What Causes Microorganisms to Grow

Factors that affect microbial proliferation rate are *extrinsic* and *intrinsic*.

Extrinsic Factors

Extrinsic factors relate to the environmental factors that affect the growth rate of microorganisms.

Temperature
Microbes have an optimum, minimum, and maximum temperature for growth. Therefore, the environmental temperature determines the proliferation rate and the genera of microorganisms that will thrive and the extent of microbial activity that occurs. For example, a change of only a few degrees in temperature may favor the growth of entirely different organisms and result in a different type of food spoilage and foodborne illness. These characteristics have been responsible for the use of temperature as a method of controlling microbial activity.

The optimal temperature for the proliferation of most microorganisms is from 14–40 °C (57–104 °F). However, some microbes will grow below 0 °C (32 °F), and other genera will thrive at temperatures up to and exceeding 100 °C (212 °F).

Microbes classified according to temperature of optimal growth include:

1. Thermophiles (high-temperature-loving microorganisms), with growth optima at temperatures above 45 °C (113 °F). Examples are *Bacillus stearothermophilus*, *Bacillus coagulans*, and *Lactobacillus thermophilus*.
2. Mesophiles (medium-temperature-loving microorganisms), with growth optima between 20 °C (68 °F) and 45 °C (113 °F). Examples are most lactobacilli and staphylococci.
3. Psychrotrophs (cold-temperature-tolerant microorganisms), which tolerate and thrive at temperatures below 20 °C (68 °F). Examples are *Pseudomonas* and *Moraxella-Acinetobacter*).

Bacteria, molds, and yeasts each have some genera that thrive in the range characteristic of thermophiles, mesophiles, and psychrotrophs. Molds and yeasts tend to be less thermophilic than do bacteria. As the temperature approaches 0 °C (32 °F), fewer microorganisms thrive, and their proliferation is slower. Below approximately 5 °C (40 °F), proliferation of spoilage microorganisms is retarded, and growth of most pathogens ceases.

Oxygen Availability

As with temperature, availability of oxygen determines which microorganisms will be active. Some microorganisms have an absolute requirement for oxygen. Others grow in the total absence of oxygen, and others grow either with or without available oxygen. Microorganisms that require free oxygen are *aerobic* microorganisms (*Pseudomonas* species is an example). Those that thrive in the absence of oxygen are *anaerobic* microorganisms (i.e., *Clostridium* species). Microorganisms that grow with or without the presence of free oxygen are *facultative* microorganisms (e.g., *Lactobacillus* species).

Relative Humidity

Microorganisms have high requirements for water to support their growth and activity. A high relative humidity can cause moisture condensation on food, equipment, walls, and ceilings. Condensation causes moist surfaces, which are conducive to microbial growth and spoilage. Furthermore, a low relative humidity inhibits microbial growth.

Bacteria require a higher humidity than do yeasts and molds. Optimal relative humidity for bacteria is 92% or higher, whereas yeasts prefer 90% or higher. Molds thrive more if the relative humidity is 85–90%.

Intrinsic Factors

Intrinsic factors that affect the rate or proliferation relate more to the characteristics of the substrates (foodstuff or debris) that support or affect growth of microorganisms.

Water Activity

A reduction of water availability will reduce microbial proliferation. The available water for metabolic activity instead of total moisture content determines the extent of microbial growth. The unit of measurement for water requirement of microorganisms is usually expressed as water activity (A_w), defined as the vapor pressure of the subject solution divided by the vapor pressure of the pure solvent: $A_w = p \div p_0$, where p is the vapor pressure of the solution and p_0 is the vapor pressure of pure water. The approximate optimal A_w for the growth of many microorganisms is 0.99, and most microbes require an A_w higher than 0.91 for growth. The approximate relationship between fractional equilibrium relative humidity (RH) and A_w is RH = $A_w \times 100$. Therefore, an A_w of 0.95 is approximately equivalent to an RH of 95% in the atmosphere above the solution. Most natural food products have an A_w of approximately 0.99. Generally, bacteria have the highest water activity requirements of the microorganisms. Molds normally have the lowest A_w requirement, and yeasts are intermediate. Most spoilage bacteria do not grow at an A_w below 0.91, but molds and yeasts can grow at an A_w of 0.80 or lower. Molds and yeasts can grow on partially dehydrated surfaces (including food), whereas bacterial growth is retarded.

pH

pH is a measurement of $\log_{10}$ of the reciprocal of the hydrogen ion concentration (g/L) and is represented as $pH = \log_{10}[H^+]$. The pH for optimal growth of most microorganisms is near neutrality (7.0). Yeasts can grow in an acid environment and thrive best in an intermediate acid (4.0–4.5) range. Molds tolerate a wider range (2.0–8.0), although their growth is generally greater with an acid pH. They thrive in a medium that is too acid for either bacteria or yeasts. Bacterial growth is optimal by near-neutral pH values. However, acidophilic (acid-loving) bacteria grow on food or debris down to a pH of approximately 5.2. Below 5.2, microbial growth is below that in the normal pH range.

Oxidation-Reduction Potential

The oxidation-reduction potential is an indication of the oxidizing and reducing power of the substrate. To attain optimal growth, some microorganisms require reduced conditions; others need oxidized conditions. Thus, the importance of the oxidation-reduction potential is apparent. All saprophytic microorganisms that are able to transfer hydrogen as H^+ and E^- (electrons) to molecular oxygen are *aerobes*. Aerobic microorganisms grow more rapidly under a high oxidation-reduction potential (oxidizing reactivity). A low potential (reducing reactivity) favors the growth of anaerobes. Facultative microorganisms are capable of growth under either condition. Microorganisms can alter the oxidation-reduction potential of food to the extent that the activity of other microorganisms is restricted. For example, anaerobes can decrease the oxidation-reduction potential to such a low level that the growth of aerobes is inhibited.

Nutrient Requirements

In addition to water and oxygen (except for anaerobes), microorganisms have other nutrient requirements. Most microbes need external sources of nitrogen, energy (carbohydrates, proteins, or lipids), minerals, and vitamins to support their growth. Amino acids and nonprotein nitrogen sources provide nitrogen. However, some microorganisms utilize peptides and proteins. Molds are the most effective in the utilization of proteins, complex carbohydrates, and lipids because they contain enzymes capable of hydrolyzing these molecules into less complex components. Many bacteria have a similar capability, but most yeasts require the simple forms of these compounds. All microorganisms need minerals, but requirements for vitamins vary. Molds and some bacteria can synthesize enough B vitamins for their needs, whereas other microorganisms require a ready-made supply.

Inhibitory Substances

Inhibitory substances affect microbial proliferation. *Bacteriostats* are substances or agents that inhibit microbial activity, whereas *bactericides* destroy microorganisms. Bacterial substances such as nitrites retard bacterial proliferation when incorporated during food processing. Most bactericides decontaminate foodstuffs or serve as a sanitizer for cleaned equipment, utensils, and rooms. (Sanitizers are discussed in detail in Chap. 10.)

Interaction Between Growth Factors

The effects that factors such as temperature, oxygen, pH, and A_w have on microbial activity may be dependent on each other. Microorganisms generally become more sensitive to oxygen availability, pH, and A_w at temperatures near growth minima or maxima. For example, bacteria may require a higher pH, A_w, and minimum temperature for growth under anaerobic conditions than when aerobic conditions prevail. Microorganisms that grow at lower temperatures are usually aerobic and generally have a high A_w requirement. Lowering A_w by adding salt or excluding oxygen from foods (such as meat) held at a refrigerated temperature dramatically reduces the rate of microbial spoilage. Normally, some microbial growth occurs when any one of the factors that controls the growth rate is at a limiting level. Microbial growth is curtailed or stopped if more than one factor becomes limiting.

Role of Biofilms

Biofilms are microcolonies of bacteria closely associated with an inert surface attached by a matrix of complex polysaccharide-like material with other trapped debris, including nutrients and microorganisms. A biofilm is a unique environment that microorganisms generate for themselves, enabling the establishment of a "beachhead" on a surface resistant to intense assaults by sanitizing agents. When a microorganism lands on a surface, it attaches itself with the aid of filaments or tendrils. The organism produces a polysaccharide-like material, a sticky substance that will cement in a matter of hours the bacteria's position on the surface and act as a glue to which nutrient material will adhere with other bacteria and, sometimes, viruses. The bacteria become entrenched on the surface, clinging to it with the aid of numerous appendages. Bacteria within a biofilm are up to 1,000 times more resistant to some sanitizers when compared to those dispersed freely in a solution.

A biofilm builds upon itself, adding several layers of the polysaccharide material populated with microorganisms, such as *Salmonella*, *Listeria*, *Pseudomonas*, and others common to the specific environment. Increased time of organism contact with the surface contributes to the size of the microcolonies formed, amount of attachment, and difficulty of removal. The biofilm will eventually become a tough plastic normally removed only by scraping. A firmly established biofilm has layers of organisms protected from the sanitizer. Biofilm buildup can be responsible for portions sheared off by the action of food or liquid passing over the surface. Because the shear force is greater than the adherence force in the topmost layers of the biofilm, chunks of the polysaccharide cement, with the accompanying microbial population transferred to the product with subsequent contamination.

There is an interest in biofilms because *Listeria monocytogenes* will adhere to stainless steel and form a biofilm. Biofilms form in two stages. First, an electrostatic attraction occurs between the surface and the microbe. The process is reversible at this state. The next phase occurs when the microorganism exudes an extracellular polysaccharide, which firmly attaches the cell to the surface. The cells continue to grow, forming microcolonies and, ultimately, the biofilm.

These films are very difficult to remove during the cleaning operation. Microorganisms that appear to be more of a problem to remove because of biofilm protection are *Pseudomonas* and *L. monocytogenes*. Heat application appears to be more effective than that of chemical sanitizers, and Teflon appears to be easier to clear of biofilm than stainless steel.

At this writing, it appears that cold plasma may offer potential as a nonthermal intervention to protect against biofilms. Cold plasma is a form of ionized gas that can be utilized near room temperature. Cold plasmas are very reactive and are made by energizing pure gases or gas mixtures with high-voltage electricity. According to Niemira (2017), the electric charge strips molecules apart, creating ions, free electrons, oxygen singlet atoms, reactive radical species, and other gas plasma products. The antimicrobial modes of action arise from chemical reactions of plasma-reactive particles and molecules with bacterial cell structures and from additional ultraviolet (UV) damage to DNA and other cellular components caused by a UV light component of the cold plasma. Cold plasma has received attention from the medical field where biofilms present a challenge such as dental and oral treatments.

Biofilms protect against the penetration of water-soluble chemicals such as caustics, bleaches, iodophors, phenols, and quaternary ammonium sanitizers without microbial destruction. A biocide may require use at 10–100× normal strength to achieve inactivation.

In tests of sanitizers, including hot water at 82 °C (180 °F); chlorine at 20, 50, and 200 parts per million (PPM); and iodine at 25 PPM, bacteria on stainless steel chips survived after 5 min immersion in the sanitizer. Baker and Riche (2015) evaluated the attachment of cells to plastic materials analyzed with a scanning electron microscope (SEM) to study particles (biological or synthetic). This principle involves a beam of electrons emitted onto a surface or sample and capturing secondary electrons by detectors inside the SEM, which ultimately transmits the signal into an image viewed on a computer. Baker and Riche (2015) found that sanitizers used to sanitize plastic materials are not sufficient in removing attached cells visible by SEM. Thus, the removal of biofilms from a surface is very difficult and not well understood.

Microbial Growth

As temperature decreases, the *generation interval* (time required for one bacterial cell to become two cells) is increased. This is especially true when the temperature goes below 4° C (40 °F). Figure 3.2 illustrates the effect of temperature on microbial proliferation. For example, freshly ground beef may contain 1 million bacteria/g. When the number of this microbial population reaches approximately 300 million/g, abnormal odor and some slime development, with resultant spoilage, can occur. This trend does not apply to all genera and species of bacteria. However, these data reveal that initial contamination and storage temperature dramatically affect the shelf life of food. The storage life of ground beef that contains 1 million bacteria/g is approximately 28 h at 15.5 °C (60 °F). At normal refrigerated storage temperature of approximately—1 °C to 3 °C (34–38 °F), the storage life exceeds 96 h.

Effects of Microorganisms on Spoilage

When it becomes undesirable for human consumption, food is spoiled. Microorganisms cause spoilage through decomposition and putrefaction. Spoilage is an undesirable change in the flavor, odor, texture, or color of food caused by growth of microorganisms and ultimately the action of their enzymes.

Physical Changes

The physical changes caused by microorganisms usually are more apparent than the chemical changes. Microbial spoilage usually results in an obvious change in physical characteristics such as color, body, thickening, odor, and flavor degradation. Food spoilage is either aerobic or anaerobic, depending on the spoilage conditions, including whether the principal microorganisms causing the spoilage were bacteria, molds, or yeasts.

Aerobic spoilage of foods from molds is normally limited to the food surface, where oxygen is available. When molded surfaces of foods such as meats and cheeses are trimmed, the remainder is generally acceptable for consumption. This is especially true for aged meats and cheeses. When these surface molds are trimmed, surfaces underneath usually have limited microbial growth. If extensive bacterial growth occurs on the surface, penetration inside the food surface usually follows, and toxins may be present.

Anaerobic spoilage occurs within the interior of food products or in sealed containers, where oxygen is either absent or present in limited quantities. Facultative and anaerobic bacteria cause spoilage and is expressed through souring, putrefaction, or taint. Souring occurs from the accumulation of organic acids during the bacterial enzymatic degradation of complex molecules. Proteolysis without putrefaction may contribute to souring. Souring causes the production of various gases. Examples of souring are milk and round sour or ham sour in meat. Meat sours, or taints, are caused by anaerobic bacteria that have been originally present in lymph nodes or bone joints or that might have gained entrance along the bones during storage and processing.

Chemical Changes

Through the activity of endogenous hydrolytic enzymes present in foodstuffs (and the action of enzymes that microorganisms produce), proteins, lipids, carbohydrates, and other complex molecules are degraded into smaller and simpler compounds. Initially, the endogenous enzymes are responsible for the degradation of complex molecules. As microbial load and activity increase, degradation subsequently occurs. These enzymes hydrolyze the complex molecules into simpler compounds, subsequently utilized as nutrient sources for supporting microbial growth and activity. The effects of microbial action depend upon oxygen availability. Availability of oxygen permits hydrolysis of proteins into products such as simple peptides and amino acids. Under anaerobic conditions, protein degradation yields a variety of sulfur-containing compounds, which are odorous and generally obnoxious. The nonprotein nitrogenous compounds usually include ammonia.

Other chemical changes include action of lipases secreted by microorganisms that hydrolyze triglycerides and phospholipids into glycerol and fatty acids and phospholipids into nitrogenous bases and phosphorus. Extensive lipolysis accelerates lipid oxidation.

Most microorganisms prefer carbohydrates to other compounds as an energy source because of their readily utilization of energy. Utilization of carbohydrates by microorganisms results in a variety of products, such as alcohols and organic acids. In many foods, such as sausage products and cultured dairy products, microbial fermentation of sugar that has been added yields organic acids (such as lactic acid), which contribute to their distinct and unique flavors.

Effects of Microorganisms on Foodborne Illness

The United States has the safest food supply of all nations. However, the CDC estimates that there are 76 million foodborne illnesses per year in the United States with approximately 325,000 annual hospitalizations and 5,000 deaths attributable to this illness. However, the actual number of confirmed cases documented by the CDC is much lower.

The development of gastrointestinal disturbances following the ingestion of food can result from any one of several plausible causes. Although the sanitarian is most interested in those related to microbial origin, other causes are chemical contaminants, toxic plants, animal parasites, allergies, and overeating. Although each of these conditions is a potential source of illness in humans, subsequent discussions will address those illnesses caused by microorganisms.

Foodborne Disease

A foodborne disease is any illness associated with or in which the causative agent is through the ingestion of food. A *foodborne disease outbreak* is "two or more persons experiencing a similar illness, usually gastrointestinal, after eating a common food, if analysis identifies the food as the source of illness." Bacterial pathogens cause approximately 66% of all foodborne illness outbreaks. Of 200 foodborne outbreaks reported each year, approximately 60% are of undetermined etiology. Unidentified causes may be from the *Salmonella* and *Campylobacter* species, *Staphylococcus aureus*, *Clostridium perfringens*, *Clostridium botulinum*, *Listeria monocytogenes*, *Escherichia coli* O157, *Shigella*, *Vibrio*, and *Yersinia enterocolitica*, which are transmitted through foods. A wide variety of home-cooked and commercially prepared foods are implicated in outbreaks, but they are most frequently related to foods of animal origin, such as poultry, eggs, red meat, seafood, and dairy products.

Foodborne Illnesses

Food poisoning is an illness caused by the consumption of food containing microbial toxins or chemical poisons. Food poisoning caused by bacterial toxins is called *food intoxication*, whereas that caused by chemicals that have gotten into food is referred to as *chemical poisoning*. Illnesses caused by microorganisms exceed those of chemical origin. Illnesses that are not caused by bacterial by-products, such as toxins, but through ingestion of infectious microorganisms, such as bacteria, rickettsia, viruses, or parasites, are *food infections*. Foodborne illnesses caused from a combination of food intoxication and food infection are *food toxicoinfections*. In this foodborne disease, pathogenic bacteria grow in the food. Large numbers are ingested with food by the host, and, when in the gut, pathogen proliferation continues, with resultant toxin production, which causes illness symptoms. Illness caused by the mind, due to one witnessing another human sick or to the sight of a foreign object, such as an insect or rodent, in a food product, is termed *psychosomatic food illness*.

To provide protection against foodborne illness, it is necessary to have up-to-date knowledge of production, harvesting, and storage techniques for accurate evaluation of the quality and safety of raw materials. Thorough knowledge of design, construction, and operation of food equipment is essential to exercise control over processing, preservation, preparation, and packaging of food products. An understanding of the vulnerability of food products to contamination will help establish safeguards against food poisoning.

Aeromonas hydrophila Foodborne Illness

Evisceration and cold storage of chickens at 3 °C (38 °F) may permit an increase in *Aeromonas hydrophila*. Chill waters and the evisceration process itself appear to be probable sources of contamination in the typical broiler processing operation and may contribute to the high efficiency of occurrence of this microorganism at the retail level. This microorganism has been isolated from raw milk, cheese, ice cream, meat, fresh vegetables, finfish, oysters, and other seafood. It is a motile, facultative anaerobic, gram-negative rod with polar flagellum. The temperature range for growth is 4–43 °C (40–115 °F) with an optimum of 28 °C (82 °F). The pH range is 4.5–9.0, and the maximum concentration of salt for growth is 4.0%. *A. hydrophila* can cause gastroenteritis in humans and infections in patients immunocompromised by treatment for cancer.

Bacillus cereus Foodborne Illness

Bacillus cereus is a gram-positive, rod-shaped, spore-forming obligate aerobe that is widely distributed. Although some strains of this microbe are psychrotrophic and able to grow at 4–6 °C (40–42 °F), most proliferate at 15–55 °C (58–130 °F) with an optimal temperature of 30 °C (85 °F). The normal habitat for *B. cereus* is dust,

water, and soil. It is in many foods and food ingredients. Because this microorganism is a spore-former, it is heat resistant. Most of the spores have moderate resistance, but some have high heat resistance. The pH range for the proliferation of this bacterium is 5.0–8.8 with a minimum A_w of 0.93.

This microorganism produces two types of gastroenteritis: emetic and diarrheal. The *diarrhetic type* has relatively mild symptoms, such as diarrhea and abdominal pain that occur 8–16 h after infection and may last for approximately 6–24 h. In the emetic form of *B. cereus* illness, the symptom is primarily vomiting (which occurs within 1–6 h after infection and endures for 24 h or less), although diarrhea may occur also. The *B. cereus* emetic toxin is in the food and is heat stable like *S. faecalis*. The production of an enterotoxin within the gut causes the emetic form, which is more severe than the diarrhetic type. Consumption of rice or fried rice served in restaurants or warmed-over mashed potatoes is associated with foodborne outbreaks. Other foods associated with this foodborne illness include cereal dishes, vegetables, minced meat, meat loaf, milk products, soups, and puddings. The number of cells required for an outbreak is 5–8 log colony-forming units (CFU) per gram of food. Effective sanitation in restaurants including holding starchy cooked foods above 50 °C (122 °F) or refrigerating at below 4 °C (40 °F) within 2 h after cooking retards growth and toxin production.

Botulism

Botulism is a foodborne illness that results from the ingestion of a toxin produced by *Clostridium botulinum* during its growth in food. This microbe is an anaerobic, gram-positive, rod-shaped, spore-forming, gas-forming bacterium found primarily in the soil. The optimal growth temperature is 30–40 °C (85–104 °F). Temperature growth ranges are normally 10–50 °C (50–122 °F) except for type E, which thrives at 3.3–45 °C (36–116 °F). There are eight different botulinum toxins rec-

Table 3.1 Type of botulinum toxin

Type	Characteristics
A	Toxin is poisonous to humans; the most common cause of botulism
B	Toxin is poisonous to humans; found more often than type A in most soils
C_1	Toxin is poisonous to waterfowl, turkeys, and several mammals, but not to humans
C_2	Toxin is poisonous to waterfowl, turkeys, and several mammals, but not to humans
D	Toxin is responsible for forage poisoning of cattle, but rarely poisonous to humans
E	Toxin is poisonous to humans; usually associated with fish and fish products
F	Toxin is poisonous to humans; only recently isolated and extremely rare
G	Toxin is poisonous but rarely found
H	Poisonous toxin isolated from infant botulism patient that produced neurotoxin type B

ognized and serologically classified and another identified, but at this writing not acknowledged by the Food and Drug Administration in their documentation (see Table 3.1). The extremely potent toxin (the second most powerful biological poison known to humans) produced by this microorganism affects the peripheral nervous system of the victim. The ingestion of 10–100 spores that germinate in the intestinal tract and produce a toxin affects infants. Death occurs in approximately 60% of the cases from respiratory failure. The characteristics, including symptoms, incubation time, involved food, and preventive measures, of botulism and other common food poisonings are in Table 3.2.

Because *C. botulinum* may occur in the soil, it is also present in water. Therefore, seafood is a more viable source of botulism than are other muscle foods. However, the largest potential sources of botulism are home-canned vegetables and fruits with a low to medium acid content. Because this bacterium is anaerobic, canned and vacuum-packaged foods are also viable sources for botulism. Canned foods with a swell are unsuitable for consumption because the swelling results from the gas produced by the organism. It is essential to heat smoked fish to at least 83 °C (180 °F) for 30 min during processing to provide additional protection.

To prevent botulism, effective sanitation, proper refrigeration, and thorough cooking are essential. This toxin is relatively heat-labile, but the bacterial spores are very heat-resistant, and severe heat treatment is required to destroy them. Thermal processing at 85 °C (185 °F) for 15 min inactivates the toxin. The combinations of temperatures and times given in Table 3.3 are required for complete spore destruction.

Campylobacteriosis

Campylobacter has become a major concern because of transmission by food, especially inadequately cooked foods and through cross-contamination. The temperature for growth ranges from 30–45.5 °C (86–110 °F) with an optimum of 37–42 °C (98.5–108 °F). It survives to a maximum sodium chloride level of 3.5%. *Campylobacter* exists as commensals of the gastrointestinal tract of wild and domesticated animals. This fastidious, facultative (microaerophilic requiring 5% O_2 and 10% CO_2), gram-negative, non-spore-forming, spiral curve-shaped rod, which is motile by means of flagella, is now the most common cause of foodborne illness in the United States. It is the causative agent of veterinary diseases in poultry, cattle, and sheep and is quite common on raw poultry. Improvement of the detection and isolation of this microorganism has incriminated it in foodborne disease outbreaks. This microbe is one of the most frequent causes of bacterial diarrhea and other illnesses, and there is a mounting body of evidence that it causes ulcers. The infective dose of *Campylobacter* is normally 400–500 bacteria, depending on individual resistance. The pathogenic mechanisms of this pathogen allow it to produce a heat-liable toxin that may cause diarrhea.

Campylobacteriosis can occur at least twice as frequently as salmonellosis. The symptoms of foodborne illness from *Campylobacter* vary. Humans with a mild case may reflect no visible signs of illness but excrete this microorganism in their feces. Symptoms of those with a severe case may include muscle pain, dizziness, headache, vomiting, cramping, abdominal pain, diarrhea,

Table 3.2 Characteristics of the more common foodborne illnesses

Illness	Causative agent	Symptoms	Average time before onset of symptoms	Foods usually involved	Preventive measures
Bacillus	Bacillus cereus	Nausea, vomiting, diarrhea, abdominal pain	1–16 h	Cooked products, pasta, fried rice, and dried milk	Sanitary handling and rigid temperature control
Botulism	Toxins produced by Clostridium botulinum	Impaired swallowing, speaking, respiration, and coordination, dizziness and double vision, weariness, weakness	12–36 h	Canned low-acid foods, including canned meat and seafood, smoked and processed fish	Proper canning, smoking, and processing procedures. Cooking to destroy toxins, proper refrigeration, and sanitation
Staphylococcal (foodborne illness)	Enterotoxin produced by Staphylococcus aureus	Nausea, vomiting, retching, abdominal cramps due to gastroenteritis (inflammation of the lining of the stomach and intestines)	2–8 h	Custard and cream-filled pastries, potato salad, dairy products, ham, tongue, and poultry	Pasteurization of susceptible foods, proper refrigeration, and sanitation
Clostridium perfringens (foodborne illness)	Toxin produced by Clostridium perfringens (infection?)	Nausea, occasional vomiting, diarrhea, and abdominal pain	8–24 h	Cooked meat, poultry, and fish held at non-refrigerated temperatures for long periods of time	Prompt refrigeration of unconsumed cooked meat, poultry, or fish; maintain proper refrigeration and sanitation
Salmonellosis (food infection)	Infection produced by ingestion of any of over 1200 strains of Salmonella that can grow in the gastrointestinal tract of the consumer	Nausea, vomiting, diarrhea, fever, abdominal pain, may be proceeded by chills and headache	8–72 h	Insufficiently cooked or warmed-over meat, poultry, eggs, and dairy products; these are especially susceptible when kept refrigerated for a long time	Cleanliness and sanitation of food handlers and equipment, pasteurization, proper refrigeration, and packaging
Shigella infection (bacillary dysentery)	Shigella species	Nausea, vomiting, water diarrhea, fever, abdominal pain and cramps, chills, and headache	1–7 days	Foods handled by unsanitary workers	Hygienic practices of food handlers
Trichinosis (infection)	Trichinella spiralis (nematode worm) found in pork	Nausea, vomiting, diarrhea, profuse sweating, fever, and muscle soreness	2–14 days	Insufficiently cooked pork and products containing pork	Thorough cooking of pork and wild game (i.e., bear and cougar) to an internal temperature of 59–77 °C (138–170 °F) (higher with microwave cooking); frozen storage of uncooked pork at −15 °C (59 °F) or lower, for a minimum of 20 days; avoid feeding pigs raw garbage

Aeromonas (foodborne illness)	*Aeromonas hydrophila*	Gastroenteritis	–	Water, poultry, red meats	Sanitary handling, processing, preparation, and storage of foods; store foods below 2 °C (33 °F)
Campylobacter (foodborne illness)	*Campylobacter* species	Diarrhea, abdominal pain, cramping, fever, prostration, bloody stools, headache, muscle pain, dizziness, and rarely death	1–7 days	Poultry and red meats	Sanitary handling, processing, preparation, and storage of muscle foods
Listeriosis	*Listeria monocytogenes*	Meningitis or meningoencephalitis, listerial septicemia (blood poisoning), fever, intense headache, nausea, diarrhea, vomiting, lesions after contact, collapse, shock, coma, mimics influenza, interrupted pregnancy, stillbirth, 30% fatality rate in infants, and immunocompromised children and adults	4 days to several weeks	Milk, coleslaw, cheese, ice cream, poultry, red meats	Avoid consumption of raw foods with contact with infected animals; store foods below 2 °C (33 °F)
Yersiniosis	*Yersinia enterocolitica*	Abdominal pain, fever, diarrhea, vomiting, skin rashes for 2–3 days, and rarely death	1–7 days	Dairy products, raw meats, seafoods, fresh vegetables	Sanitary handling, processing, preparation, and storage of foods
Escherichia coli O 157:H7 (infection)	Enterohemorrhagic *Escherichia coli* O157:H7	Hemorrhagic colitis, hemolytic uremic syndrome with 5–10% acute mortality rate, abdominal pain, vomiting, anemia, thrombocytopenia, acute renal injury with bloody urine, seizures, pancreatitis	12–60 h	Ground beef, dairy products, raw beef, water, apple cider, mayonnaise	Sanitary handling, irradiation, cooking to 65 °C (149 °F) or higher
Hepatitis	Infectious hepatitis A	Fever, abdominal pain, nausea, cramps, jaundice	1–7 weeks approx. 25 days	Raw shellfish from polluted waters, sandwiches, salads, desserts	Thorough handwashing, sanitary food handling, cooking to 70 °C (158 °F)

Table 3.3 Temperatures and times required to destroy completely *Clostridium botulinum* spores

Temperature		
(°C)	(°F)	Time (min)
100	212	360
105	220	120
110	230	36
115	240	12
120	248	4

fever, prostration, and delirium. Diarrhea usually occurs at the beginning of the illness or after fever is apparent. Blood is frequently present in the stool after 1–3 days of diarrhea. The length of illness normally varies from 2–7 days. Although death is rare, it can occur. Complications and sequelae of campylobacteriosis include relapse (5–10%), bacteremia, meningitis, acute appendicitis, urinary tract infections, endocarditis, peritonitis, Reiter's syndrome, and Guillain-Barre syndrome (Davidson 2003). This pathogen is not tolerant to environmental stresses. Most cases of campylobacteriosis are sporadic and not associated with an outbreak. Sanitary handling and proper cooking of foods from animal origin is the most effective technique for the control of *Campylobacter*.

Campylobacter is in the intestinal tract of cattle, sheep, swine, chickens, ducks, and turkeys. Because this microorganism is in fecal material, contamination of muscle foods occurs during the harvesting (slaughtering) process without effective sanitation. *Campylobacter jejuni* is in milk, eggs, and water that have been in contact with animal feces. Limited studies have shown that the incidence of *C. jejuni* on retail cuts of red meat is lower than on retail poultry cuts. Symptoms and signs of *C. jejuni* infection lack special features not differentiated from illnesses caused by other enteric pathogens. Isolation of this pathogen is difficult because it is usually present in low numbers.

Normal levels of oxygen in the air will inhibit the growth of this microorganism. The strain of *C. jejuni*, initial contamination load, and environmental conditions especially storage temperature determines the survival in raw foods. Destruction is accomplished through heating contaminated foods to 60 °C (140 °F) internal temperature and holding at this temperature for several minutes for beef and approximately 10 min for poultry. Infection reduction of this pathogen is achieved thorough handwashing with soap and hot running water for at least 18 s before food preparation and between handling of raw and prepared foods.

Campylobacter outbreaks occur most frequently in children over 10 years old and in young adults, although all age groups are affected. This infection causes both the large and small intestines to produce a diarrheal illness. Although symptoms may occur between 1 and 7 days after eating contaminated food, illness usually develops 3–5 days after ingestion of this microbe.

A garlic-derived compound, diallyl sulfide is more effective than most antibiotics in retarding the proliferation of *Campylobacter*, especially if protected by a slimy biofilm. This compound can penetrate the biofilm and destroy bacterial cells by combining with a sulfur-containing enzyme, subsequently changing the enzyme's function and ceasing cell metabolism. The total elimination of this pathogen is unlikely. The web of causation (see Chap. 5) of campylobacteriosis is so diverse that complete elimination of *Campylobacter* species from domestic animals is not currently feasible.

Clostridium perfringens Foodborne Illness

Clostridium perfringens is an anaerobic, gram-positive, rod-shaped spore-former that produces a variety of toxins as well as gas during growth. This microbe will proliferate at a temperature range of 15–50 °C (58–122 °F) with an optimal temperature of 43–46 °C (110–114 °F). The optimal pH range is 6.0–7.0, but growth can occur from pH 5.0–9.0. The minimum A_w for growth is 0.95–0.97. This microorganism has a sodium chloride maximum of 7.0–8.0% and with inhibition at 5.0%. *Clostridium perfringens* and their spores have been isolated in many foods—especially among red meats, poultry, and seafood. Numbers of these microbes tend to be higher among meat items that have been cooked, allowed to cool slowly, and subsequently held for an extended period before serving. As with *Salmonella* microorganisms, ingestion of large numbers of active bacteria causes this type of foodborne illness to occur.

The spores from various strains of this microorganism have differing resistances to heat. Some spore destruction occurs within a few minutes at 100 °C (212 °F), whereas others require from 1–4 h at this temperature for complete destruction. Control of *Clostridium perfringens* is most effect by rapid cooling of cooked and heat processed foods. Frozen storage at −15 °C (4 °F) for 35 days provides greater than 99.9% kill of this microorganism. An outbreak of foodborne illness by *C. perfringens* can usually be prevented through proper sanitation as well as appropriate holding (≥60 °C) (−76 °F) and storage (≤2 °C) (26 °F) temperatures of foods at all times, especially of leftovers. Heating leftover foods (65 °C, 150 °F) destroys vegetative microorganisms.

Escherichia coli O157:H7 Foodborne Illness

Outbreaks of hemorrhagic colitis and hemolytic uremic syndrome caused by *Escherichia coli* O157:H7, a facultative anaerobic, gram-negative, rod-shaped bacterium, have elevated this pathogen to a high echelon of concern. It is uncertain how this microorganism mutated from *E. coli*, but some scientists speculate that it picked up genes from *Shigella*, which causes similar symptoms. This microorganism is in the feces of cattle and can contaminate meat during processing. It is important to establish intervention procedures during slaughter and meat processing operations. Cooking beef to 70 °C (158 °F) ensures sufficient heat treatment to control and destroy this pathogen. A rigid sanitation program is essential to reduce foodborne illness outbreaks from this microorganism.

In 1982, the identification of *E.coli* O157:H7 (designated by its somatic (O) and flagellar (H) antigens) as a human pathogen followed two hemorrhagic colitis outbreaks. Six classes of diarrheagenic *E. coli* are recognized. They are enterohemorrhagic, enterotoxigenic, enteroinvasive, enteroaggregative, enteropathogenic, and diffusely adherent. All enterohemorrhagic strains produce Shiga toxin 1 and/or Shiga toxin 2, also referred to as *verotoxin 1* and *verotoxin 2*. The ability to produce Shiga toxin came from a bacteriophage, presumably directly or indirectly from *Shigella*. The infectious dose associated with foodborne illness outbreaks from this pathogen has been low (2000 cells or less), due to the organism's acid tolerance.

The initial symptoms of hemorrhagic colitis generally occur 12–60 h after eating contaminated food, with periods of 3–5 days reported. This bacterium attaches itself to the walls of the intestine, producing a toxin that attacks the intestinal lining. Symptoms start with mild, nonbloody diarrhea followed by abdominal pain and short-lived fever. During the next 24–48 h, the diarrhea increases in intensity to a 4–10-day phase of overtly bloody diarrhea, severe abdominal pain, and moderate dehydration.

A life-threatening complication that may occur in hemorrhagic colitis patients is hemolytic uremic syndrome, which may occur a week after the onset of gastrointestinal symptoms. Characteristics of this condition include edema and acute renal failure. It occurs most frequently in children less than 10 years old. Approximately 50% of these patients require dialysis, and the mortality rate is 3–5%. Other associated complications may include seizures, coma, stroke, hypertension, pancreatitis, and hypertension. Approximately 15% of these cases lead to early development of chronic kidney failure and/or insulin-dependent diabetes, and a small number of cases may recur.

Thrombotic thrombocytopenic purpurea is another illness associated with *E. coli* O157:H7. It resembles hemolytic uremic syndrome, except that it normally causes renal damage, has significant neurologic involvement (i.e., seizures, strokes, and central nervous system deterioration), and is restricted primarily to adults.

Ground beef has been the food most often associated with outbreaks in the United States. Dry-cured salami has been associated with an outbreak revealing that low levels of this pathogen can survive in acidic fermented meats and cause illness. Other foods associated with this pathogen are unpasteurized apple juice, apple cider, and radish and alfalfa sprouts. An outbreak in the United States involved alfalfa sprouts. Drinking water and recreational waters have been vehicles of several *E. coli* O157:H7 outbreaks.

Research has revealed that 3.2% of dairy calves and 1.6% of feedlot cattle tested were positive for *E. coli* O157:H7. Deer are a source for this pathogen, and the transmission of this microorganism may occur between deer and cattle. Fecal shedding of this pathogen is transient and seasonal. The prevalence of *E. coli* O157:H7 in feces peaks in the summer and during spring through fall on the hide.

E. coli O157:H7 can grow at 8 °C (46 °F) to 44.5 °C (112 °F) with an optimal temperature of 30–42 °C (86–108 °F). Growth rates are similar at pH values between 5.5 and 7.5 but decline rapidly under more acidic conditions even though this pathogen survives a low pH well. The minimum pH for *E. coli* O157:H7 is 4.0–4.5. Several outbreaks have been associated with low levels of this pathogen surviving in acidic foods, such as fermented sausages, apple cider, and apple juice. Research results reveal that this pathogen will survive for several weeks in a variety of acidic foods, such as mayonnaise, sausages, and apple cider.

Cooking ground beef to 72 °C (161 °F) or higher or incorporating a procedure that kills this pathogen in the manufacture of fermented sausages or the pasteurization of apple cider destroys this pathogen. The HACCP system appears to be the most effective means for systematically developing food safety protocols that can reduce infection from this pathogen. The low incidence of this pathogen limits the utility of direct microbial testing as a means of verifying the effectiveness of HACCP.

Listeriosis

Listeria monocytogenes is an especially dangerous pathogen because it can survive at refrigerated temperatures. Previously, listeriosis was rare in humans. However, foodborne outbreaks since the 1980s have increased public health concern over this pathogen. In the United States, this pathogen causes approximately 1600 illnesses every year and accounts for nearly 19% of annual foodborne-related deaths (Chaves and Brashears 2016–17). Individuals in certain high-risk groups are more likely to acquire listeriosis. Pregnant women are approximately 20 times more susceptible than other healthy adults (Duxbury 2004). *Listeria monocytogenes* is an opportunistic pathogen, as it normally does not cause severe disease in healthy individuals with strong immune systems.

This microorganism is a facultative gram-positive, rod-shaped, non-spore-forming microaerophilic (5–10% CO_2) bacterium. *L. monocytogenes*, a ubiquitous pathogen, occurs in human carriers (ca. 10% of the population) and is found in the intestinal tracts of over 50 domestic and wild species of birds and animals, including sheep, cattle, chickens, and swine, as well as in soil and decaying vegetation. Other potential sources of this microorganism are stream water, sewage, mud, trout, crustaceans, houseflies, ticks, and the intestinal tracts of symptomatic human carriers. This pathogen is in most foods, from chocolate and garlic bread to dairy products and meat and poultry. Elimination of *Listeria* is impractical and may be impossible. The critical issue is how to control its survival and proliferation.

The optimal temperature range for the proliferation of this microbe is 30–37 °C (86–98 °F); however, growth can occur at a temperature range of 0–45 °C (32–112 °F). This microorganism is a psychrotrophic pathogen, which grows well in damp environments. *L. monocytogenes* is very tolerant of environmental stresses compared to other vegetative cells and has a high vegetative cell heat resistance. It grows in over 10% salt and survives in saturated salt solutions. This pathogen will grow twice as fast at 10 °C (50 °F) as at 4 °C (40 °F) and survives freezing temperatures. Destruction of this pathogen occurs at processing temperatures above 61.5 °C (143 °F). Although *L. monocytogenes* is frequently in milk, cheese, and other dairy products, it can be present in vegetables fertilized with the manure of infected animals. This microorganism thrives in substrates of neutral to alkaline pH but not in highly acidic environments. Growth can occur in a pH range from 5.0–9.6, depending on the substrate and temperature. *L. monocytogenes* operates through intracellular growth in mononuclear phagocytes. Once the bacterium enters the host's monocytes, macrophages, or polymorphonuclear leukocytes, it can evade host defenses and grow.

Human listeriosis may be caused by any of 13 serotypes of *L. monocytogenes*, but those most likely to cause illness are 1/2a, 1/2b, and 4b (Farber and Peterkin 2000). Most cases of listeriosis are sporadic. This illness primarily affects pregnant women and infants, people over 50 year old, those debilitated by a disease, and other individuals who are in an immunocompromised state of health. Meningitis or meningoencephalitis is the most common manifestations of this disease in adults. This disease may occur as a mild illness with influenza-like symptoms, septicemia, endocarditis, abscesses, osteomyelitis, encephalitis, local lesions, or minigranulomas (in the spleen, gall bladder, skin, and lymph nodes) and fever. Fetuses of pregnant women with this disease may also be infected. These women might suffer an interrupted pregnancy or give birth to a stillborn child. Infants who survive birth may be born with septicemia or develop meningitis in the neonatal period. The fatality rate is approximately 30% in newborn infants and nearly 50% when the infection occurs in the first 4 days after birth.

Listeriosis is a dangerous infection for persons with AIDS. Because AIDS severely damages the immune system, those with the disease are more susceptible to a foodborne illness such as listeriosis. Those with AIDS can be more than 300 times as susceptible. The infectious dose for *L. monocytogenes* is unknown because of the presence of unidentified factors in persons with normal immune systems that make them less susceptible to the bacteria than immunocompromised persons. The infectious dose depends on both the strains of *Listeria* and on the individual. However, thousands or even millions of cells may be required to infect healthy animals, whereas 1–100 cells may infect those who are immunocompromised. The severe form of human listeriosis usually does not occur in the absence of a predisposing infection, although *L. monocytogenes* can cause gastroenteritis in previously healthy individuals.

Listeria monocytogenes can adhere to food contact surfaces by producing attachment fibrils, with the subsequent formation of a biofilm, which impedes removal during cleaning. The attachment of *Listeria* to solid surfaces involves two phases. They are primary attraction of the cells to the surface and firm attachment following an incubation period. A primary acidic polysaccharide is responsible for initial bacterial adhesion. This microbe adheres by producing a mass of tangled polysaccharide fibers that extend from the bacterial surface to form a "glycocalyx," which surrounds the cell of the colony and functions to channel nutrients into the cell and to release enzymes and toxins. These microbes are also potential contaminants of raw materials utilized in plants, which contribute to constant reintroduction of this organism into the plant environment. Utilization of Hazard Analysis and Critical Control Points (HACCP) and other process control practices is the most effective method of controlling this pathogen in the processing environment. The HACCP approach has helped to identify critical points and to evaluate the effectiveness of control systems through verification procedures.

Transmission of this pathogen occurs through the consumption of contaminated food, but it can also occur from person-to-person contact or by inhalation of this microorganism. For example, a person who has had direct contact with infected materials, such as animals, soil, or feces, may develop lesions on the hands and arms. This pathogen may occur in home refrigerators, suggesting the need for regular cleaning and sanitizing of this equipment.

The most effective prevention against listeriosis is to avoid the consumption of raw milk, raw meat, and foods made from contaminated ingredients. It is important for pregnant women, especially, to avoid contact with infected animals. Fail-safe procedures for the production of *Listeria*-free products do not exist. Thus, food processors must rely on a rigid environmental sanitation program and HACCP principles to establish a controlled process. The most critical areas for the prevention of contamination are plant design and functional layout, equipment design, process control operational practices, sanitation practices, and verification of *L. monocytogenes* control.

Various studies have demonstrated that *L. monocytogenes* is resistant to the effects of some sanitizers. This pathogen has resistance to the effects of trisodium phosphate (TSP), and exposure to a high (8%) level of TSP for 10 min at room temperature is required to reduce bacterial numbers by one log after a colony has grown on the surface and a biofilm has formed. Furthermore, washing skin with 0.5% sodium hydroxide has a minimal effect on the proliferation of *L. monocytogenes*. This microorganism is more resistant to the cooking process than are other pathogens, and cooking may not be a definitive means of eliminating the organism from foods. Although *L. monocytogenes* is susceptible to irradiation, it is not the final solution with regard to eliminating this pathogen from fresh meat and poultry.

Although a minimal number of listeriosis cases exist in the United States each year, a significant number of those affected die from the disease. This microorganism is a "super bacterium" that can survive environmental extremes that will eliminate other pathogenic bacteria. Thus, food processors and foodservice operators should focus on reducing the presence of this microorganism in products, even though it is nearly impossible to eliminate this pathogen from the food supply.

Salmonellosis

The reported incidence of *Salmonella* illnesses is approximately 14 cases per each 100,000 persons. Salmonellosis is a food infection because it results from the ingestion of any one of numerous strains of living *Salmonella* organisms. The infectious dose can be as low as one to ten cells. These microbes grow in a 5–47 °C (40–116 °F) (37 °C or 98 °F optimal temperature) environment and produce an endotoxin (a toxin retained within the bacterial cell) that causes the illness. The usual symptoms of salmonellosis are nausea, vomiting, and diarrhea, which appear to result from the irritation of the intestinal wall by the endotoxins. The ingestion of approximately 1 million of these microorganisms causes an infection to occur. The time lapse between ingestion and appearance of symptoms of salmonellosis is generally longer than that of staphylococcal food poisoning symptoms. Mortality from salmonellosis is generally low. Most deaths that occur are among infants, the aged, or those already debilitated from other illnesses. Salmonellosis may be especially harmful to persons with AIDS since these patients are susceptible to this foodborne illness.

Salmonella is a complicated microorganism with more than 2400 species in circulation. This pathogen is a facultative anaerobic, gram-negative non-spore-forming, oval-shaped bacterium that primarily originates from the intestinal tract. *Salmonella* generally grows at an optimum A_w of 0.86 in a pH range of 3.6–9.5 with an optimum range of 6.5–7.5. A salt concentration of over 2% will retard growth, but this microbe is very tolerant of freezing and drying. These bacteria may be present in the intestinal tract and other tissues of poultry and red meat animals without producing any apparent symptoms of infection in the animal. This microorganism is an enduring problem for fresh poultry and exists among 70% broiler carcasses.

Although *Salmonella* can be present in skeletal tissues, the major source of the infection results from the contamination of food by the handlers during processing, through recontamination or cross-contamination. *Salmonella* transferred by the fingertips are capable of surviving for several hours and still contaminating food. Thermal processing conditions for the destruction of *S. aureus* will destroy most species of *Salmonella*. Because of the origin of these bacteria and their sensitivity to cold temperature, poor sanitation and temperature abuse contribute to salmonellosis.

Shigellosis

Shigella gastroenteritis (called shigellosis or bacillary dysentery) is an infection with an onset time of 1–7 days that endures 5–6 days. Primary symptoms vary with severe cases that may result in bloody diarrhea, mucus secretion, dehydration, fever, and chills. Death may occur among immunocompromised individuals, but the mortality rate is usually low among others. Foods most associated with shigellosis are those subjected to a large amount of handling or those contaminated with waterborne *Shigella*. Foods infected with this microorganism are potato, chicken, shrimp, tuna salads, and seafood/shellfish. Most outbreaks occur in foodservice establishments such as hospital cafeterias and restaurants and are frequently attributable to ineffective handwashing after defecation.

Shigella are gram-negative, non-spore-forming rods that are weakly motile and lactose negative with low heat resistance. *Shigella* are generally not hearty and lack resistance to environmental stresses. These facultative anaerobes grow from 6 °C–48 °C with an optimum temperature of 37 °C (98 °F). This microorganism is primarily of human origin and spreads to food by carriers and contaminated water. The pH range for *Shigella* is 4.9–9.3. It requires a minimum A_w of 0.94 with a maximum salt content of 4.0–5.0%. *Shigella* is a highly infectious microorganism since the ingestion of less than 100 of these bacteria can cause illness. *Shigella* spp. elaborate a toxin that has enterotoxic, neurotoxic, and psychotoxic activities responsible for inflammatory intestinal responses.

Staphylococcal Foodborne Illness

Staphylococcus aureus, a facultative, sphere-shaped, gram-positive non-spore-forming microorganism, produces an enterotoxin that causes an inflammation of the stomach and intestines, known as *gastroenteritis*. Although mortality seldom occurs from staphylococcal food poisoning, the central nervous system is affected. Death is usually due to added stress among people with other illnesses. The bacteria causing staphylococcal food poisoning are widely distributed and can be present among healthy individuals. The pH range for *S. aureus* is 4.0–9.8 with 6.0–7.0 being optimum. It tolerates a water activity as low as 0.86 in the presence of ca. 20% salt.

Handling of improperly refrigerated food by infected individuals is one of the greatest sources of contamination. The most common foods that may cause staphylococcal food poisoning are potato salad, custard-filled pastries, dairy products (including cream), poultry, cooked ham, and tongue. With ideal temperature and high contamination levels, staphylococci can multiply enough to cause foodborne illness without noticeable changes in color, flavor, or odor. Heating *Staphylococcus aureus* organisms to 66 °C (151 °F) for 12 min destroys them, but the toxin requires heating for 30 min at 131 °C (268 °F). Therefore, the normal cooking time and temperature for most foods will not destroy the enterotoxin.

Trichinosis

Humans transmit this illness by *Trichinella spiralis*, which can infect the flesh of pork and wild game such as bear and cougar. Most humans infected by this organism are asymptomatic. Symptomatic illness includes gastroenteritis symptoms including fever, nausea, vomiting, and diarrhea. Onset time is approximately 72 h with an infection time of up to 2 weeks. Initial symptoms are followed by edema, muscle weakness, and pain when the larvae migrate encysting the muscles. Furthermore, respiratory and neurological manifestations may occur. Death may result without treatment. Prevention is possible through protection from contamination and cooking to 40 °C (104 °F) with conventional cookery (i.e., gas and electric heat) or 71 °C (160 °F) with microwave heating. Other destruction methods include irradiation or frozen storage of meat less than 15 cm (6″) thick for 6 days at −29 °C (−20 °F) or 20 days at −15 °C (5 °F).

Yersiniosis

Yersinia enterocolitica, a psychrotrophic pathogen, is in the intestinal tracts and feces of wild and domestic animals. Other sources are raw foods of animal origin and non-chlorinated water from wells, streams, lakes, and rivers. Transmission of this microorganism occurs from person to person. Fortunately, most strains isolated from food and animals are avirulent.

Y. enterocolitica will multiply at refrigerated temperatures, but at a slower rate than at room temperature. This facultative anaerobic, gram-negative, non-spore-forming rod is heat sensitive and is destroyed at temperatures over 60 °C. However, the growth range of this pathogen is −2–45 °C (26–112 °F) with an optimal temperature of 28–29 °C (82–84 °F). This pathogen grows at a pH range of 4.2–9.6 and tolerates a high pH effectively. The presence of this microbe in processed foods suggests post-heat treatment contamination. *Y. enterocolitica* has been isolated from raw or rare red meats; the tonsils of swine and poultry; dairy products such as milk, ice cream, cream, eggnog, and cheese curd; most seafoods; and fresh vegetables.

Not all types of *Y. enterocolitica* cause illness in humans. Yersiniosis can occur in adults but most frequently appears in children and teenagers. The symptoms, which normally occur from 1–3 days after ingesting the contaminated food, include fever, abdominal pain, and diarrhea. Vomiting and skin rashes can also occur. Abdominal pain associated with yersiniosis closely resembles appendicitis. In food-related outbreaks in the past, some children have had appendectomies because of an incorrect diagnosis.

The illness from yersiniosis normally lasts 2–3 days, although mild diarrhea and abdominal pain may persist 1–2 weeks. Death is rare but can occur due to complications. The most effective prevention measure against yersiniosis is proper sanitation in food processing, handling, storage, and preparation.

Foodborne Illness from *Arcobacter butzleri*

This microorganism, which is in beef, poultry, pork, and non-chlorinated drinking water, occurs in up to 81% of poultry carcasses. It is more resistant to irradiation and more tolerant of oxygen than *C. jejuni* and grows at refrigerated temperatures in atmospheric oxygen.

Cryptosporidiosis

Cryptosporidium parvum causes cryptosporidiosis through transmission via fecal contamination of water or food. Onset time is 1–2 weeks and the duration is 2 days–4 weeks. This bacterium forms oocysts that persist for long periods in the environment and are resistant to chlorine. Oocysts are susceptible to high temperatures, freezing, dehydration, and sanitizers such as ozone, hydrogen peroxide, and chlorine dioxide. Filtration removes them from water by filtration. Symptoms of cryptosporidiosis include watery diarrhea, abdominal pain, and anorexia.

Foodborne Illness from *Helicobacter pylori*

Research results suggest that this pathogen, related to *Campylobacter*, may cause gastroenteritis and is a causative agent for gastritis, stomach and intestine ulcers, and stomach cancer in humans. It is suspected that this microorganism, which is the most common chronic bacterial infection in humans, can swim and resist muscle contractions that empty the stomach during contraction. This bacterium is in the digestive tract of animals, especially pigs. It is present in 95% of duodenal and in up to 80% of human gastric ulcer cases, in addition to clinically healthy individuals, including family members of patients. Sewage-contaminated water is a source of transmission of this microorganism.

Legionellosis

Legionella pneumophila is a vibrant bacterium that causes Legionnaires' disease. This facultative gram-negative microbe is in contaminated waters in most of the environment and is becoming a widespread concern. This bacterium is able to multiply intracellularly within a variety of cells. The dominant extracellular enzyme produced by *L. pneumophila* is a zinc metallo-protease, also called a *tissue-destructive prote-ase*, *cytolysin*, or *major secretory* protein. This protease is toxic to different types of cells and causes tissue destruction and pulmonary damage, which suggests its involvement in the pathogen-esis of Legionnaires' disease.

This microorganism causes 1–5% of community-acquired pneumonia in adults, with most cases occurring sporadically. The Center for Disease Control and Prevention receives 1,000–3,000 reports of cases of Legionnaires' disease each year. Aerosol-producing devices, such as cooling towers, evaporating condensers, whirlpool spas, humidifi-ers, decorative fountains, and tap water faucets, cause most of the outbreaks.

Water is the major reservoir for *Legionella* organisms; however, this microorganism is in other sources, such as potting soil. Amoebae and biofilms, which are ubiquitous within plumbing systems, have a critical role in the amplification process of supporting the bacterial growth.

The inhalation of *Legionella* organisms as an aerosolized liquid to respirable size (1–5 μm) transmits legionellosis. Occasional transmission occurs through other routes, such as inoculation of surgical wounds with contaminated water dur-ing the placement of surgical dressings.

Vibrio spp.

Several species of *Vibrio*, such as *Vibrio para-haemolyticus*, *V. cholerae*, and *V. vulnificus*, are pathogens. This microbe is a gram-negative, non-spore-forming, straight-curved facultatively anaerobic rod. *Vibrio parahaemolyticus* grows at 13–45 °C (56–113 °F) with an optimum range of 22–43 °C (72–110 °F). It grows at pH 4.8–11.0 with an optimum range of 7.8–8.6, while the range and optimum for V. *cholerae* are 5.0–9.6 and 7.6 and for V. *vulnificus* are 5.0–10.0 and 7.8. The minimal A_w is 0.94, 0.96, and 0.97 for V. *parahaemolyticus*, V. *vulnificus*, and V. *cholerae*, respectively. The optimal amount of salt is 0.5, 2.5, and 3.0 for V. *cholerae*, V. *parahaemolyticus*, and V. *vulnificus*, respectively. The primary habi-tat for *Vibrio* is seawater.

The onset time for V. *parahaemolyticus* gas-troenteritis is 8–72 h with an average of 18 h. Symptoms include diarrhea and abdominal cramps accompanied by nausea, vomiting, and mild fever. Illness duration is 48–72 h with a low mortality rate. The number of cells required to cause illness is 5–7 logs.

Mycotoxins

Mycotoxins are compounds or metabolites pro-duced by molds that are toxic or have other adverse biological effects on humans and animals (Table 3.4). They originate from a wide range of fungi. The acute diseases caused by mycotoxins are *mycotoxicoses*. Mycotoxicoses are not com-mon in humans. However, epidemiologic evi-dence suggests an association between primary liver cancer and aflatoxin, one type of mycotoxin, in the diet. In large doses, aflatoxins are acutely toxic, causing gross liver damage with intestinal and peritoneal hemorrhaging, resulting in death. Mycotoxins may enter the food supply by direct

Table 3.4 Mycotoxins of significance to the food industry

Mycotoxin	Major[a] producing microorganism	Potential foods involved
Aflatoxin	*Aspergillus flavus* *Aspergillus parasiticus*	Cereal, grains, flour, bread, corn meal, popcorn, peanut butter
Patulin	*Penicillium cyclopium, Penicillium expansum*	Appeals and apple products
Penicillic acid	*Aspergillus* species	Moldy supermarket foods
Ochratoxin	*Aspergillus ochraceus, Penicillium viridicatum*	Cereal grains, green coffee beans
Sterigmatocystin	*Aspergillus versicolor*	Cereal grains, cheese, dried meats, refrigerated, and frozen pastries

[a]Other genera and species may produce these mycotoxins

contamination, resulting from mold growth on the food. Entry can occur by indirect contamination through contaminated ingredients in processed foods or from the consumption of foods containing mycotoxin residues.

Molds that are capable of producing mycotoxins are frequent contaminants of food commodities. Those that are important in the food industry because of potential mycotoxin production include members of the genera *Aspergillus*, *Penicillium*, *Fusarium*, *Cladosporium*, *Alternaria*, *Trichothecium*, *Byssochlamys*, and *Sclerotinia*. Most foods are susceptible to invasion by these or other fungi during some stage of production, processing, distribution, storage, or merchandising. Mold growth produces mycotoxins. The existence of mold in a food product, however, does not necessarily signify the presence of mycotoxins. Furthermore, the absence of mold growth on a commodity does not indicate that it is free of mycotoxins, because a toxin can exist after the mold has disappeared.

Of the mycotoxins, aflatoxin poses the greatest potential hazard to human health. *Aspergillus flavus* and *Aspergillus parasiticus* produce aflatoxins, which are nearly ubiquitous with spores that are widely disseminated by air currents. These molds occur among cereal grains, almonds, pecans, walnuts, peanuts, cottonseed, and sorghum. The microorganisms will normally not proliferate unless these commodities are insect damaged, not dried quickly, and not stored in a dry environment. Growth can occur by the invasion of the kernels with mold mycelium and subsequent aflatoxin production on the surface and/or between cotyledons.

The clinical signs of acute aflatoxicosis include lack of appetite, listlessness, weight loss, neurological abnormalities, jaundice of mucous membranes, and convulsions. Death may occur. Other evidence of this condition is gross liver damage through pale color, other discoloration, necrosis, and fat accumulation. Edema in the body cavity and hemorrhaging of the kidneys and intestinal tract may also occur.

Control of mycotoxin production is complex and difficult. Insufficient information exists regarding toxicity, carcinogenicity, and teratoge-

nicity to humans, stability of mycotoxins in foods, and extent of contamination. Such knowledge is required to establish guidelines and tolerances. The best approach to eliminating mycotoxins from foods is to prevent mold growth at all levels of production, harvesting, transporting, processing, storage, and marketing. Prevention of insect damage and mechanical damage throughout the entire process—from production to consumption—as well as moisture control, is essential. An A_w level above 0.83, or approximately 8–12% kernel moisture, depending on the type of grain, produces mycotoxins. Therefore, rapid and thorough drying and storage in a dry environment is necessary. The peanut industry incorporates photoelectric eyes that examine and pneumatically remove discolored kernels that may contain aflatoxins to aid in control and to avoid the difficult, tedious, and costly process of hand sorting.

Other Bacterial Infections

Other bacterial infections that occur cause illnesses with symptoms similar to food poisoning. The most common of these infections is *Streptococcus faecalis*. Infections caused by enterotoxigenic *E. coli* are the most common cause of "traveler's diarrhea," an illness frequently acquired by individuals from developed countries during visits to developing nations where hygienic practices may be substandard.

Microbial Destruction

Microorganisms are dead when they cannot multiply, even after being in a suitable growth medium under favorable environmental conditions. Death differs from dormancy, especially among bacterial spores, because dormant microbes have not lost the ability to reproduce, as evidenced by eventual multiplication after prolonged incubation, transfer to a different growth medium, or some form of activation.

Regardless of the cause of death, microorganisms follow a logarithmic rate of death, as in the accelerated death phase of Fig. 3.1. This pattern suggests that the population of microbial cells is dying at a relatively constant rate. Deviations from this death rate can occur due to accelerated effects from a lethal agent, effects due to a population mixture of sensitive and resistant cells, or with chain- or clump-forming microbial flora with uniform resistance to the environment.

Heat

Historically, application of heat has been the most widely used method of killing spoilage and pathogenic bacteria in foods. Heat processing is a way to cook food products and destroy spoilage and pathogenic microorganisms. Therefore, extensive studies have determined optimal heat treatment to destroy microorganisms. A measurement of time required to sterilize completely a suspension of bacterial cells or spores at a given temperature is the *thermal death time* (TDT). The value of TDT will depend on the nature of the microorganisms, its number of cells, and factors related to the nature of the growth medium.

Another measurement of microbial destruction is *decimal reduction time* (D value). This value is the time in minutes required to destroy 90% of the cells at a given temperature. The value depends on the nature of the microorganism, characteristics of the medium, and the calculation method for determining the D value. This calculated value is for a period of exponential death of microbial cells (following the logarithmic order of death). The D value can be determined through an experimental survivor curve.

Increased concern about pathogens of fecal origin (such as *E. coli* O157:H7) has been responsible for the investigation and implementation of hot-water spray washing of beef carcasses immediately after slaughter and dressing as a method of cleaning and decontamination. Smith (1994) identified the best combination (and sequence) of interventions reducing microbial load to be use of 74 °C (165 °F) water in the first wash and 20 kg/cm^2 (110 lbs/in^2) pressure and spray wash with hydrogen peroxide or ozone in the second wash (especially if 74 °C (165 °F) water temperature is

not incorporated in the first wash). The passage of ready-to-eat meats through a tunnel of heated coils prior to packaging and post-package pasteurization is also an effective tool for controlling L. monocytogenes surface contamination.

Chemicals

Many chemical compounds that destroy microorganisms are not appropriate for killing bacteria in or on a foodstuff. Acceptable applied sanitizing agents protect equipment and utensils that can contaminate food. As the cost of energy for thermal sanitizing has increased, the use of chemical sanitizers has grown. Chlorine disinfection may result from slow penetration into the cell or the necessity of inactivating multiple sites within the cell before death results. (Additional discussion related to this subject is in Chap. 10.) Chlorine, acids, and phosphates are potential decontaminants for microbial load on red meat and poultry carcasses.

Radiation

When microorganisms in foods are irradiated with high-speed electrons (beta rays) or with X-rays (or gamma rays), the log of the number of survivors is directly proportional to the radiation dose. The relative sensitivity of a specific strain of microorganisms subjected to specific conditions is the slope of the survivor curve. The thermal D value results from plotting the log_{10} of survivors from radiation against the radiation dosage and the radiation D or D_{10} value, which is comparable with the thermal D value. The D_{10} value is defined as the amount of radiation in rods (ergs of energy per 100 g (3.5 oz.) of material) to reduce the microbial population by 1-log (90%).

The destructive mechanism of radiation is unclear. It appears that death occurs by inactivation of cellular components through energy absorbed within the cell. A cell inactivated by radiation cannot divide and produce visible outgrowth. (Additional information related to radiation as a sanitizer is in Chap. 10.)

Electronic Pasteurization

Pasteurization is an act or process, usually involving heat, which reduces the number of bacteria in a food product without changing the chemistry or property of the food. Electron-beam accelerators provide electron pasteurization of food products through bombarding the products directly with electrons or optimizing the conversion of electron energy to X-rays and treating the product with these X-rays. For electron treatment, 10 million electron volts (meV) kinetic energy is the maximum allowed by international agreement.

Accelerators provide X-rays or electrons for treatment of food. An accelerator provides energy to electrons by providing an electric field (potential energy) to accelerate the electrons. Electrons are atomic particles, rather than electromagnetic waves, and their depth of penetration in the product is smaller. Therefore, the direct use of electrons is limited to packages less than 10 cm (4″) thick.

Pulsed Light

A potential method of microbial reduction on both packaging and food surfaces is the utilization of intense pulses of light. Pulsed light is energy released as short, high-intensity pulses of broad-spectrum "white" light that can sterilize packaging materials and decrease microbial populations on food surfaces. Reductions of more than 8 logs of vegetative cells and 6 logs of spores on packaging materials, and in beverages, and 1–3 logs on complex or rough surfaces, such as meat, may be achieved.

Compressing electrical energy into short pulses and using these pulses to energize an inert gas lamp create pulsed-light flashes. The lamp emits an intense flash of light for a few hundred microseconds. Only a few flashes are required to produce a high level of microbial kill because of multiple lamp flashing times per second. Thus, an online procedure for food processing can be very rapid.

The advantage of pulsed light is that it penetrates deeper than continuous UV light and it is faster. Depending on the characteristics of the pulsed-light device, high inactivation levels occur in seconds. Koutchma (2016) indicated that pulsed-light treatment at 8400 mJ/cm^2 did not affect the sensory quality of cooked ham, while treatments above 2100 mJ/cm^2 negatively influenced the sensory properties of bologna.

Pruett and Dunn (1994) reported that the incorporation of an acetic acid spray before pulsed-light treatment led to higher levels of pathogen kill. A potential multi-hurdle concept is the use of a hot-water spray in combination with pulsed light. Past investigations have revealed no nutritional or sensory changes attributable to pulsed light.

Microbial Growth Control

Most methods used to kill microorganisms are a milder treatment to inhibit microbial growth. Sublethal heating, irradiation, or treatment with toxic chemicals frequently causes injury to microorganisms and impaired growth without death. An increased lag phase, less resistance to environmental conditions, and greater sensitivity to other inhibitory conditions indicate injury. Synergistic combinations of inhibitory agents, such as irradiation plus heat and heat plus chemicals, can increase microbial sensitivity to inhibitory conditions. Injured cells appear to require synthesis of some essential cell materials (i.e., ribonucleic acid or enzymes) before recovery is accomplished. Inhibition of microbial growth is through maintenance of hygienic conditions to reduce debris available to support bacterial proliferation.

Refrigeration

Previous discussion addressed the effect of temperature on microbial proliferation. Freezing and subsequent thawing will kill some of the microbes. Those that survive freezing will not proliferate during frozen storage. Yet, this method of reducing the microbial load is not practical. Microorganisms that survive frozen storage will grow on thawed foods at a rate similar to those that are unfrozen. Refrigerated storage complements other methods of inhibition-preservatives, heat, and irradiation.

Chemicals

Chemicals that increase osmotic pressure with reduced A_w below the level that permits growth of most bacteria function as bacteriostats. Examples include salt and sugar.

Dehydration

Reduction of microbial growth by dehydration is another method of reducing the A_w to a level that prevents microbial proliferation. Some dehydration techniques restrict the types of microorganisms that may multiply and cause spoilage. Dehydration is most effective when combined with other methods of controlling microbial growth, such as salting and refrigeration.

Fermentation

In addition to producing desirable flavors, fermentation can control microbial growth. It functions through anaerobic metabolism of sugars by acid-producing bacteria that lower the pH of the substrate, the foodstuff. A pH below 5.0 restricts growth of spoilage microorganisms. Acid products that result from fermentation contribute to a lower pH and reduced action of microorganisms. Acidified and heated foods packed in hermetically sealed containers prevent spoilage by aerobic growth of yeasts and molds.

Biopreservation

Biopreservation encompasses food preservation techniques that range from ancient fermentation methods to modern technologies such as bacteriocins and bacteriophages. This concept incorporates the use of nonpathogenic microorganisms that antagonize or inhibit the growth of undesirable spoilage and/or pathogenic microbes through their metabolic activity or capacity to compete for nutrients or attachment niches.

Bacteriophages (also called phages) are a group of viruses that affect only bacteria and exist where bacteria are present. Bacteriophages are proven and vicious attackers of bacteria and utilize them as their hosts. Lytic cycle phages break open and destroy cells to replicate the phage virus, which then hunts down new hosts. Phages are generally stable but less active at refrigerated temperatures.

Approved phage products are for use in food processing to combat *Listeria*, *Salmonella*, and *E. coli* O157:H7. Various processing steps, including finished products, incorporate some of the phage products. Fuhrman (2014) indicated that phage specificity is a benefit but also a challenge as a specific phage needed to infect most of the pathogen strains needing to be controlled. Because bacteriophages are specific for their host bacteria, they are most suited for the control of pathogenic bacteria that fall into a relatively narrow spectrum of bacterial species. These organisms do not possess their own metabolism and therefore are sensitive to many of the same intrinsic and extrinsic factors that affect bacteria. Since more emphasis is on the reduction of antibiotic use in animal production, phages offer potential in the supplementation or replacement of antibiotics that control disease or promote growth. Furthermore, phages have a potential application in preharvest interventions through processing.

According to the World Health Organization, probiotics are live microorganisms administered in appropriate amounts that confer a health benefit on the host. Two of the well-known probiotic bacteria are *Lactobacillus* and *Bifidobacterium*. These probiotics are inhibitory by competing for nutrients or producing a metabolic product that is antimicrobial (e.g., lactic acid). Reilly (2016) suggested that probiotics act by mechanisms that follow:

1. Blanketing. Populating the food contact surface with a probiotic film and prevention of pathogen binding by occupying available

space and possible excretion of anti-adhesion molecules that change the molecular charge or the hydrophobicity of the surface preventing pathogen binding.

2. Biosurfactant production. Secretion of a biosurfactant that breaks down the existing biofilm through a change in surface tension to permit the surface to become wet and facilitate biofilm dispersion or prevention of pathogen adherence. Probiotics may compete by the development of their own biofilm, but a biosurfactant removes pathogenic biofilms that may exist in hidden niches on equipment. These biosurfactants possess a lower biodegradability than conventional synthetic surfactants discussed in Chap. 9.

3. Exopolysaccharide (EPS) production. This mechanism involves a matrix released from the cell that dampens a pathogen's ability to remain viable. EPS may modulate the expression of pathogen genes that produce biofilms or surface adhesions.

Microbial Load Determination

Various methods are available for determining microbial growth and activity in foods. The choice of method depends on the information required, tested food product, and the characteristics of the microbe(s). One of the most important factors in obtaining accurate and precise results is the collection of representative samples. Because of the large numbers and variability of microorganisms present, microbial analyses are less accurate and precise and, therefore, more subjective than are chemical methods of analysis.

Different test methodologies offer advantages, limitations, and disadvantages. Speed, accuracy, test breadth or robustness, and costs determine the overall desirability. A Fourier transform using infrared spectrometry accelerates the time to results to approximately 20–24 h, although this method can only test for a limited number of microorganisms. An excellent information source for rapid test kits is the AOAC Research Institute that has certified a large number of test kits. Technical knowledge and experience related to

microbiology and food products are essential for selection of the most appropriate method and the ultimate application of results.

Among the most advanced rapid microbial testing platforms utilized are fluorescent DNA markers to identify species power. Some culture-independent platforms utilize in situ hybridization, fluorescent microagglutination, and flow cytometry to identify pathogens rapidly (McCright 2016).

During the past, many of the microbial determination methods were culture based, with microorganisms grown on agar plates. These methods have been slow, labor intensive, and tedious to perform. Now, the food industry utilizes several rapid microbial test kits and automated systems to enable firms to detect, identify, and correct potential microbial hazards in their products before release from the plant. These technologies (usually DNA based) include immunological methods (i.e., ELISA), automated biochemical identification and optical systems (i.e., biosensors), and molecular methods (i.e., PCR and microarrays). Immunocapture techniques incorporate antibodies attached to plastic beads to facilitate recovery of pathogens from a food matrix.

Although microbial analysis may not provide precise results, it can indicate the degree of hygiene reflected through equipment, utensils, other portions of the environment, and food products. In addition to reflecting sanitary conditions, product contamination, and potential spoilage problems, microbial analysis can indicate anticipated shelf life. Because several new and improved methods are now available, it is difficult to indicate which will be the most viable in the future. Therefore, we will look at some potential methods of assessment of microbial load here. Readers interested in more information should review current technical microbiology journals.

Aerobic Plate Count Technique

This technique is among the most reproducible methods used to determine the population of microorganisms present on equipment or food products. It assesses the amount of contamination from the air, water, equipment surfaces, facilities, and food products. This technique includes swabbing of equipment, walls, and/or food products with the subsequent transfer to peptone water or a phosphate buffer according to the anticipated amount of contamination. The sample goes to a growth medium containing agar in a sterile, covered plate (Petri dish) with the diluted material transferred to a culture medium (such as standard methods agar) that nonselectively supports microbial growth.

The number of colonies that grow on the growth medium in the sterile, covered plate during an incubation period of 2–20 days (depending on incubation temperature and potential microorganisms) at an incubation temperature consistent with the environment of the product being tested reflects the number of microorganisms contained by the sample. This technique provides limited information related to the specific genera and species of the sample, although physical characteristics of the colonies can provide a clue. Special methods that permit the selective growth of specific microorganisms are available to determine their presence and quantity.

This method is reliable, but it is slow and laborious. The need for a faster response to a high-volume production environment has encouraged the investigation of more rapid methods. Slowness of "end-product" testing can retard production and does not provide an actual total count. Its use continues because of reliability and wide acceptance. Test kits such as TEMPO AC can shorten the time required to conduct this technique.

Surface Contact Technique

This method of assessment, called the *contact plate technique*, is similar to the plate count technique except for swabbing. A covered dish or rehydrated Petrifilm is opened with the growth medium (agar) pressed against the area to be sampled. The incubation process is the same as for the total plate count method. This method is easy to conduct, and less chance for error (including contamination) exists. The greatest limitation of this technique is that it is only for lightly contaminated surfaces because dilution is not possible. Press plates monitor the effectiveness of a sanitation program. The amount of growth on the media suggests the amount of contamination.

Indicator and Dye Reduction Tests

Various microorganisms secrete enzymes as a normal metabolic function of their growth, which are capable of inducing reduction reactions. Some indicator substances (such as dyes) are the basis of these tests. The rate of their reduction, indicated by a color change, is proportional to the number of microorganisms present. The time required for the complete reduction of a standard amount of the indicator is a measure of the microbial load. A modification of these methods involves a dye-impregnated filter paper applied directly onto a food sample or piece of equipment. The time required for the filter paper to change color determines the microbial load.

This method lacks utility because (1) biofilms and not all microorganisms are detected and (2) material cost. This technique does not quantify the extent of contamination. However, it is quicker and easier to conduct than the plate count technique and has become and acceptable tool for evaluation of a sanitation program effectiveness.

Direct Microscopic Count

A known volume is dried and fixed to a microscope slide, stained with a number of fields (frequently 50) counted. Although most staining techniques do not distinguish between viable and nonviable bacteria, this method estimates the number of microorganisms. Sophisticated digital cameras attached to microscopes capture images using image analysis software. These images analyze different bacteria based on size and enumerate organisms/field, thus eliminating human error. Although this method provides morphologic or specific staining information, it receives limited use because analyst fatigue can produce errors and only the limited quantity examined.

Most Probable Number

This estimate of bacterial populations involves placing various dilutions of a sample in replicate tubes containing a liquid medium. The number of tubes in each replicate set of tubes in which growth occurred (as evidenced by turbidity) compared to the number in a standard most probable number (MPN) table determines the number of microorganisms. This method measures only viable bacteria and it permits further testing of the cultures for purposes of identification.

Petrifilm Plates

Petrifilm plates contain a dehydrated nutrient medium on a film. This self-contained, sample-ready approach is an alternative method to the standard aerobic plate count (SPC) and coliform counts, as determined by violet red bile (VRB) pour plates. The most commonly used methodology for enumerating *E. coli* from broiler chicken carcasses and ground beef are rapid detection methods such as Petrifilm (3 M Co.) and SimPlate (Neogen). These methods, which are available as commercial test kits, depend on the detection of the production of an enzyme (glucuronidase) through *E. coli*.

Cell Mass

The quantifying of cell mass estimates microbial populations in certain research applications, but not normally in routine analysis since it can be more time-consuming and less practical than other methods. The measured and centrifuged fluid packs the cells, with subsequent decanting and discarding of the supernatant, or filtered through a bored asbestos or cellulose membrane and weighed.

Turbidity

Turbidity is an arbitrary determinant of the number of microorganisms in a liquid. This rarely used technique lacks utility because the food particles in suspension contribute to turbidity and inaccurate results.

Radiometric Method

With this technique, a sample goes into a medium containing a ^{14}C-labeled substrate, such as glucose. The measured amount of $^{14}CO_2$ produced relates to microbial load. Because some microorganisms will not metabolize glucose, ^{14}C-glutamate and ^{14}C-formate media are incorporated. This technique is limited to applications where data acquisition is required within 8 h and/or need for technician labor reduction. Utilization of this method has been limited.

Impedance Measurement

Impedance measurements determine the microbial load of a sample by monitoring microbial metabolism rather than biomass. Impedance is the total electrical resistance to the flow of an alternating current passed through a given medium. Microbial colonies on media produce changes in impedance that measured by the continuous passage of a small electrical current in as soon as 1 h. This technique offers potential as a rapid method of determining microbial load. Previous research has revealed a correlation of 0.96 between impedance-detecting time and bacterial counts. Impedance enumerates aerobic plate count (APC) coliforms, *E. coli*, psychrotrophs, and *Salmonella* organisms to predict shelf life and to do sterility testing.

Endotoxin Detection

The limulus amoebocyte lysate (LAL) assay is for the detection of endotoxins produced by gram-negative bacteria (including psychrotrophs and coliforms). Amoebocyte lysate from the blood of the horseshoe crab forms a gel in the presence of minute amounts of endotoxin. Due to heat stability, the detection of both viable and nonviable bacteria makes this test useful in tracing the history of the food supply. The LAL assay involves placing a sample into a prepared tube of lysate reagent, incubating 1 h at 37 °C (98 °F), and evaluating the degree of gelation.

Bioluminescence

This biochemical method, simplified for easy use, measures the presence of adenosine triphosphate (ATP) by its reaction with the luciferin-luciferase complex. It estimates the microbial load of a food sample. The bioluminescent reaction requires ATP, luciferin, and firefly luciferase—an enzyme that produces light in the tail of the firefly. During the reaction, the oxidized luciferin emits light. A luminometer measures the light produced, which is proportional to the amount of ATP present in the sample. The ATP content of the sample correlates with the number of microorganisms present because all microbial cells have a specific amount of ATP. An automated, palm-sized

luminometer can detect the presence of yeast, mold, or bacterial cells in liquid samples in as few as 5 s (Figs. 3.3 and 3.4). A computer-interfaced luminometer, which employs customized

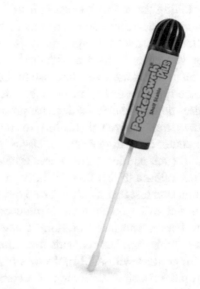

Fig. 3.3 Swab for a rapid hygiene test (Courtesy of Ecolab Inc., St. Paul, Minnesota)

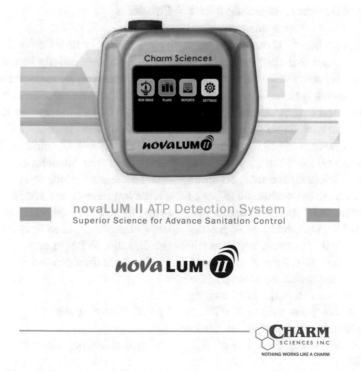

Fig. 3.4 Device for the rapid determination of hygienic conditions (Courtesy of Ecolab Inc., St. Paul, Minnesota)

software, a printer, and an automatic sampler, can analyze samples with a sensitivity of one microorganism per 200 mL (0.21 quarts). Furthermore, the detection unit illustrated in Fig. 3.4 features enhanced onboard data analysis tools including the ability to search historical results and add corrective action results with the generation of graphs for rapid analysis of pass/fail for key areas in food establishments. Use of this method has increased because of the need for more rapid results from product testing. It requires approximately 12 days for products to flow from microbial testing to the distribution center and out to retailers. Use of a rapid method, such as bioluminescence, accelerates product release to less than 24 h. A surface contamination test that requires 2 or 3 days using agar-based testing methods can be reduced to less than 1 min. The incorporation of new, highly sensitive biochemical reagents that emit light when in contact with ATP molecules has permitted rapid microbial screening to detect extremely low levels of microorganisms.

Benefits of rapid methods testing and the reduced risk of contamination have enhanced the evolution of bioluminescence technology as a reliable rapid test for microbial contamination. Although agar plate-based technology may appear to be less expensive than bioluminescence, a cost-analysis study demonstrated that rapid microbial testing offers a savings of approximately 40% over traditional testing methods. A limitation to this test is that cleaning compound residues can quench the light reaction to prevent proper response from the assay system. Many commercial bioluminescence detection kits contain neutralizers to combat the effect of detergents/sanitizers. ATP bioluminescence is ineffective in powder plants when milk powder or flour residues exist. Furthermore, naturally luminescent organisms exist in seafood plants. This increases the incidence of false-positive results on the surfaces tested. Furthermore, yeasts have up to 20 times as much ATP as bacteria, to complicate enumeration. A major advantage of this test is the detection of ATP from tissue exudates, whereas other tests do not offer this feature. Furthermore, this test identifies dirty equipment.

Previous research has involved increasing the sensitivity of bioluminescence reactions through identification of the adenylate kinase enzyme that produces ATP. This approach permits the counting of lower numbers of microorganisms present.

A colored hygiene test strip detects residues on surfaces by measurement of nicotinamide adenine dinucleotide. This technique provides an easy and rapid monitoring of cleaning measures.

A bioluminescent enzyme immunoassay (BEIA), using salmonella-specific monoclonal antibody M183 for capture and biotinylated monoclonal antibody M183 for detection, offers another alternative for the detection of salmonella. This immunoassay offers an advantage of providing a 24-h test for detecting salmonella in chicken carcass rinses.

MicroSnap is a modification of the ATP bioluminescence reaction. This novel rapid test system is capable of detecting bacteria at low levels in a variety of sample types in 6–8 h with a multifunction luminometer.

Catalase

This enzyme is in foods and aerobic bacteria. Because catalase activity increases with the bacterial population, its measurement can estimate bacterial load. A Catalasemeter utilizes the disc flotation principle to measure catalase activity in foods and can detect 10,000 bacteria/mL (0.0021 pint) within minutes. This unit, which incorporates the biochemical method of detection and enumeration, is an online monitoring device to detect contamination problems in raw materials and finished products, to control vegetable blanching and milk quality, and to detect subclinical mastitis in cows. The catalase test is applicable to fluid products.

Spiral Assay System

This equipment deposits a liquid sample in a spiral pattern onto a rotating agar plate and can create a 3-log dilution effect. The merits of this system include reduced or elimination of serial dilutions, less materials (pipettes, plates, media, and other supplies), less time and labor, and simplified plate counting. The disadvantages of this system include investment cost and required specialized equipment (i.e., plating machine and counting machine).

Direct Epifluorescence Filter Technique (DEFT)

This biophysical technique is a rapid, direct method for counting microorganisms in a sample. This method monitors milk samples and has been applied to other foods, even though it is not used routinely in the food industry. This technique incorporates both membrane filtration and epifluorescence microscopy. A sample on a polycarbonate membrane captures microorganisms. The cells, stained with acridine orange, cause the viable bacteria to fluoresce orange and the dead bacteria to fluoresce green under the blue portion of the ultraviolet spectrum. An epifluorescence microscope, which illuminates the sample with incident light, counts the fluorescing bacteria.

This technique evaluates dairy and muscle foods, beverages, water, and wastewater. A prediction of the keeping quality of pasteurized milk stored at 5 °C (40 °F) and 11 °C (52 °F) within 24 h occurs by sample preincubation and counting bacteria by DEFT. The enumeration of *L. monocytogenes* in ready-to-eat packaged salads and other fresh vegetables and in the detection of *E. coli* O157:H7 in ground beef, apple juice, and milk occurs by incorporating the antibody-direct epifluorescent filter technique (Ab-DEFT). In addition to membrane filtration of food to collect and concentrate microbial cells on the membrane surface, fluorescent antibody staining of the filter surface and epifluorescence microscopy are involved. Examination under a microscope follows placement of the added fluorescent antibody on a slide. This quantifying method for *L. monocytogenes* has demonstrated the potential of Ab-DEFT as a rapid alternative for the quantitation of *Listeria* in food. However, nonspecific reactivity of the fluorescent antibodies to indigenous microbial populations has resulted in false-positive reactions using Ab-DEFT.

Remote Inspection Biological Sensor

Biosensors provide an instantaneous indication of the presence of specific pathogens in a food sample without need for enrichment and can detect generic *E. coli* and *Salmonella*. They may provide continual feedback of pathogen loads in fluids within a plant. The remote inspection biological sensor (RIBS) uses a laser spectrographic technique with the laser beam directed onto the surface of a carcass. Based on the characteristics of the reflected light, this equipment can make a specific identification of pathogenic bacteria and give a general indication of the number of organisms present. It has a sensitivity of up to five colony-forming units (CFUs) per square centimeter (0.4″) and effectively discriminates target organisms from the background.

Microcalorimetry

Heat production measurement from a biological reaction, such as the catabolic processes occurring in growing microorganisms cultured from contaminated samples, is by a sensitive calorimeter called a *microcalorimeter*. This biophysical technique enumerates microorganisms in food. The procedure correlates a thermogram (a heat-generation pattern during microbial growth) with the number of microbial cells. The establishment of a reference thermogram permits a comparison of the reference to others obtained from contaminated samples.

Radiometry and Infrared Spectrophotometry

An inverse relationship exists between the number of microorganisms in a sample and the time required for the detection of certain levels of radioactivity by this biophysical technique. This method employs sterility testing of aseptically packaged products. Results are available in

4–5 days, compared with 10 days with conventional methods. The enumeration of microorganisms in food samples requires less than 24 h.

Hydrophobic Grid Membrane Filter System

This culturing method detects and enumerates *E. coli* in foods. An ISO-GRID hydrophobic grid membrane filter (HGMF) system is available to detect and enumerate *E. coli*. Filtering a sample through a membrane without use of an enrichment step and a complex medium (SD-39) detects the target organism. The test involves 48 h, including biochemical and serological confirmation of presumptive colonies.

Other Screening Devices

The RapidChek lateral flow device offers another screening technique. Beyond detection, whole-genome sequencing is incorporated.

Diagnostic Tests

Enzyme-Linked Immunoassay Tests

A high level of technical skill is required to perform these tests. Because of the time and skill required, several rapid methods for detecting *Salmonella* have been developed such as enzyme-linked immunosorbent assays (ELISAs), immunodiffusion methods, immunomagnetic bead ELISAs, nucleic acid hybridization methods, and polymerase chain reaction methods. Furthermore, there are automated immunodiagnostic assays. The VIDAS-SLM automated method is a rapid screening technique and a potential alternative to the time- and labor-intensive culture method. Goodridge et al. (2003) developed a rapid MPN-ELISA for the detection and enumeration of *Salmonella typhimurium* in poultry processing wastewater.

Two other developments in rapid immunoassay and molecular methods areas are the magneto immunochromatography test (MICT) and loop-mediated isothermal amplification (LAMP). MICT consists of antibody-coated superparamagnetic nanoparticles in a lateral flow immunoassay format. This test utilizes ELISA technology for pathogen detection, achieving capture of the target through antigen-/antibody-binding affinity, but utilizes magnetic nanoparticles in the detection phase versus an optical conjugate/substrate enzyme reaction to color change or fluorescence. LAMP represents a process innovation in DNA-based testing methodologies, specifically in the area of polymerase chain reaction. To achieve exponential DNA amplification and enable rapid, accurate detection of the target analyte, PCR-based methods utilize thermocycling. Temperature cycling and DNA amplification enabled by polymerase chain reaction (PCR) are conducted within a thermocycler.

Antigens are the specific constituents of a cell or toxin that induce an immune response and interact with a specific antibody, whereas antibodies are immunoglobulins that bind specifically to antigens. Immuno-based assays incorporate either monoclonal or polyclonal antibodies. Monoclonals are a single type of antibody with a high affinity for a specific target antigen epitope. A polyclonal antibody is a set of different antibodies specific for an antigen but able to recognize different epitopes of the antigen. The advantages of these assays are rapid results, increased sensitivity and specificity, and decreased costs. Enzyme-linked immunoassays have been effective in detecting pathogens and are easy to conduct. Similar competing organisms in food producing similar antigens resulting in a cross-reaction complicate this detection method.

Formatted systems described previously consist of antibodies attached to a solid support, such as the walls of a microtiter plate or a plastic dipstick. An added enrichment culture to the solid support permits antibodies to bind target antigens in the sample. An added sandwich format in which a second added enzyme-labeled antibody to the sample, followed by a reactive substrate, produces a positive color reaction. If the target antigens are not present, the labeled antibody will not attach and no color reaction occurs.

An efficient and sensitive method of analyzing samples for pathogens is immunoblotting. The common procedure involves an enrichment culture that is spotted onto a solid support (i.e., nitrocellulose paper), with the remaining protein-binding areas of the paper blocked by dipping in a protein solution such as bovine serum albumin or reconstituted dry milk. An enzyme-labeled antibody solution specific for the target pathogen is applied, and a substrate for the added enzyme after washing removes the unbound antibody. If the labeled antibody is present, due to attachment to the target antigen, a color reaction will indicate a positive sample. This modified procedure is for use in conjunction with other methods, such as the HGMF system.

Another technique for pathogen detection is the use of superparamagnetic microspheres coated with an antibody specific to a target antigen. The selectively enriched sample transferred to a test tube includes a small amount (approximately 10 mL or 0.021 pints) of the enrichment culture. The antibody-coated beads are added and gently and briefly shaken. A magnetic particle concentrator separates the beads from the sample homogenate. After reconstitution in a buffer, the beads are spread-plated onto a selective agar to observe growth of the target pathogen. Confirmation of presumptive colonies occurs if present in the original sample. These beads detect *E. coli* O157:H7 in foods.

A latex agglutination test provides quick results with an acceptable degree of specificity for *E. coli* O157 but not for H7 confirmation. An available assay uses a polyclonal O157 antibody coated onto polystyrene latex particles and an incorporated slide agglutination format to transfer a suspect culture to a paper card, followed by the addition of the antibody reagent. Agglutination indicates presence of the O157 antigen.

A lateral flow immunoprecipitate assay serves as a screen test for *E. coli* O157:H7. This assay, approved by the Association of Official Analytical Chemists (AOAC), requires an enrichment broth and incubation for 20 h at 36 °C (97 °F). Subsequently, deposition of a 0.1 mL (0.0021 pint) sample of the enrichment broth in a test window in a self-contained, single-use test device that contains proprietary reagents occurs. As lateral flow occurs across the reagent zone, the target antigen, if present, reacts with the reagents to form an antigen-antibody-chromogen complex. After approximately 10 min of incubation at room temperature, a line will form in the test window, indicating the possible presence of *E. coli* O157:H7. If no line appears, a confirmed negative test results. As flow continues through the test verification zone, all samples will react with reagents, and a line will appear, indicating proper completion of the test. A positive test does not ensure that an *E. coli* O157:H7 strain exists. A tested suspect sample further confirms the presence of the pathogen. This test, which is easy to conduct, incorporates an assay system into a single test unit.

A key difference between ELISA and PCR tests is detection limits. The detection limit difference is a longer required enrichment time for an ELISA method. To combat overgrowth of competing nontarget bacteria in a food sample, a secondary selective enrichment is typically required in an ELISA method to permit the target bacteria an opportunity to grow. The following web address provides a listing of *Salmonella* detection and determination methods: https://www.researchgate.net/publication/264275306.

RAPID ONE System

This test for *Enterobacteriaceae* relies on preformed enzymes. This one-step inoculation is easy to use. It provides results in 4 h, but a competent microbiologist is required for correct interpretation.

Crystal™ Identification Systems

This system relies on preformed enzymes. The one-step inoculation is easy to use with the inoculum suspended in lysing buffer. A 3-h computer-assisted ID match yields results; however, a competent technician is required for consistent interpretation.

Salmonella 1–2 Test

This rapid screening test for *Salmonella* is conducted in a single-use, plastic device that contains a nonselective motility medium and a selective enrichment broth. It yields a positive test by an immobilization precipitation band that forms in the motility medium from the reaction of motile *Salmonella* with flagellar antibodies.

This test incorporates a clear plastic device with two chambers. The smaller chamber contains a peptone-based, nonselective motility medium. This procedure involves the addition of the sample to the tetrathionate-brilliant green-serine broth contained in the inoculation chamber of the 1–2 test unit. After approximately 4 h of incubation, motile *Salmonella* move from the selective motility medium. As these organisms progress through the motility medium, they encounter flagellar antibodies diffused into this medium. The reaction of the motile *Salmonella* with the flagellar antibodies results in an immobilized precipitation band 8–14 h after an inoculation.

DNA-Based Microarray Assays

The emergence of new DNA-based microarray assays permits a look at DNA sequences of microorganisms, including strains within an organism, for very precise identification. DNA microarrays are a revolutionary concept in the evolution of food microbiology tests because in a single or small number of assays one can screen for a large number of microorganisms. Following the standard PCR protocol that amplifies the DNA for detection of a microbe, an analyst can use a single DNA chip to identify 40–100 species or strains of microorganisms in a single test. DNA chip technology also changes the way to approach an unknown organism in a food matrix. With conventional tests, one can only detect one pathogen per single test. Knowledge of what organisms may be in the food matrix is essential before choosing an appropriate test. DNA microarrays permit one to identify what microbe is in the food matrix.

IDEXX Bind

The IDEXX Bind for *Salmonella* incorporates genetically engineered bacteriophages. The modified bacteriophages attach to *Salmonella* receptors and insert DNA into the bacterial cells. During incubation, the modified DNA causes *Salmonella* to produce ice nucleation proteins. At a specified temperature, the ice nucleation proteins promote the formation of ice crystals. Positive samples will freeze and turn orange at this temperature, whereas negative samples will not freeze.

Random Amplified Polymorphic DNA

The random amplified polymorphic DNA (RAPD) method has achieved promising results, especially to trace *L. monocytogenes* infections in humans. Advantages are the low cost of the multiple DNA primers, discriminating nature of the test, and the ability to trace small amounts of *L. monocytogenes*. Since this assay is time-consuming, it has more utility as a research tool than as a diagnostic test for industry use.

Immunomagnetic Separation and Flow Cytometry

This technique detects less than 10 *E. coli* O157:H7 cells/g of ground beef after enrichment for 6 h. The immunomagnetic beads concentrate cells, making it easier to detect, using flow cytometry. The presence of other microorganisms does not influence the detection limit. This method is more of a research tool than as a diagnostic tool in the food industry.

Diagnostic Identification Kits

These kits are for human clinical medicine but can aid in the identification of various microorganisms. Most of these tests are for use with isolated colonies, which require 1–3 days to obtain. A Petrifilm Salmonella Express System is available as an all-in-one test and biochemical confirmation used for the detection of this microorganism in enriched foods and food process environmental samples.

CAMP Test

This test involves a suspected isolate of *L. monocytogenes* streaked adjacent to or across a streak of a second, known bacterium on a blood agar plate. At the juncture of the two streaks, the metabolic by-products of the two bacteria diffuse and result in an augmented hemolytic reaction. Hemolysis of blood cells is an important characteristic of pathogenic bacteria such as *L. monocytogenes* because it appears to be closely associated with virulence. Through this method, the virulence of *L. monocytogenes* may be determined.

Fraser Enrichment Broth/Modified Oxford Agar

This method is for *Listeria* detection using Fraser enrichment broth combined with modified Oxford agar for motility enrichment. The *Listeria* organisms are enriched in Fraser broth and held at 30 °C (85 °F) for 24 h, and 1 mL (0.0021 pint) of the enrichment broth is placed in the Fraser broth in the left arm of a U-shaped tube. The Fraser broth selectively isolates and promotes *Listeria* growth and precludes the growth of nonmotile microorganisms. The microbes migrate through the modified Oxford agar and arrive as a pure culture in the second branch of the Fraser broth. This becomes the second enrichment necessary for the identification of *Listeria*. An easier indication that *Listeria* organisms are present is the formation of a black precipitate as the bacteria move through the modified Oxford agar. When turbidity develops, the sample for DNA probe analysis confirms the presence of *Listeria*. The second enrichment step requires 12–24 h. The US Food and Drug Administration also lists a number of alternative screening tests for *Listeria*.

Crystal Violet Test

The retention of crystal violet by *Y. enterocolitica* correlates with virulence. Most *Y. enterocolitica* strains isolated from meat and poultry are avirulent. Thus, this rapid test allows the identification and rapid discarding of samples with virulent strains.

Methyl Umbelliferyl Glucuronide Test

The enzyme, glucuronidase, produced by most *E. coli* and other microbes such as *Salmonella*, splits methyl umbelliferyl glucuronide (MUG). When split, MUG becomes fluorescent under ultraviolet illumination of a specific wavelength and permits rapid identification in tubed media or on spread plates for enumeration.

Assay for *E. coli*

Several techniques exist for the rapid identification of microorganisms. Many techniques have not been available long enough to establish their efficacy or to achieve AOAC approval. Although several methods are available, most require 24–48 h for incubation of the microorganisms and may need additional testing to confirm the presence of *E. coli*. Many commercial assays for the detection of *E. coli* incorporate membrane filtration technology, and others employ a reagent/sample mixture incubated for 24–48 h to obtain a presence/absence result of total *E. coli* contamination.

An assay for a rapid, inexpensive determination of *E. coli* concentrations in aqueous environments is the IME. *Test™-EC KOUNT Assayer*. This assayer uses a reagent mixture containing an indicator compound that provides a colorimetric (bright blue) indication of *E. coli* concentration in a water-based sample, predicated on cleavage by the beta-galactosidase enzyme specific to *E. coli*. This assay provides a simple method for quantifying the concentration of viable *E. coli* in an aqueous sample in 2–10 h.

The procedure involves filling a snapping cup with a sample and introducing it to a vacuum-sealed test ampoule by snapping off the sealed tip in one of the holes in the bottom of the cup. The ampoule automatically fills with the aqueous sample. The sample, incubated at 35 °C (95 °F) and monitored for the production of a blue fluorescence, results in enzymatic cleavage of the indicator molecule, MUG. The time for the production of a bright blue color, visualized under long-wave ultraviolet light optically or via instrument, is proportional to the total *E. coli*/mL (0.0021 pint) in the sample. Based on time to positive, a comparison chart provides the corresponding *E. coli* count for the sample. Concentration and detection times are:

E. coli concentration	Detection time (h)
9.9×10^6 CFU/mL (0.0021 pint)	2
100 CFU/mL (0.0021 pint)	10

Further incubation of samples that are negative at 12 h provides a presence/absence determination after 24 h. This technique permits sampling at a remote site and return to a laboratory for analysis. The major limitation appears to be that not all of the *E. coli* bacteria react in the presence of MUG.

Micro ID and Minitek

Micro ID is a self-contained identification unit containing reagent-impregnated paper discs for biochemical testing for the differentiation of *Enterobacteriaceae* in approximately 4 h. This technique has provided reliable results. The Minitek system is another miniaturized test kit for the identification of *Enterobacteriaceae*. This kit also utilizes reagent-impregnated paper discs requiring 24 h of incubation. It is accurate and versatile. The Analytab Products, Inc. (API) strip is the most commonly used identification unit.

DNA Hybridization and Colorimetric Detection

This assay methodology combines DNA hybridization technology with nonradioactive labeling and colorimetric detection. With the appropriate specific DNA probes, enrichment, and sample preparation procedures for a particular organism, this basic assay is for the analysis of a wide variety of microbes. The assay requires approximately 2.5–3 h after 2 days of broth culture enrichment of the sample.

An application of this principle is a colorimetric assay, which employs synthetic oligonucleotide DNA probes against ribosomal RNA (rRNA) of the target organism. This approach offers increased sensitivity because rRNA, as an integral part of the bacterial ribosome, is present in multiple copies (1000–10,000) per cell. The number of ribosomes present per cell is dependent on the growth state of the bacterial culture.

Polymerase Chain Reaction

This technique detects low levels of pathogens found in food products. Polymerase chain reaction (PCR) amplifies very low DNA levels (as low as one molecule) or detectable levels of target DNA (approximately 10^6) through a series of DNA hybridization reactions and thermocycling. Various methods, such as gel electrophoreses and colorimetric or chemiluminescent assays, detect PCR products. In real-time PCR, specificity increases by the use of probes and primers designed to target conserved regions of the target genome (Lauer 2012). Selective enrichment is not required for the PCR method because of the selectivity of the probes and primers used in the assay.

Even though PCR tests are the most common alternative to traditional culturing, they rely on this step. PCR tests exist for the major foodborne bacteria, but generally require 1–5 days for results. Advanced PCR tests include quantitative tests and real-time tests.

A real-time PCR kit permits rapid control and reaction times. A liquid-handling platform speeds food pathogen testing by automating the DNA extraction and PCR plate setup for real-time PCR kits. Using gene-specific probes and primers enhances reproducibility while increasing sensitivity and specificity, and PCR internal controls ensure accurate results (Anon 2008). Ready-made reagents permit simple extraction, amplification, and detection. A genetic-based listeria assay that involves reverse transcriptase is available that delivers results in 8 h or less.

An application of genomics is molecular serotype determination from a colony isolate in 72 h. This assay utilizes targeted amplicon PCR combined with sequencing to develop a genetic profile for the colony isolate with a comparison of sequence results with the known genetic makeup of the reference database to determine the specific *Salmonella* serotype present. LAMP PCR systems also merit consideration for diagnostics.

Surface-Enhanced Raman Scattering

This technique involves the placement of a specimen on a rough surface and subsequent scanning with the Raman spectrometer's laser beam. The scattered light forms a distinct pattern known as a "Raman spectral signature." This test can differentiate between live and dead cells and antibodies and a biomarker is not required. It offers the potential of accelerating the process of pathogen testing, from sampling to results, by 2 h or less. Possible applications are evaluation of the efficacy of processing methods such as high-pressure processing, irradiation, and thermal processing.

Ribotyping

This approach utilizes restriction enzymes to digest the DNA in bacteria, creating hybridized and digitized fragments analyzed by comparison with reference organisms in a database to determine the species present. These tests are for a wide range of bacteria and require approximately 8 h for results.

Biosensors

Biosensors similar to pregnancy test kits are being developed and evaluated for rapid, reliable, and inexpensive identification and quantification of pathogenic microorganisms as well as for biosafety and biosecurity. The bioanalytical microsystem, fabricated using nanotechnology, contains a microfluidic biosensor with the desired characteristics of the black box type of pathogen sensor. Furthermore, lateral flow assays that detect pathogens based on antibodies that detect pathogens with a 10–20 min assay. Baeumner (2004) developed a lateral flow universal biosensor made specifically for any pathogen within a few minutes with no special equipment and skills. It detects pathogenic microorganisms based on their nucleic acid sequences. The lateral flow assay appears to need more development for the bioanalytical microsystem.

Rapid Method Selection

A laboratory should evaluate the needs and determine the current level of knowledge, instrumentation, and potential application. With a large number of samples evaluated consistently, the speed and costs of supplies and labor may justify an investment in automated instrumentation.

An extensive amount of effort and money has been devoted to the development of instantaneous or real-time pathogen detection techniques. It is possible to reveal plant sanitation levels quickly and to incorporate these measurements to set high standards for the involved plant. However, a pathogen-free status necessitates additional technology. Even though improved technology may not provide a pathogen-free environment, complementary strategies will contribute to improved hygiene.

Microbial Surveillance

PulseNet

This surveillance system, utilized by public health authorities for tracking down pathogens after they have left the production facility, relies on several tests. It materialized in 1996 in collaboration with the Association of Public Health Laboratories. In the United States, this system which includes state and local public health laboratories and federal food regulatory agency laboratories tracks pathogens and their subtypes. The PulseNet laboratories break down pathogens' DNA and enter it into a database. The laboratories analyze the DNA using pulsed-field gel electrophoresis, a common technique that sends an electrical pulse from varying directions through a gel that contains DNA that has been isolated from the bacteria. PulseNet utilizes "DNA fingerprinting" (Keefe 2011).

Study Questions

1. What is the difference between a microorganism and a bacterium?
2. What is a virus?
3. How does contamination affect the lag phase of the microbial growth curve?
4. What is a psychrotroph?
5. What is A_w?
6. What is a biofilm?
7. What is generation interval?
8. What is an anaerobic microorganism?
9. What is psychosomatic food illness?
10. What microorganism is most likely to cause influenza-like symptoms?
11. What is a mycotoxin?
12. What is cross-contamination?
13. What is a Petrifilm plate?
14. What is the difference between a foodborne disease and food poisoning?
15. What is the role of bacteriophages in the food industry?

References

Anon (2008). Real-timer PCR test kit provides faster time to results. *Food Saf* 14(4): 36.
Baeumner A (2004). Nanosensors identify pathogens in food. *Food Technol* 58(8): 51.
Baker CA, Ricke (2015). Biofilms, processing equipment, and efficacy of sanitization. *Food Qual Saf* 22(4): 28.
Caul EO (2000). Foodborne viruses. In *The microbiological safety and quality of food*, eds. Lund BM, Baird-Parker TC, Gould GW, 1457. Aspen Publ., Inc.: Gaithersburg, MD.
Chaves BD, Brashears MM (2016–17). Mitigation of *Listeria monocytogenes* in ready-to-eat meats using lactic acid bacteria. *Food Saf Mag* 22(6): 56.
Davidson PM (2003). Foodborne diseases in the United States. In *Food plant sanitation*, eds. Hui YH, et al., 7. Marcel Dekker, Inc.: New York.
Duxbury D (2004). Keeping tabs on listeria. *Food Technol* 58(7): 74.
Farber JM, Peterkin PI (2000). *Listeria monocytogenes*. In *The microbiological safety and quality of food*, eds. Lund BM, Baird-Parker TC, Gould GW, 1178. Aspen Publ., Inc.: Gaithersburg, MD.
Fuhrman E (2014). On the offensive. *The National Provisioner* 228(6): 72.
Goodridge CL, Goodrich D, Gottfried P, Edmonds P, Wyvill JC (2003). A rapid most-probable-number-based enzyme-linked immunosorbent assay for the detection and enumeration of Salmonella typhimurium in poultry wastewater. *J Food Prot* 66: 2302.
Keefe LM (2011). Warning light. *Meatingplace* 06(11): 51.
Koutchma (2016). Light technologies as post-lethality in meat processing. *Meatingplace* 09(16): 85.
Lauer W (2012). PCR a simple solution for a more sustainable lab. *Food Qual* 19(3): 23.
McCright (2016). Rapid microbial detection can pay big dividends. *Food Qual Saf* 23(5): 39.
Niemira BA (2017). Targeting biofilms with cold plasma: New approaches to a persistent problem. *Food Saf Mag* 23(3): 24.
Pruett WP, Dunn J (1994). Pulsed light reduction of pathogenic bacteria on beef carcass surfaces. *Proc Meat Ind Res Conf*, 93. American Meat Institute: Washington, DC.
Reilly SS (2016). Biofilm and pathogen mitigation: A real culture change. *Food Saf Mag* 22(1): 16.
Shapton DA, Shapton NF, eds. (1991). Microorganisms— An outline of their structure. In *Principles and practices for the safe processing of foods*, 209. Butterworth-Heinemann: Oxford.
Smith GC (1994). Fecal material removal and bacterial count reduction by trimming and/or spray-washing beef external fat surfaces. *Proc Meat Ind Res Conf*, 31. American Meat Institute: Washington, DC.

The Relationship of Allergens to Sanitation

4

Abstract

Allergens are substances that cause the immune system to trigger an act against itself. Normally, this condition occurs when foreign bodies enter the human body. Those involved with sanitation should be aware of how to protect foods against allergens. Allergen infestation frequently occurs through product cross-contamination of an allergen-containing product during manufacture.

An effective method for the control of allergens is the organization and implementation of an allergen control plan. Such a plan avoids inadvertent allergen cross-contamination with resultant recalls and potentially adverse and a potentially fatal reaction. Allergen contamination can be most effectively reduced through effective education, sanitation, and monitoring. Additional information about sanitary practices for the control of allergens is provided in Chapters that follow.

Keywords

Allergen(s) • Allergen control plan • Cleaning • Contamination • Labeling • Mast cell • Monitoring

Introduction

Allergens in foods have become one of the most visible and urgent issues facing the food industry. More than 170 foods can cause allergenic reactions. The Food and Drug Administration (FDA) indicated that 69 of the 229 (30.1%) food safety entries were for undeclared allergens. Allergens are protein based, smaller than bacteria, and not destroyed with thermal treatment. Currently, there is no known cure for food allergies. Thus, strict avoidance is the only way for consumers to avoid an allergenic reaction. It is essential that the food industry ensure that derivatives of common food allergens are included on labels and that manufacturing facilities and equipment do not contribute to contamination of these substances.

Knowledge of undeclared allergens that can occur in food processing and preparation is essential to the effective sanitation and the maintenance of a safe food supply. Those involved with sanitation must be knowledgeable about how to protect foods against allergens that can be devastating and even fatal to a segment of the population. It is essential that the food industry keep these chemical organisms out of the food supply.

Approximately 30,000 emergency room visits and 200 deaths each year are attributable to food allergens. According to the US Centers for Disease Control and Prevention (CDC), 1 in 25 adults and 1 in 17 children under the age of 3 in the United States are affected by food allergies with the prevalence that appears to be increasing. Most infants diagnosed with food allergies outgrow them within a few months, but some food allergies (e.g., peanuts and shellfish) are more persistent, often enduring for a lifetime. The impact of allergens is increasing dramatically as evidenced by no recalls for undeclared food allergens before 1990, but a large number now. Allergen-related recalls rose from approximately 9.7% in 1999 to over 25%. There has been increased regulatory attention given to food allergens by both state and federal regulators. The FDA has declared that the control of food allergens is a top priority.

Most allergies originate in foodservice. Over 170 foods cause allergic reactions. The "Big 8" foods that are most likely to contain allergens include (1) peanuts; (2) tree nuts such as almonds, cashews, Brazil nuts, and pistachios; (3) dairy products; (4) eggs; (5) soybeans; (6) crustacea; (7) fish; strawberries; and (8) cereals. These eight most common allergen sources account for approximately 90% of all allergenic reactions. Other potential foods that may contain allergens are cottonseed, sesame seed, poppy seed, mollusks, and other legumes. Natural common airborne allergens include grass pollen, tree pollen, mold spores, and animal dander. Allergenic substances and products include yeasts, mannitol, sorbitol, polysorbates, rice maltodextrins, citrus, bioflavonoids, lactose, artificial preservatives, artificial colors, citrus pectin, talc, soy lecithin, corn flour, gluten, soy flour, rice flour, alfalfa, potato starch, and acacia gum. Any food protein can be an allergen. The human immune system may not recognize it properly and identifies it as a foreign body (e.g., bacteria) that may attack and become an allergy.

Typical symptoms of allergenic reactions to food include nausea, vomiting, abdominal pain, diarrhea, anaphylactic shock, atopic dermatitis, rhinitis, and asthma.

An increase in food product recalls has occurred because of undeclared allergens and ingredients of public health concern (FSIS 2015). The major causes of undeclared allergen recalls are packaging, product formulation changes, new suppliers, and misprinted labels.

What Are Allergens?

Allergens are substances that cause the immune system to trigger and act against itself. Normally, this condition happens when foreign bodies such as bacteria enter the human body. However, innocent and harmless bodies (proteins) such as pollen, peanuts, milk, penicillin, etc. may not be recognized by the immune system and continue to function as a harmful foreign body. Yet, wasps and other insects produce allergens as a defense mechanism.

Food allergies can occur when the human immune system reacts to proteins consumed. In an attempt to protect the body, the immune system produces antibodies to that food. Those antibodies cause mast cells (allergy cells in the body) to release chemicals, such as histamine, into the bloodstream. Histamine acts on the eyes, nose, throat, skin, lungs, or gastrointestinal tract causing symptoms of the allergic reaction.

A food allergy occurs when a natural substance is mistaken for a hostile invader causing immune systems to mobilize to repel the invader. IgE antibodies to proteins—a characteristic shared with other allergens such as hay fever (an acute allergic nasal condition) and wasp sting reactions—mediate food allergies. The severity of food allergy symptoms varies from life-threatening reactions, when exposed to food proteins that are allergens, sensitized, to less severe reactions such as skin irritation and breathing difficulty. Since no cure is available for food allergies, avoidance is the only preventive measure available to allergic consumers.

Allergenic Reaction(s)

An allergenic reaction is a reaction between an antigen and an antibody. Allergenic reactions are due to an inappropriate immunological response to an otherwise harmless food. In a sensitizing step, the first time a body is exposed to a foreign substance, it produces antibodies. This reaction requires 5–15 days with a small amount of antibodies produced. Through memory, the second exposure to the same antigen elicits antibiotic production more rapidly than the first exposure. The incorporation of vaccines in the sensitizing step and the memory of the immune system are responsible for vaccine efficiency. With the exposure to the same antigen after vaccination or initial exposure, the body's immune system mounts a rapid and vigorous attack. Furthermore, each time with body exposure to the same antigen, the immune response induces aggressive attacks.

Specific defenses involve the production of antibodies in response to a foreign substance. A specifically produced antibody attacks a specific antigen and binds it to remove the antigen.

When an allergenic reaction occurs, the body incorrectly views an ingested protein of a food as an antigen, and the immune system hosts an immune reaction by the production of IgE antibodies that bind to the antigen and a mast cell, forming a complex. An immediate hypersensitivity reaction by IgE-antigen-mast cell complex formation causes a systemic or localized attack (Baldus et al. 2009). Either reaction may occur between minutes and a few hours after initial sensitization and subsequent ingestion of the offending food(s). Systemic reactions are abdominal cramps, vomiting, diarrhea, respiratory distress, and, in severe cases, anaphylactic shock. Localized reactions center on the skin and cause rashes, hives, eczema, and itching.

Common sites for allergenic reactions are those areas where mast cells are concentrated. Examples are swelling of the lips or tongue, breathing restrictions, asthma, rhinitis, nausea, vomiting, diarrhea, hives, rash, and itching.

Causes of Allergen Contamination

Possible processing errors that result in allergen-containing product contamination include:

- Cross-contamination through inadequate cleaning of equipment used for the manufacture of non-allergen-containing products produced after allergen-containing foods
- Changing of ingredients without an allergen assessment of the new materials
- Use of reworks
- Formulation errors
- Incorrect labeling

The cause of a true food allergy is the protein in a food item, typically the primary protein. These proteins are heat stable and not eliminated by cooking or thermal processing. When an allergenic individual contacts this protein, the body has an immune-mediated response because of the necessity for elimination of the identified foreign substance. A release of histamine can cause symptoms that may range from itchy skin or eyes to nausea or difficulty breathing and potentially fatal anaphylaxis.

Allergen Control

Preventive control rules in the Food Safety Modernization Act specify that plants must have an allergen control program, making allergen control a fundamental prerequisite for safe food processing. Identification of allergen control practices is essential in a written set of allergen control program standard operating procedures and in the documentation of execution available for review. An essential label development procedure should include a checklist completed during label development. Proof of labels and label receiving into plants require thorough review for accuracy.

Sanitation is the first line of defense for preventing allergen cross contact in a food plant. Ineffective sanitation fails to remove potential allergen residues and cross contact. Thus, food plants need documented standard operating procedures (SOPs) and sanitation standard operating procedures (SSOPs), as discussed in Chap. 7, to provide direction for employees to perform their tasks correctly. An effective sanitation program involves thorough training about allergen awareness and control for all workers, including seasonal and/or temporary employees.

An effective technique for allergen control is the organization and implementation of an allergen control plan (ACP). Such a plan can avoid inadvertent allergen cross-contamination with resultant recalls and potentially adverse or possibly fatal physiological reactions from consumers. ACP is a systematic method in a food processing facility that identifies and controls allergens from the incoming ingredients to the final packaged product. Corporate managers, plant managers, and management employees involved in quality assurance, quality control, production, sanitation, and purchasing should all accept the responsibility for the development, implementation, and maintenance of an ACP.

An allergen control plan is an ancillary program to a manufacturing plant's Hazard Analysis and Critical Control Points (HACCP) plan. The two major components of an ACP are:

1. Allergen assessment as part of the hazard analysis (a chemical hazard).
2. After allergen identification, as a raw ingredient or contained within a roll ingredient, control steps should be established if the product is not run on a separate line or a complete wet cleaning is performed between allergen- and non-allergen-containing products.

Some states have initiated independently, or in cooperation with the FDA, allergen inspections and analysis of products selected randomly from grocery stores with a resultant increase in product recalls. An allergen control program should address:

Employee Education Education of employees includes instruction about the handling of materials that may contain allergens. Training includes the incorporation of good manufacturing practices instruction and documentation through employee signature, date, and materials covered.

Supplier Monitoring Product or ingredient formulations, specification sheets, and certificates of analysis from suppliers of raw materials are necessary. Testing to verify the quantity of an allergen present can determine essential precautions necessary during production. Verification that suppliers have an ACP is essential as well as letters of guarantee that allergens are not present.

Cleaning The allocation of adequate sanitation time, including inspection, promotes effective cleaning. Allergen control through the reduction of cross-contamination in a manufacturing plant involves the production of allergen-containing foods as the last product on the production line followed by a wet cleaning program. Since the protein component within a food is responsible for the immunological symptoms of an allergenic reaction in humans, complete removal of these proteins is important. An allotment of 24 h between regular and gluten-free food preparation permits flour particles to settle and then be removed.

Special cleaning attention adopted from Kochak (2016–17) involves:

1. Food surfaces with an allergen
2. Containers that transport allergens
3. Cleaning utensils that clean production equipment in contact with allergens such as brushes, scrubbers, rags, and dust collectors
4. Push-through products for cleanout prior to running a product containing an allergen
5. Reworks if not from a like product
6. Sampling devices that draw samples from a run containing an allergenic ingredient
7. Final clean-in-place rinse

The allergen, product, and processing equipment dictate the appropriate cleaning method and protocols. The potential cleaning methods are wet cleaning, dry cleaning, and/or use of sanitizing agents. Opting for the wet cleaning method necessitates assessing the food items processed with shared equipment. Each product may contain a different allergen, requiring the implementation of scheduling or cleaning procedures. Cleaning operations not performed between allergen- and non-allergen-containing products, a parts-per-million analysis, establish the safety of products that do not list allergens on the label. A label declaration may be sufficient for allergen control if all products contain the same allergen.

Product Changeover Cross contact occurs during product changeover when manufacturing transitions from one product to another. During this stage, adequate time allocation ensures effective cleaning, label monitoring, and verification and documentation of all changeovers as they occur. If possible, it is important that the same equipment manufacture products with similar allergens. When production lines are in close proximity, physical barriers separate allergenic and non-allergenic materials and mitigate the risk of allergen cross contact. If personnel on non-allergenic production lines do not work on allergenic areas, they are less likely to carry allergenic residues on their clothing or hands.

Scheduling Processing Segregation of allergenic and non-allergenic products minimizes allergen cross contact. Cleaning immediately after the production of foods containing allergenic ingredients reduces allergen contamination.

Raw Material Storage All raw materials and foods that contain allergens should be stored in an area secluded or removed from non-allergenic materials. Incoming palletized materials should be shrink-wrapped to prevent cross-contamination from potential leakage. Partially used bags or other containers of allergen-containing materials should be sealed and stored in segregated areas. Label all materials that contain allergens accordingly with a color-coded tag. For easy identification by plant personnel, place color-coding charts in the production area, especially above wall-mounted equipment and near storage areas. Allergen-containing materials should be stored on the bottom of racks or nearest to the floor to prevent spillage on other items. Dedicated scoops and storage containers for specific materials maintain separation of allergens.

Plant Layout Product flow may determine if allergen-containing materials contact other foods with resultant contamination. A potential example is exposure through overhead conveyors that cross one another or over exposed products provide separate food preparation zones and storage areas. Incorporate controlled airflow that minimizes the deposition of airborne particles on prepared or processed foods.

Color-Coding of Utensils Color-coding provides an easier method to keep different materials, utensils, and equipment separate.

Incorporation of Reworks Add only like foods to reworked products. Label reworked products to indicate which products contain allergens. Reworked products containing allergenic ingredients must be stored in areas separate from those that do not contain such products. Color-code containers with allergen-containing products that do not contact non-allergen-containing products. If feasible, incorporate reworked products into the same production run.

Label Review Develop a system for maintaining labels placed on foods containing allergens in easy-to-identify areas. Conduct a thorough review and matching of the current formulations. Provide documentation for all material specifications, formulations, and finished product labels. When a raw material ingredient statement changes, provide a cross-reference with the finished product labels to comprehend affected products and labels by the change.

The word "Contains" followed by the name of the food allergen or a parenthetical statement within the list of adjuncts is a labeling option for allergenic ingredients. When an allergen is in the food contained within an allergenic group such as "tree nuts" or "seafood," list the specific nutmeat or seafood. This practice is necessary because some people may be allergic to one tree nut or seafood but not all within the grouping.

Documentation   Review   of   Activities
Documentation proves the specific activity. Production schedule and sanitation check-off sheets should be filled out and reviewed by a supervisor (signed and dated) to complete the records for allergen control.

Evaluation of Program Effectiveness Changes in customers, suppliers, and raw materials necessitate the need for continuous reevaluation of the effectiveness of an allergy control program. A key component in the continuous verification and success of an ACP is the incorporation of routine auditing practices for suppliers and in-plant operations. Allergen plans should be reviewed as determined appropriate and especially during an annual HACCP plan validation. Review internal audits placed on the agenda and review during monthly HACCP meetings should be conducted. During internal audits, review documentation to ensure that all practices written within the allergen policy are performed.

The following sources suggested by Bush (2015) provide information and/or assistance with the control of allergens:

1. *The International Food Information Council (IFIC) Foundation*
2. *AllergyHome.org*
3. *SnackSafety.com*
4. *Food Allergy & Anaphylaxis Connection Team (FAACT)*

Food Allergen Tests

Accurate and affordable testing for allergens is essential. Initially, tests ensure that processing equipment was free of allergens. However, expanded testing examines all aspects of the manufacturing process.

The food industry has relied on two methods for allergy testing. They are the ELISA and polymerase chain reaction (PCR). PCR is a fast and inexpensive method to identify DNA. According to Cowan-Lincoln (2013), this method amplifies, or copies, small segments of DNA until a large enough grown sample determines if an allergen is present. Although this technique can identify the DNA of milk, soy, peanuts, hazelnuts, walnuts, fish, and crustaceans, it can fail to find all allergens because it detects the presence of DNA but not proteins. Cowan-Lincoln (2013) indicated that egg whites and milk, which are significant allergens, contain little or no DNA, but a large quantity of protein. Consequently, this method is unreliable for these foods.

The enzyme-linked immunosorbent assay (ELISA) method can detect antibodies in a sample that indicate the presence of allergens. However, this technique requires a separate kit for each allergen, which is an expensive approach. With utilization of the ELISA method, most firms do not test products for the presence of all possible allergens, but incorporate a cost-effectiveness analysis and select the top 1–3 allergens most likely to be present. ELISA tests are widely incorporated that give food processors quick, simple, and accurate tools to check for traces of certain allergenic foods on manufacturing equipment or in food processed on shared equipment. However, this test method does not work properly with heat-treated products, hydrolyzed proteins, and fermented products (Cook 2011). The basis for the immunoassay is protein binding to specific enzyme-labeled antibodies to permit detection and quantification by comparison to standard curves. These are primarily laboratory tests. Available low-cost kits utilized in a manufacturing plant by workers require approximately 30 min.

Strip tests incorporate the formation of complexes between anti-allergen antibody-coated colored beads with allergenic proteins in the sample and anti-allergen antibodies on the test strip. These complexes give rise to a colored test line on the strip, indicating a positive (allergen-containing) sample. A formed colored control band confirms correct conduct of the test. This test is easy to conduct, inexpensive, and rapid (a few min to conduct) and can be utilized in the field since instrumentation is not required. Only analyzed single samples detect single allergens at one time when strip tests are incorporated.

According to Schag (2009), allergen test kits detect milk, soy, and egg residue in addition to foodborne bacteria and parasites, within 2½ h. These differently sized commercial kits include lower-volume applications and larger-volume laboratories that may utilize some degree of automation. Commercial kits are available for the detection and quantification of gluten at very low concentrations in cooked and uncooked foods and in environmental samples. An assay incorporates extracted and diluted food samples with a specially formulated buffer. Furthermore, a developed and validated test incorporates the DNA screening to detect most nuts.

Another technology for allergen detection is mass spectrometry (MS). This process identifies proteins and peptides with a high level of accuracy. MS directly detects allergens by breaking them down into peptides or short strings of amino acids that link together to form larger proteins. Enhanced reliability occurs by the detection of peptides instead of entire protein structures because processing degrades proteins and recognition failure of an altered structure when an assay is looking for an allergen. Shorter peptides are more likely to remain intact after processing and may be detectable by MS. Since MS detects more than one peptide per allergen, if one is degraded, detection of the other occurs through at least one of the peptides (Cowan-Lincoln 2013). Increased mass spectrometer accuracy occurs through the direct detection of allergen components instead of indirect detection through DNA or antibodies as with ELISA or PCR. Mass spectrometers can multiplex, detecting all of the eight main allergens in one test making this approach easier, faster, and less expensive, when testing for multiple allergens, than incorporating a series of ELISA assays. A limitation of MS is the equipment cost. Most testing laboratories own mass spectrometers, which makes this detection technique available to those that cannot justify purchasing this equipment.

Biosensors such as surface plasma resonance (SPR)-based biosensors have become increasingly accepted tools for allergen detection (Bremer 2009). This detection method relies on changes in the refractive index at the surface of a sensor chip, caused by the binding of an analyte to an immobilized ligand. Immobilized specific antibodies are on the chip surface since allergens are high molecular weight compounds. Monitoring of the binding of allergens in samples occurs in real time. The calculated sample concentration from a calibration curve occurs from a signal change. This automated technique provides results in only minutes. Yet, this approach is expensive due to equipment costs. Furthermore, testing of only a single sample at one time occurs, and trained laboratory personnel are required.

Assays based on flow cytometry detection utilize sets of differently colored micron-size beads. Coupled antibodies against different colored allergenic compounds occur to each color-coded set of beads. Specific, fluorescently labeled, second antibodies visualize the binding of allergens to the beads. For analysis, simultaneously added different bead sets to a sample in a microtiter will detect different allergens. The beads drawn into a fluidic tube causes the microspheres to line up in a single file before passing through the detection chamber. In the chamber, one laser identifies each bead and categorizes it into the appropriate bead set (based on which detected allergen), while another laser checks the beads for the quantity of fluorescently labeled antibodies per bead and determines the concentration of the detected allergen. The detection of multiple allergens occurs simultaneously in a sample. These assays provide simultaneous detection of multiple allergens from small-volume samples in seconds, and the equipment costs are low-priced when compared to biosensors. However, labor requirements are similar to ELISAs.

When testing, it is important to incorporate an official testing method recognized by a standard organization or government agency and internal validation of the methodology. If the utilization of an external third-party laboratory occurs, it is necessary to confirm that the testing methodology is recognized by a standard organization or government agency and that the method has been validated in that laboratory. Furthermore, the results of the selected method must permit the level of detection needed.

Allergen Labeling

To protect consumers against the eight major allergens listed by the FDA, congress passed the Food Allergen Labeling and Consumer Protection Act in 2004, which contains several requirements for food manufacturers. Primary provisions of the act are the requirement of easy-to-understand labeling of the eight major allergen ingredients (which together cause 90% of allergenic reactions to food in the United States) on food packages; declaration of allergens present in flavoring, coloring, or incidental additives; and a report to congress by the Secretary of Health and Human Services detailing:

- Analysis of how foods are unintentionally contaminated with allergens during manufacturing
- Advice on industry best practices that can be employed to prevent cross-contamination
- Description of advisory labeling (such as "may contain") incorporated by food manufacturers
- Statement of the number of food facilities inspected in the past two years with a description of the agents handling a number of nonconforming facilities, the nature of the violations, and the number of voluntary recalls or assurances of proper labeling
- Proposal of rules to define and permit the use of the term "gluten-free" on labeling
- Improved collection and presentation of data on the prevalence of food allergies, clinical significance or serious adverse events, and modes of treatment for food allergies
- Recommendations on research activities related to food allergies
- Pursuance of Food Code revisions to provide guidance for the preparation of allergen-free foods
- Provisions for technical assistance to state and local emergency medical services for the treatment and prevention of food allergy responses

This act provides a change in the way that foods are labeled, increased inspection by government agencies, and the likelihood of more regulations involving handling and production of foods in environments that handle allergenic agents. It is essential that the food industry will need to develop the discipline to implement an effective allergen control and labeling management strategy.

Allergen Management

The primary responsibility to provide safe foods free from allergen cross-contamination belongs to food manufacturers. Because of variations in plant layout, ingredients, and products, it may be necessary to incorporate different allergen management strategies. Incorporation of the following into food manufacturing and foodservice operations protects against allergens:

- Adopt a "zero tolerance" protection program against allergen cross-contamination.
- All personnel should be trained in allergen management strategy.
- Ensure that incoming ingredients are clearly labeled, and the labels are reviewed periodically to confirm that suppliers have not changed ingredients without notice.
- Develop an allergen storage policy including a procedure for the cleanup of spills.
- Control airflow to minimize airborne particles from landing on gluten-free food.
- Permit at least 24 h between regular and gluten-free food preparation to allow flour particles to settle and be subsequently cleaned away.
- Utilize equipment to facilitate cleaning and the prevention of allergen harborage niches.
- Manufacture different shaped gluten-free products to avoid mixing with other foods.
- Conduct an allergen risk assessment as part of or in addition to the HACCP program.
- Adopt a comprehensive rework policy, including clear identification of work-in-process materials and reworks.
- Reject in-process materials or finished products suspected of cross-contamination.

- Review labels prior to use and confirm that the correct labels are incorporated in the process.
- Conduct internal audits or use a third-party auditor to assess the allergen management strategy.

In foodservice, arrange dishes to prevent spillage and splattering of foods onto those that are gluten-free, and ensure that these foods differ in shape and other appearances served with plates, bowls, and napkins of differing colors.

Evaluate and track consumer complaints involving allergen issues, and designate a trained person to respond to consumer inquiries regarding allergens.

Study Questions

1. What is an allergen?
2. Why is allergen contamination a major problem for food manufacturing firms?
3. What are the two major components of an allergen control plan?
4. What are the three most important components for the control of allergen contamination?
5. How can a plant layout affect allergen contamination?
6. What precautions are essential for allergen control with the incorporated reworks in product manufacture?
7. What are three tests available for allergen detection in foods?
8. What are the advantages and disadvantages of available allergen tests?
9. What are the "Big 8" food allergens?

References

Baldus K, Werrenon K, Deibel V (2009). Adverse reactions to allergens. *Food Saf Mag* 14(6): 16.

Bremer M (2009). Selecting a suitable food allergen detection method. *Food Saf Mag* 15(3): 16.

Bush T (2015). Food companies & food allergies: Unite! *Food Saf Mag* 21(2): 22.

Cook J (2011). Get a handle on allergens. *Food Qual* 18(5): 25.

Cowan-Lincoln M (2013). The future of allergen testing with mass spectrometry. *Food Qual Saf* 20(3): 48.

Food Safety and Inspection Service, United States Department of Agriculture (2015). FSIS compliance guidelines: Allergens and ingredients of public health concern: Identification, prevention and control, declaration through labeling.

Kochak JW (2016–17). Getting ready for FSMA's allergen guidelines. *Food Saf Mag* 22(6): 18.

Schag AK (2009). Protect your company from food allergens. *Food Qual* 16(2): 22.

Food Contamination Sources

5

Abstract

Food products are rich in nutrients required by microorganisms and may become contaminated. Major contamination sources are water, air, dust, equipment, sewage, insects, rodents, and employees.

Contamination of raw materials can also occur from soil, sewage, live animals, external surfaces, and the internal organs of meat animals. Additional contamination of animal foods originates from diseased animals, although advances in health care have nearly eliminated this source. Contamination from chemical sources can occur through accidental mixing of chemical supplies with foods. Ingredients can contribute to additional microbial or chemical contamination. Contamination can be reduced through effective housekeeping and sanitation, protection of food during storage, proper disposal of garbage and litter, and protection against contact with toxic substances.

Keywords

Contamination • Contamination sources • Cross-contamination • Foods • Infection

Introduction

Most foods provide an ideal nutrition source for microorganisms and generally have a pH value and water activity in ranges needed to contribute to growth and proliferation. During growing, harvesting, transporting, processing, distribution, and preparation, food is contaminated with soil, air, and waterborne microorganisms. Extremely high numbers of microorganisms are found in meat animals' intestinal tracts, and some of these find their way to the carcass surfaces during harvesting. Some apparently healthy animals may harbor various microorganisms in the liver, kidneys, lymph nodes, and spleen. These microorganisms and those from contamination through the slaughter process can migrate to the skeletal muscles via the circulatory system. When carcasses and cuts are subsequently handled through food distribution channels, where they are reduced to retail cuts, they are subjected to an increasing number of microorganisms from the cut surfaces. The fate of these microorganisms

© Springer International Publishing AG, part of Springer Nature 2018
N.G. Marriott et al., *Principles of Food Sanitation*, Food Science Text Series,
https://doi.org/10.1007/978-3-319-67166-6_5

and those from other foods depends on several important environmental factors, such as the ability of the organisms to utilize fresh food as a substrate at low temperatures. In addition, oxygenated conditions and high water activity will segregate those microorganisms most capable of rapid and progressive growth under these conditions.

Refrigeration, one of the most viable methods for reducing the effects of contamination, is widely applied to foods in commercial food processing, food retailing, foodservice, and food distribution. Its use has prevented outbreaks of foodborne illness by effectively controlling microbial growth. When the correct techniques for cooling food and cold storage of these products are not followed, organisms that are present will grow. The growth rate of microorganisms may sustain a large increase in an environment slightly above the minimal temperature required for growth. Generally, foods cool slowly in air, and the cooling rate decreases with increased container size, so it is very difficult to properly cool large volumes of food. Many *Clostridium perfringens* foodborne illness outbreaks have been caused when large containers of food or broth were allowed to cool slowly.

Identification of contamination sources in a food production facility impacts directly the ultimate effectiveness of an establishment's sanitation control strategies. Both direct and indirect food contact surfaces, water, air, and personnel are primary areas of concern as contamination sources in a food facility. Food products may transmit certain microorganisms, causing foodborne illness in several ways, including through infections, intoxications, or toxicoinfections (toxin-mediated infection) as indicated by Knechtges (2012):

1. An infection is caused when a pathogenic bacterium present in a food is ingested and then multiplies, as is true for *Salmonella*, *Campylobacter*, *Listeria*, and some enteropathogenic *Escherichia coli*. Most foodborne illnesses are infections caused by bacteria, viruses, and parasites.
2. An intoxication is caused when certain toxin-producing microorganisms present in foods, multiply, sporulate, or lyse, releasing the toxin in the food. The food is then ingested, causing illness. Examples of such infections are caused

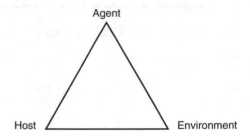

Fig. 5.1 The epidemiologic triangle (CDC 2011)

by *S. aureus*, *C. botulinum*, and *Bacillus cereus* (emetic or vomiting syndrome).
3. A toxicoinfection (toxin-mediated infection) is caused when some pathogenic organisms, capable of producing a toxin in the body, are ingested. Organisms causing toxicoinfections include *C. perfringens*, *Bacillus cereus* (diarrheal syndrome), and Shiga toxin-producing *E. coli* (Knechtges 2012).

Transfer of Contamination

When describing how a foodborne illness (or any infectious disease) is caused, a simple model called the epidemiologic triangle or triad is often used to illustrate the concept (CDC 2011) (Fig. 5.1).

The triangle consists of an external agent (a pathogenic microorganism that must be present), a susceptible host (a human who can get the illness), and an environment that brings the agent and host together. The environment consists of the external factors, such as physical, biological, and socioeconomic factors, that affect the agent and the opportunity for exposure (CDC 2011). The goal is to break at least one side of the triangle to disrupt the connection between these components and prevent a disease from occurring. From the epidemiologic triangle, specific transmission of an agent occurs through sequence of events called the chain of infection (CDC 2011) that will be described below.

The Chain of Infection

A chain of infection is a series of related events or factors that must exist or materialize and be

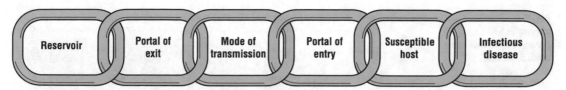

Fig. 5.2 The chain of infection

linked together before an infection will occur. These links can be identified as the reservoir, the portal of exit (CDC 2011), the mode of transmission, the portal of entry, host, and the infectious disease and are illustrated in Fig. 5.2 (CDC 2011).

The essential links in the chain of infection are all necessary for the transmission of an infectious disease and include the following:

1. Reservoir. This is the location where the pathogenic microorganism usually lives, grows, and multiplies. Examples of reservoirs include animals and raw foods of animal origin, other foods, plants, soil, water, a biofilm or microbial niche, and the human body (in fecal material, respiratory secretions, etc.).
2. Portal of Exit. This is where the microorganism leaves the reservoir and includes places like the nose, mouth, in respiratory secretions, the intestinal tract, a biofilm, or microbial niche.
3. Mode of Transmission. This is the mechanism where an agent may be transmitted from its reservoir to a susceptible host by either direct transmission (from reservoir to a susceptible host) or by indirect transmission (through the air), through vehicles like food and water and by inanimate objects (contaminated surfaces of equipment, utensils, etc.) or through vectors such as insects.
4. Portal of Entry. This is the way that pathogens enter the susceptible host. For foodborne illnesses, the primary portal of entry is the ingestion of pathogen-contaminated food (often via the "fecal-oral" route), and for non-foodborne infections primary portals of entry include the respiratory tract, mucous membranes, and blood.
5. Susceptible Host. This is a person who is at risk for developing an infection. There are several factors that make a person more susceptible to disease including age (the very

young and the elderly), chronic diseases, specific immunity, medical conditions that weaken the immune system, certain types of medications, malnutrition, and alcoholism.
6. Infectious Disease. An example is, a microorganism that is capable of causing illness, including bacteria, viruses, and parasites. Organisms can cause an infection based on their virulence (ability to multiply and grow), invasiveness (ability to enter tissues), and pathogenicity (ability to cause disease) (CDC 2011).

More specifically, the causative factors that are necessary for the transmission of a bacterial foodborne disease have been described by Bryan (1979) and are listed below:

1. The causative agent must be in the environment in which the food is produced, processed, or prepared.
2. A source (or reservoir) of the agent.
3. Transmission of the agent from the source to a food.
4. The food must support the growth of the microorganism.
5. The food must be kept in a temperature range for a sufficient time to permit growth to a level capable of causing infection or intoxication.
6. The susceptible host consumes the contaminated food.

Conditions such as required nutrients, water activity, pH, oxidation-reduction potential, lack of competitive microorganisms, and lack of inhibitors must also exist for bacterial pathogens to survive and grow.

Foodborne illnesses caused by viruses, parasites, and chemicals require only factors 1, 2, 3, and 6. Illnesses caused by plant toxicants or toxic animals require only factors 1, 2, and 6 (Bryan 1979).

Fig. 5.3 Hygienic
zoning in food
processing facilities for
environmental
monitoring sampling
and sanitation (Courtesy
of the Food Safety
Preventive Controls
Alliance (FSPCA 2015))

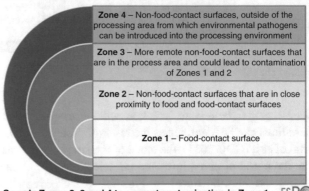

The chain of infection emphasizes the multiple causation of foodborne diseases. The presence of the disease agent is indispensable, but all of the steps are essential in the designated sequence before a foodborne disease can result. For an illness to develop, each link of the chain must be connected, but when any link of the chain is broken, the transmission is stopped (CDC 2011).

Contamination of Foods

A viable way for the identification of contamination sources in food facilities is to incorporate the concept of "hygienic zoning" to environmental monitoring (ICMSF 2002) that has been advanced by the former Kraft Foods Company, adopted by many other food companies (Slade 2002) and noted in the Food Safety Modernization Act (FSMA) training program on Preventive Controls for Human Food (FSPCA 2015).

This concept is an effective way to identify areas in a food processing environment that have different levels of risk and can be used to select sites for a plant environmental sampling program. It can also be used to maintain effective sanitation control strategies through targeting specific areas of concern in the facility. Hygienic zoning is used to identify areas of the highest risk (Zone 1) to the lowest risk (Zone 4) and differentiate sanitation requirements in different areas of the facility to minimize product contamination (Fig. 5.3).

The hygienic zoning approach is designed as a target with the center circle (or bull's eye) designated as Zone 1 representing the zone of highest risk and most critical areas for environmental monitoring and cleaning and sanitizing. Zone 1 represents direct food contact surfaces that include, but are not limited to, production equipment, utensils, and container conveyors, tables, racks, pumps, valves, slicers, filling and packaging machines, etc. (ICMSF 2002; FSPCA 2015). The second circle (Zone 2) of the target includes the areas adjacent to food contact surfaces. These are considered indirect food contact surfaces and include the exterior of equipment, equipment panels, bearings, aprons, or other surfaces that are in close proximity to the product flow in Zone 1 and could indirectly lead to product contamination (ICMSF 2002; FSPCA 2015). Zone 3 includes all other items in the food processing area of the facility such as floors, walls, ceilings, drains, and other equipment. Zone 4 includes the non-production areas of a facility such as hallways, employee locker rooms, cafeteria, maintenance shops and equipment, and areas further away from the production area (ICMSF 2002; FSPCA 2015).

One of the most viable contamination sources is the food product itself. Waste products that are not handled in a sanitary way become contaminated and support microbial growth. ATP bioluminescence and protein test kits are nonmicrobial tests that can be used to rapidly detect soil and organic material that are left on a surface and

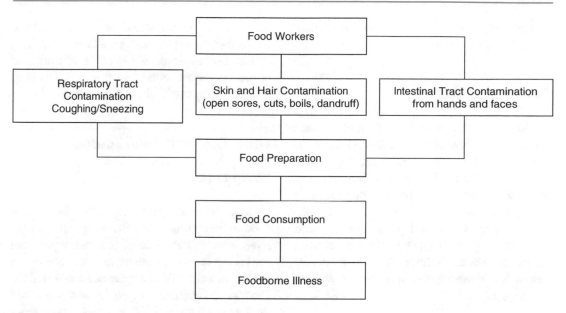

Fig. 5.4 Potential contamination of food by humans

cannot be seen by the naked eye. ATP bioluminescence test kits detect any cells that contain
ATP and are often used as a validation of the
effectiveness of the cleaning and sanitizing process (Powitz 2007). Protein tests are also used to
detect surface contamination and identify protein
in soils, which is an indicator of contamination
such as fecal material. Figure 5.4 illustrates
potential contamination by humans.

Dairy Products

Equipment with extensively designed sanitary
features to improve the hygiene of milk
production and to eliminate disease problems in
dairy cows has contributed to more wholesome
dairy products, although contamination can
occur from the udders of cows and milking
equipment. The subsequent pasteurization in processing plants has further reduced milk-borne
disease microorganisms. Nevertheless, dairy
products are vulnerable to cross-contamination
from items that have not been pasteurized.
Because not all dairy products are pasteurized,
the presence of pathogens (including *Listeria
monocytogenes*) in this industry has increased.

(Additional discussion related to contamination
of dairy products is presented in Chap. 16.)

Red Meat Products

The muscle tissues of healthy living animals are
nearly free of microorganisms. Contamination of
meat occurs from the external surface, such as
hair, skin, and the gastrointestinal and respiratory
tracts. The animal's white blood cells and the
antibodies developed throughout their lives effectively control infectious agents in the living body.
These internal defense mechanisms are destroyed
when blood is removed during slaughter.

Initial microbial inoculation of meat results
from the introduction of microorganisms into the
vascular system when contaminated knives are
used for exsanguination. The vascular system
rapidly disseminates these microorganisms
throughout the body. Contamination subsequently occurs by the introduction of microorganisms on the meat surfaces in operations
performed during the slaughter process, hide
removal, cutting, processing, storage, and distribution of meat. Other contamination can occur
by contact of the carcass with the hide, feet,

manure, dirt, and visceral contents from punctured digestive organs.

Poultry Products

Poultry is vulnerable to contamination by bacteria such as *Salmonella* and *Campylobacter* during processing. The processing of poultry, especially defeathering, evisceration, and chilling, permits an opportunity for the distribution of microorganisms among carcasses. Contaminated hands and gloves and other tools of processing plant workers also contribute to the transmission of salmonellae. A more detailed discussion of meat and poultry plant sanitation is in Chap. 17.

Seafood Products

Seafoods are excellent substrates for microbial growth and are vulnerable to contamination during harvesting, transportation, processing, distribution, and marketing. They are excellent sources of proteins and amino acids, B vitamins, and a number of minerals required for bacterial nutrition. Seafoods are handled extensively from harvesting to consumption, and since they are sometimes stored without proper refrigeration, contamination and growth of spoilage microorganisms and microbes of public health concern can occur. (Chap. 18 provides additional discussion related to seafood contamination.)

Ingredients

Ingredients (especially spices) are potential vehicles of harmful or potentially harmful microorganisms and toxins. The amounts and types of these agents vary with the origin of the spices and harvest method, type of food ingredient, processing technique, and handling. The food plant management team should be aware of the hazards connected with individual incoming ingredients. Only supplies and materials from approved sources, which meet company specifications and

are produced (or processed) in accordance with recognized good practices, should be used. This requirement also applies to control of testing of critical materials, either by the manufacturing firm, receiving establishment, or both.

Other Contamination Sources

Equipment

Contamination of equipment occurs during production, as well as when the equipment is idle. Even with sanitary design features, equipment can collect microorganisms and other debris from the air, as well as from employees and materials. Product contamination of equipment is reduced through improved sanitary design and more effective cleaning and sanitizing according to master sanitation schedule.

Employees

Of all the viable means of exposing microorganisms to food, employees are the largest contamination source. Employees who do not follow sanitary practices contaminate food that they touch with spoilage and pathogenic microorganisms that they come in contact with through work and other parts of the environment. The hands, hair, nose, and mouth harbor microorganisms that can be transferred to food during processing, packaging, preparation, and service by touching, breathing, coughing, or sneezing. Because the human body is warm, microorganisms proliferate rapidly, especially in the absence of good personal hygiene and sanitary practices.

After the chain of infection is broken, the spread of bacteria from one location to another can be prevented. Generally, the mishandling of food by people perpetuates the chain of infection until someone becomes ill or dies before corrective actions were taken to prevent additional outbreaks (Chao 2003). If every person that works with food could achieve appropriate personal hygiene and perform sanitary practices regularly and routinely, food contamination could be minimized. Every employee who works with food can play a vital role in preventing food contamination.

Air and Water

Water is used in production and preparation of fresh fruits and vegetables, is an ingredient added in the formulation of various processed foods, and serves as a key component of the cleaning and sanitizing operation. It can also serve as a source of contamination. If excessive contamination exists, another water source should be obtained, or the existing source should be treated with chemicals after other methods to assure its safety.

Contamination can result from airborne microorganisms in food processing, packaging, storage, and preparation areas. This contamination can result from unclean air surrounding the food plant or from contamination through improper sanitary practices. The most effective methods of reducing air contamination are through sanitary practices, filtering of air entering the food processing and preparation areas, and protection from air by appropriate packaging techniques and materials.

Sewage

Raw, untreated sewage can contain pathogens that have been eliminated from humans and animals, as well as other materials of the environment. Microorganisms present in raw sewage can cause typhoid and paratyphoid fevers, dysentery, and infectious hepatitis. Sewage has contaminated food and equipment through faulty plumbing in food facilities.

If raw sewage drains or flows into potable water lines, wells, irrigation ponds, rivers, lakes, and ocean bays, the water and living organisms such as seafood are contaminated. To prevent this contamination, large animal production operations, privies, and septic tanks should be sufficiently separated from wells, streams, and other bodies of water. Raw sewage should not be applied to fields where fruits and vegetables are

grown. (Additional discussion related to sewage treatment is presented in Chap. 12.)

Insects and Rodents

Flies and cockroaches are associated with homes, eating establishments, food processing facilities, and food warehouses, as well as with toilets, garbage, and other filth. These pests transfer filth from contaminated areas to food through their waste products; mouth, feet, and other body parts; and with flies, during regurgitation of filth onto clean food and equipment. To stop contamination from these pests, food processing, preparation, and serving areas should be protected against their entry, and if they do find entry, then eradication is necessary. In addition, other stored product pests can also contaminate foods resulting in infestation, contamination, and spoilage.

Rats and mice transmit filth and disease through their feet, fur, and intestinal tract. Like flies and cockroaches, they transfer filth from garbage dumps, sewers, and the environment where they live to food and food facilities. (Discussion about the control of rodents, insects, and other pests is provided in Chap. 13.)

Protection Against Contamination

The Environment

Ready-to-eat foods should not be touched with bare hands when consumed raw or after cooking, and food workers should also minimize bare hand contact with foods that are not in a ready to eat form. Contact with hands can be reduced by the use of clean, intact disposable plastic gloves, utensils, or deli paper during food processing, preparation, and service. A processed or prepared food, either in storage or ready for serving or holding, should be covered with a close-fitting clean cover that will not collect loose dust, lint, or other debris. If the nature of the food does not permit this method of protection, it should be placed in an enclosed, dust-free cabinet at the appropriate temperature. Foods in small modular wrappers or containers, such as milk and juice, should be dispensed directly from those packages. If foods are served from a buffet, they should be presented on a steam table (or other warming unit) or ice tray (or refrigerated unit), depending on temperature requirements, and should be protected during display by a transparent shield over and in front of the food. The shield will protect the food against contamination from the serving area (including ambient air), from handling by those being served, and from sneezes, coughs, or other employee- and customer-originated contamination. Any food that has touched any unclean surface should be cleaned thoroughly or discarded. Equipment and utensils for food processing, packaging, preparation, and service should be cleaned and sanitized between uses. Foodservice workers should be instructed to handle dishes and eating utensils in such a way that their hands do not touch any surface that will be in contact with food or the consumer's mouth.

Storage

Storage facilities should provide adequate space with appropriate control and protection against dust, insects, rodents, and other extraneous matter. Organized storage layouts with appropriate stock rotation can frequently reduce contamination and facilitate cleaning and can contribute to a tidier operation. In addition, storage area floors should be swept or scrubbed and shelves and/or racks cleaned with appropriate cleaning compounds and subsequent sanitizing. (Chaps. 9 and 10 discuss appropriate cleaning compounds and sanitizers.) Trash and garbage should not be permitted to accumulate in a food storage area.

Litter and Garbage

The food industry generates a large volume of wastes including used packaging materials, containers, and waste products. To reduce contamination, refuse should be placed in appropriate containers for removal from the food area.

The preferred disposal method (required by some regulatory agencies) is to use containers for garbage that are separated from those for disposal of litter and rubbish. Clean, disinfected receptacles should be located in work areas to accommodate waste food particles and packaging materials. These receptacles should be seamless, with close-fitting lids that should be kept closed except when the receptacles are being filled and emptied. Plastic liners are inexpensive, provide added protection, and can be removed quickly. All receptacles should be washed and disinfected regularly and frequently, usually daily. Containers in food processing and food preparation areas should not be used for garbage or litter, other than that produced in those areas.

Toxic Substances

Poisons and toxic chemicals should not be stored near food products. In fact, only chemicals required for cleaning and sanitizing should be stored in the same premises. Cleaning compounds should be clearly labeled and, when possible, be stored in their original containers. Only cleaning and sanitizing compounds, supplies, utensils, and equipment approved by regulatory or other agencies should be used in food handling, processing, and preparation.

Study Questions

1. What is the chain of infection?
2. What is the major contamination source of food?
3. Which microorganism is most likely to cause foodborne illness if large quantities of food have been stored in slowly cooling containers?
4. Which pathogenic microorganism may be found in unpasteurized dairy products that have become cross-contaminated?
5. What is the best way to reduce contamination from food equipment?
6. How can sewage-contaminated water, if consumed, affect humans?

References

Bryan FL (1979). Epidemiology of foodborne diseases. In *Food-borne infections and intoxications*, eds. Riemann H, Bryan FL. 2nd ed., 4. Academic Press: New York. p. 3.

CDC (2011). *Principles of epidemiology in public health practice. An introduction to applied epidemiology and biostatistics.* 3rd ed. https://www.cdc.gov/ophss/csels/dsepd/ss1978/lesson1/section8.html

Chao TS (2003). Workers' personal hygiene. In *Food plant sanitation*, eds. Hui YH, et al. Marcel Dekker, Inc.: New York.

Food Safety Preventive Controls Alliance (FSPCA) (2015). *FSPCA Preventive Controls for Human Food Training Curriculum.* 1st ed. 16 Chapters and 6 Appendices.

International Commission on Microbiological Specifications for Foods (ICMSF) (2002). *Microorganisms in foods 7. Microbiological testing in food safety management.* Kluwer Academic/Plenum Publishers: New York, NY.

Knechtges PL (2012). *Food safety: Theory and practice.* Jones and Bartlett Learning: Burlington, MA.

Powitz RW (2007). ATP systems help put clean to the test. *Food Saf Mag* June/July. http://www.foodsafetymagazine.com/magazine-archive1/junejuly-2007/atp-systems-help-put-clean-to-the-test/

Slade PJ (2002). Verification of effective sanitation control strategies. *Food Saf Mag* 8(1): 24. http://www.foodsafetymagazine.com/magazine-archive1/februarymarch-2002/verification-of-effective-sanitation-control-strategies/

Personal Hygiene

6

Abstract

Food workers are potential sources of microorganisms that cause illness and food spoilage. *Hygiene* is a word used to describe sanitary principles for the preservation of health. *Personal hygiene* refers to the cleanliness of a person's body. Parts of the body that contribute to the contamination of food include the skin, hands, hair, eyes, mouth, nose, nasopharynx, respiratory tract, and excretory organs. These parts are contamination sources as carriers, through direct or indirect transmission, of detrimental microorganisms.

Management must select clean and healthy employees and ensure that they practice good personal hygiene. Employees must be held responsible for personal hygiene so that the food that they work with remains safe and wholesome.

Keywords

Bacteria • Contamination • Disease transmission • Employees • Handwashing • Hygiene • Skin

Introduction

Humans are a major source of food contamination, and those who work with food can transmit a variety of pathogens to food that can cause illness. Their hands, breath, hair, and perspiration, as well as their unguarded coughs and sneezes, can contaminate food and cause illness. Transfer of human and animal excreta by workers is also a potential source of pathogenic microorganisms that can also invade the food supply.

The food industry and regulatory agencies are focusing more on employee education and training and emphasizing that supervisors and workers be familiar with the principles of food protection. In multiunit chain operations, the negative effects of public opinion often spiral outward to uninvolved units. This recently occurred when a highly publicized series of foodborne outbreaks occurred in multiple locations across the country in the same restaurant chain that severely affected their brand image, reputation, and business.

© Springer International Publishing AG, part of Springer Nature 2018
N.G. Marriott et al., *Principles of Food Sanitation*, Food Science Text Series,
https://doi.org/10.1007/978-3-319-67166-6_6

In an analysis of 816 reported outbreaks from 1997 to 2006 associated with infected food workers, it was found that over 61% of these outbreaks came from foodservice facilities and catered events and another 11% of them were attributed to schools, day care centers, and health-care institutions. The two most frequently reported risk factors associated with these implicated food workers were bare hand contact with food and failure to properly wash hands (Greig et al. 2009). Researchers at the Centers for Disease Control and Prevention (CDC) (Green et al. 2007) found that handwashing was more likely to occur in restaurants whose food workers received food safety training, had more than one handwashing sink, and had a handwashing sink in the observed worker's sight. This research suggests that improving food worker hand hygiene requires not only food safety education, but appropriate facilities as well.

Personal Hygiene

The word *hygiene* is used to describe an application of sanitary principles for the preservation of health, while personal hygiene refers to the cleanliness of a person's body. The health and hygiene of workers both play a vital role in food sanitation and the safety of foods. People are potential sources of microorganisms including pathogenic bacteria, viruses, and parasites that can contaminate the foods that they work with and cause illness.

Workers who are ill should not come in contact with food or equipment and utensils used in the processing, preparation, and serving of food. Human illnesses that may be transmitted through food include those associated with intestinal disorders, dysentery, typhoid fever, infectious hepatitis (hepatitis A), and norovirus. In many illnesses, the disease-causing microorganisms may remain with the person after recovery. A person with this condition is known as a *carrier*, and the role of carriers will be discussed later in this chapter. Food workers should report to their supervisor or person in charge of the operation information about their health and activities that relate to diseases or infec-

tions transmissible through food (US FDA 2013). For example, food workers with symptoms of nausea, vomiting, jaundice, and sore throat with fever or who have a lesion containing pus (such as a boil or infected wound) that is open and draining should report these to their supervisor and not work with food. In addition, if a worker has an illness that has been diagnosed by a health practitioner due to *Norovirus*, hepatitis A virus, *Shigella* species, Shiga toxin-producing *E. coli*, *Salmonella typhi*, or nontyphoidal *Salmonella*, they should also be excluded from working with food (US FDA 2013).

When food workers become ill, their potential as a source of contamination increases. Staphylococci are normally found in and around infected cuts and burns, acne, boils, carbuncles, and eyes and ears. A sinus infection, sore throat, nagging cough, and other symptoms of the common cold are further signs that microorganisms are increasing in number. The same principle applies to gastrointestinal ailments, such as nausea, vomiting, diarrhea, or an upset stomach. Even when evidence of illness passes, some of the causative microorganisms may remain as a source of recontamination. For example, salmonellae may persist for several months after a worker has recovered from a *Salmonella* infection. The virus responsible for hepatitis A has been found in the intestinal tract over 5 years after the disease symptoms have disappeared. To explain the importance of employee hygienic practices, it is beneficial to look at different parts of the human body in terms of their potential sources of bacterial contamination.

Skin

This living organ provides four major functions: protection (against abrasion, invasion of microorganisms, dehydration, and ultraviolet radiation), sensation (pain, heat, cold, and pressure), regulation (raises or lowers body temperature as necessary), and secretion (perspiration and oils). Protection is an important function in terms of personal hygiene. There are three layers of skin including the epidermis (outermost layer) that provide a waterproof barrier; the dermis (beneath the epidermis) that contains tough connective tissue, hair follicles, and sweat glands; and the subcutaneous layer (also called the subcutis or hypodermis) that consists of fat and connective tissue (Page 2017; Grice and Serge 2011; Brodell and Rosenthal 2008). The two outer layers are tough, pliable, elastic layers that provide resistance to damage from the environment. The epidermis is less subject to damage than other parts of the body because it does not contain nervous tissue or blood vessels. The outermost layer of the epidermis is called the *stratum corneum* that consists of 25–30 rows of cells. They tend to be flatter and softer than most other cells and function through the formation of a layer that is impermeable to microorganisms. This layer is important to the distribution of transient and resident microbial flora. These tissues are replaced with newly created cells from the underlying layers every 4–5 days as they wear away. These dead cells are 30×0.6 µm in diameter and are easily dislodged in clothing or disseminated into the air. The dermis, an underlying layer of the skin, is composed of connective tissue, elastic fibers, blood and lymph vessels, nervous cells and fibers, muscle tissue, sweat glands, hair roots, and ducts. The glands of the dermis secrete perspiration and oil. The skin functions as a working organ through constant deposition of perspiration, oil, and dead cells on the outer surface. When these materials mix with environmental substances such as dust, dirt, and grease, they form an ideal environment for bacterial growth. Thus, the skin becomes a potential source of bacterial contamination. The normal flora of the skin is characterized by a wide diversity of microorganisms with over 200 species identified (Brodell and Rosentahl 2008), with most of them being harmless and suppressing the growth of pathogens (Brodell and Rosenthal 2008; Grice and Segre 2011). As the secretions build up and the bacteria continue to grow, the skin may become irritated. Food workers may rub and scratch the area, thereby transferring bacteria to food. Improper handwashing and infrequent bathing increase the amount of microorganisms dispersed with the dead cell fragments. Contamination results in shortening the product's shelf life if they are spoilage organisms or in foodborne illness if they are pathogens.

Foodborne illness may occur if a food worker is a carrier of *Staphylococcus aureus*, a bacterial species normally present on the skin. These organisms are present in the hair follicles and in the ducts of sweat glands. They are capable of causing abscesses, boils, and wound infections following surgical operations. As secretions occur, perspiration from the eccrine gland and sebum (a fatty material secreted by the sebaceous gland) contain bacteria that are deposited to the skin surface, with subsequent reinfection.

Certain genera of bacteria do not grow on the skin because the skin acts as a physical barrier and also secretes chemicals that can destroy some of the microorganisms that are foreign to it. This self-disinfectant characteristic is most effective when the skin is clean.

The epidermis contains cracks, crevices, and hollows that can provide a favorable environment for microorganisms. Bacteria can also grow in hair follicles and in the sweat and sebaceous glands. Because hands are very tactile, the opportunity for cuts, calluses, and contact with a wide variety of microorganisms is evident. Hands are in association with so much of the environment that contact with contaminating bacteria is unavoidable.

Bacteria on the skin, particularly the hands, can be classified into two categories, the resident and the transient flora. Resident bacteria of the skin, which occur naturally, live in microcolonies that are usually buried deep in the pores of the skin and protected by fatty secretions of the sebaceous glands. They can also be found on the surface of the skin (CDC 2002; WHO 2009). The microorganisms in the resident group include *S. epidermidis* (the most dominant species), propionibacteria, corynebacteria, dermobacteria, and micrococci (WHO 2009). The resident flora does not normally cause disease but may enhance infection in sterile body cavities, the eyes, or on non-intact skin (WHO 2009). The transient microbiota colonize the superficial layers of the skin and are picked up by the hands of food workers from a variety of contaminated sources and are transient in that they reside on the hands only temporarily (e.g., *S. aureus*, *Escherichia coli*, etc.). These transient microorganisms can be removed by routine handwashing. Poor care of the skin and skin disorders, aside from detrimental appearance, may cause bacterial infections, such as boils and impetigo. *Boils* are severe local infections that result when microorganisms penetrate the hair follicles and skin glands after the epidermis has been broken. This damage can occur from excess irritation of clothing. Swelling and soreness result as microorganisms such as staphylococci multiply and produce an exotoxin that kills the surrounding cells. The body reacts to this exotoxin by accumulating lymph, blood, and tissue cells in the infected area to counteract the invaders. A restraining barrier is formed that isolates the infection. A boil should never be squeezed. If it is squeezed, the infection may spread to adjoining areas and cause additional boils. Such a cluster is called a *carbuncle*. If staphylococci gain entrance to the bloodstream, they may be carried to other parts of the body, causing meningitis, bone infection, or other undesirable conditions. Employees with boils should exercise caution if they must handle food because the boil is a prime source of pathogenic staphylococci. An employee who touches a boil or a pimple should thoroughly wash hands and use an alcohol-based hand sanitizer for disinfection. Cleanliness of the skin and wearing apparel is important in the prevention of boils.

Impetigo is an infectious disease of the skin that is caused by members of the staphylococci group. This condition appears more readily in young people who fail to keep their skin clean. The infection spreads easily to other parts of the body and may be transmitted by contact. Keeping the skin clean also helps to prevent impetigo.

Fingers and Hands

Bacteria may be picked up when the hands touch raw animal products, dirty equipment, garbage, contaminated food, clothing, or other areas of the body. When this occurs, the employees should wash their hands thoroughly and use a hand-dip sanitizer to reduce transfer of contamination. Clean, intact, disposable plastic gloves may be a solution (although their use has been considered controversial by some sanitation experts who maintain that their use may allow contamination). They help prevent the transfer of pathogenic bacteria from the fingers and hands to food and have a favorable psychological effect on those observing the food being handled in this way.

The use of gloves offers both benefits and liabilities. A clean contact surface may be attained initially, and bacteria that are sequestered on and in the skin are not permitted to enter foods as long as the gloves are not torn or punctured. However, the skin beneath the gloves is occluded, and heavily contaminated perspiration builds up rapidly between the internal surface of the glove and skin. Furthermore, gloves tend to promote complacency that is not conducive to good hygiene. Gloves must be changed frequently, especially when moving between raw and ready-to-eat foods and changing tasks while preparing foods. The hands should always be thoroughly washed and dried before clean gloves are worn.

Fingernails

One of the easiest ways to spread bacteria is through dirt under the fingernails, and employees with dirty fingernails should never handle any food. Fingernails should be trimmed and cleaned regularly to prevent dirt buildup and contamination. Frequent washing with soap and water is effective in reducing the microbiota on the hands.

Jewelry

To promote both food safety and personal safety, jewelry should not be worn in food facilities. The FDA Food Code (US FDA 2013) allows for a plain ring such as a wedding band while preparing food, but it should be remembered that food debris and bacteria can accumulate in, around,

Fig. 6.1 High-speed photograph of a human sneeze (Courtesy of the CDC Public Health Image Library 2017)

and under jewelry providing a source of food contamination. Decorative jewelry can also break and fall into food causing a physical hazard. In addition, jewelry can also be a personal safety hazard if worn around equipment and machinery in food operations.

Hair

Microorganisms (especially staphylococci) are found on hair. Employees should wash their hair regularly and use an effective hair restraint to reduce the risk of hair falling into food. Employees should minimize contact with their hair while working with food and wash their hands thoroughly if they do so. The necessity for wearing hair coverings in food processing and preparation areas should be considered a condition of employment for all new employees and should be made known at the time that they are hired. Disposable hair covers should also be worn beneath hard hats in food processing environments. Facial hair should also be covered with disposable beard nets and/or mustache snouts.

Eyes

The eyes are normally free of bacteria but mild bacterial infections may develop. Bacteria can then be found on the eyelashes and at the indentation between the nose and eye. By rubbing the eyes, the hands are contaminated and can, in turn, contaminate foods that are being processed or prepared.

Mouth

Many bacteria are found in the mouth and on the lips. Various disease-causing bacteria, as well as viruses, can be found in the mouth, especially if an employee is ill. These microorganisms can be transmitted to other individuals, as well as to food products, when one sneezes. During a sneeze, microorganisms are propelled at high speed into the air and can contaminate exposed foods that are in the area. Figure 6.1 is a high speed photograph of a human sneeze that illustrates this concept. Eating and drinking in a food environment is also discouraged, since organisms from the lips and mouth can easily be transferred to food if this practice occurs. The Food Code (US FDA 2013) allows food workers to drink from a closed beverage container if it is properly handled to prevent contamination. In addition, smoking should be prohibited while working with food, so companies should provide designated smoking areas for those workers who do smoke.

Regular oral hygiene including brushing the teeth prevents the buildup of bacterial plaque on the teeth and reduces the degree of contamination that might be transmitted to a food product if an employee gets saliva on the hands or sneezes.

Nose, Nasopharynx, and Respiratory Tract

The nose and throat have a more limited microbial population than does the mouth. This is because of the body's effective filtering system. Particles larger than 7 μm in diameter that are inhaled are retained in the upper respiratory tract. This is accomplished through the highly viscid mucus that constitutes a continuous membrane overlying the surfaces within the nose, sinuses, pharynx, and esophagus. Approximately half of the particles that are 3 μm or larger in diameter are removed in the remaining tract, and the rest penetrate the lungs. Those particles that do penetrate and lodge themselves in the bronchi and bronchioles are destroyed by the body's defenses. Viruses are controlled through virus-inactivating agents found in the normal serous fluid of the nose.

Occasionally, microorganisms do penetrate the mucous membranes and establish themselves in the throat and respiratory tract. Staphylococci, streptococci, and diphtheroids (nonpathogenic corynebacteria) are frequently found in these areas. Other microorganisms occasionally inhabit the tonsils. The *common cold* is one of the most prevalent of all infectious diseases and is frequently caused by rhinoviruses and caronaviruses. The initial viral attack is generally followed by the onset of a secondary infection because the initial disease lowers the resistance of the mucous membranes in the upper respiratory tract. The secondary infection may be caused by a variety of agents, including bacteria. Bacteria, especially from employees with a cold, can be easily transmitted from the nose to hands to food.

Sinus infection results from the infection of the membrane of the nasal sinuses. The mucous membranes become swollen and inflamed, and secretions accumulate in the blocked cavities. Pain, dizziness, and a runny nose result from the pressure buildup in the cavities. Precautions should be taken if employees with nasal discharges must handle food products. An infectious agent is present in the mucous discharge, and other organisms, such as S. aureus, could be present. For this reason, employees should wash and disinfect their hands after blowing their noses, and all sneezes should be blocked.

A *sore throat* is sometimes caused by a species of streptococci. The primary source of pathogenic streptococci is the human being, who carries this microbe in the upper respiratory tract. "Strep throat," *laryngitis*, and *bronchitis* are spread by the mucous discharge of carriers. Streptococci are also responsible for scarlet fever, rheumatic fever, and tonsillitis. These conditions may be spread through employees with poor hygienic practices.

Influenza, commonly referred to as *flu*, is an acute infectious respiratory disease caused by a virus that occurs in small to widespread epidemic outbreaks. It gains entrance to the body through the respiratory tract. Death may result from secondary bacterial infections by staphylococci, streptococci, or pneumococci.

Most of these ailments are highly contagious, so employees infected with any of them should not be permitted to work with food. They endanger the products that they handle, fellow employees, and consumers of the food. All coughs and sneezes contain atomized droplets of mucous containing the infectious agents and should be blocked by the elbow or shoulder and the hands should be washed. The hands should be kept as clean as possible by thorough and frequent washing and through the use of alcohol-based hand sanitizers to prevent contamination of the infectious microorganism. As mentioned earlier, food workers who are experiencing persistent sneezing, coughing, or a runny nose that causes discharges from the eyes, nose, or mouth should not work with exposed food, clean equipment, utensils, and linens or unwrapped single-service articles (US FDA 2013). After coughing, sneezing, or nose blowing, food workers must wash hands thoroughly, otherwise bacteria can be transferred to the food being prepared or processed. The Food Code provisions on employee health are aimed at removing highly infectious food employees from the work place.

Excretory Organs

Intestinal discharges are a prime source of bacterial contamination. It is estimated that thousands of species of bacteria and 100 trillion microorganisms are present in the adult human intestine

(Round and Mazmanian 2009). *Enterococcus faecalis* and staphylococci are generally the only bacteria found in the upper part of the small intestine; however, the species and individual organisms become more numerous in the large intestine. Particles of feces collect on the hairs in the anal region and are spread to clothing. When employees go to the washroom, they may pick up some of the intestinal bacteria, and if their hands are not washed properly, these organisms will be spread to food products. A lack of good personal hygiene is responsible for this type of fecal-oral contamination. For this reason, employees should thoroughly wash their hands with soap before leaving the washroom and should use a hand-dip sanitizer or sanitizer gel before working with food. According to the Food Code, after using certain types of hand sanitizers, the hands must be rinsed in clean water before hand contact with food or by the use of gloves (US FDA 2013).

Both viruses and pathogenic bacteria can be spread through fecal contamination to food products. Unlike bacterial contaminants, viruses such as *Norovirus* and hepatitis A (infectious hepatitis) cannot multiply in food. In this case, the food serves as a vehicle or carrier of the virus from one human host to another where the virus replicates and causes illness.

The intestinal tracts of humans and animals carry the most common forms of bacteria, which, when multiplied sufficiently, can cause illness. The infections range from slight to severe and may even result in death. *Salmonella*, *Shigella*, and enterococci bacteria that cause different types of intestinal disorders are the most common.

Personal Contamination of Food Products

The intrinsic factors that affect microbial contamination by people are as follows:

1. *Body location*. The composition of the normal microbial flora varies depending on the body area. The face, neck, hands, and hair contain a higher proportion of transient microorganisms and a higher bacterial density. The exposed areas of the body are more vulnerable to contamination from environmental sources. When environmental conditions change, the microbial flora adapts to the new environment.
2. *Age*. The microbial population changes as a person matures. This trend is especially true for adolescents entering puberty. They produce large quantities of lipids known as *sebum*, which promotes the formation of acne caused by *Propionibacterium acnes*.
3. *Hair*. Because of the density and oil production, the hair on the scalp enhances the growth of microbes such as *S. aureus* and *Pityrosporum*.
4. *pH*. The pH of the skin is affected through the secretion of lactic acid from the sweat glands, bacterial production of fatty acids, and diffusion of carbon dioxide through the skin. The approximate pH value for the skin (5.0) is more selective against transient microorganisms than it is against the resident flora. Factors that change the pH of the skin (soap, creams, lotions, etc.) alter the normal microbial flora.
5. *Nutrients*. Perspiration contains water-soluble nutrients (i.e., inorganic ions and some acids), whereas sebum contains lipid (oil)-soluble materials such as triglycerides, esters, and cholesterol. The role of perspiration and sebum in the growth of microorganisms continues to be investigated.

Humans are the most common contamination source of food, and people transmit diseases as carriers. A carrier is a person who harbors and discharges pathogens but does not exhibit the symptoms of the disease. There are several types of carriers including:

1. *Convalescent carriers* are those who have recovered from an illness, continue to harbor the causative organism for a variable length of time, and are capable of transmitting it to others.
2. *Chronic carriers* are those who continue to harbor the infectious organism indefinitely, although they do not show symptoms of the disease.
3. *Contact carriers* are those who acquire and harbor a pathogen through close contact with

an infected person but do not acquire the disease.

People harbor a number of organisms, including:

- *Streptococci*. These organisms, commonly harbored in the human throat and intestines, are responsible for a wider variety of diseases than other bacteria. They are also frequently responsible for the development of secondary infections.
- *Staphylococci*. The most important single reservoir of staphylococci infection of humans is the nasal cavity. Equally important to the food industry are those who possess the pathogenic varieties of the organism as part of their natural skin flora. These people are a constant threat to consumer safety if they are allowed to handle food products.
- *Intestinal microorganisms*. This group of organisms includes *Salmonella*, *Shigella*, Shiga toxin-producing *Escherichia coli*, hepatitis A (infectious hepatitis), *Norovirus*, infectious intestinal amoebas, and parasites. These microorganisms are of public health concern because they can contribute to serious illness.

Handwashing

The first line of defense against disease is frequent and effective handwashing by those who work with food (Taylor 2000). A large percentage of foodborne disease outbreaks are spread by contaminated hands. Appropriate handwashing practices can reduce the risk of foodborne illness, and studies have shown that handwashing can reduce the risk of respiratory infections by 16% (Rabie and Curtis 2006). The most effective method to ensure effective handwashing is through proper education and training, as well as motivation, reinforcement, incentives, and modeling by supervisors and managers who practice a proper handwashing technique. Handwashing is conducted to break the transmission route of the microorganisms from the hands to another source and to reduce transient bac-

teria. It has been shown that microorganisms such as *Pseudomonas aeruginosa*, *Klebsiella pneumoniae*, *Serratia marcescens*, *E. coli*, and *S. aureus* can survive for up to 90 min when artificially inoculated on the fingertips (Filho et al. 1985). Figure 6.2 illustrates bacteria on the hands, in fingernail scrapings, on the nose, lips and hair, and the effects of handwashing.

Handwashing for 20 s with soap and water, which act as emulsifying agents to solubilize grease and oils on the hands, will remove transient bacteria. Increased friction through rubbing the hands together can reduce the number of bacteria more than quick handwashing. There are a number of viable ways of drying the hands and other skin surfaces. Paper roll and sheet towels are acceptable and should be deposited in a waste container after use. Electric blow dryers should be used only in restrooms to avoid temperature rise in other areas. The location of this equipment in processing areas is unacceptable since dust and microorganisms can be spread to food contact surfaces. The proper procedure for handwashing is detailed below and illustrated in Fig. 6.3:

1. Wet hands under running water
2. Apply soap
3. Lather by rubbing hands together, with friction, for at least 20 s, paying special attention to the palms, fingertips, areas between the fingertips, areas between the fingers, and under the fingernails, as well as exposed areas of the forearms
4. Rinse hands under running water
5. Dry hands and arms with a single-use paper towel
6. If using a single sink, turn water off using a paper towel and also use a paper towel to open the restroom door
7. Discard the paper towel in a waste container

Alcohol-based hand sanitizers used after handwashing provide an additional 10- 100-fold reduction (Anon 2002). Instant hand sanitizers should be considered when washing is not possible, but they do not have a lasting effect (Taylor 2000). The key elements of improved handwashing are

HANDS
Let's see how handwashing affects the number of bacteria present

An unwashed hand that looks clean is touched to the agar.

After washing hands for 15 seconds with hot water and soap, bacteria are reduced in number.

Bacteria grow on gelatin-like food (agar) in covered, sterile, plastic plates (petri dishes).

The plate is incubated at 98.6°F for 24 hours. The heavy growth of white colonies indicates that this hand was not very clean and that millions of bacteria were present.

Washing hands with soap and water for another 15 seconds reduces the bacteria even more.

The bacteria grow rapidly when kept at a warm temperature (98.6°F). 24 to 48 hours later, small colonies or clumps of bacteria approximately the size of a pinhead or larger can be seen in the agar.

Fig. 6.2 Bacteria on the body and the effects of handwashing (*Source*: Reprinted with permission from Gravani (1995))

This plate shows what happens when the lips and nose are pressed against agar in a petri dish.

Each sneeze contains between 10,000 and 100,000 bacteria, and they are moved through the air at more than 200 mph.

This picture shows the bacteria present in fingernail scrapings.

Hair is also a source of bacteria and has no place in food. It is unappetizing, unappealing, and adds bacteria to food. This picture shows the bacteria associated with human hair.

Fig. 6.2 (continued)

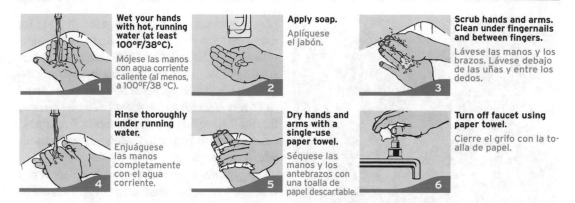

Fig. 6.3 Recommended handwashing procedure (Courtesy of Ecolab, Inc., St. Paul, Minnesota)

motivation, effective training techniques, proper equipment that is fully stocked with soap and disposable towels, and positive reinforcement. Although a certain amount of education is essential, a multidisciplinary framework should target institutional or organizational change so a culture of food safety and sanitation is achieved. Training should be risk based with the consequences of improper handwashing clearly expressed and positive reinforcement provided when food workers consistently perform this task properly and regularly.

Because proper handwashing is essential to attain a sanitary operation, automated handwashing units are being used (Fig. 6.4) in some food facilities. A typical unit is located in the processing area, so when workers enter the area, they must use the washing unit. This equipment is responsible for increased handwashing frequency by 300%. The user inserts the hands into two wash cylinders, passing a photo-optic sensor, which activates the cleansing action. High-pressure jet sprays within each cylinder spray a mixture of antimicrobial cleansing solution and water on the hands, followed by a potable water rinse. The 12-second wash-sanitize-rinse, massage-like cycle has been clinically proven to be 60% more effective at removing pathogenic bacteria from the hands than the manual handwashing (Anon 1997) and reduced water costs. Independent testing has shown that the 12-s cycle

is equivalent to a full minute wash. The high-pressure, low-volume spray uses approximately 2 L (2.1 quarts) of water per wash cycle, one-third of the amount spent in most manual handwashing methods. Up to a 300% increase in washing frequency is accomplished because this equipment provides an easy-to-use, massaging effect on the hands and is nonirritating. Also, this process can remove contamination from gloves and can accomplish hand- or glove washing with approximately 2 L (2.1 quarts) of water or only one-third of the amount used in most manual handwashing methods.

Antimicrobial soaps and lotions have been used for many years in the food industry and also by consumers and were thought to be effective in reducing the bacterial load on the hands, decreasing the possibility of cross-contamination. In September, 2016, the US Food and Drug Administration (FDA) issued a final rule establishing that over-the-counter (OTC) consumer antiseptic wash products containing certain active ingredients can no longer be marketed (US FDA 2016). Information from research studies have suggested that long-term exposure to certain active ingredients used in antibacterial products such as triclosan (liquid soaps) and triclocarban (bar soaps) could pose health risks, such as bacterial resistance or hormonal effects. This final rule applies to consumer antiseptic wash products containing one

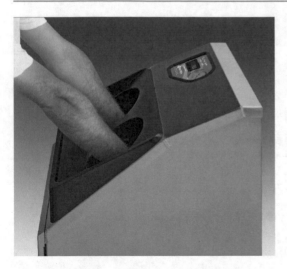

Fig. 6.4 Automated handwashing system (Courtesy of Meritech Handwashing Systems, Golden, Colorado)

or more of 19 specific active ingredients, such as the most commonly used ingredients including triclosan and triclocarban (US FDA 2016). These products are intended for use with water and are rinsed off after use. Companies will no longer be able to market antibacterial washes with these ingredients because manufacturers did not demonstrate that the ingredients are both safe for long-term daily use and more effective than plain soap and water in preventing illness and the spread of certain infections. Some manufacturers have already started removing these ingredients from their products. This rule does not apply to consumer hand sanitizers, hand wipes, or antibacterial products used in health-care settings, such as hospitals and nursing homes (US FDA 2016).

Alcohol hand rub, gel, or rinse sanitizers that contain at least 60% alcohol have been incorporated as a disinfection step after washing hands with soap and water. The alcohol present evaporates in approximately 15–20 s. This is an effective sanitizer that improves personal hygiene and does not contribute to the emergence of microbial resistance. Use of this hand sanitizer before handling food is generally considered to be a safe practice. Figure 6.5 shows a wall-mounted hand sanitizer that can be used to reduce the contamination of workers after handwashing. Ethanol is more effective at destroying viruses than isopro-

panol; however, both alcohols are effective for the destruction of bacteria, fungi, and some viruses (CDC 2016). The CDC recommends that alcohol-based sanitizers can be used "in addition" to handwashing, never as a substitute. Hand sanitizers can reduce the spread of some viruses, like the flu, but they are largely ineffective against norovirus, so it is best to thoroughly wash hands with soap and water.

To provide a barrier between bare hands and ready-to-eat foods, clean, intact gloves should be put on after the hands are washed and thoroughly dried. If the hands are not dry, residual moisture forms an incubation environment for bacteria under the gloves. Although clean, intact gloves provide a barrier between hands and ready-to-eat foods, there are some important facts to consider. Gloves may tear, get punctured, or leak, and natural rubber latex gloves may cause allergenic reactions in sensitized people. Non-latex gloves should be considered when working with ready-to-eat foods. Workers should be reminded that soil on gloves is not as easy to feel as on the bare hands, so changing gloves frequently, especially when they become soiled, tear, or are used for different tasks (between raw and RTE foods), is very important to prevent contamination of foods.

Foodborne Outbreaks Caused by Poor Personal Hygiene

The following examples provide evidence of how poor handwashing and poor personal hygiene have caused major foodborne illness outbreaks.

On a 4-day Caribbean cruise, 72 passengers and 12 crew members had diarrhea, and 13 people had to be hospitalized. Stool samples of 19 of the passengers and two of the crew contained *Shigella flexneri* bacteria. The illness was traced to German potato salad prepared by a crew member that carried these bacteria. The disease spread easily because the toilet facilities for the galley crew were limited (Lew et al. 1991).

Over 3,000 women who attended a 5-day outdoor music festival in Michigan became ill with gastroenteritis caused by *Shigella sonnei*.

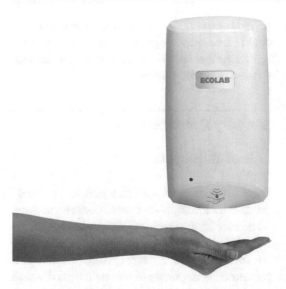

Fig. 6.5 Illustration of a wall-mounted hand sanitizer to reduce microbial contamination of workers (Courtesy of Ecolab, Inc., St. Paul, Minnesota)

The illness began 2 days after the festival ended, and patients were spread all over the United States before the outbreak was recognized. An uncooked tofu salad served on the last day caused the outbreak. Over 2,000 volunteer food handlers prepared the communal meals served during the festival. Before the festival, the staff had a smaller outbreak of shigellosis. Sanitation at the festival was mostly acceptable, but access to soap and running water for handwashing was limited. Good handwashing facilities could have prevented this explosive outbreak of foodborne illness (Lee et al. 1991).

Shigella sonnei caused an outbreak of foodborne illness in 240 airline passengers on 219 flights to 24 states, the District of Columbia, and four countries. The outbreak was identified only because it involved 21 of 65 professional football team players and coaches. Football players and coaches, airline passengers, and flight attendants with the illness all had the same strain of *S. sonnei*. The illness was caused by cold sandwiches served on the flights that had been prepared by hand at the airline flight kitchen. The flight kitchen should have minimized hand contact when preparing cold foods or eliminated them from the menu (Hedberg et al. 1992).

Methods of Disease Transmission

Direct Transmission

Many diseases are transmitted through direct transfer of the microorganisms to another person through close contact. Examples are diphtheria, scarlet fever, influenza, pneumonia, smallpox, tuberculosis, typhoid fever, dysentery, norovirus, and venereal diseases. Respiratory diseases may be transferred via atomized particles expelled from the nose and mouth when a person talks, sneezes, or coughs. When these particles become attached to dust, they may remain suspended in the air for an indefinite length of time. Other people may then become infected upon inhaling these particles.

Indirect Transmission

The host of an infectious disease may transfer organisms to vehicles such as water, food, and soil. Lifeless objects (fomites), other than food, capable of transmitting infections are doorknobs, handles, switches, elevator buttons, telephones, pencils, books, washroom fittings, clothing, money, knives, and many other commonly handled or touched objects. Intestinal and respiratory diseases such as salmonellosis, norovirus, dysentery, and diphtheria may be spread by indirect transmission. To reduce the transfer of microorganisms by indirect transmission, sinks should have foot-operated controls instead of hand-operated faucets, and doors should be self-closing.

Requirements for Hygienic Practices

Management must establish a protocol to ensure hygienic practices by employees. Supervisors and managers should set an example for employees by their own high levels of hygiene and good health while conveying the importance of these practices to the employees. When applicable, they should provide proper laundry facilities or services for maintenance of cleanliness of uniforms through clean dressing rooms, services, and welfare facilities. All employees who work with food should regularly report signs of illness, infection, and other unhealthy conditions.

These practices should be conducted to ensure personal hygiene:

1. Physical health should be maintained and protected through practice of proper nutrition and physical cleanliness.
2. Illness should be reported to the employer before working with food so that work adjustments can be made to protect food from the employee's illness or disease.
3. Hygienic work habits should be developed to eliminate potential food contamination.
4. During the work shift, the hands should be washed after using the toilet; handling garbage or other soiled materials; handling uncooked muscle foods, egg products, or dairy products; handling money; and smoking, coughing, or sneezing.
5. Personal cleanliness should be maintained by daily bathing and use of deodorants, washing hair frequently, trimming and cleaning fingernails regularly, using a hat or hair restraint while working with food, and wearing clean underclothing and uniforms.
6. Employee hands should not touch ready-to-eat foods, service equipment, and utensils with bare hands. Clean intact, and frequently changed disposable gloves should be used when contact is necessary.
7. Rules such as "no smoking" should be followed, and other precautions related to potential contamination should be taken.

Employers should emphasize hygienic practices of employees as follows:

1. Employees should be adequately trained in personal hygiene and the principles of safe food preparation.
2. A regular inspection and observation of employees and their work habits should be conducted. Deficiencies should be immediately corrected.
3. Incentives for superior hygiene and sanitary practices should be provided.

Food workers should be responsible for their own health and personal cleanliness. Employers should be responsible for making certain that the public is protected from unsanitary practices that could cause public illness. Personal hygiene is a basic step that should be taken to ensure the production of wholesome food.

Sanitary Food Handling

Role of Employees

Food processing and foodservice firms should protect their employees and consumers from workers with diseases or other microorganisms of public health concern that can affect the wholesomeness or sanitary quality of food. This precaution is important to maintain a good image and sound operating practices consistent with regulatory organizations. In most communities, local health codes prohibit employees having communicable diseases or those who are carriers of such diseases from handling foods or participating in activities that may result in the contamination of food or food contact surfaces. Responsible employers should exercise caution in selecting employees by screening unhealthy individuals. Selection of employees should be predicated upon these facts:

1. Absence of reportable illnesses.
2. Applicants should not exhibit evidence of a sanitary hazard, such as open sores or presence of excessive skin infections or acne.
3. Applicants who display evidence of respiratory problems should not be hired to handle food or to work in food processing or food preparation areas.
4. Applicants should be clean and neatly groomed and should wear clothing free of unpleasant odor.
5. Applicants should successfully complete a sanitation course such as those provided by a number of local regulatory agencies and organizations.

Required Personal Hygiene

Food organizations should establish personal hygiene rules that are clearly defined and uniformly and rigidly enforced. These rules should be documented, posted, and/or clearly spelled out in all training programs. Policy should address personal cleanliness, working attire, acceptable food-handling practices, and the use of tobacco and other prohibited practices.

Facilities

Hygienic food handling requires appropriate equipment and supplies. Food-handling and food processing equipment should be constructed according to regulations of the appropriate regulatory agency. Restroom and locker facilities should be clean, neat, well lighted, and conveniently located away from production areas. Restrooms should have self-closing doors. It is also preferred that handwashing stations have motion sensor and foot- or knee-operated faucets that supply water at 100 °F (38 °C) (US FDA 2013). Remotely operated liquid soap dispensers are recommended, and disposable towels are best for drying hands. The consumption of snacks, beverages, and other foods, as well as smoking, should be confined to specific areas, which should be clean and free of insects.

Employee Supervision

Employees who handle food should be subjected to the same health standards used in screening prospective employees. Supervisors should observe employees daily for infected cuts, boils, respiratory complications, and other evidence of infection. Many local health authorities require foodservice and food processing firms to report an employee who is suspected to have a contagious disease or to be a carrier.

Employee Responsibilities

Although the employer is responsible for the conduct and practices of employees, responsibilities should be assigned to employees at the time employment begins.

- Employees should maintain a healthy condition to reduce respiratory or gastrointestinal disorders and other physical ailments.
- Injuries, including cuts, burns, boils, and skin eruptions, should be reported to the employer.
- Abnormal conditions, such as respiratory system complications (e.g., head cold, sinus infection, and bronchial and lung disorder), and intestinal disorders, such as diarrhea, should be reported to the employer.
- Personal cleanliness that should be practiced includes daily bathing, regular hair washing at least twice a week, daily changing of undergarments, and maintenance of clean fingernails.
- Employees should tell a supervisor if items such as soap or towels in washrooms should be replenished.
- Habits such as scratching the head or other body parts should be avoided.
- The mouth and nose should be covered during coughing or sneezing, and the hands should be washed afterward .
- The hands should be washed after visiting the toilet, using a handkerchief, smoking, and handling soiled articles, garbage, or money.
- The hands should be kept out of food. Food should not be tasted from the hand, nor should it be consumed in food production areas.
- Food should be handled in utensils that are not touched with the mouth.
- Rules related to use of tobacco should be enforced.

Study Questions

1. What is hygiene?
2. What is a chronic carrier?
3. What is the difference between direct and indirect transmission of diseases?
4. What is a contact carrier?
5. What are resident bacteria?
6. Which microorganisms cause the common cold?
7. What are transient bacteria?
8. What are the four major functions of the skin?
9. What are the two most predominant bacterial species normally present on the skin?
10. What is a carbuncle?

References

Anon (1997). Hands-on hygiene. *Food Qual III* 19: 56.

Anon (2002). Handwashing and hand drying effectiveness. *Food Qual* 11(5): 49.

Brodell LA, Rosenthal KS (2008). Skin structure and function: The body's defense against infection. *Infect Dis Clin Pract* 16(2): 113.

Centers for Disease Control and Prevention (CDC) (2002). Guidelines for hand hygiene in health care settings: Recommendations of the healthcare infection control practices advisory committee and the HICPAC/SHEA/APIC/IDSA Hand Hygiene Task Force. MMWR. 51 (No. RR-16): 1.

Centers for Disease Control and Prevention (CDC) (2016). Handwashing save lives. Show me the science – When & how to use hand sanitizer. https://www.cdc.gov/handwashing/show-me-the-science-hand-sanitizer.html

Filho GPP, Stumpf M, Cardoso CL (1985). Survival of gram negative and gram positive bacteria artificially applied on the hands. *J Clin Microbiol* 21: 652.

Gravani RB (1995). *Safe food preparation: It's in your hands*. Cornell Cooperative Extension Publication: Ithaca, NY.

Green LR, Radke V, Mason R, Bushnell L, Reiman DW, Mack JC, Motsinger MD, Stigger T, Selman CA (2007). Factors related to food worker hand hygiene practices. *J Food Prot* 70(3): 661.

Greig JD, Todd ECD, Bartleson C, Michaels B (2009). Outbreaks where food workers have been implicated in the spread of foodborne disease. Part 6. Transmission and survival of pathogens in the food processing and preparation environment. *J Food Prot* 72: 202.

Grice EA, Segre JA (2011). The skin microbiome. *Nat Rev Microbiol* 9(4): 244.

Hedberg CH, Levine WC, White KE. (1992). An international foodborne outbreak of shigellosis associated with a commercial airline. *JAMA* 268(22): 3208.

Lee LA, Ostroff SM, McGee HB, Johnson DR, Downes FP, Cameron DN, Bean NH, Griffin PM (1991). An outbreak of shigellosis at an outdoor music festival. *Am J Epidemiol* 133(6): 608.

Lew JF, Swerdlow DL, Dance ME, Griffin PM, Bopp CA, Gillenwater MJ, Mercatante T, Glass RI (1991). An outbreak of shigellosis aboard a cruise ship caused by a multiple-antibiotic resistant strain of shigella flexneri. *Am J Epidemiol* 134(4): 413.

Page EH (2017). The structure and function of skin. *Merck Manual*. https://www.merckmanuals.com/home/skin-disorders/biology-of-the-skin/structure-and-function-of-the-skin

Rabie T, Curtis V (2006). Handwashing and risk of respiratory infections: A qualitative systematic review. *Trop Med Int Health* 11(3): 258.

Round JL, Magnanian SK (2009). The gut microbiome shapes intestinal immune response during health and disease. *Nat Rev Immunol* 9(5): 313.

Taylor AK (2000). Food protection: New developments in handwashing. *Dairy Food Environ Sanit* 20(2): 114.

U.S. Food and Drug Administration (2013). Food code 2013. https://www.fda.gov/downloads/Food/GuidanceRegulation/RetailFoodProtection/FoodCode/UCM374510.pdf

U.S. Food and Drug Administration (2016). FDA issues final rule on safety and effectiveness of antibacterial soaps. https://www.fda.gov/newsevents/newsroom/pressannouncements/ucm517478.htm

World Health Organization (2009). WHO guidelines on hand hygiene in health care. Booklet. 270 pps. http://apps.who.int/iris/bitstream/10665/44102/1/9789241597906_eng.pdf

The Role of HACCP in Sanitation

7

Abstract

Hazard Analysis and Critical Control Points (HACCP), a preventive approach to safe food production, applies the principles of prevention and documentation. The essential components for HACCP development are HACCP team assembly, description of the food and its intended use, identification of the consumers of the food, development and verification of a process flow diagram, hazard analysis, identification of critical control points (CCPs), establishment of critical limits, monitoring, corrective actions for deviations, procedures for verification, and record keeping.

Good manufacturing practices (GMPs) are essential building blocks of HACCP, and Sanitation Standard Operating Procedures (SSOPs) are the cornerstones for the plan. Documentation needed for an effective plan includes descriptions of HACCP team-assigned responsibilities, product description and intended use, flow diagram with identified critical control points (CCPs), details of significant hazards with information concerning preventive measures, critical limits, monitoring to be conducted, corrective action plans in place for deviations from critical limits, procedures for verification of the plan, and record-keeping procedures. Periodic auditing is necessary for validation and evaluation of the program. Although parallels exist between HACCP and the Preventive Controls Rule for Hazard Analysis and Risk-Based Preventive Controls (HARPC), a challenge is likely as to how it will interface with HACCP.

Keywords

Control • Development • Documentation • HACCP • Hazards • Implementation • Prevention • Validation • Verification • HARPC

Introduction

Development of an effective HACCP program is crucial to enhancing safety within food manufacturing and distribution. HACCP is a preventive approach to promote consistently safe food production. This program consists of two important concepts of safe food production-prevention and documentation. The major thrusts of HACCP are to determine how and where food safety hazards may exist and how to prevent their occurrence. The important documentation concept is essential to verify that potential hazards are controlled. HACCP has been recommended and/or required for use throughout the food industry and is the basis for federal food inspection in the United States.

This proactive, prevention-oriented program incorporates sound science. HACCP focuses on the prevention or control of food safety hazards that fall in the three main categories of biological, chemical, and physical hazards. The program focuses on safety and not quality and should be considered separate from or a supplement to quality assurance. The objective of HACCP is to ensure that effective sanitation and hygiene and other operational considerations be conducted to produce safe products and to provide proof that safety practices have been followed.

What Is HACCP?

The HACCP concept originated in the 1950s through the National Aeronautics and Space Administration (NASA) and Natick Laboratories for use in aerospace manufacturing under the name "failure mode and effect analysis." The Pillsbury Company, NASA, and the US Army Natick Laboratories developed jointly this rational approach to process control in 1971 to apply a zero-defect program to the food processing industry. HACCP ensured that food consumed in the US space program would be 100% free of bacterial pathogens. This concept provides a simple but very specific method to identify hazards with the implementation of appropriate controls to prevent potential hazards. The US Department

of Agriculture (USDA) Food Safety and Inspection Service (FSIS) identified HACCP as a tool to prevent food safety hazards during meat and poultry production. Several scientific groups embrace and recommend the HACCP concept. These include the National Academy of Sciences (NAS) Committee on the Scientific Basis of the Nation's Meat and Poultry Inspection Program and the NAS Subcommittee on Microbiological Criteria of the Committee on Food Protection. These two committees recognized HACCP as a rational and improved approach to food production control that can determine those areas where control is most critical to the manufacturing of safe and wholesome food.

This technique, which assesses the flow of food through the process, provides a mechanism to monitor these operations frequently and to determine the points that are critical for the control of foodborne disease hazards. A *hazard is the potential to cause harm to the consumer. A critical control point is an operation or step by which preventive or control measures can be exercised that will eliminate, prevent, or minimize a hazard (hazards) that has (have) occurred prior to this point*. The HACCP concept is a valuable program for process control is a harbinger of the trend toward more sophistication in food sanitation and inspection. Governmental regulators legitimized HACCP, and progressive food companies have adopted this prevention program.

The HACCP concept has two parts: (1) hazard analysis and (2) determination of CCPs. Hazard analysis requires a thorough knowledge of food microbiology, which microorganisms may be present, and the factors that affect their growth and survival.

The HACCP evaluation process describes the product and its intended use and identifies any potentially hazardous food items subject to microbial contamination and proliferation during food processing or preparation with subsequent process observation. Hazard analysis is a procedure for conducting risk analysis for products and ingredients by diagramming the process to reflect the manufacturing and distribution sequence, microbial contamination, survival, and proliferation capable of causing foodborne illness. Critical control points are determined from a flow chart with the correction of deficiencies. Established monitoring steps evaluate effectiveness. The HACCP program, implemented by the food industry and monitored by regulatory agencies, provides the industry with tools and monitoring points and protects the consuming public effectively and efficiently.

The HACCP concept provides a more rational approach to the control of microbial hazards in foods. Although HACCP originated over 60 years ago, this concept did not catch on with other products until 1985, when the NAS recommended HACCP for food processing operations. Later NAS studies supported HACCP for the inspection of meat and poultry products and seafood inspection. HACCP continues to evolve especially the verification and validation concepts. Although HACCP is the current trend in the food industry, this concept may evolve to a portion of a more complete program for total quality management in the future.

HACCP relates to a quality assurance (QA) function and is a systematic approach to hazard identification, risk assessment, and hazard control in a food processing and/or foodservice facility and distribution channel to ensure a hygienic operation. Potential product abuse can occur at each stage of the process and examined as an entity and in relationship to other stages. The analysis should include the production environment as it contributes to microbial and foreign material contamination.

HACCP offers benefits to the regulator, processor, and consumer. The regulator and processor receive a history of the operations and can concentrate on components related to controlling hazards. Through monitoring of CCPs, both can evaluate the effectiveness of the control methods. Furthermore, the processor can control the operation on a continuous basis and prevent hazards, instead of reacting to what has already happened. Ultimately, the consumer benefits through access to a product manufactured under conditions with identified and controlled hazards.

Monitoring must encompass systematic observation, measurement, and recording of the significant factors for the prevention or control of hazards. Follow-up is essential to correct any out-of-control processes or to bring the product back into acceptable limits before startup or during the operation. The procedures should define the acceptable performance of a process and describe the handling of process deviations. Bauman (1987) suggested that because specifications for producing a product will contain points critical to safety and some critical to quality, it is important that these not be blended together so that plant people will not confuse them.

The food industry implemented HACCP, but regulatory agencies monitor this program. The Food and Drug Administration (FDA) has adopted the HACCP philosophy because this systems approach allows it to utilize its resources more efficiently. This program provides management with tools to protect the consumer's health.

A major target of HACCP is *Listeria monocytogenes*. HACCP can help prevent the growth of *L. monocytogenes* because it requires steps to confirm the effectiveness of this concept. Samples taken from the food facility environment and product lots confirm that the control measures are effective. *L. monocytogenes* appears to provide the greatest hazard through environmental contamination. Therefore, most sampling is from environmental sources. Environmental samples should include those from ceilings, floors, floor drains, water hoses, equipment surfaces, and other areas on a random basis. Essential routine testing includes floor drains, which can carry microorganisms from a large area, using a rapid microbial method such as immunoassay technology.

HACCP Development

Common prerequisite programs may include, but are not limited to:

1. Facilities. The facilities should be located, constructed, and maintained according to sanitary design principles.
2. Supplier control. Continuing supplier guaranty and supplier HACCP system verification.

3. Specifications. Written specifications for all ingredients, products, and packaging materials.
4. Production equipment. Constructed and installed according to sanitary design principles with preventive maintenance and calibration schedules that are established and documented.
5. Cleaning and sanitation. Conforming to written procedures.
6. Personal hygiene. All personnel entering the manufacturing area should follow the requirements for personal hygiene.
7. Training. All employees should receive training in personal hygiene, GMPs, cleaning and sanitation procedures, personal safety, and their role in the HACCP program.
8. Chemical control. Adopt documented procedures to ensure the segregation and proper use of nonfood chemicals (i.e., cleaning compounds, fumigants, pesticides, and rodenticides) in the plant.
9. Receiving, storage, and shipping. Raw materials and products should be stored under sanitary conditions.
10. Traceability and recall. Raw materials and products should be lot coded and a recall system developed to facilitate rapid and complete traces if recalls are necessary.
11. Pest control. Implementation of an effective past control system.

Essential steps for the development of a HACCP plan are:

1. Assembly of a HACCP team, including the person responsible for the plan. Selections should include employees with expertise in sanitation, quality assurance, and plant operations. It is desirable to have expertise in marketing, personnel management, and communications. HACCP should be a part of a firm's quality assurance program.
2. Description of the food and its distribution. The name and other descriptors including storage and distribution requirements are essential in addition to the listing of all raw materials and adjuncts.

3. Identification of the intended use and consumers of the food. It is especially important to identify intended consumers if infants and other immunocompromised people are the targeted customers.
4. Development of a flow diagram (discussed later under this topic).
5. Verification of the flow diagram. The HACCP team should inspect the operation to verify the accuracy and completeness of the flow diagram. Modifications are appropriate as necessary.
6. Conduction of a hazard analysis:
   (a) Identify steps in the process where the hazards of potential significance occur.
   (b) List all identified hazards associated with each step.
   (c) List preventive measures to control hazards.
7. Identification and documentation of the CCPs in the process.
8. Establishment of critical limits for preventive measures associated with each identified CCP.
9. Establishment of CCP-monitoring requirements, including monitoring frequency and person(s) responsible for the specific monitoring activities.
10. Establishment of corrective action taken when monitoring reveals that a deviation from an established critical limit exists. The action should include the safe disposition of affected food and the correction of procedures or conditions that caused the out-of-control situation.
11. Establishment of procedures for verification that the HACCP system is working correctly. Responsible company personnel should conduct verification of compliance with the HACCP plan on a scheduled basis.
12. Establishment of effective record-keeping procedures that document the HACCP system and update the HACCP plan when a change of products, manufacturing conditions, or evidence of new hazards occurs.

Steps 6 through 12 are the *seven HACCP principles* to be discussed later.

A food and its raw materials are the categories that follow:

- Step 1. Risk assessment as accomplished through examination of the food for possible hazards
- Step 2. Assignment of hazard categories through identification of general food hazard characteristics

Determination of CCPs is also part of the development process. Not all steps in a process are critical, and it is important to separate critical from noncritical points. A practical approach to determining CCPs consists of utilizing a HACCP worksheet with the following headings:

1. Description of the food product and its intended use
2. Flow diagram with the following components:
   - Raw material handling
   - In-process preparation, processing, and fabrication steps
   - Finished product packaging and handling steps
   - Storage and distribution
   - Point-of-sale handling

It is easy to identify CCPs from the flow diagram. A CCP can be a location, practice, procedure, or process, and, if controlled, it can prevent or *minimize contamination*. Monitoring of CCPs ensures that the steps are under control. Monitoring may include observation, physical measurements (temperature, pH, A_w), or microbial analysis and most often encompasses visual and physiochemical measurements because microbial testing is often too time-consuming. Possible exceptions are microbial analysis of the raw materials. Microbial testing may be the only acceptable monitoring procedure when the microbial status of the raw material is a CCP. Microbial testing determines directly the presence of hazards during processing and in the finished product. They can indirectly monitor effectiveness of control points for cleaning and employee hygiene. Yet, this use of microbiology is a check and does not have to be an ongoing process.

Critical limits are essential for each monitoring procedure.

Monitoring verified by laboratory analysis ensures that the process is working. The HACCP concept has been effective because:

1. Cooperation existed between the government and industry to develop monitoring procedures for CCPs.
2. Education of processors is required.
3. Government agencies foster HACCP.

The following is important for HACCP to function effectively:

1. Food processors and regulators must be educated about HACCP.
2. Technical sophistication applied by plant personnel is essential.
3. Avoid overuse of HACCP items that are not hazardous.

A sequence of events must occur in the implementation of HACCP strategy that analyzes potential error and provides an approach for implementing effective risk mitigation strategies (Anon 2014). A viable tool for developing a HACCP plan is the failure mode and effect analysis (FMEA), which provides a weighted metric to apply to a HACCP. FMEA is a method that identifies and quantifies method constraints and steps with potential process variation to quantify risk in each test method. It is a calculation of the total number of steps in a process, the total times the sample is in contact with the operator, and the weighing of the risk associated with the failure mode. Weighing includes factors such as severity of the risk to the overall outcome of the result, frequency at which the error may occur, as well as the likelihood of an operator detecting the error or defect and intervening. These determine a risk priority, where a lower number indicates a lower risk related to that particular method. The application of a scale for risk impact is viable. Such a scale enables a manufacturer to make an informed choice as to which methods best meet the criteria for its facility. Through combining an effective HACCP plan with FMEA and monitoring program effectiveness, a superior product is achievable with lower risk (Anon 2014).

HACCP Program Implementation

A sequence of events must occur in the implementation of HACCP as will discussed briefly.

HACCP Team Assembly

Initial program development involves the designation of a HACCP team, consisting of members with specific knowledge and expertise appropriate to the product and process. Selection criteria should emphasize production and quality assurance knowledge; however, marketing and communication expertise may be appropriate if these employees have an appreciation and understanding of the product and process. The team should include employees who are involved in daily manufacturing as they are more familiar with the variability and limitations of the operation. Furthermore, those involved with the process should be involved to foster a sense of ownership among those who must implement the plan.

Involvement from experts outside of the organization may be beneficial to provide additional expertise, but they must have the support of production employees. Experts who are knowledgeable about the product and process may serve more effectively in verification of the completeness of the hazard analysis and the HACCP plan. These individuals should have the knowledge and experience to correctly (1) identify potential hazards, (2) assign levels of severity and risk, (3) provide direction for monitoring verification and corrective actions when deviations occur, and (4) assess the success of the HACCP plan.

Food Description and Distribution Method

A separate HACCP plan is essential for each food product manufactured in the plant. Product description should include the name, formulation, method of distribution, and storage requirements.

Intended Use and Anticipated Consumers

If the food is for a specific segment of the population, such as infants, immunocompromised people, or those in other categories, identification of the intended group is necessary.

Flow Diagram Development Describing the Process

A simple description of the operation for each step that occurs is important. This diagram is essential for hazard analysis and assessment of CCPs. The diagram serves as a record of the operation and a future guide for employees, regulators, and customers who must understand the process for verification. The flow diagram should include steps that take place before and after the process that occurs in the plant and should contain words rather than engineering drawings.

Flow Diagram Verification

The HACCP team should check the operation to verify the accuracy and comprehensiveness of the flow diagram. Modifications are appropriate if and when necessary. Verification of the effectiveness of sanitizing has received additional attention during the past decade because of pathogens that cause foodborne illness. Thus, additional emphasis is being placed on ensuring that cleaning with cleaning compounds is followed by sanitizing, a lethal step to eradicate remaining invisible microorganisms or debris on surfaces and equipment.

CGMPs: The Building Blocks for HACCP

The Current Good Manufacturing Practice (CGMP) regulations provide criteria for complying with provisions of the Federal Food, Drug, and Cosmetic Act, which mandates that all human foods be free from adulteration. This requirement includes the prevention of product contamination from direct and indirect sources.

Good manufacturing practices are the minimum sanitary and processing requirements necessary to ensure the production of wholesome food. They are broad and general in nature and can explain tasks that are part of many jobs. Good Manufacturing Practices apply to each of the following areas:

1. Personnel. These practices include direction for disease control, cleanliness, education and training, and supervision.
2. Buildings and facilities. The building surrounding grounds, plant construction design, and sanitary operations are included.
3. Equipment and utensils. The hygienic design of all plant equipment and utensils facilitates adequate cleaning and maintenance.
4. Production and process control. Sanitation practices for production-related functions, i.e., inspection, storage, and cleaning of raw material ingredients, and procedures for processing operations.
5. Records and reports. Records should include filing and maintaining for suppliers, processing/ production, and distribution.
6. Defect action levels. These levels are defect limits at which the FDA will take action. The levels are set to avoid health hazards.
7. Miscellaneous. These include other guidelines such as visitor rules.

Sanitation regulations promulgated by the USDA contain identical or similar requirements. Included is a summary of responsibilities for plant management regarding plant personnel. Criteria for disease control, cleanliness (personal hygiene and dress requirements), education, and training are also included. These requirements prevent the spread of disease among workers in the food processing area and from workers to the food itself. A competent supervisor should ensure compliance by all personnel.

Good manufacturing practices are a prelude to HACCP implementation. The application of CGMPs is essential to an effective HACCP program. Furthermore, CGMPs are the foundation for the development of sanitation standard operating procedures (SSOPs). Compliance with specific CGMPs should be included as part of a HACCP program for meat and poultry plants, as CGMP regulations and the USDA sanitation regulations address some biological, chemical, and physical hazards associated with food production. A CGMP compliance program should contain documented plans and procedures.

Good manufacturing practices and SSOPs are interrelated and an important part of process control. CGMPs are the minimum sanitary and processing requirements necessary to ensure the production of wholesome food. The areas addressed through CGMPs are personnel hygiene and other practices, buildings and facilities, equipment and utensils, and production and process controls. CGMPs should be broad in nature.

Sanitation Standard Operating Procedures: The Cornerstones of HACCP

Although SSOPs are interrelated with CGMPs, they detail a specific sequence of events necessary to perform a task to ensure sanitary conditions. SOPs are either SSOPs or manufacturing SOPs. CGMPs should guide the development of SSOPs. SSOPs contain a description of the procedures that an establishment will follow to address the elements of preoperational and operational sanitation relating to the prevention of direct product contamination.

Federally and state-inspected meat and poultry plants are required to develop, maintain, and adhere to written SSOPs. This requirement was because the USDA FSIS concluded that SSOPs were necessary in the definition of each establishment's responsibility to follow effective sanitation procedures and to minimize the risk of direct product contamination or adulteration.

In meat and poultry plants, SSOPs cover daily preoperational and operational sanitation procedures that establishments implement to prevent direct product contamination or adulteration. Establishments must identify the officials who monitor daily sanitation activities, evaluate whether the SSOPs are effective, and take appropriate corrective action when needed. Daily records that reflect completion of the procedures in the SSOPs are required. Deviations and corrective actions taken must be documented and maintained for a minimum of six months and must be made available for verification and monitoring. Corrective actions (1) include procedures to ensure appropriate disposition of contaminated products, (2) restore sanitary conditions, and (3) prevent the recurrence of direct contamination or product adulteration, including the appropriate reevaluation and modification of the SSOPs and the procedures specified therein.

Written SSOPs contain a description of all cleaning procedures necessary to prevent direct contamination or adulteration of products. The frequency with which each procedure in the SSOPs is included along with a designation of the employee(s) responsibility for the implementation and maintenance through actual performance of such activities or that of the person responsible for ensuring that the sanitation procedures are executed.

SSOPs implementation in meat and poultry plants is ensured by the signature and dating by one with overall authority on site or by a higher-level official of the establishment. Furthermore, a signature is required for initiation or any modification. The establishment must evaluate and modify SSOPs, as necessary, to reflect changes in the establishment facilities, personnel, or operations to ensure that they remain effective in the prevention of direct product contamination and adulteration.

HACCP Interface with GMPs and SSOPs

Sanitation SOPs are a prelude to HACCP. The intent of a HACCP plan is to ensure safety at specific CCPs within specific processes. Sanitation SOPs transcend specific processes. Sanitation SOPs are the cornerstones for a HACCP plan and can serve as a preventive approach to direct product contamination and/or adulteration.

HACCP Principles

HACCP is a systematic approach to food production as a means to ensure food safety. The basic principles that underlie the HACCP concept include an assessment of the inherent risks that may be present from harvest through ultimate consumption. It is necessary to establish critical limits at each CCP, appropriate monitoring procedures, corrective action if a deviation occurs, record keeping, and verification activities. The following discussion indicates the seven basic principles of HACCP and gives a brief description of each.

1. Conduct a hazard analysis through the identification of hazards and assessment of their severity and risks by listing the steps in the process where significant hazards occur and describing preventive measures This step provides for a systematic evaluation of a specific food and its ingredients or components to determine the risk from hazardous microorganisms or their toxins. Hazard analysis can guide the safe design of a food product and identify the CCPs that eliminate or control hazardous microorganisms or their toxins at any point during production. Hazard assessment is a two-part process, consisting of characterization of a food according to six hazards followed by the assignment of a risk category based upon the characterization.

The hazard assessment procedure occurs after the development of a working description of the product, establishment of the types of raw materials and ingredients required for preparation of the product, and preparation of a diagram for the food production sequence. The two-part assessment of hazard analysis and assignment of risk categories is as will be described.

Hazard Analysis and Assignment of Risk Categories

Food hazard characteristics are A through F, using a plus symbol (+) to indicate a potential hazard. The number of pluses determines the risk category. If a product falls under hazard class A, it is a risk category VI. A description of the six hazards follows. Hazards are chemical or physical hazards:

- *Hazard A*: This hazard applies to a special class of nonsterile products designated and intended for consumption by at-risk populations, e.g., infants or older, infirm, or immunocompromised individuals.
- *Hazard B*: Products that fit this hazard contain "sensitive ingredients" in terms of microbial hazards.
- *Hazard C*: Manufactured foods in this group do not contain a controlled microbial destruction step.
- *Hazard D*: Foods that fit this hazard are subject to recontamination after processing and before packaging.
- *Hazard E*: With this hazard, there is substantial potential for abusive handling in distribution or in consumer handling that could render the product harmful when consumed.
- *Hazard F*: Foods in this group have not been subjected to a terminal heat process after packaging or when cooked at home.

The following risk categories are according to hazard characteristics:

- *Category O*—No hazard.
- *Category I*—Food products subject to one of the general hazard characteristics.
- *Category II*—Food products subject to two of the general hazard characteristics.
- *Category III*—Food products subject to three of the general hazard characteristics.
- *Category IV*—Food products subject to four of the general hazard characteristics.
- *Category V*—Food products subject to all five of the general hazard characteristics: hazard classes B, C, D, E, and F.
- *Category VI*—A special category that applies to nonsterile products designated and intended for consumption by at-risk populations, e.g., infants, aged, infirm, or immunocompromised individuals. All hazard characteristics apply.

2. Determine CCPs that control the identified hazards A *CCP* is "a point, step, or procedure at which control can be applied and a food safety hazard can be prevented, eliminated, or reduced to an acceptable level." A CCP must be established where control can be exercised. Control of identified hazards must occur at some point(s) in the food production sequence, from growing and harvesting raw materials to the ultimate consumption of the prepared food.

Critical control points are located at any point in a food production sequence where it is essential to destroy or control hazardous microorganisms. An example of a CCP is a specified heat process at a given time and temperature implemented to destroy a specified microbial pathogen. Another temperature-related CCP is refrigeration required to prevent hazardous organisms from growing or the adjustment of the pH of a food to prevent toxin formation. CCPs are not those points that do not control safety. A *control point* differs from a CCP, in that it is "any point, step, or procedure in a specific food production operation at which biological, physical, or chemical factors can be controlled." Figure 7.1 presents a CCP decision tree recommended by the National Advisory Committee on Microbiological Criteria for Foods (1997) to assist in the identification of CCPs.

Information developed during the hazard analysis serves as a guideline to identify the steps in the process that are CCPs. CCPs are located at any point where hazards require prevention, elimination, or reduction to acceptable levels. Examples of CCPs may include but are not limited to specific sanitation procedures, cooking, chilling, product formulation, and cross-contamination prevention.

The number of established CCPs should be minimal to simplify monitoring and documentation and to avoid dilution of the HACCP program effectiveness. CCPs must be carefully developed and documented and address only product safety.

Food operations may differ in the risk of hazards and the points, steps, or procedures that are CCPs. Differences such as the process, layout, equipment, products manufactured, and

HACCP: Principles and Application

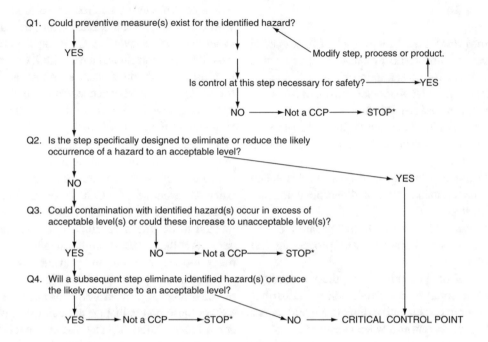

Fig. 7.1 CCP decision tree (*Source*: Pierson and Corlett 1992)

ingredients incorporated determine whether a CCP exists. Although general HACCP plans may serve as a guide, each operation necessitates scrutiny before the assignment of CCPs and development of a HACCP plan. A CCP may not be as appropriate as an SOP or SSOP to prevent a hazard. However, if a CCP exists, an SOP or SSOP may not be an acceptable substitute.

3. Establish critical limits for preventive measures associated with each identified CCP A *critical limit* is one or more prescribed tolerances to ensure that a CCP effectively controls a microbial health hazard to an acceptable level. Information about critical limits is essential for safe control of a CCP. Each preventive measure has associated with it critical limits that serve as boundaries of safety for each CCP. The critical limits for preventive measures are time, temperature, physical dimensions, pH, A_w, etc. Development of these critical limits may require

determination of probable maximum numbers of microorganisms in the product, as well as sources such as regulatory standards and guidelines. The food industry is responsible for engaging competent authorities to validate that the critical limits will control the identified hazard.

4. Establish procedures to monitor CCPs Scheduled testing or observation of a CCP and its limits occurs through monitoring. Documentation of results obtained from monitoring is essential. From a monitoring standpoint, failure to control a CCP is a critical defect. A critical defect may result in hazardous or unsafe conditions for those who use or depend on the product. Monitoring procedures must be very effective because of the potentially serious consequences of a critical defect.

Monitoring is a planned sequence of observations or measurements to assess whether a CCP is under control and to produce an accurate record for future use in verification. Monitoring is essential to food safety management because it tracks the system's operation. If monitoring reveals that there is a trend toward loss of control, i.e., exceeding a target level, action is essential to bring the process back into control before a deviation occurs. Monitoring identifies a loss of control or a deviation at a CCP, such as exceeding the critical limit and the need for corrective action. Furthermore, monitoring provides written documentation for use in verification of the HACCP plan.

If feasible, monitoring should be continuous. It is possible to attain continuous monitoring of pH, temperature, and humidity with recorders. Insufficient control, as recorded on the chart, suggests a process deviation. When it is impractical to monitor a critical limit continuously, a monitoring interval must be established that will be reliable enough to indicate that the hazard is under control. This can be accomplished through a statistically designed data-collection program or sampling system. Statistical procedures are useful for measuring and reducing the variation in manufacturing equipment and measurement devices.

Monitoring procedures for CCPs require rapid results because insufficient time exists for time-consuming analytical testing. Microbial testing is also normally unsatisfactory for monitoring CCPs because of the amount of time involved. Physical and chemical measurements are more viable because they are rapid and can indicate microbial control of the process. Physical and chemical measurements for monitoring include measurements of pH, time, temperature, and moisture which are preventive measures for cross-contamination and specific food handling procedures.

Random checks supplement the monitoring of certain CCPs. They check incoming pre-certified supplies and ingredients, to assess equipment and environmental sanitation, airborne contamination, and cleaning and sanitization of gloves. Random checks include physical and chemical testing and microbial tests, as appropriate.

Certain foods, microbially sensitive ingredients, or imports may not have alternative to microbial testing. However, a sampling frequency that is adequate for reliable detection of low levels of pathogens is seldom possible because of the large number of samples needed. Microbial testing has limitations in HACCP but is valuable as a means of establishing and randomly verifying the effectiveness of control of the CCPs.

5. Establish corrective measures when there is a deviation from an established critical limit Specific corrective actions must demonstrate that the CCPs are under control. Acceptance by the appropriate regulatory agency is essential for documented deviation procedures prior to approval of the plan. If a deviation occurs, the production facility should place the product on hold until appropriate corrective actions and analyses are completed.

6. Establish procedures for verification that the HACCP plan is working correctly The concept of verification was a late addition to HACCP and is the most complicated principle. Verification occurs through methods, procedures, and tests to determine compliance with the HACCP plan. Verification confirms the identification of all hazards in the HACCP plan. It may be accomplished through chemical and sensory methods and testing for conformance with microbial criteria when established. This activity may include, but is not limited to:

1. A scientific or technical process to ensure that critical limits at CCPs are satisfactory.
2. Establishment of appropriate verification inspection schedules, sample collection, and analyses.
3. Documented periodic revalidation independent of audits or other verification procedures that must be performed to ensure accuracy of the HACCP plan; revalidation includes a documented on-site review and verification of all flow diagrams and CCPs in the HACCP plan.
4. Governmental regulatory responsibility and actions to ensure that the HACCP plan is functioning satisfactorily.

The basic role of verification is to ensure that the Food Safety Management Systems (FSMS) or HACCP plan is functioning as designed and is effective.

Four major types of verification activities are:

1. A non-audit review of documents such as CCP records to ensure that a specific lot of product complies with the HACCP plan
2. Conducting various measurements and assessment activities to ensure that a prerequisite program (PRP) or the product is within defined parameters such as the collection of environmental microbial swabs to ensure that the cleaning and sanitizing program is compliant with the internal specifications
3. Conducting assessments to determine if other components of the FSMS are operating within defined parameters such as determination of effective training, inspection of equipment calibration, and conducting mock recalls
4. Internal and external audits to provide an unbiased assessment of the FSMS

7. Establish effective record-keeping procedures that document the HACCP plan The HACCP plan must be on file at the food establishment to provide documentation relating to CCPs and to any action on critical deviations and production disposition. It should clearly designate records that are available for government inspection.

The HACCP plan should contain the following documentation:

1. Listing of the HACCP team and assigned responsibilities
2. Product description and its intended use
3. Flow diagrams of the entire manufacturing process with the CCPs identified
4. Descriptions of hazards and preventive measures for each hazard
5. Details of critical limits
6. Descriptions of monitoring conducted
7. Description of corrective action plans for deviations from critical limits

8. Description of procedures for verification of the HACCP plan
9. Listing of record-keeping procedures

HACCP Implementation

A designated HACCP plan addresses a specific process or product. The plan should include the objective of the analysis, whether it be safety, spoilage, or foreign control. Documentation should include the objective(s); job title of each employee involved; flow charts of the operations involved, with the CCPs highlighted; hazards and details, with control options, cross-references to equipment maintenance and cleaning schedules, and procedures or CGMPs that apply to the process; and summary and conclusions, including action to be taken as a result of the analysis. The HACCP report forms the record of the plan and is readily available and discernable to anyone who needs to use the report. It is an important resource when any changes are proposed for the process or specification concerned. A matrix has been suggested (Shapton and Shapton 1991) to allow this feature, with these suggested column headings:

1. CCP number
2. Process/storage state of this CCP
3. Description of the state
4. Hazards associated with the state
5. Hazards controlled
6. Control limits
7. Deviations and how they have been or may be corrected
8. Planned improvements

To ensure success, employees should be educated, trained, and retrained in the use of HACCP. Employee turnover rate necessitates a need for continuous education so that plant personnel will understand HACCP and the need for various controls. This approach can be responsible for the reduction of foodborne illness outbreaks and the replacement of costly crisis management with cost-effective control.

Effective implementation of the HACCP concept encompasses education of employees, especially workers in the production areas where problems can occur. An effective approach contains the following steps:

1. *Management education.* Quality assurance personnel and higher management need to understand the HACCP concept so that an effective program can be instituted that relies on total commitment of all personnel. Training courses for the management team are important to create the awareness of the entire program. Furthermore, plant managers and supervisors should set a positive example.
2. *Operational steps.* The plant design and operating procedures may require change to avoid interference with a hygienic operation. It is important to assign experienced, properly trained personnel to critical operations.
3. *Employee motivation.* Improvement of working conditions can be a motivating force in the implementation of HACCP. Task redesign may be a helpful tool in attaining success. All employees must feel a sense of personal responsibility for the quality and safety of food products.
4. *Employee involvement.* To ensure commitment of the workers, they must be involved in problem solving. Consultation groups (quality circles) and their recommendations offer an approach toward more employee involvement. Management should guide, not administer, the HACCP program. Furthermore, this program requires total commitment to a long-term undertaking by all levels of management and production employees.

Because HACCP represents a structured approach to control the safety of food products, effective organization and management is essential to ensure that the plan is operating correctly and perseveres in the future. The most viable source for leadership to organize and implement HACCP is the quality assurance group within a firm or a similar group with past and/or present responsibilities for the food safety functions in the organization. The challenges that ensure proper implementation include execution of the 12 developmental steps discussed previously.

Two deficiencies in a HACCP plan most frequently detected are (1) documentation of the HACCP plan (insufficient "background" documentation on decision making and inadequate documentation of actual processes) and (2) management of the HACCP program. Ineffective management is most likely to cause failure to ensure that (1) a comprehensive plan is in place to yield safe products and (2) adequate review mechanisms to ensure success of the HACCP plan.

The plan success depends on management commitment, detailed planning, appropriate resources, and employee empowerment. A corporate statement of support for HACCP is an effective tool for communicating the importance of HACCP to all employees. Furthermore, management should establish specific objectives and implementation schedules for additional support.

Management and Maintenance of HACCP

One person within an organization should be responsible for the maintenance of HACCP. This responsibility includes coordination of input from others, monitoring of activities, review, validation, verification, and documentation. Furthermore, the coordinator should ensure that the HACCP team has access to the variety of information required to conduct the various assignments. Reporting structures and the relationships of those involved should be determined, and the appropriate forms must be developed and provided for employees.

Frequent evaluation and revision as needed enhances the viability of a HACCP plan. Evaluation should involve the review and interpretation of results and verification and validation of the plan. A mandatory evaluation process guarantees that a systematic evaluation will be made of any changes in the process, thus assuring evaluation of any revisions affecting product safety before implementation. Evaluation and needed revisions result from the verification principle of the HACCP plan.

Most food establishments have instituted environmental sampling to meet HACCP prerequisites and/or validate the process. Sampling strategies evaluated in advance with the particular operation being evaluated and consistent with the goals of the plant's daily preoperational sanitation policy and procedures is essential. Sampling strategies validate the process in concert with statistical process control and other environmental monitoring to ensure that the manufactured products are not getting too dirty throughout the production shift. The detection of contaminated products and a quality assurance investigation should identify the problem origin and corrective actions necessary for eradication of the problem. The stringency of environmental sampling within preoperational sanitation is by the zone or "shell" concept under which a map of sampling sites is prepared, followed by routine testing of three of these identified sites. Samples taken occasionally ad hoc at non-mapped sites avoid surprises. Identified sampling sites result from considering where the team or plant crew has observed an accumulation of food, biofilms, and potential bacteria.

Maintenance of effective HACCP (process) depends upon regularly scheduled verification activities. The FDA and FSIS do not set reassessment time requirements. However, regulatory authorities have suggested plan updates and revisions as needed. Good management practices dictate appropriate reassessment protocol. A manufacturing facility is responsible for establishing verification tables. The process, procedure, ingredients, CGMPs, and decisions must be correct and have documentation of scientific data and history to support this correctness.

HACCP Auditing and Validation

Within the first year of HACCP plan implementation, auditing determines its effectiveness. Verification reviews those activities, other than monitoring, that determine the adequacy of and compliance with the plan. Verification confirms adherence to requirements and procedures, whereas validation is proof that the control measures work. Validation requires scientific proof that a process should work in theory and does work in practice. Changes in the process, such as ingredients, suppliers, procedures, etc., dictate that the process may need to be revalidated.

Auditing occurs through the HACCP team, management, or a consultant and/or food scientist. It should include a comprehensive review of the entire plan with evaluation and documented observations, conclusions, and recommendations. Auditing serves as a report card for the plan and provides future direction. Furthermore, auditing contributes to validation of the plan. Validation, as defined by the National Advisory Committee on Microbiological Criteria for Foods (1997), is that element of verification that focuses on the collection and evaluation of information to determine whether the HACCP plan, when implemented properly, will effectively control the significant hazards.

Electronic HACCP

With the continued use of the computer in food production, scientists have investigated the development of a viable electronic HACCP system called eHACCP. This program manages the operational aspects of HACCP including data collection, validation, and reporting results. Surak and Cawley (2009) indicated that an effective eHACCP system must contain the following components:

1. Electronic data collection—to eliminate the manual clipboard to reduce labor cost and improve data quality
2. Database that is secure and can provide immediate access
3. Alarm system—to warn when a critical value is violated
4. Analytics—to provide real-time operational understanding of the process
5. Reporting—to provide documents that meet the requirements of regulators, customers, and auditors
6. Security—to provide electronic security and comply with electronic regulatory record-keeping requirements

Successful transition from paper to electronic HACCP depends upon factors such as the KISS (keep it simple stupid) approach, identification and solving of potential glitches, competent employees, and updating of new technology. If properly implemented, this concept can be a valuable tool in conducting HACCP.

HACCP Vulnerability Assessments

HACCP is the basis of Vulnerability Assessment and Critical Control Points (VACCP). VACCP outline is a method to protect food products from fraud and potential adulteration. The international version is Threat Assessment and Critical Control Points (TACCP). VACCP is the application of the HACCP-type system specifically to the unique attributes of a food fraud incident. Labs (2016) suggested that VACCP is unlike food defense covered by TACCP that protects against malicious tampering to cause harm. VACCP and TACCP cover the entire food fraud spectrum.

Hazard Analysis and Risk-Based Preventive Controls (HARPC)

At this writing, the main compliance date is September 2017 for the Preventive Controls Rule for Human Food. As with HACCP, HARPC requires hazard analysis to assess the significance of food safety hazards. To meet the Preventive Control Rule, all assessed food safety hazards are without considering any current controls.

The incorporation of a preventive control (PC) recognizes when a hazard (i.e., contamination) is present and corrects the error or stops product release with positive confirmation that the hazard has, or has not, occurred. Consequences of this change may be significant, requiring a different approach to ensure system manageability. Marsh (2016) suggested that food establishments have a major task of trying to amend their current safety systems to meet new requirements.

HARPC consists of the following steps:

1. Define the assessment scope
2. Hazard identification
3. Conduct a hazard analysis
4. Preventive controls addition
5. Monitoring systems implementation
6. Corrective actions and corrections addition
7. System verification
8. System reanalysis

The Preventive Control Rule does not indicate how it should be organized and documented nor does it address a portion of the practices from HACCP such as scope, product description, intended use and user, and process flow. However, the introduction of PCs as an additional tier of control can be advantageous. With proper incorporation, these tools enhance food safety.

Study Questions

1. What is a hazard?
2. What is a critical control point?
3. What are CGMPs?
4. What is the meaning of sanitation SOPs?
5. What are the seven HACCP principles?
6. What are the five steps necessary to develop a HACCP plan prior to conducting a hazard analysis?
7. What is monitoring?
8. What is a control point?
9. What is critical limit?
10. What is HACCP verification?
11. What is HACCP plan validation?
12. What is the justification for auditing?
13. What is VACCP?
14. What is TACCP?
15. What is HARPC?

References

Anon (2014). Applying HACCP strategies to pathogen detection methods. *Food Saf Mag* 20(4): 40.

Bauman HE (1987). The hazard analysis critical control point concept. In *Food protection technology*, ed. Felix CW, 175. Lewis Publishers: Chelsea, MI.

Labs W (2016). VACCP: HACCP for vulnerability assessments. *Food Eng* February. p. 53.

Marsh K (2016). How to combine your HACCP and HARPC plans. *Food Qual Saf* 23(5): 16.

National Advisory Committee on Microbiological Criteria for Foods (1997). *Hazard analysis and critical control point principles and application guidelines*.

Pierson MD, Corlett Jr DA (1992). *HACCP principles and applications*. Van Nostrand Reinhold: New York.

Shapton DA, Shapton NF, eds. (1991). Establishment and implementation of HACCP. In *Principles and practices for the safe processing of foods*, 21. Butterworth-Heinemann: Oxford.

Surak JG, Cawley JL (2009). Moving from paper to electronic HACCP records. *Food Saf Mag* 15(1): 20.

Quality Assurance for Sanitation

8

Abstract

Product wholesomeness and uniformity can be more effectively maintained through a quality assurance (QA) program that incorporates available scientific and mechanical tools. Quality is considered to be the degree of acceptability by the user. These characteristics are both measurable and controllable. The major ingredients needed for a successful quality assurance (QA) program are education and cooperation. An important component of a QA program is third-party auditing, Safe Quality Food (SQF) certification, and ISO Accreditation 9000. The Hazard Analysis Critical Control Point (HACCP) approach can be incorporated in a QA program because it applies to a zero defects concept in food production. Effective surveillance of a QA program can detect unsanitary products and variations in production.

Statistical Quality Control (SQC) techniques make inspection more reliable and eliminate the cost of 100% inspection. The principal tool of a statistical QC system is the control chart. Trends of control charts provide more information than do individual values. Values outside the control limits indicate that the production process should be closely observed and possibly modified. Control limits should be determined not only by natural variability in the process but also based on quality and safety specifications and optimization of the process.

Keywords

Quality assurance • Total quality management • Quality control • Auditing • Statistical quality control • Control limits • Control charts

Introduction

One key function of QA and QC is to evaluate the effectiveness of a sanitation program. A QA program provides the avenue to establish checks and balances in the areas of food safety, public health, technical expertise, and legal matters affecting food processors. Activities related to food sanitation include sanitation inspections, product releases and holds, packaging sanitation, and product recalls and withdrawals. When cleaning

© Springer International Publishing AG, part of Springer Nature 2018
N.G. Marriott et al., *Principles of Food Sanitation*, Food Science Text Series,
https://doi.org/10.1007/978-3-319-67166-6_8

and sanitation occur, it will be documented and testing will be conducted to evaluate the effectiveness of the sanitation program. Quality assurance also implies that equipment is properly calibrated, tests are performed properly, positive and negative controls are run, and laboratory results are documented accurately. QC will conduct testing to ensure that sanitation is effective at enhancing food safety without negatively impacting product quality. A specific example of this relationship can be demonstrated through a process step in a poultry processing plant. Broilers are commonly placed in a chill tank after harvesting to cool the carcasses down to refrigeration temperatures within 2 h postmortem. Peracetic acid is used in the chill tank to control *Salmonella* contamination. Quality control personnel will monitor peracetic acid concentration and pH to ensure that *Salmonella* is controlled and meat proteins are not denatured. This demonstrates how quality control is related to food safety, sanitation, and quality.

The food industry emphasizes an organized sanitation program that monitors the microbial load of raw ingredients in production plants and the wholesomeness and safety of the finished products, in an effort to maintain or upgrade the acceptability of its food products. As a result of increased consumer sophistication, desire for clean labels, increased globalization, and information available on the internet, it is even more vital for the food industry to develop an effective QA and sanitation program. Cleaning and sanitizing are the two most important elements that comprise a food sanitation program, and both should be performed in tandem to successfully achieve food safety and quality assurance goals.

Since the middle of the 1990s, additional emphasis on sanitation, food safety (including Hazard Analysis and Critical Control Points (HACCP)), and consumer and customer pressure has placed an increasing onus on food processors to expand their testing initiatives, utilize rapid testing methodology, and employ emerging technologies to reshape and enhance their testing programs. Food scientists have also had a positive impact on QA programs because many of these professionals have joined various companies in the food industry. Their efforts have been instrumental in the adoption and/or upgrading of QA programs for the organizations that they represent. In its initial stages, QA was primarily a QC function, acting as an arm of manufacturing. It has now evolved to a formidable force within the executive structure of large food firms and has emerged into a broad spectrum of activities.

A QA program that emphasizes sanitation is vital to the growth of a food establishment. If foods are to compete effectively in the marketplace, established hygienic standards must be strictly maintained. However, it is sometimes impractical for production personnel to measure and monitor sanitation while maintaining a high level of productivity and efficiency. Thus, an effective QC program should be available to monitor, within established priorities, each phase of the operation. All personnel should incorporate the team concept to attain established sanitary standards, ensuring that food products in the marketplace are safe.

The development of an effective testing program requires a commitment to the many aspects of a food processor's operation. It must be decided whether what kind of and how much in-plant testing is appropriate. Other decisions are what testing and how much should be outsourced to a contract laboratory. Additional requirements include the implementation of a laboratory quality assurance program that defines best practices and operations, personnel, and instrumentation. Furthermore, it must be decided whether to accredit the laboratory and what kind of proficiency testing program is most appropriate. It is essential to explore new methods and technologies to increase the accuracy and meaning of results that are obtained.

All processors, regardless of industry segmentation, should view regulatory guidelines as a basis for establishing testing programs and strive to exceed prescribed testing requirements to protect their products and consumers. Many meat processors that manufacture ready-to-eat (RTE) products are embracing a proactive stance and take more microbial samples that are required for regulatory compliance. For example, a meat company may produce a cooked RTE meat product, and their testing demonstrates that there is no *Listeria monocytogenes* or other pathogens on their finished product. However, they may have determined that their desired shelf life of 90 days is not achieved under refrigeration conditions and modified atmospheric packaging (MAP). Quality control would need to conduct testing to determine if the shelf life is due to microbial growth, chemical deterioration, and/or sensory off-flavors. Once the mechanism of spoilage is determined, the process can be evaluated to determine the point in the process that contamination occurs. For example, if it is microbial spoilage, the MAP gas conditions could be off target, or bacterial spoilage may happen somewhere in the process. The processing steps would need to be evaluated to determine the cause. Were there high microbial counts of raw materials? Did contamination occur between heat processing and packaging? What are the sanitation changes that need to occur to enhance control of spoilage bacteria?

It is imperative for food testing laboratories to ensure that access to hazardous biological and chemical agents is controlled so that they cannot be used in criminal or terrorist acts in food and water. Several leading testing organizations, including the American Council of Independent Laboratories, are urging plant food testing laboratories to aggressively implement food defense programs as part of their quality assurance initiatives and verify the stringency of their efforts through independent assessments conducted by reputable auditing organizations.

It is important to recognize that QA is an investment. A company with a QA program can offset the cost through an improved image, reduced likelihood of product liability suits, consumer satisfaction with a uniform and wholesome product, and improved sales. In practical terms, it makes good sense to have a QA program.

The Role of Total Quality Management

An effective sanitation program is a segment of total quality management (TQM), which must be applied to all aspects of the operation within an organization. Total quality management applies the "right first time" approach. The most critical aspect of TQM is food safety. Thus, sanitation is an important segment of TQM. Additional discussion of TQM will be provided in Chap. 22.

The successful implementation of TQM requires that management and production workers be motivated to improve product acceptability. All involved must be skilled and understand the TQM concept. Computer software is available for training, implementation, and monitoring of TQM programs.

Quality Assurance for Effective Sanitation

Quality is the degree of acceptability. Component characteristics of quality are both measurable and controllable.

An effective sanitation QA program can achieve the following goals:

- Identify raw material suppliers that provide a consistent and wholesome product
- Make possible stricter sanitary procedures in processing to achieve a safer product, within given tolerances
- Segregate raw materials on the basis of microbial quality to allow the greatest value at the lowest price

By tradition, the food industry has applied QA principles to ensure effective sanitation practices, including inspection of the production area and equipment for cleanliness. If evidence of poor cleanup is reported, necessary action is taken to correct the problem. More sophisticated operations frequently incorporate the use of a daily sanitation survey with appropriate checks and forms. Visual inspection should include more than a superficial examination, because a film buildup that can harbor spoilage and food poisoning microorganisms can occur on equipment.

Major Components of Quality Assurance

The following tasks should be included as components:

1. Clear delineation of objectives and policies
2. Establishment of sanitation requirements for processes and products
3. Implementation of an inspection system that includes procedures
4. Development of microbial, physical, and chemical product specifications
5. Establishment of procedures and requirements for microbial, physical, and chemical testing
6. Development of a personnel structure, including an organizational chart for a QA program
7. Development, presentation, and approval of a QA budget for required expenditures
8. Development of a job description for all positions
9. Setup of an appropriate salary structure to attract and retain qualified QA personnel
10. Constant supervision of the QA program with written results in the form of periodic reports
11. Compliance with the requirements Food Safety Modernization Act
12. Food defense plan and implementation

The Major Functions of Quality Assurance and Quality Control

The major thrust of a QA organization is one of education and surveillance to ensure that regulations and specifications that are defined by the organization are implemented. Those involved with the QA program should be responsible for checking the wholesomeness and uniformity of raw materials assigned to manufacturing and for informing production personnel of these results. Further monitoring involves checks for good manufacturing practices and the finished products to ensure that they comply with specifications established under the QA program and are agreed upon previously by those involved with production or sales. If compliance is not attained, QA personnel should inform those who can implement corrections.

Quality assurance is generally a function of corporate management, which sets the policies, programs, systems, and procedures to be executed by those assigned to quality control. The major internal responsibility is working with the various functional departments of the company.

Quality control is closely related to manufacturing activities at the plant level. A QC program consists of measures and procedures pertaining to physical, chemical, or organoleptic attributes of food products to ensure the cost-effective production of uniform and consistent products. Those assigned to QC normally report to QA. Sometimes QC employees report to manufacturing, but they should never be totally independent of QA. Regardless of the organization structure, QA should have the ultimate responsibility for implementing and maintaining an effective sanitation program. The QA organization should be responsible for improving the sanitation program to keep current with trends, new regulations, and technical expertise. All QC procedures should be formulated and followed precisely. Quality control differs from TQM because it is only a segment of the latter and is not a comprehensive management approach.

The basic elements of QC programs serve as a way for food processors to achieve both quality assurance and safety requirements. Implementation of new in-process intervention technologies reduces the incidence of microbial, chemical, and physical contaminants and improves processing equipment design and placement within facilities and automated data monitoring systems. These technologies favorably position processors to ensure a high degree of confidence that products are produced, packaged, and distributed and reach consumers in a high quality and safe state (Bricher 2003).

Organization for Quality Assurance

Large-volume plants should place enough emphasis on process control to form a QA department. Those involved with QA have the obligation to respond to technical requests, interpret results in practical and meaningful terms, and assist with corrective actions. A QA department should be structured as a corporate function so that it is directly responsible for the establishment, organization, execution, and supervision of an effective QA program that is integrated into corporate strategy.

Major Responsibilities of a Sanitation Quality Assurance Program

Before a QA program is implemented, these requirements must be established:

1. Criteria for measuring acceptability (e.g., microbial levels) should be determined
2. Appropriate control checks should be selected
3. Sampling procedures (e.g., sampling times, numbers to be sampled, and measurements) should be determined
4. Analysis methods should be selected

The major responsibilities of sanitation QA are:

- Perform facility and equipment sanitation inspections at least daily
- Prepare sanitation specifications and standards
- Develop and implement sampling and testing procedures
- Implement a microbial testing and reporting program for raw products and manufactured products
- Evaluate and monitor personnel hygiene practices
- Evaluate compliance of the QA program with regulatory requirements, company guidelines and standards, and cleaning equipment
- Inspect production areas for hygienic practices

- Evaluate performance of cleaning compounds, equipment, and sanitizers
- Implement a waste product handling system
- Report and interpret data for the appropriate area so that corrective action, if necessary, can be taken
- Incorporate microbial analyses of ingredients and the finished product
- Educate and train plant personnel in hygienic practices, sanitation, and quality assurance
- Collaborate with regulatory officials on technical matters when necessary

A significant risk is involved if microbial testing is conducted inside a food plant, especially if pathogen testing is being conducted. It is necessary for a laboratory to enrich and culture large quantities of pathogenic microorganisms to perform analytical tests. Although products are usually negative for pathogens, a well-managed in-plant laboratory must use positive controls on a daily basis. Thus, a serious risk is involved because the plant may be cross-contaminated through laboratory activities. Food companies that perform on-site pathogen testing should have properly trained personnel, laboratory facilities that are separate from the manufacturing area to reduce the possibility of cross-contamination, and enough volume to support additional costs and resources required to conduct pathogen testing. Other laboratory requirements include an air-handling system designed to produce a negative air pressure in the laboratory and to remove biological agents (filtered air), a well-qualified microbiologist with two or more years of laboratory experience, adherence to the Center for Disease Control (CDC) safety requirements for a Biosafety Level 2 laboratory, a pathogen monitoring program to assess risk of cross-contamination of tests and the food plant, and the use of a known positive culture to verify recovery, consequently requiring strict adherence to these other requirements.

In-plant laboratories frequently promote personnel that are not well versed in basic laboratory techniques, aseptic sampling, equipment calibration, or safety training. Hazards can occur anytime in food laboratories. So, technicians must be properly trained on how to protect themselves and the facility on serious damage. Ongoing training for laboratory technicians on good laboratory practices and safety procedures should be an integral part of all QA initiatives. Companies should consider enrolling laboratory personnel in proficiency sample testing programs in choosing testing procedures that have Association of Official Analytical Chemists (AOAC) International official methods certification. This certification assures that analytical results can withstand regulatory and legal scrutiny.

Another concern of company owned operations is the safe disposal of biohazardous waste generated from pathogen tests. Pouring enrichment broth down sinks or the addition of bleach prior to discarding waste materials is an inappropriate disposal method as well as being illegal in some states.

The Role of ISO Accreditation

Because of risks involved, many food companies use contract laboratories to conduct their pathogen testing to reduce the risk of cross-contamination from positive controls on-site and existing issues surrounding biosecurity. In response to this trend, more contract laboratories are pursuing accreditation by their country's member body of the International Standards Organization (ISO) to provide their customers additional confidence in the validity and accuracy of test results.

ISO accreditation exists through a member body, one-member body per country, in over 150 countries. For example, the American National Standards Institute (ANSI) is the ISO member body in the United States that provides ISO accreditation. All laboratories worldwide are accredited to ISO work to the same internationally recognized standard, reinforcing the integrity and consistency of the testing or calibration that they undertake. ISO covers every aspect of laboratory management such as sample preparation, proficiency testing, record keeping, and reports while ensuring that analytical results can withstand regulatory and legal scrutiny in the event of a dispute in the United States or other countries.

Achieving ISO accreditation is a long, intensive, and expensive process involving quality system verification, internal audits, proficiency programs, equipment calibration, staff assessments, and corrective actions. Although some contract laboratories have chosen not to incur accreditation, some large food companies have obtained ISO accreditation for their internal laboratories.

The Role of Management in Quality Assurance

The success or failure of a sanitation program is attributed to the extent to which it is supported by management. Management can be the major impetus or deterrent to a QA program. Managers are often uninterested in QA because it is considered a long-term program. Because quality assurance programs reflect a cost and dividends cannot always be accurately measured in terms of increased sales and profits, they are not consistently supported by management. Frequently, lower and middle management are unable to convey the importance of QA when top management does not fully comprehend the concept.

Some of the more progressive management teams have been enthusiastic about QA. They have recognized that a QA program can be used in promotional efforts and can improve sales and product stability. Other managers have improved sales and product stability, and some have enhanced the image of their organization through sanitary practices and QA laboratories.

One of the limitations of viewing quality as conformance to specifications is its effect on management. When all specifications are met, the perception is that all is well and that management is not compelled to take immediate corrective action through the issue of orders down the hierarchy until results are obtained. This management style leads to a "fire-fighting" approach to problem solving and consumes valuable resources, is very costly, and frustrates people because problems at best, go away only temporarily. Safe Quality Food (SQF) certification is an important mechanism for preventing a fire-fighting approach and pushing for continuous improvement.

Quality Assurance and Job Enrichment

Because many employees, including managers and supervisors, fail to recognize the importance of QA, all employees must be made aware of the importance of their responsibilities. Through effective management, QA can be glamorized and made exciting. Although it is beyond the scope of this text to provide specific guidelines for the implementation of a job enrichment program for QA, it is suggested that this concept be considered. An effective job enrichment program can ensure that employee responsibilities are more interesting and rewarding. This program also includes employees more as a part of the operation and can actually be more demanding of personnel through assignment of more responsibilities. If more information regarding this concept is desired, the reader is referred to a management textbook or technical journals related to management.

Quality Assurance Program Structure

Before organizing a QA program, it is important to determine who is responsible for QA and how the chain of command will operate. In the most successful efforts, the QA program is part of top management, not under the jurisdiction of production. Under this arrangement, the QA people report directly to top management and are not responsible to production management. However, a close working relationship must be maintained between QA and the production departments. The QA organization is responsible for ensuring that deviations in sanitation practices are corrected, in addition to checking the final product and determining the stability or keeping quality. Figure 8.1 illustrates areas of responsibility of the administrator of the QA program.

Responsibility for the daily functions of a QA program related to sanitation should be delegated to a designated sanitarian, who should be provided with the time and means to keep abreast of

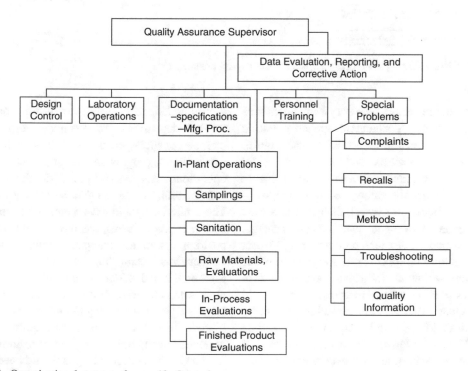

Fig. 8.1 Organizational structure for specific QA tasks

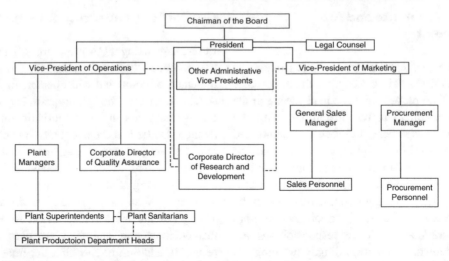

Fig. 8.2 Chart reflecting status of a plant sanitarian in a large organization

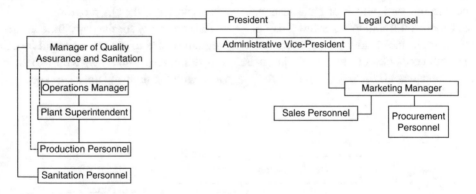

Fig. 8.3 Chart reflecting status of a plant sanitarian in a small organization

methods and materials necessary to maintain sanitary conditions. A sanitarian applies scientific knowledge to recognize, evaluate, and control environmental hazards and to preserve and improve environmental factors for the achievement of the health, safety, comfort, and well-being of humans. In general, their role in a food company is to evaluate, plan, design, manage, organize, and implement programs that protect public health. The specific role and position of the sanitarian within an individual processing firm should be made clear to all personnel. Management should clearly define parameters of responsibility by a written job description and an organizational chart. The sanitarian should report to the level of management with authority over general policy. This position should be equal to that of managers

of production, engineering, purchasing, and comparable departments to command respect and maintain adequate status to administer an effective sanitation program. Although smaller-volume operations may necessitate a combination of responsibilities, they should be clearly defined. The sanitarian should have a clear understanding of the appropriate responsibilities and how the position fits in the company structure so that assignments can be performed properly. Figures 8.2 and 8.3 show examples of how the plant sanitarian should fit in the QA program of large and small processing organizations.

A high-caliber QA program requires one or more technically trained employees to administer it. The QA director or manager should have experience in food processing and/or preparation.

Some of the QA staff can come from the ranks, provided that they show interest, leadership, and initiative. Workshops, short courses, and seminars are often available to help train new workers.

Establishment of a Quality Assurance Program

The establishment of an effective microbial food testing program requires a commitment to the use of good laboratory practices (GLPs) including an equipment calibration program, use of positive controls, adherence to CDC laboratory safety requirements, incorporation of rapid and automated techniques to identify microbial pathogens, choice of AOAC validated methods to strengthen the testing program, assignment of qualified microbiology and chemistry professionals in the laboratory through ensuring their proficiency (including continued training to reduce the potential for costly errors and raise the credibility of attained results), and the correct disposal of biohazardous wastes (McNamara and Williams 2003). It is important that food processors stay abreast of new advances in laboratory quality assurance and technology and invest in the development of testing programs that will ensure the safety and wholesomeness of the products that they manufacture. Preparation of personnel for a more unified system of control requires a change in attitude, which must be handled diplomatically. To reduce resistance, all personnel should be told why the changes are being made. Company philosophy should be developed as part of the program to help establish the new attitude and new responsibilities that personnel need to attain the desired goals.

Elements of a Total Quality Assurance System

For each production area, one person should be responsible for the controls or inspection. This can be either a plant employee or an outside contractor. The frequency of the control or inspection check must be noted, as must the records to be kept. An outline can be converted into a written format, as though it were a set of instructions for plant employees. It can serve as the operating manual for the persons responsible for conducting QA.

Sanitation Inspection Procedure

A procedure to check the overall sanitation of plant facilities and operations, including outside adjacent areas and storage areas on plant property, should be included in a total QA program.

In a total QA program, a designated plant official should make the sanitation inspection and record the results. If sanitation deficiencies are discovered, a plan for corrective action is necessary. Corrective action may include additional cleaning or closing off an area until a repair is completed. Frequent and systematic sanitation inspection procedures should be used when product contamination is possible, such as from container failure, moisture dripping, or grease escaping from machinery onto the product or surfaces that come into contact with the food.

New Employee Training

Instruction should include basic information that any new employee needs to know about food handling and cleanliness. Employees should be informed of the importance of hygienic practices. A list of all of the items that need to be covered in employee orientation should be developed, as well as how and when the orientation will be performed. Employee training should include an ongoing program to remind employees continuously of the importance of good sanitation.

Hazard Analysis and Critical Control Point (HACCP) Approach

HACCP should be incorporated as a QA function and as a systematic approach to hazard identification, risk assessment, and hazard control in a food processing and/or foodservice facility and distribution channel to ensure a sanitary operation.

Potential product abuse should be considered, and each stage of the process should be examined as an entity and in relation to other stages. Analysis should include the production environment as it contributes to microbial and foreign material contamination. Additional information on HACCP may be found in Chap. 7.

Program Evaluation

It is essential to evaluate the sanitary phase of a QA program through reliance either on the senses or on microbial techniques. Most inspectors rely on appearance as an evaluation technique for cleanliness. To the average inspector, a production area with walls, floors, ceilings, and equipment that looks clean, feels clean, and smells clean is satisfactory for production. But an effective QA program must use more than the human senses. It should incorporate a concrete method to evaluate hygienic conditions. To more objectively evaluate sanitation effectiveness, microbial testing methods should be incorporated to detect and enumerate microbial contamination. Also, knowledge of the quantity and genera of microorganisms is important in the control of product wholesomeness and spoilage.

Various techniques are available to evaluate the degree of cleanliness of equipment and foodstuffs and the effectiveness of a sanitation program. However, QA specialists do not always accurately determine or interpret results. Selection of the most appropriate technique should be based on the desired accuracy and precision, desired results, and the amount of effort and expenses available. Generally, the less complicated techniques are less accurate and precise. However, many measurements need not be exceptionally accurate and precise, as long as the degree of sanitation can be determined. Sanitation can be evaluated by the use of contact plates. However, various thermally processed products may require very sensitive techniques for determining the amount and genera of microorganisms present throughout the finished product and on the processing equipment.

Assay Procedures for Evaluation of Sanitation Effectiveness

The availability of more sensitive, accurate, and rapid test methods and systems, especially for microbial analyses, has introduced efficiencies into food testing programs. The concern for pathogens has necessitated the use of rapid microbial tests and the food laboratory as an important element for the implementation of an effective testing program. Laboratory methods are an important part of the entire scenario. Because these methods play an important role, they should be:

- Accurate
- Reproducible
- Clearly described
- Safe
- Easy to conduct
- Rapid (in turnaround time)
- Efficient
- Available commercially (all components)
- Officially recognized (i.e., AOAC, Food and Drug Administration (FDA), United States Department of Agriculture (USDA))

A brief discussion of the the most viable assay procedures follows, according to category. Additional information about microbial determination is discussed in Chap. 3.

Direct Contact Contamination Removal

With this method, plates that contain agar are pressed against a surface to determine the amount of contamination. Variation is reduced through swabbing several locations. Modifications of the contact plate method include the agar slice (syringe extrusion and agar sausage) and the use of selective and differential media. Another assay technique is the impression method. This technique involves a piece of sterile cellophane tape, which functions as a replicator to transfer cells to a growth-support agar, with subsequent incubation and counting. This approach serves as only an approximate representation of the contamination and does not distinguish between particulate contamination containing one cell or more.

Surface Rinse Method

This method uses elution of contamination by rinsing to permit a microbial assay of the resultant suspension. A sterile fluid is manually or mechanically agitated over an entire surface. The rinse fluid is then diluted and subsequently plated. When applicable, it is more precise than the swab method because a larger surface can be tested. The membrane filter is an aid to the surface rinse method if contamination is not excessive. The membrane, bearing microorganisms, can be incubated on a nutrient pad, stained in 4–6 h, and examined under a microscope with 8–100× magnification. Although the surface rinse method is more accurate and precise than the direct contact method and has a higher recovery rate (approximately 70%) and the flexibility of interfacing with the membrane fiber, it is restricted to horizontal surfaces and usually limited to container-type equipment.

Direct Surface Agar Plating (DSAP) Technique

This technique has utility for examining contamination on surfaces in situ. Eating utensils may be tested through pouring a melted medium into a cup and allowing the agar to solidify. The agar is transferred aseptically to sterile culture plates, with subsequent over-layering and incubation. The agar slab may be protected by a cover and read after 28–48 h.

ATP Detection

Presence of adenosine triphosphate (ATP) is an indicator of organic material that remains on surface after sanitation. ATP concentration correlates with sanitation effectiveness and provides a near-instant measurement of organic material left on a surface.

Interpretation of Data from QA Tests

Microbial tests to evaluate hygienic conditions of equipment and foodstuffs are discussed in Chap. 3. Additional information on the tests that can be performed is discussed in the following paragraphs.

Importance of a Monitoring Program

A monitoring program should be established and implemented to provide an internal method of evaluating the overall wholesomeness of the finished product and the degree of sanitation. The main purpose is to avoid problems related to product safety and acceptability. The development of a program should include determination of objectives, techniques, and evaluation procedures. In testing, the overall effectiveness of sanitation, not just the quantification of microorganisms on food contact surfaces, should be considered.

Products and surfaces to be tested should be determined by type of products produced, production steps, and the importance of the designated surface to sanitation practices and to the safety and/or overall acceptability of the food product. The monitoring program should be based on desired accuracy, time requirements, and costs. The type of food contact surface to be tested should also be considered in determination of the monitoring technique.

To reduce the possibility of incorrectly interpreting results, the monitoring program should be designed so that data can be statistically analyzed. Further misunderstanding can be avoided through a thorough comprehension of the benefits and limitations of the test procedures, for example, recognizing that bacterial clumps from the contact method of sampling should yield lower counts than the swab method, which breaks up cell clumps. The incorporation of 0.5% Tween 80 and 0.07% soy lecithin into media for RODAC plates is suggested if sampling is to be conducted on surfaces previously treated with a germicide.

In addition to analyzing data, the monitoring program should include a means of evaluating the information generated by the sampling technique. Acceptable and unacceptable guidelines should be determined under practical operating conditions. Repeated monitoring of given surfaces over time under given conditions (such as after cleaning and sanitizing and during manufacture) can provide a trend. The QA manager can use this information to establish realistic guidelines for the production operation. The guidelines specifying the amount of contamination should be predicted based on the stage of production, amount of food surface exposed, and the length of contact time between the food and surface. Graphs that display daily counts of microorganisms and the established guidelines can be posted for review by the supervisors and employees and can be used to stress the importance of monitoring and conforming to guidelines.

Microbial monitoring of food contact surfaces with techniques that have been discussed can be an effective tool to measure and evaluate the effectiveness of a QA program. Furthermore, a monitoring program can isolate potential problem areas in the production operation and serve as a training device for the sanitation crew, supervisors, and QA employees.

Auditing Considerations

In-house and third-party food safety audits have become a common practice and are required by most major food and foodservice retailers to ensure that they receive safe food products and to limit their liability if a foodborne illness outbreak occurs. These audits provide accurate assessments of a supplier's plant operations, written programs, and records as they relate to food safety.

In 2001, the National Food Processors Association launched a Supplier Audit for Food Excellence (SAFE). The objective of the SAFE program was to create an industry-wide standard audit. The audit checklist was developed by the SAFE council, which included representatives from 30 prominent companies. This program has been well received by the industry as evidenced by approximately 1000 of these audits being conducted within the first 2 years after implementation.

Since 2009, the Safe Quality Food Institute (SQFI) has offered a Safe Quality Food (SQF) certification program that integrates food safety and food quality into one management program. Many companies require their suppliers to have SQF certification to ensure that their food safety and food quality control systems have been implemented effectively and are monitored continuously, and verification is conducted. SQF certification links agricultural production, suppliers, food manufacturing, and distribution so that all sectors of the food industry are using the same safety and quality management program. Certification is conducted by third parties, which are often companies or consultants that are certified by SQFI to conduct company and plant audits and provide SQF certification. Many of these companies also provide good manufacturing practice (GMP) certification, other food safety audits, and assistance with HACCP plan development, implementation, and verification.

There are three levels of SQF certification:

- SQF Level 1—Focused on food safety fundamentals, this is the most basic level with the fewest requirements. It is most appropriate for low risk operations and does not include the HACCP approach. This level is not recognized by Global Food Safety Initiative (GFSI).
- SQF Level 2—This level includes food safety fundamentals and a HACCP approach to managing risks and hazards. It is comparable to the ISO 22000, Food Safety System Certification (FSSC) 22000, and British Retail Consortium (BRC) standards and registration schemes.
- SQF Level 3—The level 3 requirements include quality requirements in addition to the food safety requirements and are appropriate for the organization that wants to have an integrated system for food safety and food quality. This would be similar to having an ISO 22000 system integrated with an ISO 9001 system.

Bjerklie (2003) suggested some broad categories that should be considered during the preparation for a plant audit. They are:

1. Food safety and quality organization and responsibilities
2. Food safety, quality policies, and procedures
3. Specific training goals and programs for management and operating personnel
4. Identified HACCP team and effective HACCP plan
5. Comprehensive recall plan and procedures
6. Regulatory compliance standard
7. Document and records management
8. Change management and emergency management programs

9. Documentation to tracking effectiveness of policies
10. Management awareness and commitment to food safety and quality

An audit can be a positive learning experience for a food processor. Auditors can play an important role since they want to know how a processing plant controls the processes inside the in the plant. The entire auditing procedure is designed to provide that answer.

Preparation for an Audit

The most effective way the food plant can prepare for an audit is to determine the audit criteria, especially what the company is going to be compared against. Once that is known, the plant can plan appropriate preparations. When plant management knows that an audit is impending, they should conduct a self-evaluation of their facility against the audit criteria or audit themselves. Another important task that a plant should address prior to an audit is to prepare a small but convenient workspace for the auditor and be ready to provide assistance as needed. Management's interface with an auditor can reveal as much as the audit itself about how a plant is managed and how the company conducts its business (Bjerklie 2003).

Recall of Unsatisfactory Products

Product recall is bringing back merchandise from the distribution system because of one or more unsatisfactory characteristics. Every food business is susceptible to a potential product recall. A satisfactory public image of businesses can be preserved during a recall if a well-organized plan is implemented.

During a recall, products are recovered from distribution as a result of voluntary action by a business firm or involuntary action due to the Food and Drug Administration (FDA) action. The reasons for recall are best described in the FDA recall classifications:

Class I As a result of a situation where there is a reasonable probability that the use of or exposure to a defective product will cause a serious public health hazard including death

Class II As a result of a situation where the use of or exposure to a defective product may cause a temporary adverse health hazard or where a serious adverse public health hazard (death) is remote

Class III As a result of a situation where the use of or exposure to a defective product will not cause a public health hazard

An example of a Class I product recall would be contamination with a toxic substance (chemical or microbial). A Class II product recall involves products contaminated with food infection microorganisms. A Class III example is products that do not meet a standard of identity.

An effective way to prevent a recall is through an effective HACCP plan and instilling a food safety mindset among all employees. Some plants conduct mock recall exercises. A processor's public relations team should be a part of the entire recall. A recall plan for unsafe products caused from poor sanitation should:

1. Collect, analyze, and evaluate all information related to the product
2. Determine the imminence of the recall
3. Notify all company officials and regulatory officials
4. Provide operating orders to company staff needed to execute the recall
5. Issue an immediate embargo on all further shipments of involved product lots
6. If determined appropriate, issue news releases for consumers on the specifics of the product
7. Notify customers
8. Notify and assist distributors in tracking down the product
9. Return all products to specific locations and isolate them
10. Maintain a detailed log of recall events
11. Investigate the nature, extent, and causes of the problem to prevent recurrence

12. Provide progress reports to company and regulatory officials
13. Conduct an effectiveness check to determine the amount of questionable product recalled
14. Determine the ultimate disposition of the recalled product

Sampling for a Quality Assurance Program

An effective sampling plan is an essential component of testing for a food safety program. With an ineffective sampling plan, a test result that is negative provides a false sense of security. To obtain meaningful data, an understanding of the testing involved in the context of the sampling plan is essential. The types of swabs (individual or composite), the number of swabs, and the sites in the plant that are sampled will impact the test results.

A sample is part of anything that is submitted for inspection or analysis that is a representative of the whole population. For the sample to be appropriate, it must be statistically *valid*. Validity is achieved by selecting the sample so as to ensure that each unit of material in a lot being sampled has an equal chance of being chosen for examination. This process is called *randomization*.

A sample must be representative of the population to ensure integrity of results. A suggested sample number is the square root of the total number that would be sampled. Representative samples are not only random samples, but must constitute a proportionate amount of each part of the population. A major concern of the QA organization should be the collection, identification, and storage of samples for inspection and/or analysis. A statistically valid sample is important because:

• A sample is the basis for establishing the condition of the entire item or lot. A larger sample size increases the integrity that can be placed on findings.
• Submitting the entire item or lot for inspection is expensive and usually impractical.
• Sampling is used for the establishment of data for the development of standards and product acceptance.

• The integrity of collected samples is diminished by inaccurate and incomplete information. Forms should contain all of the information necessary for sampling and subsequent type of analysis. Sample cases should be insulated to ensure temperature maintenance during the period of transit to the point of inspection or analyses.

Samples must be held at 0 °C–4.5 °C. Sealed refrigerants, which come in several temperature ranges, are available. If maintenance in the zero to subzero temperature range is essential, dry ice should be used.

During the past decade, a limited amount of environmental testing and monitoring was conducted in food plants. However, food companies now recognize that the control of the in-plant environment is critical to the production of safe food. Tests are being performed on-site, outsourced, or a combination of both. Contamination from the processing environment is one of the most common sources of microbial contamination of the finished product. The implementation and maintenance of a rigid environmental monitoring program can be beneficial in identifying areas that can serve as growth niches on plant equipment are in the plant environment. Environmental testing is a preventive step that may lead to the recognition of a contamination problem before it becomes a source for finished product contamination. An environmental testing program can verify that the sanitation controls are effective in minimizing hazards such as foodborne pathogens, especially *Listeria monocytogenes* in wet or refrigerated environments and *Salmonella* in dry processing operations.

Sampling Procedures

An example of defined sampling procedures for solid, semisolid, viscous, and liquid samples follows:

1. Identify and collect only representative samples
2. Record product temperature, where applicable, at the time of sampling

3. Maintain collected samples at the correct temperature. Nonperishable items and those normally at ambient temperature may be maintained without refrigeration. Perishable and normally refrigerated items should be held at 0 °C–4.5 °C; normally frozen and special samples should be maintained at -18 °C or below

4. After collection, protect the sample from contamination or damage. Do not label certain plastic sample containers with a marking pen; ink can penetrate the contents

5. Seal samples to ensure their integrity

6. Submit samples to the laboratory in the original unopened container whenever possible

7. When sampling homogeneous bulk products or products in containers too large to be transported to the laboratory, mix, if possible, and transfer at least 100 g of the sample to a sterile sample container, under aseptic conditions. Frozen products may be sampled with the aid of an electric drill and 2.5-cm auger

Basic QA Tools

Depending on the food product area, items from the following equipment and supplies should be considered for sampling and product evaluation.

Measurement Apparatus
These include a centigrade thermometer, headspace gauge, vacuum gauge, titration burettes, filtering apparatus, and 0.1–10.0-mL sterile disposable pipettes.

Lab Supplies
Suggested sanitation-related supplies include petri dishes or petrifilm, glass microscope slides, can opener, record forms, marking tape, pencils, pens, aluminum foil, sterile cotton swabs, paper towels, microbial media, Bunsen burner, forceps, spoons, knives, and inoculation tubes.

Clerical Supplies
Supply list depends on what tests are being conducted. Necessary basic sanitation QA tools are:

1. Ingredient specifications
2. Approved supplier list
3. Product specifications
4. Manufacturing procedures
5. Monitoring program (analyses, records, reports)
6. Good manufacturing practice (GMP) requirements
7. Cleaning and sanitizing program
8. Recall program

Role of Statistical Quality Control

Statistical quality control is the application of statistics in controlling a process. Measurements of acceptability attributes are taken at periodic intervals during production and are used to determine whether or not the particular process in question is under control, that is, within certain predetermined limits. A statistical QA program enables management to control a product. This program also furnishes an audit of products as they are manufactured.

The samples taken for analysis are destroyed; thus, only SQC is practical for monitoring food safety. The greatest advantage of an SQC program is that it enables management to monitor an operation continuously and to enhance operating a closely controlled production process.

Sample selection and sampling techniques are the critical factors in any QC system. Because only small amounts (usually less than 10 g or 0.35 oz) of a product are used in the final analysis, it is imperative that this sample be representative of the lot from which it was selected.

Statistical quality control, also referred to as *operations research*, *operations analysis*, or *reliability*, is the use of scientific principles of probability and statistics as a foundation for decisions concerning the overall acceptability of a product (Marriott et al. 1991). Its use provides a formal set of procedures in order to conclude what is important and how to perform appropriate evaluations. Various statistical methods can determine which outcomes are most probable and how much confidence can be placed in decisions.

Table 8.1 Central tendency values

Data	Mean	Mode	Median
12,13,13,14,15,16,17,18,19	15.2	13	16

Central Tendency Measurements

Three measurements are commonly used to describe data collected from a process or lot. These are the arithmetic mean or average, mode or modal average, and median. The mean is the sum of the individual observations divided by the total number of observations. The mode is the value of observations that occurs most frequently in a data set. The median is the middle value present in collected data. By using these values, the manufacturer can represent characteristics of central tendencies of the measurements taken. Table 8.1 illustrates calculated values for the mean, mode, and median from a collection of sample data.

The equation for the mean is as follows:

$$x = \frac{x_1 + x_2 + x_3 + x_4 + x_5 \cdots + x_n}{n} = \frac{\Sigma x_i}{n}$$

Variability

There must be a uniformity and minimal variation in microbial load or other characteristics between the products manufactured. Two measures of variation are the range and standard deviation. Measuring variability by means of the range is accomplished by subtracting the lowest observation from the highest.

$$R = X_{max} - X_{min}$$

From Table 8.1 the calculation would be

$$R = 20 - 11 = 9$$

Because the range is based on just two observations, it does not provide a very accurate picture of variation. As the number of samples increases, the range tends to increase because there is an increased chance of selecting an extremely high or low sample observation. The standard deviation is a more accurate measurement of how data

are dispersed because it considers all of the values in the data set. The formula for calculating the standard deviation is

$$S = \sqrt{\frac{\left(x - \bar{x}\right)^2 + \left(x_2 - \bar{x}\right)^2 + \cdots + \left(x_n - \bar{x}\right)^2}{n-1}}$$

Although this formula is more complicated than the range calculation, it can be determined easily by using a spreadsheet program on a personal computer. As the standard deviation increases, it reflects increased variability of the data. To maintain uniformity, the standard deviation should be kept to a minimum. Using the formula above, the standard deviation for the data set from Table 8.1 is 2.44. The relative standard deviation (RSD) is also often calculated to verify and demonstrate that variability is not too great.

The RSD is calculated with the following equation:

$$\text{Relative Standard Deviation} = \text{RSD} = \frac{S}{\bar{x}} \times 100$$

For our example,

$$\text{RSD} = \left(2.44 / 15.2\right) \times 100 = 16.0\%$$

An RSD that is less than or equal to 5% is generally considered acceptable for analytical measurements.

Displaying Data

It is beneficial to represent data in a frequency table, especially when a large sample of numbers must be analyzed. A frequency table displays numerical classes that cover the data range of sampling and list the frequency of occurrence of values within each class. Class limitations are selected to make the table easy to read and graph. The frequency table of microbial load from raw materials (Table 8.2) displays how data are divided into each class. To help visualize how these data are arranged, one can graph it in the form of a histogram. Figure 8.4 takes the information from Table 8.2 and displays it graphically.

Table 8.2 Frequency table for microbial load (CFUs/g)

Class in CFUs	Frequency
0–100	5
100–1000	10
1000–10,000	22
10,000–100,000	13
100,000–1,000,000	3

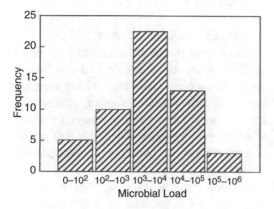

Fig. 8.4 Histogram of microbial load (CFUs/g)

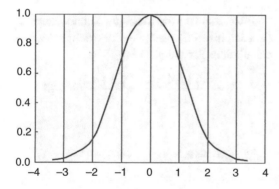

Fig. 8.5 Normal curve

The histogram in Fig. 8.4 depicts an important curve common to statistical analysis—the normal curve or normal probability density function. Many events that occur in nature approximate the normal curve. The normal curve has the easily recognizable bell shape and is symmetrical about the center (see Fig. 8.5). The area underneath the curve represents all the events described by the frequency distribution.

From Fig. 8.5, the mean is the highest point on the curve. The variation of the curve is represented by the standard deviation. It can be used to determine various portions underneath the curve.

This is illustrated in the figure where one standard deviation to the right mean represents roughly 34% of the sample values. Consequently, 68.27% of the values fall within ±1 standard deviation from the mean. Similarly 95.45% fall within ±2 standard deviations. Virtually all of the area (99.75) is represented by ±3 standard deviations. The information thus far can be used to establish control limits in order to determine whether a process is in a state of statistical control.

For Fig. 8.4, there is a large amount of variability in the bacterial counts. In general, a product will be in a state of statistical control if it is within 3 standard deviations of the mean. However, product specifications may state that the raw materials should not have more than 4-log colony-forming units (CFUs). Therefore, the normal curve distribution of bacterial counts would need to be reduced in the samples such that samples with 10^3–10^4 CFU would be the right tale of the normal curve, samples with 10^2–10^3 CFU would be the peak of the curve, and samples with 0–10^2 CFU would be the left tale of the curve since anything over four logs would not meet specifications. This generic example and the selection of four-log CFUs were arbitrarily chosen.

Control Charts

Control charts offer an excellent method of attaining and maintaining a satisfactory level of acceptability. The control chart is a widely used industry technique for online examination of materials produced. In addition to providing a desired safety level, it can be useful in improving sanitation and in providing a sign of impending trouble. The primary objective is to determine the best methodology, given the available resources, and then to monitor control points. This variation can be classified as either chance-cause variation or assignable-cause variation.

In chance-cause variation, the end products are different because of random occurrences. They are relatively small and are unpredictable in occurrence. There is a certain degree of chance-cause variation present. Assignable-cause variation is just what the name implies. Cause can be "assigned" to a contributing factor, such as a difference in microbial load of raw materials, process and machine aberration, environmental factors, or operational characteristics of individuals involved along the production line. This variation, once determined, is controlled through appropriate corrective action. When a process shows only variation due to chance causes, it is "under control." Quality control charts were developed in order to differentiate between the two types of variation and to provide a method to determine whether a system is under control. Figure 8.6 illustrates a typical control chart for a quality characteristic. The y-axis represents the characteristic of interest plotted against the x-axis, which can be a sample number or time interval. The center line represents the average or mean value of the quality trait established by the manufactured product when the process is under control. The two horizontal lines above and below the center line are labeled so that as long as the process is in a state of control, all sample points should fall between them. The variation of the points within the control limits can be attributed to chance cause, and no action is required. An exception to this rule would apply if a substantial number of data points fall above or below the center line instead of being randomly scattered. This would indicate a condition that is possibly out of control and would warrant further investigation.

If a point falls above or below the out-of-bounds lines, one can assume that a factor has been introduced that has placed the process in an out-of-control state, and appropriate action is required.

Control charts can be divided into two types:

1. Control charts for measurement
2. Control charts for attributes

Measurement Control Charts

Measurement of variable control charts can be applied to any characteristic that can be measured. The X chart is the most widely used chart for monitoring central tendencies, whereas the R chart is used for controlling process variation. The following examples show how both of these control charts are used in a manufacturing environment.

A food manufacturer may monitor the pH value of finished products to satisfy safety concerns. Five samples may be pulled every hour during an 8h shift and analyzed for pH as noted in Table 8.3.

First calculate the average (X) and range (R) for each inspection sample. For example, sample calculations for sample 1 are

$$X = \frac{4.6 + 4.4 + 4.1 + 4.8 + 4.5}{5} = 4.48$$

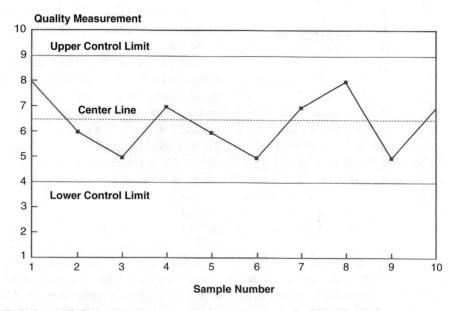

Fig. 8.6 Typical control chart

Table 8.3 X and R values for pH measurements

Sample	pH measurement					X	R
1	4.6	4.4	4.1	4.8	4.5	4.48	0.7
2	4.1	4.2	4.3	4.6	4.6	4.36	0.5
3	4.6	4.6	4.3	4.2	4.5	4.44	0.4
4	4.7	4.8	4.5	4.5	4.3	4.56	0.5
5	4.1	4.1	4.0	4.6	4.9	4.32	0.8
6	4.2	4.2	4.6	4.6	4.9	4.50	0.7
7	4.6	4.5	4.6	4.7	4.7	4.62	0.2
8	4.0	3.9	4.8	4.4	4.4	4.30	0.9
					Average	4.4475	0.5875

R is the highest value minus the smallest value of the five samples.

After all of the sample Xs and Rs are calculated, take the average of the Xs and Rs to obtain X and R.

$$\bar{X} = \frac{\text{Sum of all } \bar{X}s}{\text{number of sample lots}} = \frac{35.58}{8} = 4.4475$$

$$\bar{R} = \frac{\text{Sum of all Rs}}{\text{number of sample lots}} = \frac{4.7}{8} = 0.5875$$

From the calculation, the center line for the X and R chart can be defined to be

$$X \text{ Chart center line} = 4.4475$$

$$R \text{ Chart center line} = 0.5875$$

In order to calculate the upper control limits (UCL) and lower control limits (LCL), the standard deviation for each sample lot must be determined. Rather than perform the lengthy calculation needed for this value, another method can be used to determine these values. The control limits for the previous charts were represented by

$$\text{UCL} = \bar{X} + 3\partial$$
$$\text{LCL} = \bar{X} - 3\partial$$

By substituting a factor (A_2) from a statistical table into the above equation for UCL and LCL, the needed values for the control point can be obtained. In this example, the value for (A_2) for a sample size of 5 is 0.58. The new equation becomes

$$\text{UCL} = \bar{X} + A_2\bar{R}$$
$$\text{LCL} = \bar{X} - A_2\bar{R}$$

substituting

$$\text{UCL} = 4.475 + 0.58(0.5875) = 4.7883$$
$$\text{LCL} = 4.4475 - 0.58(0.5875) = 4.1067$$

The control limits for the R chart are determined similarly, using factors D_4 and D_3 from the statistical reference table.

$$D_4 = 2.11, \ D_3 = 0$$
$$\text{UCL} = D_4\bar{R} = 2.11(0.5875) = 1.2396$$
$$\text{LCL} = D_3\bar{R} = 0(0.5875) = 0$$

Once these calculations are complete, the values can be plotted on an X-Y chart to obtain the X and R charts for pH measurements. Figures 8.7 and 8.8 illustrate complete control charts from the sample data. Both graphs show a process currently under control, with all data points lying within the boundaries of the control limits and an equal number of points above and below the center line.

The example above displays the correct mathematical methodology for looking at this pH data. However, there may be specifications that are important to the safety and quality of products. For example, the company purchasing a product may have pH specifications between 4.2 and 4.6. The pH of 4.2 may be due to a negative effect on quality. The pH of 4.6 may be due to a requirement for food safety for it to be a shelf-stable acidified food. Based on this knowledge, the upper control limit may need to be set at 4.55, and the lower control limit may need to be set at 4.25. If this was the case, it would be important to control and lower the variability in the process such that UCL = X + 3∂ = 4.55 or less and the LCL = X − 3∂ is 4.2 or greater so that there is over 99% confidence that the samples are within specification. This is a hypothetical situation that was chosen arbitrarily but is important because variability in processes needs to be minimized so that upper and lower control limits meet product specifications. In addition, companies can also use this variability data from their process or samples to demonstrate to customers or management that in some cases the specifications that are in place are not feasible and need to be reevaluated.

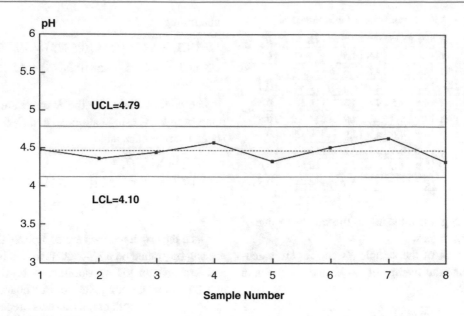

Fig. 8.7 X chart for pH measurements

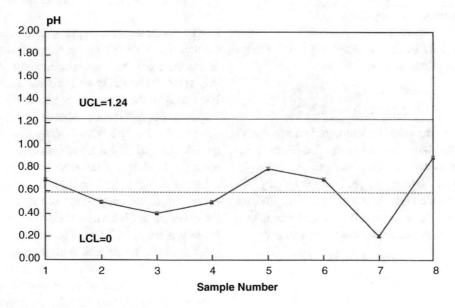

Fig. 8.8 R chart for pH measurements

Attribute Control Charts

Attribute control charts differ from measurements charts in that one is interested in an acceptable or unacceptable classification of products. The following charts are commonly used for attribute testing:

1. p charts
2. np charts
3. c charts
4. u charts

p Charts

The p chart, one of the more useful attribute control charts, is used for determining the unacceptable (p) fraction. It is defined as the number of unacceptable items divided by the total number of items inspected. For example, if a producer examines five samples per hour (for an 8-h shift) from the production line and finds a total of eight unacceptable units, p would be calculated as follows:

$$\text{Total number of unacceptable} = 9$$
$$\text{Total number of inspected} = 5(8) = 40$$
$$p = \frac{\text{number of unacceptable}}{\text{total number inspected}} = \frac{8}{40} = 0.20$$

Sometimes this value is represented as percentage unacceptable. In this example, percentage defective would be

$$0.20 \times 100 = 20\%$$

An attribute control chart can be constructed from a sampling schedule by obtaining an average fraction unacceptable (p) value from a data set and using the formula $p \pm 3\delta$ or the desired control limits. Because attribute testing follows a binomial distribution, the standard deviation would be calculated:

$$\delta = \sqrt{\frac{\bar{p}(1-\bar{p})}{n}},$$

where n is the number of items in a sample. Control limits would be obtained by

$$\text{UCL} = \bar{p} + 3\delta$$
$$\text{LCL} = \bar{p} - 3\delta$$

When these data are plotted and no points are outside of the control limits, it can be assumed that the process is in a state of statistical control, and any variation can be attributed to natural occurrences.

The UCL = 0.2 + 3 × the square root of ((0.2 × 0.8)/40) = 0.39, and the UCL equals 0.01 so the 99.75% confidence interval for this small sample would be (0.01, 0.39).

np Charts

np charts can be used to determine the number of unacceptable instead of the fraction defective, and the sampling lots are constant. The formula for the number of unacceptable (np) is

$$\text{number of unacceptable} (np) = n \times p,$$

where n is the sample size and p is the unacceptable fraction defective. If one value is known, the other can be easily calculated. For example, if a sample lot of 200 is known to be 1.5% unacceptable, the number of unacceptable should be

$$np = 200 \times 0.015 = 3$$

The calculation for determining the control limits would be the same as for the p chart, except that the standard deviation would be

$$\delta = \sqrt{np(1-\rho)}$$

The standard deviation for this example would be the square root of ((200 × 0.015) × (1−0.015)) = 2.96.

c Charts

These charts are used when the concern is the number of defects per unit of product. They are not as frequently incorporated as the p and np charts but can be effective if applied correctly. Assume that a manufacturer examines 10 lots and discovers 320 defects. The equations for the average (c) and standard deviation required for a c chart are

$$\bar{c} = \frac{320}{10} = 32$$
$$\delta = \sqrt{c} = \sqrt{32} = 5.66$$

The control limits would be

$$\text{UCL} = \bar{c} + 3\sqrt{c} = 32 + 3(5.66) = 48.97$$
$$\text{LCL} = \bar{c} - 3\sqrt{c} = 32 - 3(5.66) = 15.03$$

u Charts

Sometimes, a constant lot size may not be attainable when examining for defects per unit area. The *u* chart is used to test for statistical control. By establishing a common unit in terms of a basic lot size, one can determine equivalent inspection sample lot sizes from unequal inspection samples. The number of equivalent common basic lot sizes (*k*) can be calculated as

$$k = \frac{\text{size of sample lot}}{\text{size of common lot}}$$

The *u* statistic can be determined from *c*, the number of defects of a sample lot, and the *k* value defined in the above equation.

$$u = \frac{c}{k}$$

From these values, the upper and lower control limits for the *u* chart can be defined as

$$\text{UCL} = \bar{u} + 3\sqrt{\frac{\bar{u}}{k}}$$

$$\text{LCL} = \bar{u} - 3\sqrt{\frac{\bar{u}}{k}}$$

In addition to charting, a manufacturer may introduce other statistical analyses, such as modeling, variable correlations, regression, analysis of variance, and forecasting to the production area. These techniques provide additional statistical methods for examining processes in order to ensure maximum production efficiency.

For example, a company may have a rush during a certain season of the year when the demand for a product is greater than other times of the year. Therefore, instead of receiving the common lot size of 5000 for a raw material, the current lot size is 10,000. This would make $k = 10{,}000/5000 = 2$. For this example, assume that there are ten sample defects in this lot, so $u = 10/2 = 5$. We know the average $u = 6$ based on previous data. Therefore, the UCL = 6 + 3x the square root of (6/2) = 11.2 and the LCL = 0.8 for this example.

Sample Size and Sampling

When choosing sample size for a specific lot in attribute sampling, the military sampling plan can be used. For example, a company receives lots of 1000 cooking trays for a new venture. A company is providing these cooking trays for this new application. The company decides to test 32 for statistical validity, based on the lightest inspection from the military sampling plan. This holds true whether your accepted level of defects (AQL) is 0.5, 1.0, or 2.0%. The recommended sample size would be $n = 80$ if this analysis was conducted for a product that could still be shipped out. However, it is unrealistic for a product such as a cooking tray that cannot be sold if it is retained for testing. The 95% confidence statement would be as follows for an acceptable level of 0.5% defective trays (0.5% AQL). If the lot is accepted (0 defective out of 32), there is 95% confidence that there are less than 9.3% defective. If the lot is rejected (1 or more defectives out of 32), there is 95% confidence that there are more than 0.25% defective. For the 1.0% AQL level, there is 95% confidence that there are less than 12.0% defective in the lot if one accepts the lot (either 0 or 1 defective out of 32) or greater than 0.75% defective if the lot is rejected (2 or more defectives out of 32). For the 2.0% AQL level, there is 95% confidence that there are less than 16.2% defective in the lot. If the lot is rejected (2 or more defectives out of 32), there is 95% confidence that there are more than 1.91% defective.

Sample Size

When determining the sample size for statistical testing, two problems must be addressed: how to compare treatments and the statistical power of the test to be carried out. It is general practice to have enough samples to have a sufficiently low alpha (0.05), which is the Type I error or risk, and beta (<0.10), the Type II error or risk. Beta is not often calculated when conducting analyses or reporting data, but it should always be considered when statistical analysis is utilized to determine differences between treatments such as comparing the effectiveness of two sanitizers for an application. Though the explanation of sample size, alpha (Type I error), beta (Type II error), and power determination are not addressed in the current text, information can be found on this topic in Schilling (2014).

Summarizing, the following should be considered before carrying out sampling:

- Clarify the objective and the characteristics to be measured.
- Identify how test samples should be prepared.
- Estimate the level of Type I and Type II risks together with the number of replicates.

Explanation and Definition of Statistical Quality Control Program Standards

The following terms apply to maintenance of standards:

- *Standard*: The level or amount of a specific attribute desired in the product.
- *Quality attribute*: A specific factor or characteristic of the food product that determines a proportionate part of the acceptability of the product. Attributes are measured by a predetermined method, and the results are compared against an established standard and lower and upper control limits to determine if the product attribute is at the desired level in the food product.
- *Retained product:* A product that is not to be used in production or sold until corrective action has been taken to meet the established standards. Retained products should not be released for production or sales use until the problem is corrected.

Rating Scales

Two rating scales have been devised for evaluation of attributes:

1. *Exact measurement:* For attributes that can be measured in precise units (bacterial load, percentage, parts per million, etc.).
2. *Subjective evaluation:* Used when no exact method of measurement has been developed. The evaluation must be conducted through sensory judgment (taste, feel, sight, smell). This is usually described numerically. Two scales have been developed for evaluating acceptability:

Scale 1	Scale 2
7—Excellent	4—Extreme
6—Very good	3—Moderate
5—Good	2—Slight
4—Average	1—None
3—Fair	
2—Poor	
1—Very poor	

The number of samples to conduct at any point during production to evaluate the sanitation operation also depends on the variations of analysis of the samples. A minimum of 3–5 samples of approximately 2 kg each should be selected and pooled from each lot of incoming raw material. After a sufficient number of samples have been analyzed, control charts can be constructed for each raw material.

Sampling of the finished product should be conducted at a special step in the production sequence, such as at the time of packaging. Sampling at this stage does not need to be conducted on individual products for inspection or regulatory purposes because it is directed at monitoring process control, not individual product analysis. However, to be familiar with the wholesomeness and overall acceptability of each product, the preferred procedure is to analyze and maintain control charts on all products.

Sample size usually consists of 3-5 specimens that serve as a representative of the population sampled. Another guideline for sample size is the square root of the total units, and, for large lots, an acceptable size may be the square root of the total units divided by 2. Daily sampling is necessary to monitor process control effectively.

Action limits for finished products should be as outlined under the analysis program and should be used in determining whether the process conforms to the designated specifications. If three consecutive samples exceed the maximum limit for contamination, production should cease, with further cleaning.

Cumulative Sum (CUSUM) Control Charts

Data can be plotted where greater sensitivity in detecting small process changes is required by use of the CUSUM chart. This chart is a graphic plot of the running summation of deviation from a control value. These differences are totaled with each subsequent sampling time to provide the CUSUM values. This monitoring technique can be incorporated in sanitation operations that require a higher degree of precision than obtained from a regular statistical QC chart. The CUSUM chart gives a more accurate account of real changes, faster detection and correction of deviation, and a graphical estimation of trends. It enhances an optimum process control for various applications. The CUSUM chart was not developed for multiple levels and is not practical for use on production processes that drift over an extended period of time. If used, it is important that the results of the CUSUM system be kept current so that immediate corrective action may be taken.

A personal computer can rapidly perform the statistical computations and identify the points that require corrective action, thus reducing the burden of processing large quantities of data. These data can be available to promptly expedite corrective actions, project future performances, and determine when and where preventive QC procedures are necessary.

Standard Curves

Instrumental analyses such as spectroscopy, gas chromatography (GC), and high-performance liquid chromatography (HPLC) require the use of a standard curve to determine unknown concentrations. A standard curve is sometimes needed in quality control to determine the concentration of a compound that is in a food product. Simple linear regression with concentration as the x-variable and either absorbance or peak area as the y-variable is used to determine $y = mx + b$. The slope or **sensitivity** is m, x is the concentration, and b is the y-intercept.

Example (that may be applied to sanitation) A laboratory measures vitamin D concentration in milk to make sure that it does not decrease in concentration over storage time using an HPLC method.

The initial range of the vitamin D standards is between 0 and 0.8 IU/ml BSA. Simple linear regression was run on the samples with concentration as the x (explanatory variable) and absorbance as the y (response variable).

Standard concentration (IU/ml)	Absorbance (AU)
0	0
0.1	0.14
0.2	0.26
0.3	0.41
0.4	0.53
0.5	0.66
0.6	0.78
0.7	0.89
0.8	0.99

Figures of merit were determined and include R^2 **(coefficient of determination)**, **limit of detection (LOD)**, **linear range**, and **sensitivity**. It is a general rule that a standard curve should have a minimum R^2 of 0.99. If $R^2 = 0.99$, this indicates that 99% of the variability in the absorbance (y) is explained by the concentration (x) and the remaining 1.0% is random variability. The R^2 for the current data is 0.997. The **limit of detection (LOD)** can be calculated for analytical instruments and an assay. For instruments, the LOD is 3 × the noise of the sample when a blank is run in the analytical instrument. The LOD for the assay is dependent on the random variability of the regression line and the slope and is calculated as either 3 × standard error (s.e.)/slope (m) or 3.3 × s.e./m. We used 3 × s.e./slope (m) to calculate LOD for our examples since it is what is done in our lab, but some literature reports using 3.3 × s.e./m to calculate LOD. Any unknown sample that has a concentration below the LOD based on the standard curve should be reported as not detectable. The LOD for our example data is 0.042 IU/ml.

The limit of quantitation (LOQ) has been previously reported as either equal to the LOD or as 9 × s.e./m. The third figure of merit is **sensitivity** or the slope and is defined as the change in response (absorbance) due to a 1 unit increase in concentration. A high sensitivity assay should be used to determine small differences between samples, and a low sensitivity assay such as the biuret analysis is sufficient for samples with higher concentrations or treatment samples that have large differences in concentration. The fourth and final figure of merit is the **linear range**. The linear range starts at either the LOD or LOQ and ends at the highest concentration that is used in the final regression analysis. Therefore, for our example, the linear range would be 0.042 IU–0.8 IU/ml.

Determining Unknowns

Example Triplicate samples are run for milk samples. The samples yielded absorbance values of 0.512, 0.524, and 0.532 AU. To understand this data, the following protocol was followed: (1) verification that the concentration was within the linear range; (2) means, standard deviations, relative standard deviations, and confidence intervals were calculated; and (3) data were reported as significant digits. To determine the unknown concentrations, the $y = mx + b$ equation was used to solve for x, in which $x = (y - b)/m$. Our regression equation is y = 1.25*AU/(IU/ml)*x + 0.018, x = (0.512 AU–0.018 AU)/1.25 AU/(IU/ml); (0.524 AU–0.018 AU)/1.25 AU/(IU/ml); and (0.532 AU–0.018 AU)/1.25 AU/(IU/ml), which is 0.395 IU/ml, 0.405 IU/ml, and 0.411 AU/ml. The mean, standard deviation, and relative standard deviation were determined for the concentrations and are 0.40 IU/ml, 0.0081, and 2.1%, respec-

tively. The final concentration was reported as two significant digits because there was confidence that the 4 was correct, slight confidence that the 0 was correct, but no confidence that the number in the thousandths place was correct. Detailed information on standard curves and confidence intervals for unknown samples can be found in Schilling (2014).

Study Questions

1. What is quality?
2. What is total quality management?
3. Why should QA personnel not be placed under the supervision of production management?
4. What is SQC?
5. What is CUSUM?
6. What are Class I, II, and III recalls?
7. What are quality control charts?
8. What is the difference between quality assurance and quality control?
9. How does one know that a process or product is in a state of statistical control?
10. What factors should be considered when determining sample size?

References

Bjerklie S (2003). How to survive an audit. *Meat Process* 42(6): 58.
Bricher J (2003). Top 10 ingredients of a total food protection program. *Food Saf* 9: 30.
Marriott NG, Boling JW, Bishop JR, Hackney CR (1991). *Quality assurance manual for the food industry.* Virginia Cooperative Extension, Virginia Polytechnic Institute & State University, Blacksburg. Publication no. 458-013.
McNamara AM, Williams J Jr (2003). Building an effective food testing program for the 21st century. *Food Saf* 9: 36.
Schilling MW (2014). Chemical analysis: Sampling and statistical requirements. In *Encyclopedia of meat sciences*, eds. Devine C, Dikeman M. 2nd ed., Vol. 1. Elsevier: Oxford. p. 187.

Cleaning Compounds

9

Abstract

Knowledge of soil deposits and use of an appropriate and versatile cleaning compound for the specific cleaning application are essential for an effective sanitation program. Soil characteristics determine the most appropriate cleaning compound. Generally, an acidic cleaning compound is most effective for the removal of inorganic deposits, an alkaline cleaner for removing non-petroleum organic soils, and a solvent-type cleaner for the removal of petroleum soil.

The major function of cleaning compounds is to lower the surface tension of water so that soils may be loosened and flushed away. Detergent auxiliaries are included in cleaning compounds to protect sensitive surfaces or to improve the cleaning properties. Knowledge of how to handle cleaning compounds is essential to reduce the potential for injury of employees. If a worker is accidentally splashed with a cleaning compound, the affected area must be flushed with a large amount of water immediately.

Keywords

Cleaning compounds • Emulsification • Saponification • Sequestrant • Soil • Surfactant

Introduction

An essential tool for effective food sanitation is cleaning compounds. They are formulated specifically for performing certain jobs such as washing equipment, floors, and walls, use in a high-pressure washer, cleaning-in-place (CIP), and other purposes. Viable cleaners must be economical, nontoxic, noncorrosive, non-caking, non-dusting, easy to measure or meter, stable during storage, and easily and completely dissolved.

Requirements of cleaning compounds vary according to the area and equipment to be cleaned. Proper selection of compounds for blending to form a satisfactory cleaner requires technical knowledge. Major considerations in cleaning compound selection are the nature of the soil to be cleaned, water characteristics, application method, and area and kind of equipment to be cleaned.

Soil Characteristics

Soil should be recognized as "matter out of place." Matter can be debris that is associated with the specific operation. Soil may be divided into solids such as lipids, proteins, starch, wax, coagulated colloids, dust, inorganic salts, and metal oxides, whereas liquid soils include animal, vegetable, and mineral oil.

Chemical Contamination

Potential contamination sources in food production and preparation areas include cleaning compounds in addition to sanitizers, insecticides, rodenticides, and air fresheners. Cleaning compounds and other substances may contaminate equipment, utensils, or surfaces, serving as a vehicle for transfer of the contaminants to food. This statement can be verified through those who have drunk from a glass or cup that imparts a distinct taste of dishwashing soap. Other potential chemical contaminants could be particulate rather than soluble chemicals.

The most effective protection against chemical contamination is by establishing rigid housekeeping methods to be used by production and sanitation employees. Such contamination can be reduced or even eliminated if carelessness and sloppy personal habits of all employees are abolished.

Physical Characteristics

Soil present in food establishments consists of dirt and dust materials with discrete particles in three dimensions, organic materials with discrete particles in three dimensions, and organic materials that could be encountered in a foodservice or processing facility. Examples of soil are fat deposits on a cutting board, lubricant deposits on a moving conveyor belt, and other organic deposits on processing equipment.

Soils can be classified according to the method of removal from the object to be cleaned.

Soils Soluble in Water (or Other Solvents) Containing No Cleaner These soils will dissolve in tap water and in other solvents that do not contain a cleaning compound. They include many inorganic salts, sugars, starches, and minerals. These soils present a minimal technical problem because their removal is merely a dissolving action.

Soils Soluble in a Cleaning Solution that Contains a Solubilizer or Detergent Acid-soluble soils are soluble in acidic solutions with a pH below 7.0. Deposits include films of oxidized iron (rust), zinc carbonates, calcium oxalates, metal oxides (iron and zinc) on stainless steel, waterstone (reaction between various alkaline cleaners and chemical constituents of water having non-carbonate hardness), hard-water scale (calcium and magnesium carbonates), and milkstone (a waterstone and milk film interaction, precipitated by heat on a metal surface). *Alkali-soluble soils* are basic media with a pH above 7.0. Fatty acids, blood, proteins, and other organic deposits are solubilized by an alkaline solution. Under alkaline conditions, a fat reacts with the alkali to form soap. This reaction is called *saponification*. The soap formed from the reaction is soluble and will act as a solubilizer and dispersant for the remaining soil.

Soils Insoluble in a Cleaning Solution These soils are insoluble throughout the range of normal cleaning solutions. However, they must be loosened from the surface on which they are attached and subsequently suspended in the cleaning media.

A soil that falls into one class for one type of cleaning compound may fit in another class if another cleaner is applied. For example, sugar is soluble in water when an aqueous detergent system is used but it is insoluble in the organic solvents used in the dry cleaning industry and, therefore, falls into another class. It is important to select the appropriate solvent and the correct cleaning compound for removing a specific soil. Table 9.1 summarizes the solubility characteristics of various kinds of soil. Soils are further classified as inorganic soils. An acid cleaning

Table 9.1 Solubility characteristics of various soils

Type of salt	Solubility characteristics	Removal ease	Surface heating effects
Monovalent salts	Water soluble, acid soluble	Easy to difficult	Interaction with other constituents of removal difficulty
Sugar	Water soluble	Easy	Caramelization and removal difficulty
Fat	Water insoluble, alkali soluble	Difficult	Polymerization and removal difficulty
Protein	Water insoluble, slightly acid soluble, alkali soluble	Very difficult	Denaturation and difficulty in removal

Table 9.2 Classification of soil deposits

Type of soil	Soil subclass	Deposit examples
Inorganic soil	Hard-water deposits	Calcium and magnesium carbonates
	Metallic deposits	Common rust, other oxides
	Alkaline deposits	Films left by improper rinsing after the use of an alkaline cleaner
Organic soil	Food deposits	Food residues
	Petroleum deposits	Lubrication oils, grease, and other lubrication products
	Nonpetroleum deposits	Animal fats and vegetable oils

Table 9.3 Types of cleaning compounds for soil deposits

Type of soil	Required cleaning compound
Inorganic soil	Acid-type cleaner
Organic soil	
(Nonpetroleum)	Alkaline-type cleaner
(Petroleum)	Solvent-type cleaner

compound is most appropriate for the removal of inorganic deposits. An alkaline cleaner is more effective in removing organic deposits. So, "like cleans like." If these classes are subdivided, it is easier to determine the specific characteristics of each type of soil and the most effective cleaning compound. Table 9.2 gives a breakdown of soil subclasses, with examples of certain deposits.

Soil deposits are characteristically complex in nature and are frequently complicated by organic soils being protected by deposits of inorganic soils, and vice versa. Therefore, it is important to identify correctly the type of deposit and to use the most effective cleaning compound or combination of compounds to effectively remove soil deposits. It is frequently essential to utilize a two-step cleaning procedure that contains more than one cleaning compound to remove a combination of inorganic and organic deposits. Table 9.3 illustrates the types of cleaning compounds applicable to the broad categories of soil previously discussed.

Chemical Characteristics

Surface attachment is influenced by the chemical and physical properties of soil, such as surface tension, wetting power, and chemical reactivity with the surface of attachment, and by physical characteristics, including particle size, shape, and density. Some soils are held to a surface by adhesion forces, or *dispersion forces*. Certain soils are bonded to the surface activity of the adsorbed particles. Adsorption forces must be overcome by a surfactant that reduces surface energy of the soil and subsequently weakens the bond between the soil and surface of attachment.

Physical characteristics of soil can also affect adhesion strength, which is related directly to environmental humidity and time of contact. Adhesion forces are also dependent on geometric shape, particle size, surface irregularities, and plastic properties. Mechanical entrapment in irregular surfaces and crevices contributes to the accumulation of soils on equipment and other surfaces.

Effects of Surface Characteristics on Soil Deposition

Surface characteristics should be considered when selecting a cleaning compound and cleaning method (Table 9.4). Equipment and building

Table 9.4 Characteristics of various surfaces of food processing plants

Material	Characteristics	Precautions
Black metals	Rust may be promoted by acidic acid chlorinated detergents	Because these metals are prone to rust, they are often tinned or galvanized. Neutral detergents should be used in cleaning these surfaces
Tin	May be corroded by strong alkaline and acid cleaners	Tin surfaces should not come in contact with foods
Cement	May be etched by acid foods and cleaning compounds	Concrete should be dense, acid resistant, and non-dusting. Acid brick may be used in place of concrete
Glass	Smooth and impervious; may be etched by strong alkaline cleaning compounds	Glass should be cleaned with moderately alkaline or neutral detergents
Paint	Surface quality depends on the method of application etched by strong alkaline cleaning compounds	Certain edible paints are satisfactory for food plants
Rubber	Should be nonporous, non-spongy; not affected by alkaline detergents; is attacked by organic solvents and strong acids	Rubber cutting boards can warp, and their surface dulls knife blades
Stainless steel	Generally resistant to corrosion; smooth surfaced and impervious (unless corrosion occurs); resistant to oxidation at high temperatures; easily cleaned; nonmagnetic	Stainless steel is expensive and may be less plentiful in the future. Certain varieties are attacked by halogens (chlorine, iodine, bromine, and fluorine)

materials incorporated affect soil deposition and cleaning requirements.

Sanitation specialists should be thoroughly familiar with all finishes used on equipment and areas in the food facility and should know which cleaning chemicals will attack surfaces. If the local management team is unfamiliar with the cleaning compounds and surface finishes, a consultant or reputable supplier of cleaning compounds should be sought to provide technical assistance, including recommending chemicals and sanitation procedures.

Soil Attachment Characteristics

Soils deposited in cracks, crevices, and other uneven areas are difficult to remove—especially in hard-to-reach areas. Ease of soil removal from a surface depends on surface characteristics such as smoothness, hardness, porosity, and wettability. Soil removal from a surface consists of three sub-processes.

First is *separation of the soil from the surface*, material, or equipment to be cleaned. Soil separation can occur through mechanical action of high-pressure water, steam, air, and scrubbing,

through alteration of the chemical nature of soil (e.g., reaction of an alkali with a fatty acid to form a soap), or through the absence of alteration of the chemical nature of the soil (e.g., surfactants that reduce surface tension of the cleaning medium, such as water, to allow more intimate contact with the soil).

The soil and surface must be thoroughly wet for a cleaning compound to aid in separating soil from the surface. Cleaning compounds reduce the energy binding soil to a surface, enhancing soil loosening and separation. The effectiveness of energy reduction and reduced binding may be increased through increased temperature of the cleaning compound and water or high-pressure spray, which can aid in cutting heavy soil deposits from the surface.

The second sub-process is *soil dispersion in the cleaning solution*. Dispersion is the dilution of soil in a cleaning solution. Soil that is soluble in a cleaning solution is dispersed if an adequate dilution of cleaning medium is maintained and if the solubility limits of the soil in the media are not exceeded. The use of fresh cleaning solution or the continuous dilution of the dispersed solution with fresh solution will increase dispersion.

Some soils that have been loosened from the surface being cleaned will not dissolve in the cleaning media. Dispersion of insoluble soils is more complicated. It is important to reduce soil to smaller particles or droplets with transport away from the cleaned surface. In this application, mechanical energy supplied by agitation, high-pressure water, or scrubbing is needed to supplement the action of cleaning compounds in breaking down the soil into small particles. A synergistic action of energy reduction activity of the cleaning compound and mechanical energy can break the soil into small particles and separate it from the surface.

The third sub-process is the *prevention of redeposition of dispersed soil*. Redeposition can be reduced by removal of the dispersed solution from the surface being cleaned. Other reduction methods are continued agitation of the dispersed solution while still in association with the surface to stop the settling of the dispersed soil; prevention of any reaction of the cleaning compound with water on the soil (note that soft water containing sequestering agents will reduce the possibility of forming hard-water deposits from soap present in the cleaning compound or formed through fat saponification); elimination of any residual solution and dispersed soil that may have collected on the surface by flushing or rinsing the cleaned surface; and maintenance of soil in a finely dispersed condition to avoid further entrapment on the cleaned surface.

Adsorption of surface-active agents on the surface of soil particles causes similar electrical charges to be imparted to the particles. This condition prevents aggregation of larger particles because like-charged particles repel each other. Surface redeposition is minimized because a similar repulsion exists between surfactant-coated particles and the surfactant-coated clean surface.

A systematic approach to cleaning encompasses equipment for mechanical energy, cleaning compounds to reduce the energy holding the soil to the surface and sanitizing compounds to destroy microbial contamination associated with

soil deposits. Successful soil removal depends on cleaning procedures, cleaning compounds, water quality, high-pressure application of the cleaning media, mechanical agitation, and temperature of cleaning compounds and media.

The soil removal processes depend upon the following.

Chemical Removal Caustic solubilization of soil occurs through peptization of proteins, hydrolysis of starch and triglycerides, and acid solubilization of metal oxides.

Physical Removal Solid-solid, solid-liquid, liquid-liquid, and air-liquid interface adsorption of surfactants and sequestrants by van der Waals attractions occur.

Sequestration Sequestrants penetrate the soil. Through soil loosening, removal efficiency occurs.

The Role of Cleaning Media

Water is the cleaning medium most frequently used for soil removal. However, this medium is not "pure" since it contains dissolved and suspended materials. Rainwater dissolves carbon dioxide from the air which yields carbonic acid. This dilute acid percolates through the soil dissolving alkaline metals such as calcium and magnesium to cause water hardness. Also, foods such as a thin film of milk can add enough calcium to increase water hardness.

Other cleaning media may include *air* for removal of packaging material, dust, and other debris where water is not an acceptable cleaning medium. Additional media may include *solvents*, which are incorporated in the removal of lubricants and other similar petroleum products. The primary water requirements for food processing operations are that it must be free from disease-producing organisms, toxic metal ions, and objectionable odors and taste. Since food processing establishments do not normally have an ideal water supply, cleaning compounds must be tailored to the individual water supply and type of operation.

The major functions of water as a cleaning medium include:

- Prerinsing for the removal of large soil particles
- Wetting (or softening) of soils on the surface where removal is essential
- Transport of the cleaning compound to the area to be cleaned
- Suspension of soil to be removed
- Transport of suspended soil from the surface being cleaned
- Rinsing of the cleaning compound from the area being cleaned
- Transport of a sanitizer to the cleaned area

To complement cleaning compounds, water should be relatively free of microorganisms, clear, colorless, noncorrosive, and free of minerals (known as *soft water*). Hard water, which contains minerals, may interfere with the action of some cleaning compounds, thereby limiting their ability to perform effectively (although some cleaning compounds can counteract the adverse effects of hard water). The hardness of water affects cleaning compound consumption and may cause the formation of films, scale, or precipitates on equipment surfaces.

Cleaning Compound Characteristics

Food particles and other debris provide the nutrients required for microorganisms to proliferate. Soil must be removed thoroughly through the use of mechanical energy and chemical energy of cleaning compounds to provide a clean environment.

How Cleaning Compounds Function

The major functions of a cleaning compound are to *lower the surface tension of water so that soils may be dislodged and loosened and to suspend soil particles for subsequent flushing away*. To complete the cleaning process, a sanitizer is applied to destroy residual microorganisms that exist after cleaning.

One of the oldest and best-known cleaning compounds is plain soap. A basic soap contributes to cleaning through the removal of fats, oils, and greases by suspending particles of these water-insoluble materials, although a residual film will exist. The suspension process of water-insoluble materials through interaction with soap is called *emulsification*. However, soap has limited utility in food processing and foodservice units and is rarely used because it does not clean well and reacts with hard water to form an insoluble curd (such as a bathtub ring).

In emulsification, the cleaning compound interacts with water and the soil. Figure 9.1 illustrates that the hydrophilic portion of a cleaning compound molecule is soluble in water. The hydrophobic portion is soluble in the soil. A suspended soil particle results through micelle formation (Fig. 9.2) when a cleaning compound surrounds soil.

Factors Affecting Cleaning Compound Performance and Efficiency

Remus (1991) suggested that cleaning performance is enhanced through these factors:

Time: contact time on the surface being cleaned

Fig. 9.1 Anionic surfactant molecule

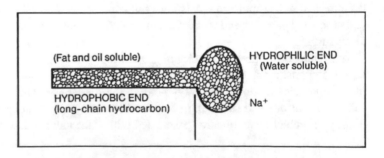

(Fat and oil soluble)

HYDROPHILIC END
(Water soluble)

HYDROPHOBIC END
(long-chain hydrocarbon)

Na+

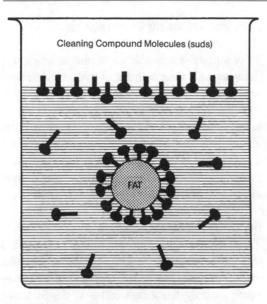

Fig. 9.2 Soil particle suspended by micelle formation

Action: physical force exerted onto the surface (velocity or flow)

Concentration: amount of cleaner used

Temperature: amount of energy (as heat) used in the cleaning solution

Water: used to prepare cleaning solution

Individual: worker performing clean-up operation

Nature: composition of the soil

Surface: material being cleaned

These factors spell out the acronym *TACT WINS* and describe important principles, involved in cleaning.

The most desirable properties of a cleaning compound for optimal cleaning as modified from Hui (2015) are (1) emulsifying, deflocculating, sequestering, dispersing, or suspending action, (2) effective rinsing properties, (3) dissolving action on soil, (4) cost-effectiveness, (5) noncorrosive, (6) surface penetration and wetting action, (7) minimal to no toxicity, and (8) quick and thorough solubility. Cleaning compound formulation can be varied to adjust to the specific cleaning operation.

Cleaning Compound Terminology

To further understand the properties of cleaning compounds, the following terms are important:

- *Chelating agent (frequently called* sequestering agent *or* sequestrant): An additive used in cleaning compounds that prevents hardness constituents and salts of calcium and magnesium from depositing on equipment surfaces by binding these salts to their molecular structure or the binding of other ions.
- *Detergent*: A compound that cleans or purges.
- *Emulsification:* A complex action consisting of a physical breakdown of fats and oils into smaller particles that are dispersed throughout the medium. The soil is still present but is reduced in physical size.
- *Peptizing:* A process that involves the formation of a colloidal solution from a material that is partially soluble, by the action of alkaline materials on protein soils.
- *Rinsibility:* The ability of a cleaning compound to be removed easily from a surface with minimal residue.
- *Saponification:* The action of an alkaline material on an insoluble soil (i.e., animal fat or vegetable oil) to produce a soluble, crude soap.
- *Sequestrant (sometimes called* chelating agent): An inorganic ingredient that is blended with cleaning compounds to prevent the precipitation of unstable salts (i.e., those that contain calcium, magnesium, and iron) that contribute to water hardness. These unstable salts will break down in the presence of alkaline compounds or at a high temperature. Many alkaline cleaning compounds are more effective with an elevated temperature; however, a high-temperature cleaning solution contributes to precipitation of calcium and magnesium carbonates, commonly known as a *scale*. A sequestrant is a chemical agent that ties up calcium and magnesium ions in a solution to prevent the ions from forming insoluble curds with the cleaning detergent, which results in precipitation deposits.
- *Soap*: A detergent, since it cleans or purges.
- *Surfactant:* A complex molecule that, when blended with a cleaning compound, reduces the surface tension of water to permit closer contact between the soil deposit and cleaning medium.
- *Suspension:* A process by which a cleaning compound loosens, lifts, and holds soil particles in a solution.
- *Water hardness:* The amount of salts such as calcium chloride, magnesium chloride,

sulfates, and bicarbonates present in water. *Permanent hardness* is frequently used when referring to calcium, magnesium chlorides, and sulfates in the water. These salts are rather stable and soluble under most conditions, causing minimal problems with cleaning. *Temporary hardness* is caused by the presence of calcium and magnesium bicarbonates, which are relatively soluble but unstable. The unstable condition of calcium and magnesium bicarbonates contributes to white deposits on equipment, heat exchangers, and water utensils. The combined amount of permanent and temporary hardness is referred to as *total hardness*.

• *Water softening:* A condition caused by the removal or inactivation of the calcium and magnesium ions in water. This is accomplished by chelation (precipitating calcium and magnesium as insoluble salts through a precipitating agent such as trisodium phosphate) and by ion exchange involving replacement of calcium and magnesium, as is accomplished by commercial water softeners. Past experiences have indicated that it is frequently less expensive to soften the hard water supply than to purchase cleaning compounds with a sequestrant.

• *Wetting (penetration):* Caused by the resultant action of a surfactant that, due to its chemical structure, is capable of wetting or penetrating the soil deposit to start the loosening process from the surface.

Classification of Cleaning Compounds

Most cleaning compounds that are used in the food industry are classified as blending products. Ingredients are combined to produce a single product with specific characteristics that performs a given function for one or more cleaning applications. The following classes of cleaning compounds are most frequently used in connection with foodservice facilities and processing plants.

Alkaline Cleaning Compounds

To describe the nature of a cleaning solution, pH (a logarithmic measurement of hydrogen ion concentration), is frequently used. A pH ranging from 0–7 is acidic. Acidity decreases from 0–7, with 7 being a neutral pH. As pH increases from 7–14, alkalinity increases. Alkaline cleaners are divided into subclasses with characteristics as discussed. Generally, fats, oils, greases, and proteins require alkaline cleaners with a pH of 11 or higher.

Strongly Alkaline Cleaners
These cleaners have strong dissolving powers and are very corrosive. They can burn, ulcerate, and scar skin. Prolonged contact may permanently damage tissue. Inhalation of the fumes or mist may cause respiratory tract damage. Mixing strong alkaline cleaners with water causes an exothermic reaction; the heat generated may cause the solution to boil or vaporize. Such explosive boiling may spray nearby personnel with the caustic compound.

Examples of strongly alkaline compounds are sodium hydroxide (caustic soda) and silicates having high $N_2O:SiO_2$ ratios. The addition of silicates tends to reduce the corrosiveness and improves the penetrating and rinsing properties of sodium hydroxide. These cleaners are used to remove heavy soils, such as those from commercial ovens and smokehouses, and have little effect on mineral deposits. Caustic soda, which has highly germicidal activity, protein dissolution, and deflocculation/emulsifying properties, is used for removing heavy soils. Because of its potential damage to humans and equipment, caustic soda is not used as a manual cleaner.

Heavy-Duty Alkaline Cleaners
These compounds have moderate dissolving powers and are generally slightly corrosive or noncorrosive. Prolonged contact with body parts may remove necessary oils from the skin, leaving it vulnerable to infections. The active ingredients of these cleaners may be sodium metasilicate (a good buffering agent), sodium hexametaphosphate, sodium pyrophosphate, sodium carbonate,

and trisodium phosphate (known for its good soil emulsification activity). Phosphates exhibit properties that make them flexible, i.e., efficacy at low concentrations, solubility in various media (e.g., concentrated acidic, alkaline, and neutral products), temperature stability, corrosion inhibition, low toxicity, stability to chlorine, and a stabilizer of peroxides.

The addition of sulfites tends to reduce the corrosion attack on tin and tinned metals. These cleaners are frequently used with high-pressure or other mechanized systems. They are excellent for removing fats but have no value for mineral deposit control. Sodium carbonate, which is one of the oldest alkaline cleaners, functions primarily as a buffering agent. Borax may be added as a buffering agent. Sodium carbonate, which is relatively low in cost, is used as a buffering agent in many formulations and has a wide range of uses in heavy-duty and manual cleaning applications. Chelators and wetting agents are normally added to tie up minerals and enhance free rinsing, respectively.

Mild Alkaline Cleaners

Mild cleaners frequently exist in solution and are used for hand cleaning of lightly soiled areas. Examples of mild alkaline compounds are sodium bicarbonate, sodium sesquicarbonate, tetrasodium pyrophosphate, phosphate water conditioners (sequesters), and alkyl aryl sulfonates (surfactants). These compounds have good water-softening capabilities but exhibit no value for mineral deposit control.

Table 9.5 summarizes cleaning characteristics of commonly used alkaline cleaners. Comparisons of emulsifying properties, detergency, and corrosiveness are also provided.

Chlorinated Alkaline Cleaners

Hypochlorite is added to these cleaners to peptize the proteins for easier removal. These cleaners are well adapted to cleaning-in-place (CIP) of pipes, tanks, and vats and remove effectively fats, oils, grease, and proteins.

A detergent sanitizer is expected to have adequate detergency and antibacterial activity; however, these cleaners have minimal sanitizing activity because of the reaction of chlorine with the soil being removed. Examples of detergent sanitizers are (1) alkaline detergents with quaternary ammonium compounds (quats) and a nonionic wetting agent and (2) acid detergents with an iodophor—usually phosphoric or sulfamic—and ampholyte compounds which ionize as either cationic or anionic compounds, depending on the pH.

Acid Cleaning Compounds

Acid cleaners constitute a group of materials having a pH <7.0 diluted in water. Acidity converts organic soils into hydrophobic materials; therefore, the use of acids is not the preferred way to perform detergency. Surfactants can be added to

Table 9.5 Characteristics of commonly used alkaline cleaning compounds

Alkaline detergent	pH of 0.5% solution	Detergency[a]	Corrosiveness[a]	Emulsifying property[a]
Sodium hydroxide (caustic soda)	12.7	2.5	3.5	2.0
Sodium orthosilicate	12.6	3.0	4.0	3.0
Sodium sesquisilicate	12.6	2.0	3.2	2.5
Sodium metasilicate	12.0	3.8	0.8	4.0
Trisodium phosphate	11.8	3.5	4.0	3.5
Sodium carbonate (soda ash)	11.3	1.5	4.0	2.8
Tetrasodium pyrophosphate	10.1	3.5	3.0	0.0
Sodium sesquicarbonate	9.7	1.3	3.2	2.5
Sodium tripolyphosphate	8.8	2.0	2.0	0.0
Sodium tetraphosphate	8.4	3.0	1.0	0.0
Sodium bicarbonate	8.2	1.5	2.3	1.5

[a]Based on a 4.0 scale, where 0 = no property and 4 = excellent property

counteract hydrophobicity and enhance penetration of the soil to allow acids to perform detergency. However, detergency strength of acids is weak and has limited application for such fresh contamination. These compounds, especially blends of acids such as phosphoric, nitric, sulfuric, and sulfamic (frequently used for the removal of blancher scale), remove encrusted surface materials and dissolve mineral scale deposits including those formed from using alkaline cleaning compounds or other cleaners. A portion of the minerals found in water may be deposited when heated to 80 °C (176 °F) or higher, adhere to metal surfaces, and appear as a rusty or whitish scale.

The hydrogen ion, which is the active ingredient of acid cleaners, is very corrosive to metals such as galvanized iron. Heterocyclic nitrogen compounds, arylthioureas, and some surface-active agents are effective corrosion inhibitors in organic acid cleaning formulations. Activity of acid cleaners is expressed through chemical action with minerals found in deposits, making them water soluble and easy to remove.

Phosphoric and nitric are the most commonly utilized acids in cleaning detergency either separately or together when they form phospho-nitric detergents. Organic acids are weaker than mineral ones. Glycolic acid is especially effective on proteic soil and is similar to phosphoric acid in removing proteins (Stanga 2010). Since organic compounds do not conduct electricity, their conductivity is attained by adding a mineral acid such as sulfamic acid.

Organic acids, such as citric, tartaric, sulfamic, and gluconic acid, are also excellent water softeners, rinse easily, and are not corrosive or irritating to the skin. Although inorganic acids are effective for removing and controlling mineral deposits, they can be extremely corrosive and irritating to the skin. Acid cleaning compounds are a specialized type of cleaner and are not recognized as effective, all-purpose cleaning compounds. They are not nearly as effective against soil caused by fats, oils, and proteins, which acts as a binder, as are alkaline cleaning compounds. Alkaline cleaning compounds chemically attack the binder of organic soils,

which releases the retaining or tenacious forces. Acid cleaning compounds are not capable of this function.

Strongly Acid Cleaners

These compounds are corrosive to concrete, most metals, and fabrics. Some of these cleaners, when heated, produce corrosive, toxic gases, which can ulcerate lungs. Strongly acid cleaners are used in cleaning operations to remove the encrusted surface matter and mineral scale frequently found on steam-producing equipment, boilers, and some processing equipment. When the solution temperature is too high, the mineral scale may redeposit and form a tarnish or whitish film on the equipment being cleaned.

Strongly acid agents used for cleaning operations in food plants are hydrochloric (muriatic), hydrofluoric, sulfamic, sulfuric, and phosphoric acids. Nitric and sulfuric acids are not used in manual cleaners because of their corrosive properties. Corrosion inhibitors, such as potassium chromate for nitric acid solutions or butylamine for hydrochloric acid detergents, may be added.

Phosphoric acid and hydrofluoric acid both clean and brighten certain metals. However, hydrofluoric acid has limited cleaning activity and is corrosive to stainless steel and dangerous to handle because of the tendency toward hydrogen evolution during use. Therefore, hydrofluoric acid use should be limited to cleaning applications where other acids are ineffective. Examples are floor maintenance and removal of tenacious rust and silicate deposits. Phosphoric acid is widely used in the United States. It is relatively low in corrosive properties, is compatible with many surfactants, and is used in manual and heavy-duty formulations.

Mildly Acid Cleaners

These compounds are mildly corrosive and may cause allergenic reactions. Some acid cleaners attack the skin and eyes. Examples of mildly acid cleaning compounds are levulinic, hydroxyacetic, acetic, and gluconic acids. Wetting agents and corrosion inhibitors (i.e., 2-naphtoquinoline, acridine, 9-phenylacridine) may be added. Organic acids, which are used as manual cleaning products, are higher in cost than are the other acid cleaning compounds. These mild compounds can also function as water softeners.

Cleaners with Active Chlorine

Cleaners containing active chlorine, such as sodium or potassium hypochlorite, are effective in the removal of carbohydrate and/or protein-aceous soils because they aggressively attack such materials and chemically modify them to render them more susceptible to interaction with the balance of the components. Active chlorine-containing products are especially valuable when cleaning a surface in which the soil is derived from a food source comprised of some form of starch or protein. Also, they are effective in removing molds from surfaces.

Because of a form of chemical bonding known as *cross-linking*, many carbohydrates are such that a large number of the "big" molecules are bonded together. In this instance, they cannot dissolve, which makes cleaning them from a surface very difficult. Heating of carbohydrate-containing materials increases the number or cross-links and complicates cleaning. Active chlorine-containing cleaners have the ability to break chemical bonds, leading to the formation of smaller, more soluble molecules and an increase in cleaning speed and efficacy.

Active chlorine, such as hypochlorite, attacks the large, complex carbohydrate molecules and degrades them to smaller, more soluble and readily removed derivatives. Because active chlorine acts quickly, only portions of the molecules need to be modified for the change in ease of removability to occur. Small amounts of active chlorine give effective cleaning results.

In the reaction of sodium hypochlorite with carbohydrates, the former can reduce the molecular weight of starch and increase its solubility. As with most cases, the reaction rates increase with elevated temperature. Because hypochlorite is an effective biocide at pH values lower than 8.5, the cleaning reaction rate of this compound is faster at a pH of 8 than at 10. A lower pH accounts for more of the hypochlorite in the form of hypo-

chlorous acid, which diffuses into bacteria and carbohydrate residues faster than the hypochlorite ion, to increase the cleaning reaction rate.

Proteins are cross-linked by chemical bonding and bonds that tie the large molecules together. Hydrogen bonding occurs because certain atoms in the molecule have a stronger attraction for electrons than do others. This reaction generates an electrostatic interaction, which complicates the removal of proteins by conventional means. Furthermore, proteins can interact through hydrogen bonding to decrease their solubility. Active chlorine-containing cleaners react with the insoluble proteins and render them soluble and/or readily dispersible through degradation by rapid oxidation of sulfide cross-links that are present. Because the degradation need not be complete for solubilization to occur, a small amount of hypochlorite will remove a relatively large quantity of protein.

Hydrogen atoms attached to nitrogen in amides are replaced by chlorine when such molecules are allowed to react with hypochlorite. This reaction appears to occur with proteins. Thus, the replacement of nitrogen-bonded hydrogen with chlorine will reduce hydrogen bonding and will improve solubility. This further explains why active chlorine degrades proteins to render them soluble and to enhance their removal from soiled surfaces or at least modifies them enough for accelerated interaction with and removal by the rest of the cleaning components. However, cleaners that contain hypochlorite should be applied soon after they are made up as they lack stability during storage.

Synthetic Detergents

The major components of synthetic detergents serve essentially the same function as soap—emulsification of fats, oils, and greases—except that there is no reaction to cause a curd formation. The hydrophilic end of soap curds in hard water, whereas this end of a synthetic detergent surfactant does not have this characteristic. Synthetic detergents are effective because their addition lowers the surface tension of the solution, promotes wetting of particles, and deflocculates and suspends soil particles. The properties of synthetic cleaning compounds are influenced by the water-soluble portion of the molecule (hydrophile) and by the water-insoluble segment.

Wetting agents may be divided into three major categories:

1. *Cationic wetting agents* (such as quaternary ammonia) are normally considered sanitizers instead of wetting agents. They produce positively charged active ions in an aqueous solution. Detergents in this category are poor wetting agents, although they are strong bactericides.
2. *Anionic wetting agents* have a negatively charged active ion when in solution. They are the most commonly used wetting agents in cleaning compounds because of their compatibility with alkaline cleaning agents and effective wetting qualities. Anionic agents differ from cationic agents by not being associated with any bactericidal properties.
3. *Nonionic wetting agents* have no charge associated with them when in aqueous solution. Therefore, they are effective under both acid and alkaline conditions. Wetting agents are also responsible for suds formation produced by a detergent. Their main problem is that they produce foam, which can cause complications in drainage and sewage systems. A cleaning compound does not have to foam to be an effective cleaner. One advantage of nonionic wetting agents is that they are not affected by water hardness.

Wetting agents serve an important function as components in cleaning compounds. Most have strong emulsifying, dispersion, and wetting capabilities. They are noncorrosive, nonirritating, and rinsed easily from equipment and other surfaces.

Alkaline Soaps

Soaps, created by the reaction of an alkali compound with a fatty acid, are considered to be alkaline salts of carboxylic acids. Most are made from lauric (C_{12}) to stearic (C_{18}) of the fatty acid

series, naphthenic acids, rosin, and the monovalent alkalies (such as sodium, potassium, ammonium), or amine salts. Soaps are not popular in industrial cleaning because they are less effective in hard water and are generally inactivated by acid solutions.

Enzyme-Based Cleaners

Enzymes function by lowering the activation energy of a reaction. Because of their bacterial attachment characteristic, enzyme-based cleaners merit consideration because they break soil down into smaller pieces and aid in its removal by destroying its attachment sites. They are classified as proteases because they break peptide bonds that link amino acids together in a polypeptide chain and work best on the alkaline side at 60 °C (104 °F) or lower. They can lower the pH of the effluent. These cleaners offer potential because they contain no chlorine or phosphates and are less corrosive than chlorine sanitizers.

The disadvantages of enzyme-based cleaners are that liquid detergents require injection equipment with two-part system activation and they are not effective on all soils. Enzymes acting on one substrate will not work on another. Thus, protease or peptidase breaks the peptide bonds in proteins but does not act on starch. Amylase breaks down starch and lipase hydrolyzes fats.

Phosphate Substitutes for Laundry Detergents

Phosphates in laundry detergents have been prohibited in certain areas of the United States. Some of the substitutes for phosphates approved for use, such as carbonates and citrates, have provided less acceptable results. Unbuilt liquids and phosphate-built powders are more effective in soil removal and whiteness retention than are the carbonate-built powders. Carbonate-built detergents, although less expensive, tend to give less acceptable results because of deposit buildup on washed materials and on parts of the washer, especially with hard water.

Solvent Cleaners

Solvent cleaners are normally used on petroleum-based soils and greases in the maintenance area. Their use should be strictly controlled. Solvent cleaners are ether- or alcohol-type materials capable of dissolving soil deposits. These compounds are most frequently used to clean soils caused by petroleum products, such as lubricating oils and greases. These cleaners may contain a foaming agent to aid in the application and cleaning. Unlike alkaline cleaners that digest organic materials, solvents "melt" or break down these compounds. Because most organic soils are saponified through alkaline cleaners, an alkaline or a neutral cleaning compound is more frequently used. However, solvent cleaners are frequently used if large amounts of petroleum deposits exist. A solvent-type cleaner is frequently required to remove this type of soil deposit from the equipment. This type of soil will not usually be found directly on processing equipment surfaces but rather in the general area.

Solvent cleaners are derived from various volatile materials from the petroleum industry and combined with wetting agents, water softeners, and other additives. Heavy-duty solvent cleaners are immiscible with water and frequently form an emulsion when water is added. Heavy-duty solvent cleaners are manufactured for use without water, whereas some solvent cleaners with low solvent content can be combined with water and still exhibit the grease-cutting action expected from a solvent.

Detergent Auxiliaries

Detergent auxiliaries are additives included in cleaning compounds to protect sensitive surfaces or to improve the cleaning properties of the compound.

Protection Auxiliaries

Acid Compounds

Acids may be used with synthetic cleaning compounds for cleaning alkaline-sensitive surfaces—for example, surfaces coated with alkaline-sensitive paints or varnishes—and light metal cleaning. The following acids are useful in protecting sensitive surfaces:

- Phosphoric acid, used to clean metals before painting, because it removes rusts and metal scales and subsequently passivates the surface
- Oxalic acid, which effectively removes iron oxide rust without attacking the metal, although precautionary steps are necessary because this acid can react with hard-water constituents to form calcium oxalate, a poisonous precipitate
- Citric acid, which does not produce toxic compounds but is not as efficient as oxalic acid in rust removal
- Gluconic acid, which removes alkali and protein films through sequestering power without a toxic effect and may be used as a water conditioner
- Sodium bisulfate, a low-cost course for heavy-duty powdered acid cleaners

Protective Colloids and Suspending Agents

Hydrophilic colloids that prevent particle redeposition on the cleaned surface are commonly referred to as *protective colloids*, *thickeners*, and *suspending agents*. Examples are gelatin, glue, starch, sodium cellulose sulfate, hydroxyethyl cellulose, and carboxymethyl cellulose. Other agents with protective properties are:

- Low-alkali, high-silica compounds, such as glassy or colloidal silicates, metasilicates, and sodium chromates (and gelatin), which inhibit tin and aluminum spangling
- Sodium chromate or dichromate, borax, and sodium nitrate in neutral detergent systems, which are efficient inhibitors of steel and iron corrosion
- Metasilicates and colloidal silicates, which protect glass and enamel surfaces from caustic etching
- Sodium sulfite, sodium fluorosilicate, and metabisulfite, which are reducing agents in the detergent system and protect tin and tin-plated surfaces by removing dissolved oxygen from the wash solution

Cleaning Auxiliaries

Various cleaning auxiliaries protect sensitive surfaces or improve the cleaning properties of a compound. Some are described below.

Sequestrants

These cleaning auxiliaries (also called *chelating agents* and *sequestering agents*) chelate by complexing with magnesium and calcium ions to produce compounds. This action effectively reduces the reactivity of water hardness constituents. Other actions of sequestrants are wetting, defoaming, emulsifying properties, and foam gel generation. The major sequestrants consist of polyphosphates or organic amine derivatives. Phosphates differ in heat stability, wetting and rinsing properties, water conditioning, hardness, and sequestering power.

Although detergency occurs because of several interconnected reactions such as wetting, solubilization, emulsification, peptization, adsorption, desorption, suspension, dispersion, and anti-redeposition, most of these cannot occur without sequestrants. Compounds that disturb cleaning (i.e., ionic precipitants and metals) are controlled by sequestrants.

Stoichiometric and threshold behavior are the primary distinction among sequestrants. A stoichiometric sequestrant, when excluding the coordination bonds, can react with a number of positive charges such as metals equal to the sum of its negative valences. If the amount of metal causes this number to be exceeded, all complexes are inactivated, and the sequestrant precipitates and contributes to scale. A sequestrant is defined as threshold if it keeps the multivalent cations soluble in a quantity exceeding the simple sum of its valences (Stanga 2010).

Sequestering compounds with carbon, hydrogen, and oxygen in carboxylic and hydroxyl groups are known as sequestering carbohydrates. Those that have application in the cleaning of food facilities are hydroxyacetic acid, citrate, tartrate, sorbitol, heptonate, and gluconate.

Phosphinopolycarbonate (PPC) is considered to be effective against calcium sulfate, barium sulfate, and calcium carbonate in water treatment. However, PPC has been considered an alternative to sodium tripolyphosphate because combinations of phosphonates and polyacrylates provide an equivalent and more economical performance.

Long-chain saccharides and aluminosilicates are referred to as polysaccharides and bentonites. The polysaccharide group consists of alginates, carrageenans, xanthans, galactomannans, starches, and cellulose derivatives (e.g., carboxymethyl [ethyl] cellulose). The bentonite group includes minerals with 60–80:10–20 ratios between SiO_2 and Al_2O_3 and traces of other minerals. These adjuncts are characterized by their ability to keep surfaces clean through their sequestering reaction, flocculation, dispersion, absorption, and anti-redeposition.

Cleaning detergents consist of a surfactant and a builder. *Builders* increase the effectiveness of a cleaner by controlling properties of the cleaning solution that tend to reduce the surfactant's effectiveness. Phosphates are considered excellent builders, especially for heavy-duty cleaning compounds. Phosphates serve as builders in cleaning compounds by providing:

- Enhancement of the wetting effect and resultant cleaning efficiency of cleaning compounds
- Sufficient alkalinity necessary for effective cleaning without being hazardous
- Maintenance of the proper alkalinity in the cleaning solution through buffering ability
- Emulsification of oily, greasy soil by degradation, and subsequent release from the surface to be cleaned
- Loosening and suspension of soil with the ability to prevent redeposition on the clean surface
- Water softening by keeping minerals dissolved to prevent settling on what is being cleaned
- Reduction in numbers of bacteria associated with a clean surface

There are a number of polyphosphates of special significance. *Sodium acid pyrophosphate* has excellent buffering and peptizing properties, with limited capability for sequestering water hardness constituents. *Tetrasodium pyrophosphate*, which does not sequester calcium as the higher phosphates, is very stable above 60 °C (140 °F) and in alkaline solutions.

Sodium tripolyphosphate and *sodium tetraphosphate* have calcium-sequestering power superior to that of tetrasodium pyrophosphate but tend to revert to orthophosphate and pyrophosphate when held above 60 °C (140 °F) or in the alkalinity of pH 10 or higher. *Sodium hexametaphosphate* (Calgon) is an effective calcium sequestrant with limited magnesium-sequestering power. *Amorphous phosphates* are complex glassy phosphates with excellent calcium-sequestering power.

Organic chelating agents, used in the formulation of water conditioners, are more efficient than the phosphates in sequestering calcium and magnesium ions and in minimizing scale buildup. According to Stanga (2010), aminopolycarboxylates are the most investigated class of sequestrants. Ethylenediaminetetraacetic acid (EDTA) is considered to be the most important and frequently used of the aminopolycarboxylic sequestrants. The success of EDTA in food sanitation is its capacity to compete with strong precipitants. Most organic agents are salts of EDTA. These chelating agents are stable above 60 °C (140 °F) and in solution for extended periods of storage. The chelating properties for EDTA salts improve as pH increases. They may be used in conveyor lubricant formulations.

Surfactants

These surface-active agents function to facilitate the transport of cleaning and sanitizing compounds over the surface to be cleaned. Surfactants are known to "make the water wetter." Surface wettability is important for cleaning rough and porous surfaces because of enhancing liquid penetration, air displacement, and cleaning agent transport into the interstices. An efficient wetting agent reduces the solid-water interfacial tension by rapid diffusion from the bulk solution to the interface and forming an oriented interfacial film. Although the major functions of surfactants are wetting and penetrating, detergency characteristics, such as emulsification, deflocculation, and suspension of particles, contribute to their effectiveness.

Surfactants are classified as synthetic detergents because of their numerous properties. As auxiliaries, they are also classified in the same three groups, according to their wetting properties and active components in solution. These auxiliaries are classified as cationic surfactants, which ionize in solution to produce active positively charged ions and serve as excellent bactericidal agents and ineffective detergents. Anionic surfactants, which ionize in solution to produce active negatively charged ions, are generally excellent detergents and ineffective bactericides; and nonionic surfactants with no positive and negative ions in solution or bactericidal properties have excellent wetting and penetrating characteristics. In addition, the *amphoteric* surfactants have a positive or negative charge, depending on the pH of the solution.

The general structure for anionic surfactants is Q–X⁻M⁺, where Q is the hydrophobic portion of the molecule, X⁻ is the anionic or hydrophilic portion, and M⁺ is the counterion in the solution. The hydrophobic portion of the molecule is normally a hydrocarbon chain of the form C_nH_{2n+1}, which is usually designated as R. Q may represent an alkyl-substituted aromatic molecule, amide, ether, fatty acid, oxyethylated alcohol, phenol, amine, or olefin. The two most familiar anionic surfactants are soaps and linear alkylbenzene sulfonates.

The hydrophobic group forms a part of the cation dissolved in water in the cationic surfactants, whereas the hydrophobic portion of an anionic surfactant forms a part of the anion in aqueous solution. A cationic compound is formed by reacting a tertiary amine with an alkyl halide to form a quaternary ammonium salt, $R_1R_2R_3 + R_{4X} \rightleftharpoons R_1R_2R_3R_4N^+ + X^-$. At least one of the R substituents is a hydrophobic group, such as dimethylammonium chloride, a germicidal agent.

The hydrophilic portion of nonionic surfactants often is composed of one or more condensed blocks of ethylene oxide. The hydrophobic portion can be any of several groups, including those named for the anionic types. The bond between the hydrophobe and the hydrophile may be an ether grouping or an amide or ester grouping. Other nonionic surfactants are alkanolamides and amine oxides.

The behavior of amphoteric surfactants is a result of two different functional groups in the molecule. The principal amphoteric surfactants are alkyl betaine derivatives, imidazole derivatives, amine sulfonates, and fatty amine sulfates.

Surfactants exhibit certain characteristics, such as:

- Solubility in at least one phase of a liquid system
- Amphipathic structure with opposing solubility tendencies, i.e., hydrophilic, lipophilic, or hydrophobic
- Orientation of monolayers at phase interfaces formed by ions of surfactant molecules
- Equilibrium concentration of a surfactant solute at a phase interface greater than the concentration in the bulk of either of the solutions
- Micelle formation when the concentration of the solute in the bulk of the solution exceeds a limiting value that is a fundamental characteristic of each solute-solvent system
- One or more functional properties such as detergency, wetting, foaming, emulsifying, solubilizing, dispersion, demulsifying, and defoaming

Scouring Compounds

Scouring compounds, also known as *chemical abrasives*, are normally manufactured from inert or mildly alkaline materials. These abrasives are generally compounded with various soaps and are provided for scouring with brushes or metal sponges. Neutral scouring compounds are frequently compounded with acid cleaners for removal or alkaline deposits and encrusted materials. Abrasive cleaning compounds should be used carefully when cleaning stainless steel to avoid scratching.

Slightly Alkaline Scouring Compounds

Scouring compounds that are made from mildly alkaline materials and used for light deposits of soil are borax and sodium bicarbonate. These compounds have limited detergency and emulsifying capabilities.

Neutral Scouring Compounds

These compounds are made from earth, including volcanic ash, seismotites, pumice, silica flours, and feldspar. They may be found in cleaning powders or pastes used in manual scrubbing and scouring operations.

Chemical-Free (Green) Cleaning

During the past, chemical-free cleaning has attained limited application in food production and retailing. The efficacy of this concept was not easily validated and uses were limited because of not being cost-efficient or cost-effective.

However, Powitz (2014) suggested that the chemical-free cleaning industry has made a quantum leap to legitimize newly and sometimes redesigned technologies for application in nontraditional uses within the food industry. Three new technologies considered to be chemical-free and "green" are electrochemically activated water, dry steam, and dry ice blasting. The major limitation of electrochemically activated water, especially the single or mixed steam technology and dry steam application, is the validation of microbicidal claims (Powitz 2014).

Ultrasonic Cleaning

Ultrasonic cleaning typically involves an immersion tank containing a 66 °C (150 °F) cleaning solution, an ultrasonic generator which produces waves with a frequency of 30,000–40,000 cycles/s, and ceramic transducers for

Table 9.6 US Geological Survey definitions for water hardness

Hardness	Parts per million
Very hard	>180
Hard	120–180
Moderately hard	60–120
Soft	0–60

converting ultrasonic energy to mechanical vibrations. Unless equipment is very difficult to clean, this technique is not practical for a food processing or foodservice cleaning operation.

Water Quality Considerations

The chemical properties of water should be considered as this is a cleaning medium basic to most cleaning compounds. Water with varying amounts of calcium, magnesium, and other alkali metals (hard water) interferes with the effectiveness of cleaning compounds (especially bicarbonates), contributing to precipitate formation. Precipitates serve as sites for accumulation of debris and microorganisms and make effective sanitation more difficult. The US Geological Survey (USGS) definitions for water hardness are provided in Table 9.6.

If hard water exists, it may be more economical to use a water softener than to include chelators that mitigate the problem. With few exceptions, hot water causes less scale formation than cold water. However, where hard water is used, maximum scale formation occurs at 82 °C (180 °F).

Cleaning Compound Selection

Cleaning compound selection can be challenging. There is no general purpose cleaner that is appropriate for all soils. The type of soil, cleaning compound function, and water characteristics determine which cleaning compound can be used most effectively. It is important to recognize the rule of thumb which states that "like cleans like". Selection of a cleaning compound depends on its capability of soil separation from the substrate,

soil dispersion into the cleaning medium, prevention of soil redeposition, and compatibility with equipment construction materials. In general, organic soils are most effectively removed through alkaline, general purpose cleaning compounds. Heavy deposits of fats and proteins require a heavy-duty alkaline cleaning compound. Mineral deposits and other soils that are not successfully removed by alkaline cleaners require acidic cleaning compounds. The most frequently used types of cleaner-sanitizers are phosphates complexed with organic chlorine. A discussion of other factors that are also important in determining which cleaning compound is most effective will follow. Table 9.7 illustrates appropriate compound application and the prevention of various soils.

Because cleaning compounds are formulated for different applications, large-volume food production plants may require up to five different cleaners such as a (an):

(1) General purpose cleaner that contains an alkaline adjunct such as sodium hydroxide or sodium carbonate blended with a sequestrant such as a polyphosphate, a low foaming wetting agent to enhance wetting and penetration of soils present, and a silicate to inhibit corrosion
(2) Strongly alkaline cleaner to remove soils that are firmly attached
(3) Acid cleaner with an organic acid to remove mineral deposits, a heterocyclic nitrogen compound to inhibit corrosion, and a wetting agent for penetration
(4) CIP cleaner formulated as in (1) above but with a nonfoaming wetting agent
(5) Cleaner for washing of utensils with less alkalinity and a high foaming wetting agent

Soil Deposition

The amount of soil to be removed affects the alkalinity or acidity of the cleaning compound used and determines which surfactants and sequestrants may be needed. The extent of soil deposition and the selection of an appropriate cleaning compound affect the degree of cleaning.

Table 9.7 Examples of common detergent properties

Ingredients	Emulsification	Saponification	Wetting	Dispersion	Suspension	Water softening	Mineral deposit control	Rinsibility	Noncorrosive	Non-irritating
Basic alkalies										
Caustic soda	C	A	C	C	C	C	D	D	D	D
Sodium metasilicate	B	B	C	B	C	C	C	B	B	D
Soda ash	C	B	C	C	C	C	D	C	C	D
Trisodium phosphate	B	B	C	B	B	A	D	B	C+	C−
Complex phosphates										
Sodium tetraphosphate	A	C	C	A	A	B	B	A	AA	A
Sodium tripolyphosphate	A	C	C	A	A	A	B	A	AA	B
Sodium hexametaphosphate	A	C	C	A	A	B	B	A	AA	A
Tetrasodium pyrophosphate	B	B	C	B	B	A	B	A	AA	B
Organic compounds										
Chelating agents	C	C	C	C	C	AA	A	A	AA	A
Wetting agents	AA	C	AA	A	B	C	C	AA	A	A
Organic agents	C	C	C	C	C	A	AA	B	A	A
Mineral acids	C	C	C	C	C	A	AA	C	D	D

Note: A-high value, *B*-medium value, *C*-low value, *D*-negative value

The kind of soil deposit also dictates which class of cleaning compounds should be used. Soil characteristics also indicate which protection auxiliaries and cleaning auxiliaries are needed, which ultimately determines the degree of cleaning.

Temperature and Concentration of Cleaning Compound Solution

As the temperature and concentration of the cleaning compound solution increase, the activity of the compound increases. The chemical effects of cleaning and disinfection increase with temperature in a linear relationship and approximately double for every 10 °C (50 °F) rise. For fatty acids and oily soils, temperatures above their melting point permit breakdown and emulsification and aid in their removal. However, an extreme temperature (above 55 °C or 132 °F) and concentration exceeding recommendations of the manufacturer or supplier may be uneconomical, damaging to equipment, dangerous and can cause water condensation and protein denaturation of the soil deposits, with an ultimate reduction in the effectiveness of soil removal.

Cleaning Time

As the length of time that the cleaning compound is in direct contact with the soil increases, the surface becomes cleaner and efficiency may increase. When extended exposure to cleaning compounds can be incorporated, e.g., soak tank cleaning, other energy input reduction (e.g., cleaning compound concentration, temperature, and mechanical action) is linear with respect to cleaning time and follows first-order reaction kinetics. The method of cleaning compound application and characteristics of the cleaning compound affect this exposure time.

Mechanical Force Used

The amount of mechanical energy in the form of agitation and high-pressure spray will affect the penetration of the cleaning compound and the physical separation of soil from the surface. The amount of agitation also helps in soil removal. Chapter 11 discusses further the role of mechanical energy (cleaning equipment) in soil removal.

Handling Precautions

Careless use of cleaning compounds is a health hazard and safety threat. Sanitors should be trained for the proper use of these chemicals and supplied with appropriate safety clothing (gloves, boots, glasses, etc.). Furthermore, US safety regulations require that Material Safety Data Sheets (MSDS) be available to all employees involved in these operations.

Most cleaners, except the liquid materials, are classified as hygroscopic in nature. They will absorb moisture when left exposed; thus, the product will deteriorate or cake in the container. Containers must be resealed properly after use to prevent contamination and to keep these materials free from moisture.

The use of an inventory sheet is recommended as an aid for reordering and pointing out irregularities in product consumption. The control of these cleaning materials should be assigned to one person appointed by the facility's management to minimize product waste and ensure that sufficient quantities of each cleaning material are available when required. This worker should be familiar with each cleaning operation so that he or she can instruct other employees in the correct techniques of any specific cleaning operation or use of cleaning equipment.

Suppliers of cleaning compounds can provide specific directions for both the compound and its use. Clear instructions will ensure that the product is used effectively without damaging the surface being cleaned. Supplier instructions for cleaning specific equipment with commercial cleaning compounds should be reviewed. Compounds from different suppliers should not be mixed.

Various areas in food plants require different cleaning mixtures. Large plants normally purchase basic cleaning compounds and blend them into concentrated batch lots. Many processing plants may devise 12–15 formulations to do specific jobs around the plant. Smaller facilities frequently purchase formulated cleaners in drum lots.

Regardless of how cleaning compounds are procured and blended, these materials should be handled with caution. Strong chemical cleaners can cause burns, poisoning, dermatitis (inflammation of the skin), and other problems to workers handling them. Since the use of stronger compounds has become prominent, there has been an increase in vulnerability to injuries.

Alkali Hazards

Strong alkaline cleaning compounds, in both solid form and in solution, have a corrosive action on all body tissue, especially the eyes. Irritation from exposure to the material is usually evident immediately. Damage frequently includes burns and deep ulceration, with ultimate scarring. Prolonged contact with dilute solutions may have a destructive effect on tissue. Dilute solutions may gradually degrease the skin, leaving vulnerable tissue exposed to allergens or other dermatitis-promoting substances. It is important to be aware that dry powder or particles can get inside a glove or a shoe and cause a severe burn. Inhalation of the dust or concentrated mist of alkaline solutions can cause damage to the upper respiratory tract and lung tissue.

Many alkaline materials react violently when mixed with water. The heat of reaction upon mixing may elevate the temperature above the boiling point, and large amounts of a hazardous mist and vapor may erupt.

Acid Cleaner Hazards

Sulfamic Acid

This compound, one of the safer acid cleaners, is a crystalline substance that can be stored easily with a minimal hazard from decomposition. However, it should be stored in a location protected from fire because it emits toxic oxides of sulfur when heated to decomposition.

Acetic Acid

This acid attacks the skin and is especially hazardous to the eyes. It presents a greater fire hazard than do many other common acids used in cleaners and should be stored in areas designed for flammable materials.

Citric Acid

This compound is one of the safer acids. Although allergenic reactions may be anticipated from prolonged exposure, it presents only a slight fire hazard. However, acid fumes are emitted when it is heated to decomposition.

Hydrochloric Acid (Muriatic Acid)

Misuse of this acid can easily result in injury. The maximum allowable concentration of vapor in air for an 8 h exposure period has been previously reported as 5 parts per million (PPM). After a short exposure, 35 PPM will cause throat irritation. This acid is frequently used in cleaners intended for descaling metal equipment because it reacts with tin, zinc, and galvanized coatings. It loosens the outer layers of material and carries soil and stain away. Hydrochloric acid will roughen the surface of concrete floors through an etching effect to produce a slip-resistant surface. When heated or contacted by hot water or steam, this acid will produce toxic and corrosive hydrogen chloride gas.

Sodium Acid Sulfate and Sodium Acid Phosphate

These cleaners will cause skin irritation or chemical burns with prolonged exposure. Water solutions of these compounds are strongly acidic and will damage the eyes if flushing is not immediate.

Phosphoric Acid

This acid is used in metal cleaners and metal brighteners. In a concentrated state, it is extremely corrosive to the skin and eyes. Phosphoric acid and sulfuric acid remove water from tissues. When heated, phosphoric acid emits toxic fumes of oxides of phosphorous. When compounded with other chemicals for use as a metal cleaner, only small amounts should be used to minimize the hazard.

Hydrofluoric Acid

Aluminum can be cleaned effectively with small amounts of this ingredient. In its pure state, hydrofluoric acid is extremely irritating and corrosive to the skin and mucous membranes. Inhalation of the vapor may cause ulcers of the respiratory tract. This material, even in very dilute amounts, should be used with caution. When heated, it emits a highly corrosive fluoride vapor, and it will react with steam to produce a toxic and corrosive mist. Ordinarily, it is used in small amounts because larger quantities can cause hydrogen evolution if in contact with metal containers. It must be stored in a safe environment, such as those used for flammable liquids.

Acid cleaners of this nature do not always attack the skin or eyes as quickly as do alkaline cleaning compounds. A severely exposed person may not realize the extent of injury until serious damage has occurred. This acid can penetrate the oil barrier of the skin to the point at which washing and flushing the area may be of little value. Hydrofluoric acid is especially hazardous because it gives little warning of injury until extensive damage has been done. Inhaled fluoride can cause damage to bones. This acid should not be confused with other acids because its action and indicated medical treatment are specific.

Soaps and Synthetic Detergents

Chemical builders used to increase the cleaning effectiveness of these substances in mixtures are usually alkaline compounds. Alkalies and alkaline substances are sometimes called *caustics* but are more correctly designated by the general term *bases*. They emulsify fats, oils, and other types of soil, which can then be washed away. Soaps and detergents for household cleaning use generally have a pH of 8–9.5. Continuous exposure to them can cause harmful degreasing of the skin, but they are safe in ordinary use. Detergents can either remove the natural oils from the skin or set up a reaction with the oils of the skin to increase susceptibility to chemicals that ordinarily do not affect the skin. Some slightly acid cleaners with a pH of 6 (the pH of the skin) are used for removing

heavy, adherent grime from the body. These hand soaps usually contain solvents that suspend greasy soil without materially degreasing the skin.

Protective Equipment

Sanitation workers should wear waterproof, knee-high footwear to maintain dry feet. Trouser legs should be worn on the outside of the boots to prevent entry of powdered material, hot water, or strong cleaning solutions. Strap-top boots are recommended where trouser legs may be worn inside boots.

Protective equipment requirements vary with the strength of solution and method of use. Where cleaning materials are dispersed through spray and brush form for overhead cleaning, protective hoods, long gloves with gauntlets turned back to prevent the cleaner from running up the arms, and long aprons should be worn. Proper respiratory protective devices approved for the specific exposure should be worn where mists or gases are encountered during mixing or use. Supervisors should be made aware of the proper size and type of respiratory equipment and must ensure that this equipment is used and maintained properly.

Chemical goggles or safety glasses should be used when handling even mild cleaning compounds. Cleaning compounds of the strength of hand soaps can cause severe eye irritation (even though these materials are considered relatively mild) as their average pH is 9.0. Constant contact with even milder cleaning solutions can cause dermatitis due to chemical reaction, degreasing effects on the skin, or both. A person wearing contact lenses should not work in any area where dangerous chemicals are handled.

Mixing and Using

An apron, goggles, rubber gloves, and dust respirator must be worn when mixing or compounding dry ingredients. Cleaners should be mixed and dispensed only by experienced, well-trained personnel. The sanitation supervisor should have knowledge of chemical fundamentals of cleaning ingredients and should provide workers with the knowledge required to prevent accidents. They should know the hazards of each individual compound and how compounds are likely to react when mixed. Safety information on new compounds put in use should be made available. Workers should be instructed that cleaning compounds are not simply soaps but strong and potentially dangerous chemicals that require protective measures. Protective equipment must be cleaned after use.

Most cleaning solutions should be compounded with cold water only. A few must be mixed with hot water to go into solution. These materials must be limited to those that do not produce a heat reaction during mixing with water. Cold water should be added during mixing to keep the solution below the boiling point or the point at which obnoxious vapors are emitted.

All cleaning compounds should be used in recommended concentrations. Once a dry cleaner is mixed or compounded, it should be stored in an identified container indicating its commonly used name, ingredients, precautions, and recommended concentration. Proper supervision is essential. Sanitation workers are frequently prone to take the attitude that "if a little is good, a lot is better." The result is concentrations that are too strong for safe use. Workers must be impressed with the importance of not mixing cleaning ingredients once they are compounded. They should be warned not to place small amounts of dry chemicals back in a barrel or to blend them with unknown chemicals.

Storage and Transport

Cleaning compounds should be stored in the area remote from normal plant traffic, with dry floors, moisture-free air, and moderate temperature (to prevent freezing of liquid products). This area should be equipped with pallets, skids, or storage racks to keep the containers off of floors and should be locked to prevent theft.

Cleaning ingredients and batches of compounded cleaners must be kept in locked storage and dispensed only with supervision. A system of inventory

control should be maintained to aid in supervision and to discover deficiencies in dispensing.

Bulk storage of cleaning ingredients should be in areas designated for whatever hazard might be characteristic of that material. Reactive, basic, and acidic materials should be segregated. All bulk materials should be stored in fire-safe areas. Lids should be tightly in place, especially if the containers are stored under automatic sprinklers. Special chemicals should bear their own particular warnings that should be observed.

Containers of alkaline material should be kept tightly sealed because these materials generally take up water from the air. They should be closed as soon as possible after opening to protect the material from atmospheric moisture.

First Aid for Chemical Burns

Whenever an employee is splashed with cleaning chemicals, *flush the individual with a large amount of water immediately. Keep flushing for 15–20 min. Do not use materials of opposing pH to neutralize contaminated skin or clothing.* Such material may merely aggravate the condition through effects of its own properties.

Workers can carry a buffered solution for the eyes, which is sold in sealed containers. If water is unavailable, this liquid can be used to dilute and wash away chemicals from the eye. This emergency measure must be followed at once by washing the eyes for approximately 15–20 min as soon as the worker can be reached. After injury, the worker's eyes should be examined by a physician. Instead of or in addition to the buffered solution, a plastic squeeze bottle of sterile water may be carried. Although these emergency measures are available, workers should not be allowed to regard eye contact accidents lightly. *The use of eye protection devices should be firmly enforced, especially where flushing water is not readily available.*

An injured employee should not be released from first aid or medical treatment until the chemical is removed. Speed is the most important factor in first aid for chemical exposures. An employee who is severely burned may act confused and need help. *Prompt flushing of chemicals from the skin, including the removal of contaminated clothing, is the most important factor in the handling of such chemical burns.* Insufficient flushing with water is very little better than none at all. Sources of water such as chemical burn showers or eye wash stations are best. However, any other source of water, regardless of its cleanliness, should be used for speed. An ample supply of water must be available near all locations where workmen may be exposed to corrosive chemicals. An ordinary shower head or garden hose spray nozzle does not supply water at a fast enough rate to flush a chemical. A flood of water is required. A satisfactory type of shower bath is one with a quick-opening valve that operates as soon as a person steps on a platform or works some other type of readily accessible control.

Everyone concerned with the chemical exposure problem should be thoroughly familiar with the following steps:

1. A worker that is exposed to a concentrated chemical should be assisted by others.
2. Flush the employee immediately at the nearest source of water. A shower is best, but any source will do. The eyes should be held open, and an extensive amount of water should be thrown into the eyes if necessary.
3. Remove all clothing.
4. After preliminary flushing, if a better source of water is near, get to it quickly and continue flushing all parts of the body thoroughly for at least 15 min. Secondary first-aid treatments, after flooding the victim's injury with water, should be kept to a minimum. Laymen should not attempt treatments with which they are not familiar or which they are not authorized to give.
5. An injured person who is confused or in shock should be immobilized immediately and given warm clothing; then cover and transfer the individual to a medical facility by stretcher.
6. All but the most minor chemical burns should be treated by a medical doctor with specific knowledge of such burns. Some chemicals may have an internal toxic action, and the danger of bacterial infection exists when the skin has been eroded by a chemical.

Dermatitis Precautions

The industrial physician has the primary responsibility for determining whether an individual may be predisposed to skin irritations and for recommending suitable placement on the basis of these findings. When dermatitis suddenly develops among individuals on a job, the affected employees should be sent immediately to an experienced physician for examination and tests to determine whether they have acquired sensitivity to the substance or substances being handled. If sensitivity has developed, the physician may decide that the affected worker should be removed from the exposure. Chemical compounds used in the cleaning operation should be listed and posted with the suggested treatment for exposure in the first-aid and supervisor's offices. Area physicians and medical centers should be listed.

Study Questions

1. What does soil mean to those involved with cleaning a food facility?
2. How does a cleaning compound function?
3. What is emulsification?
4. What is a chelating agent?
5. What does suspension mean to those cleaning a food facility?
6. What is a surfactant?
7. What is a sequestrant?
8. What is a builder?
9. What are cleaning auxiliaries?
10. Which two acid cleaning compounds are considered to be among the safest to use?
11. What treatment should be given to an employee who is splashed with cleaning chemicals?
12. What three words state a rule of thumb in cleaning compound selections?
13. What are the three steps in soil removal during cleaning?
14. What substitutes are being used for laundry detergents compounded with phosphates?
15. What potential does ultrasonic cleaning offer for the food industry?

References

Hui YH (2015). *Plant sanitation for food processing and food service*, 354. CRC Press: Boca Raton, FL.

Powitz RW (2014). Chemical-free cleaning: Revisited. *Food Saf Mag* 20(5): 20.

Remus CA (1991). When a high level of sanitation counts. *Beverage World* March: 63.

Stanga M (2010). *Sanitation cleaning and disinfection in the food industry*, 83, 180. Wiley-VCH: Weinheim, Germany.

Sanitizers

10

Abstract

The application of sanitizers is essential to reduce pathogenic and spoilage microorganisms present in food facilities and equipment. Soils must be completely removed for sanitizers to function properly.

The major classifications of sanitizers are thermal, radiation, and chemical. Thermal and radiation techniques are less practical for food production facilities than chemical sanitizing. Of the chemical sanitizers, the chlorine compounds tend to be the most effective and the least expensive. However, they are known to be more irritating and corrosive than the iodine compounds or quaternary ammonium compounds (quats). Bromine compounds are more beneficial for wastewater treatment than for sanitizing cleaned surfaces, although bromine and chlorine are synergistic when combined. The quats are more restrictive in their activities but are effective against mold growth and have residual properties. They do not kill bacterial spores but can limit their growth. Acid-quat and chlorine dioxide sanitizers are considered to be effective for the control of *L. monocytogenes*, and ozone is a potential chlorine substitute. Silver has been identified as an effective antimicrobial agent. Glutaraldehyde and fatty imidazoline compounds can be incorporated as a sanitizer for conveyor lubricants used in food operations. Various tests are available to determine the concentration of sanitizing solutions.

Keywords

Acid sanitizers • Chloramines • Hypochlorite • Iodophor • Ozone • Pasteurization • Quats • Sanitizer

Introduction

Residual soil on food processing equipment after use is usually contaminated with microorganisms nourished by the nutrients of soil deposits and provides a medium for microbial proliferation. A sanitary environment is obtained through removing soil deposits with subsequent destruction of residual microorganisms by sanitizing. However, sanitizing does not replace thorough equipment and facility cleaning. Sanitation employees handling sanitizers need to be trained to properly handle, formulate, and dispense these chemicals.

An antimicrobial agent does not exist that is complete and universal enough to combine all of the properties required of a sanitizer. However, there are various methods for the destruction of pathogenic and spoilage microorganisms.

A *sterilant* is an agent that destroys or eliminates all forms of microbial life. Chemical sterilants include ethylene oxide, glutaraldehyde, and peroxyacetic acid. Heat, both dry heat ovens and moist heat such as steam under pressure or autoclaving, is a sterilization process.

A *disinfectant* is an agent that kills infectious vegetative bacteria, although not necessarily bacterial spores, on inanimate surfaces. Disinfection is a less lethal process than sterilization. A surface is disinfected when the total viable count is not strictly zero but has a value compatible with safe food production. Disinfectants do not kill spores on hard or inanimate surfaces. Good manufacturing practices (GMPs) arbitrarily require the absence of pathogens and harmful microorganisms. General disinfectants are frequently used in households, swimming pools, and water purifiers.

A *sanitizer* is a substance that reduces, but not necessarily eliminates microbial contaminants on inanimate surfaces to levels that are considered to be safe from a public health standpoint. A sanitizer is effective in destroying vegetative cells. Sanitizers and disinfectants are not interchangeable and are intended for different purposes. Sanitizers are regulated by the Environmental Protection Agency (EPA) and require stringent laboratory test data and registration. They are categorized as no-rinse food contact surface sanitizers and nonfood contact surface sanitizers. Food contact sanitizers include sanitizing rinses for equipment, utensils, and containers used in dairy processing plants, food processing and beverage plants, and eating and drinking establishments. Biocides provide microbial control of a process (fogging disinfection, disinfection of an aseptic line, or biofilm removal). These compounds are classified as oxidative sanitizer biocides (various halogens), hydrogen peroxide-based biocides (peracetic acid, peracids, chlorine dioxide, and ozone), and surfactant-based biocides (acid anionic sulfonic acid, sulfonated fatty acids, and quaternary ammonium compounds). Others are chlorhexidine gluconate, phenolics, and aldehydes (glutaraldehyde and formaldehyde).

Sanitizing Methods

Thermal

Thermal sanitizing is relatively inefficient because of the energy required. Its efficiency depends on the humidity, temperature required, and length of time a given temperature must be maintained. Microorganisms can be destroyed with the correct temperature if the item is heated long enough and if the dispensing method and application design, as well as equipment and plant design, permit heat to penetrate to all areas. Temperature should be measured with accurate thermometers located at the outlet pipes to ensure effective sanitizing. The two major sources for thermal sterilization are steam and hot water.

Steam

Sanitizing with steam is expensive and frequently ineffective. Workers often mistake water vapor for steam; therefore, the temperature usually is not high enough to sterilize that which is being cleaned. If the surface being treated is highly contaminated, a cake may form on the organic residues and prevent sufficient heat penetration to kill the microbes. Experience in the industry has shown that steam is not amenable to continuous sanitizing of conveyors. Condensation from this operation and other steam applications has complicated cleaning operations.

Hot Water

Immersion of small components (i.e., knives, small parts, eating utensils, and small containers) into water heated to 80 °C (176 °F) or higher is another thermal method of sterilization. The microbicidal action is thought to be the denaturation of some of the protein molecules in the cell. Pouring "hot" water into containers is not a reliable sterilizing method because of the difficulty of maintaining a water temperature high enough to ensure adequate sterilization. Hot water is an effective, nonselective sanitizing method for food contact surfaces; however, spores may survive more than an hour at 100 °C (212 °F). Hot water is frequently used for plate heat exchangers and eating utensils.

Water temperature determines the time of exposure needed to ensure sterilization. An example of time-temperature relationships would be combinations adopted for various plants that utilize 15 min of exposure time at 85 °C (185 °F) or 20 min at 80 °C (176 °F). A shorter time requires a higher temperature. The volume of water and its flow rate will also influence the time taken by the components to reach the required temperature. If water hardness exceeds 60 mg/L (0.000192 oz./1.05 quarts), water scale is frequently deposited on surfaces being sanitized unless the water is softened.

However, hot water is readily available and nontoxic. If incorporated, sanitizing can be accomplished either by pumping the water through assembled equipment or by immersing equipment in the water.

Radiation

Radiation at a wavelength of approximately 2,500 Å in the form of ultraviolet (UV) light or high-energy cathode or gamma rays will destroy microorganisms. For example, ultraviolet light has been used in the form of low-pressure mercury vapor lamps to destroy microorganisms in hospitals and homes. UV activity appears to be pH and temperature independent and produces no taste or odor in treated water. It has been found to produce few, if any, undesirable by-products and little or no mutagenic activity or halogenated by-products. Ultraviolet light units are now commonly used in Europe to disinfect drinking and food processing waters and are being installed in the United States. The effective killing range for microorganisms through use of ultraviolet light is short enough to limit its utility in food operations even though its activity is independent of pH and temperature.

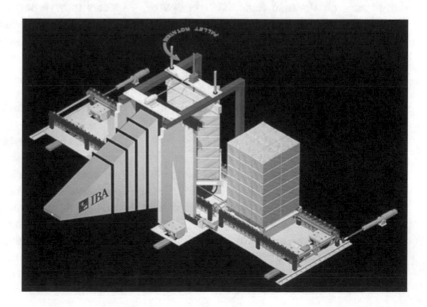

Fig. 10.1 Irradiation equipment for palletized foods

There are three different sources of ionizing radiation available for the treatment of food products. They are electron beam (e-beam), gamma rays, and X-rays. E-beam radiation has the shortest penetration range of approximately 7.5 cm (3″), whereas gamma and X-rays can penetrate one or more meters (40″) (Zammer 2004). Figure 10.1 illustrates how palletized foods may be irradiated.

Bacterial resistance determines the lethal exposure time. Light rays must actually strike the microorganisms. Radiation does not penetrate well, and its use as an antimicrobial agent should be restricted to microorganisms on sources in the air, or in clear liquids. Liquids that may be treated with UV light include beverage plant water, brine solutions, vegetable product transfer water, clean-in-place (CIP) rinse water, heating and cooling water, cheese curd rinse water, and wastewater effluents. They may be absorbed by dust, thin films of grease, and opaque or turbid solutions. Also, radiation controls the infestation of insects, regardless of the stage of their life cycle. The effectiveness of UV lamps depends on the spectral characteristics of the bulb, time of exposure, distance from the light source, and any interfering substances that interfere with light. Because the process uses glass bulbs and quartz reaction chambers, there is a risk of breakage that makes a protective shield essential. Since UV intensity dissipates with distance from the light source, there is need to minimize the distance from the lamp of the material or surface being treated.

Safety is a major concern since UV radiation can cause severe eye damage and skin irritation of exposed individuals. Furthermore, when exposed to visible light, bacterial cells injured by UV light can repair themselves.

High Hydrostatic Pressure (HHP)

This technique is applied to foods, which can be liquid or solid, packaged or unpackaged, that are subjected to high pressure (which varies depending upon application) usually for 5 min or less. HHP can be used on many foods such as raw and cooked meats, fish and shellfish, fruit and vegeta-

ble products, cheeses, salads, dips, grains and grain products, and liquids including juices, sauces, and soups. High pressure does not destroy the food, because it is applied evenly from all sides. Microorganisms living on the surface and in the interior of the food are inactivated. Inactivation is accomplished by affecting the molecular structure of chemical compounds necessary for metabolic metabolism.

HHP is equally effective on molds, bacteria, viruses, and parasites and has achieved some success in treating bacterial spores, which are resistant to many biocidal processing treatments.

The decomposition of proteins and lipids (which may result from enzymes of microbial contamination) from many foods that have active enzyme systems of their own contributes to product spoilage at refrigerated temperatures. HHP has resulted in the inactivation of certain enzymes that result in the deterioration of food.

Vacuum/Steam-Vacuum

A process exists that exposes solid food products to vacuum, steam, and vacuum again. Saturated steam is incorporated to capitalize on the large latent heat of condensation relative to the sensible heat transferred due to temperature difference in cooling superheated steam. This process appears to have potential for the destruction of pathogenic microorganisms in fresh meat and poultry, processed meats, seafood, and fruits and vegetables.

Chemical Sanitizing

Chemical sanitizers available for use in food processing and foodservice operations vary in chemical composition and activity, depending on conditions. To be effective, chemical agents must find, bind to, and transverse microbial cell envelops before they reach their target site prior to initiating the reactions that destroy microorganisms. Generally, the more concentrated a sanitizer, the more rapid and effective its action. The individual characteristics of each chemical sanitizer must be known and understood so that the most appropriate sanitizer for a specific application can be selected. Because chemical sanitizers lack penetration ability, microorganisms present in cracks, crevices, pockets, and mineral soils may not be totally destroyed. For sanitizers to be effective when combined with cleaning compounds, the temperature of the cleaning solution should be 55 °C (131 °F) or lower, and the soil should be light. The efficacy of sanitizers (especially chemical sanitizers) is affected by physical-chemical factors such as:

Exposure Time Death of a microbial population follows a logarithmic pattern, indicating that, if 90% of a population is killed in a unit of time, the next 90% of the remaining is destroyed in the next unit of time, leaving only 1% of the original number. Microbial load and the population of cells having varied susceptibility to the sanitizer due to age, spore formation, and other physiological factors determine the time required for the sanitizer to be effective. The EPA-registered label for appropriate contact time should be noted. When a sanitizer is applied via a central sanitizing system or spray application, which is generally used to sanitize exterior equipment surfaces or for environmental sanitizing, it should be used at the maximum concentration permitted on the EPA product label as a no-rinse food contact surface sanitizer. This approach is necessary to compensate for inadequate manual cleaning—especially in difficult to clean areas—and to compensate for the natural dilution that may occur because of the presence of condensation or residual rinse water from cleaning.

Temperature The growth rate of the microorganisms and the death rate due to chemical application will increase as temperature elevates. A higher temperature generally lowers surface tension, increases pH, decreases viscosity, and creates other changes that may help bactericidal action. An exception is the iodophors that vaporize above 50 °C (122 °F). These chemicals are more aggressive to surfaces, especially elastomers and gasketing materials, as the temperature rises. Thus, chemical sanitizers should be applied at ambient

temperatures, ideally 21–38 °C (70–100 °F). Generally, the degree of sanitation greatly exceeds the growth rate of the bacteria, so that the final effect of increasing temperature is to enhance the rate of destruction of the microorganisms.

Concentration Increased sanitizer concentration enhances the rate of destruction of the microorganisms.

pH The activity of antimicrobial agents occurring as different species within a pH range may be dramatically influenced by relatively small changes in the pH of the medium. Chlorine and iodine compounds generally decrease in effectiveness with an increase in pH.

Equipment Cleanliness Hypochlorites, other chlorine compounds, iodine compounds, and other sanitizers can react with the organic materials of soil that have not been removed from equipment and other surfaces. Failure to clean surfaces properly can reduce the effectiveness of a sanitizer. Oxidizing chemicals react with organic materials, such as soils, reducing their effectiveness against target microorganisms.

Water Hardness A sanitizer is affected by water composition, which can make the sanitizer chemically inactive or buffer the pH and diminish effectiveness. Quaternary ammonium compounds are incompatible with calcium and magnesium salts and should not be used with over 200 parts per million (PPM) of calcium in water or without a sequestering or chelating agent. As water hardness increases, the effectiveness of these sanitizers decreases.

Microbial Population All sanitizers are not equally effective against all microorganisms. Cells in the spore state or in a biofilm are more resistant than those in the vegetative and freely suspended state. Beverage plants with yeasts and molds as their primary contaminants may need a different sanitizer than fluid milk plants which are primarily concerned with psychrotrophic spoilage bacteria. Since a sanitizer can only reduce the number of bacteria, the higher the initial number present, the higher the amount of possible survivors. High numbers can overwhelm the sanitizer.

Bacterial Attachment   The attachment of certain bacteria to a solid surface provides an increased resistance to chlorine. With attachment, the resultant resistance to chlorine is increased.

Most of these factors are interrelated, and one can normally compensate by adjusting another. For example, if one can only prepare the ideal sanitizer in cold water, it may be possible to increase the contact time or the concentration to obtain effectiveness comparable to warm temperature in shorter contact time or lower concentration.

Desired Sanitizer Properties

The ideal sanitizer should have the following properties:

- Microbial destruction properties of uniform, broad-spectrum activity against vegetative bacteria, yeasts, and molds to produce rapid kill
- Environmental resistance (effective in the presence of organic matter [soil load], detergent and soap residues, and water hardness and pH variability)
- Good cleaning properties
- Nontoxic and nonirritating properties
- Water solubility in all proportions
- Acceptability of odor or no odor
- Stability in concentrated and use dilution
- Ease of use
- Ready availability
- Inexpensive
- Ease of measurement in use solution

A standard chemical sanitizer cannot be effectively utilized for all sanitizing requirements. The chemical selected as a sanitizer should pass the Chambers test (also referred to as the *sanitizer efficiency test*): Sanitizers should produce 99.999% kill of 75–125 million *Escherichia coli* and *Staphylococcus aureus* within 30 s after application at 20 °C (68 °F). The pH at which the compound is applied can influence the effectiveness of the sanitizer. Chemical sanitizers are normally divided according to the agent that kills the microorganisms.

Chlorine Compounds

Liquid chlorine, hypochlorites, inorganic chloramine and organic chloramines, and chlorine dioxide function as sanitizers. Their antimicrobial activity varies. Chlorine gas may be injected slowly into water to form the antimicrobial form, hypochlorous acid (HOCl). Liquid chlorine is a solution of sodium hypochlorite (NaOCl) in water. Hypochlorous acid is 80 times more effective as a sanitizing agent than an equivalent concentration of the hypochlorite ion. The amount of HOCl is dependent on the pH of the solution. A lower pH enhances HOCl formation but stability decreases. However, as pH decreases below 4.0, increasing amounts of toxic and corrosive chlorine gas are formed. Chlorine is more stable at a high pH but is less effective.

The activity of chlorine as an antimicrobial agent has not been fully determined. Hypochlorous acid, the most active of the chlorine compounds, appears to kill the microbial cell through inhibiting glucose oxidation by chlorine-oxidizing sulfhydryl groups of certain enzymes important in carbohydrate metabolism. Aldolase was considered to be the main site of action, owing to its essential nature in metabolism.

Other modes of chlorine action that have been proposed are (1) disruption of protein synthesis; (2) oxidative decarboxylation of amino acids to nitrites and aldehydes; (3) reactions with nucleic acids, purines, and pyrimidines; (4) unbalanced metabolism after the destruction of key enzymes; (5) induction of deoxyribonucleic acid (DNA) lesions with the accompanying loss of DNA-transforming ability; (6) inhibition of oxygen uptake and oxidative phosphorylation, coupled with leakage of some macromolecules; (7) formation of toxic N-chlor derivatives of cytosine; and (8) creation of chromosomal aberrations.

Vegetative cells take up free chlorine but not combined chlorine. Formation of chloramines in the cell protoplasm does not cause initial destruction. Use of ^{32}P in the presence of chlorine has suggested that there is a destructive permeability change in the microbial cell membrane. Chlorine impairs cell membrane function, especially transport of extracellular nutrients. Chlorine-releasing compounds are known to stimulate spore germination and subsequently inactivate the germinated spore.

Granular chlorine sanitizers are based on the salts of an organic carrier that contains releasable ions. Chlorinated isocyanurate is a highly stable, rapidly dissolving chlorine carrier that releases one of its two chloride ions to form NaOCl in aqueous solution. Buffering agents, which are mixed with the dry chlorine carrier in these products, control the rate of antimicrobial activity, corrosion characteristics, and stability of solutions of the sanitizers by adjusting the solution to an optimal use pH.

The chemical properties of chlorine are such that, when liquid chlorine (Cl_2) and hypochlorites are mixed with water, they hydrolyze to form hypochlorous acid, which will dissociate in water to form a hydrogen ion (H^+) and a hypochlorite ion (OCl^-), according to the reactions shown below. When sodium is combined with hypochlorite to form sodium hypochlorite, the following reactions apply:

$$Cl_2 + H_2O \rightarrow HOCl + H^+ + Cl^-$$
$$NaOCl + H_2O \rightarrow NaOH + HOCl$$
$$HOCl \rightleftharpoons H^+OCl^-$$

Chlorine compounds are more effective antimicrobial agents at a lower pH where the presence of hypochlorous acid is dominant. As pH increases, the hypochlorite ion, which is not as effective as a bactericide, predominates. Another chlorine compound, chlorine dioxide, does not hydrolyze in aqueous solutions. Therefore, the intact molecule appears to be the active agent.

Chlorine is known to be effective as a sanitizer for mechanically polished stainless steel, unabraded electropolished stainless steel, and the polycarbonate surfaces. This sanitizer is less effective on abraded electropolished stainless steel and mineral resin surfaces.

Hypochlorites, the most active of the chlorine compounds, are also the most widely used. Calcium hypochlorite and sodium hypochlorite are the major compounds of the hypochlorites.

These sanitizers are effective in deactivating microbial cells in aqueous suspensions and require a contact time of approximately 1.5–100 s. A 90% reduction in cell population for most microorganisms can be attained in less than 10 s, with relatively low levels of free available chlorine (FAC). Bacterial spores are more resistant than vegetative cells to hypochlorites. The time required for a 90% reduction in cell population can range from approximately 7 s to more than 20 min. The concentration of FAC needed for inactivation of bacterial spores is approximately 10–1,000 times as high (1,000 PPM, compared with approximately 0.6–13 PPM) for vegetative cells. *Clostridium* spores are less resistant to chlorine than *Bacillus* spores. These data suggest that in sanitizing applications, where the concentration of hypochlorous acid is low and the contact time is short, there is limited effect on bacterial spores. Although 200 PPM is effective for numerous surfaces, 800 PPM is suggested for porous areas.

The following example indicates how to formulate a 200-PPM solution of chlorine in a 200-L (52.5 gallons) tank. This calculation assumes that the chlorine contains 8.5% NaOCl:

$$8.5\% \ NaOCl = 85,000 \ PPM \left(0.085 \times 1,000,000\right)$$
$$1 \ L = 1,000 \ mL$$
$$200 \ L = 200,000 \ mL$$
$$\frac{X}{200,000 \ mL} = \frac{200 \ ppm}{85,000 \ ppm}$$
$$85,000 \ X = 40,000,000 \ mL$$
$$X = 470 \ mL \ of \ 8.5\% \ NaOCl$$

Calcium hypochlorite, sodium hypochlorite, and brands of chlorinated trisodium phosphate may be applied as sanitizers after cleaning. The hypochlorites may also be added to cleaning compound solutions to provide a combination cleaner-sanitizer. Organic chlorine-releasing agents, such as sodium dichloroisocyanurate and dichlorodimethylhydantoin, can be formulated with cleaning compounds.

Molecular hypochlorous acid is present in highest concentration near pH 4, decreasing rapidly as pH increases. At a pH higher than 5, hypochlorite (OCL^-) increases, whereas, at pH < 4, chlorine gas increases. Furthermore, the formation of Cl_2 is a safety issue. Because there are substantial amounts of hypochlorous acid present when the pH exceeds 6.5, sanitizing operations are normally executed in the pH range of 6.5–7.0.

The reaction time of chlorine-based sanitizers is temperature dependent. Up to 52 °C (126 °F), the reaction rate doubles for each 10 °C (50 °F) increase in temperature. Although hypochlorites are relatively stable, Cl_2 solubility decreases rapidly above 50 °C (122 °F).

A buffered sodium hypochlorite solution is effective against bacterial contamination and will reduce the presence of *Salmonella enteritidis*. No adverse effects on protein functionality, lipid oxidation, and starch degradation after exposure of food products to the sanitizing solution have been identified. Furthermore, this sanitizer is non-film forming without residual activity.

Active chlorine solutions are very effective sanitizers, especially as free chlorine and in slightly acid solutions. These compounds appear to act through protein denaturation and enzyme inactivation. Chlorine sanitizers are effective against gram-positive and gram-negative bacteria and conditionally against certain viruses and spores. These sanitizers tolerate a low temperature; however, the available chlorine from hypochlorite and other chlorine-releasing chemicals reacts with and is inactivated by residual organic matter. If the recommended volume and sufficient concentration is applied, a sanitizing effect can still be achieved. Only freshly prepared solutions should be used. Storage of used solutions may result in a decline in strength and activity. Concentration of active chlorine can be easily measured by use of test kits to ensure application of the desired concentration. Liquid chlorine, which is a solution of sodium hypochlorite in water, can be applied to processing and cooling waters to prevent bacterial growth and slime formation.

Inorganic chloramines are compounds formed from the reaction of chlorine with ammonia nitrogen; *organic chloramines* are formed through the reaction of hypochlorous acid with amines, imines, and imides. Bacterial spores and vegetative cells are more resistant to chloramine than to the hypo-

chlorites. Chloramine T apparently releases chlorine slowly, and, as a result, its lethal effects are slow when compared with the hypochlorites.

Other chloramine compounds are as effective as, or more effective than, the hypochlorites in deactivating microorganisms. However, these compounds release chlorine slowly and produce a slower kill rate. Sodium dichloroisocyanurate is more active than sodium hypochlorite against *E. coli*, *S. aureus*, and other bacteria.

Less is known about the antimicrobial effects of *chlorine dioxide* than about the other chlorine compounds; however, interest in this compound has increased. New chemical formulations of this compound allow it to be shipped to areas of use (rather than being generated on-site); consequently, it is being used more in the food industry. *Chlorine dioxide* (ClO_2) is known to have 2.5 times the oxidizing power of chlorine. This compound is not as effective as chlorine at pH 6.5, but at pH 8.5, ClO_2 is the most effective. Thus, ClO_2 appears to be less affected by alkaline conditions and organic matter than hypochlorites, making it a viable agent for sewage treatment.

Examples of how chlorine dioxide sanitizers are produced are indicated by the reactions that follow:

$$5NaClO_2 + 4HCl \rightarrow 4ClO_2 + 5NaCl + 2H_2O$$
$$NaOCl + HCl \rightarrow NaCl + HOCl$$
$$HOCl + 2NaClO_2 \rightarrow ClO_2 + 2NaCl + H_2O$$

Foam generation of ClO_2 is being used in cleaning and sanitizing. This sanitizer can be produced through combining chlorine salt and chlorine or hypochlorite and acid, followed by the addition of chlorite. Biodegradable foam containing 1–5 PPM of ClO_2 can be produced and is effective with a shorter contact time than the quats or hypochlorites. Chlorine dioxide is effective against a broad spectrum of microorganisms, including bacteria, viruses, and spore-formers. As a chemical oxidant, the residual activity significantly inhibits microbial redevelopment. It is active over the broad pH range normally encountered in food facilities and more tolerant of organic matter than chlorine. This compound is less corrosive than other chlorine sanitizers because of the low concentration necessary to be effective and produces less "undesirable" chlorinated organics. The major disadvantages of chlorine dioxide include cost, difficulty to handle, sensitivity to light and temperature, and potential safety and toxicity limitations.

The US Food and Drug Administration (FDA) approved the use of stabilized chlorine dioxide for sanitizing of food processing equipment. Anthium Dioxcide is a compound with 5% aqueous solution of stabilized chlorine dioxide supplied with a pH of 8.5–9.0. Free ClO_2 is the potential biocidal agent in the solution. Although Anthium Dioxcide does exhibit bacteriostatic properties, it is not nearly as effective as free ClO_2. The active biocide is free ClO_2, even though the stabilized ClO_2 at pH 8.5 is mildly bacteriostatic. The Anthium Dioxcide complex is a combination of oxygen and chlorine joined as ClO_2 in aqueous solution, which provides a longer residual effect than other chlorine sanitizers. Industrial applications include a no-rinse sanitizer at 100 PPM, poultry chill tanks at 3–5 PPM, and drinking water treatment.

Oxine differs from generated ClO_2 as it is formulated from scratch, using a proprietary process, as opposed to being converted from chlorite. Increased microbial kill is possible by adjusting the ratio of chlorite and chlorine dioxide, and of other oxychlorine species, through the formation of oxine. Oxine is stabilized through dissolving it into a proprietary aqueous solution and essentially converting it into its "salt" form. An activator, such as food-grade acid, is needed for this binary product to lower the pH and retrieve the gas. The major application of this compound is as a surface sanitizer that is effective against biofilms. Oxine will destroy *E. coli* O157:H7 at 6 PPM.

Acidified sodium chlorite (ASC), an antimicrobial agent generated by mixing concentrated sodium chlorite solution with a generally-recognized-as-safe acid at sodium chlorite concentrations of 500–1,200 PPM, is FDA approved for use with poultry, red meat, comminuted meat products, and processed fruits and vegetables to reduce bacterial contamination. It is also approved by EPA as a pesticide for use on food contact surfaces. This sanitizer can also be incorporated in water or ice at concentrations of 40–50 PPM for thawing, washing, rinsing, transporting, and/or storage of seafood.

When chlorine compounds are used in solutions or on surfaces where available chlorine can react with cells, these sanitizers are bactericidal and sporicidal. Vegetative cells are more easily destroyed than are *Clostridium* spores, which are killed more easily than are *Bacillus* spores. Chlorine concentrations of less than 50 PPM lack antimicrobial activity against *Listeria monocytogenes*, but exposure to more than 50 PPM effectively destroys this pathogen. This lethal effect of most chlorine compounds is enhanced, with an increase in free available chlorine, a decrease in pH, and an increase in temperature. However, chlorine solubility in water decreases, and corrosiveness increases with a higher temperature, and solutions with a high chlorine concentration and/or low pH can corrode metals. Chlorine compounds offer the following advantages:

- This broad-spectrum biocide is effective against a variety of bacteria, fungi, and viruses.
- They include fast-acting compounds that will pass the Chambers test at a concentration of 50 PPM in the required 30 s.
- They are the cheapest sanitizer (if inexpensive chlorine compounds are used).
- Equipment does not have to be rinsed if 200 PPM or less is applied.
- They are available in liquid or granular form.
- They are unaffected by hard-water salts (except when slight variations, due to pH, exist).
- High levels of chlorine may soften gaskets and remove carbon from rubber parts of equipment.
- Toxic by-products are not produced.
- They are less corrosive than chlorine.

However, they have some disadvantages:

- They are unstable when exposed to heat or contamination with organic matter.
- Their effectiveness decreases with increased solution pH.

- They are corrosive to stainless steel and other metals.
- They must be in contact with food handling equipment, especially on any type of dishes, for only a short time to prevent corrosion.
- They deteriorate during storage when exposed to light or to a temperature of above 60 °C (140 °F).
- Solutions at a lower pH can form toxic and corrosive chlorine gas (Cl_2).
- Concentrated in the liquid form, they may be explosive.
- Chlorine is irritating to the skin and mucous membranes.
- The environmental impact is questionable because of the formation of potentially toxic organochlorine by-products because chlorine reacts with natural-occurring organic materials, primarily humic acids and water, which result in the formation of suspected carcinogenic trihalomethane compounds.

Iodine Compounds

The mode of antibacterial action of iodine has not been studied in detail. It appears that diatomic iodine is the major active antimicrobial agent, which disrupts bonds that hold cell proteins together and inhibit protein synthesis. Generally, free elemental iodine and hypoiodous acid are the active agents in microbial destruction. The major iodine compounds used for sanitizing are iodophors, alcohol-iodine solutions, and aqueous iodine solutions. The two solutions are normally used as skin disinfectants. The iodophors have value for cleaning and disinfecting equipment and surfaces and as a skin antiseptic. Iodophors are also used in water treatment.

The iodophor complex releases an intermediate triiodide ion, which, in the presence of acid, is rapidly converted to hypoiodous acid and diatomic iodine. Both the hypoiodous acid and diatonic iodine are the active antimicrobial forms of an iodophor sanitizer.

Ionic surface-active agents (surfactants) are compounds composed of two principal functional groups—a lipophilic portion and a hydrophilic portion. When placed in water, these molecules ionize, and the two groups induce a net charge to

the molecule, which results in either a positive or a negative charge for the surfactant molecule. Cationic and anionic sanitizers have similar modes of action.

When elemental iodine is complexed with nonionic surface-active agents such as nonyl phenolethylene oxide condensates or a carrier such as polyvinylpyrrolidone, the water-soluble complexes known as *iodophors* are formed. Iodophors, the most popular forms of iodine compounds used today, have greater bactericidal activity under acidic conditions. Thus, these compounds are frequently modified with phosphoric acid. Complexing iodophors with surface-active agents provides detergent properties and qualifies them as detergent-sanitizers. These compounds are bactericidal and, when compared to aqueous and alcoholic suspensions of iodine, have greater solubility in water, are nonodorous, and are nonirritating to the skin.

To prepare the surfactant-iodine complex, iodine is added to the nonionic surfactant and heated to 55 °C–65 °C (130–150 °F) to enhance the solution of iodine and to stabilize the end product. The exothermic reaction between the iodine and surfactant produces a rise in temperature, dependent on the type of surfactant and the ratio of surfactant to iodine. If the iodine level does not exceed the solubilizing limit of the surfactant, the end product will be completely and infinitely soluble in water.

The amount of free available iodine determines the activity of iodophors. The surfactant present does not determine the activity of iodophors but can affect the bactericidal properties of iodine. Spores are more resistant to iodine than are vegetative cells, and the lethal exposure times noted in Table 10.1 are approximately 10–1,000 times as long as for vegetative cells. Iodine is not as effective as chlorine in spore inactivation. Iodine-type sanitizers are somewhat more stable in the presence of organic matter than are the chlorine compounds. Because iodine complexes offer low toxicity and are stable at a very low pH, they may be incorporated at a very low concentration of 6.25 PPM and are frequently used at 12.5–25 PPM. Iodine sanitizers are more effective than other chemicals on viruses. Only

Table 10.1 Inactivation of bacterial spores by iodophors

Organism	pH	Concentration (PPM)	Time for a 90% reduction (min)
Bacillus cereus	6.5	50	10
	6.5	25	30
	2.3	25	30
Bacillus subtilis	–	25	5
Clostridium botulinum type A	2.8	100	6

Note: All tests were conducted in distilled water at 15–25 °C (59.5–77 °F)

6.25 PPM is required to pass the Chambers test in 30 s. Nonselective iodine compounds kill vegetative cells, over a broader pH range than chlorine, and many spores and viruses.

Iodophor sanitizers, used in recommended concentrations, usually provide 50–70 mg/L (110–155 lbs/1.05 quarts) of free iodine and yield pH values of 3 or less in water of moderate alkaline hardness. Excessive dilution of iodophors with highly alkaline water can severely impair their efficiency because acidity is neutralized. Solutions of this sanitizer are most effective at a pH of 2.5–3.5.

In a concentrated form, formulated iodophors have a long shelf life. In solution, however, iodine may be lost by vaporization. This loss is especially rapid when the solution temperature exceeds 50 °C (122 °F) because iodine tends to sublime. Plastic materials and rubber gaskets of heat exchangers absorb iodine, with resultant staining and antiseptic tainting. Iodine stain can be advantageous because most organic and mineral soil stains yellow, thus indicating the location of inadequate cleaning. The amber color of iodine solutions provides visible evidence of the presence of the sanitizer, but color intensity is not a reliable guide to iodine concentration.

Because iodophor solutions are acidic, they are not affected by hard water and will prevent accumulation of minerals if used regularly. Yet, existing mineral deposits are not removed through the application of iodine sanitizers. Organic matter (especially milk) inactivates the iodine in iodophor solutions, with a subsequent fading of the amber color. Iodine loss from solutions is slight, unless excessive organic soils are present. Because iodine loss increases during storage, these solutions should be checked and adjusted to the required strength.

Iodine compounds cost more than chlorine and may cause off-flavor in some products. Other disadvantages of iodine compounds are that they vaporize at approximately 50 °C (122 °F), are less effective against bacterial spores and bacteria phage than are chlorines, have poor low-temperature efficacy, are very sensitive to pH changes, and stain porous and plastic materials. Iodine sanitizers are effective for sanitizing the hands because they do not irritate the skin. They are recommended for hand-dipping operations in food plants, and even though they produce excessive foam with CIP applications, these sanitizers are used frequently on food handling equipment.

Bromine Compounds

Bromine has been used alone or in combination with other compounds and more in water treatment than as a sanitizer for processing equipment and utensils. At a slightly acid to normal pH, organic chloramine compounds are more effective in destroying spores (such as *Bacillus cereus*) than are organic bromine compounds, but chloramine with bromine tends to be less affected by an alkaline pH of 7.5 or higher. The addition of bromine to a chlorine compound solution can synergistically increase the effectiveness of bromine and chlorine.

Blending hydrogen bromide with sodium hypochlorite (bleach) forms hypobromous acid, which can be efficiently and quickly mixed on-site. Hydrogen bromide is the only known acid (mineral or organic) to react with chlorine or sodium hypochlorite to form a nonhazardous product (hypobromous acid). At concentrations of less than 2%, a noxious gas is not produced. This compound can serve as a sanitizer for the application to destroy pathogens and other microorganisms on carcasses (especially beef) in an abattoir.

Quaternary Ammonium Compounds

The quaternary ammonium compounds, frequently called the *quats*, are used most frequently on floors, walls, furnishings, and equipment. They are good penetrants and have value for porous surfaces. They are natural wetting agents with built-in detergent properties and are referred to as *synthetic surface-active agents*. Thus, they can be applied through foaming. The most common agents are the cationic detergents, which are poor detergents but effective germicides. Quaternary ammonium compounds are very effective sanitizers for the destruction of *L. monocytogenes* and in reducing mold growth. They are stable with a long shelf life.

The quats are ammonium compounds in which four organic groups are linked to a nitrogen atom that produces a positively charged ion (cation). In these quaternary ammonium compounds, the organic radical is the cation, and chlorine is usually the anion. The mechanism of germicidal action is not fully understood, but it is hypothesized that the surface-active nature of the quat surrounds and covers the cell's outer membrane, causing a failure of the wall, which consequently causes leakage of the internal organs and enzyme inhibition. The general formula of the quaternary ammonium compound is

$$R_1 \overset{\overset{R_2}{|}}{\underset{\underset{R_4}{|}}{N}} R_3 \quad Cl^- \text{ or } BR^-$$

The quats act against microorganisms differently than do chlorine and iodine compounds. They form a residual antimicrobial film after being applied to surfaces. Although the film is bacteriostatic, these compounds are selective in the destruction of various microorganisms. The quats do not kill bacterial spores but can inhibit their growth. Quaternary ammonium compounds are more stable in the presence of organic matter than are chlorine and iodine sanitizers, although their bactericidal effectiveness is impaired by the presence of organic matter. Stainless steel and polycarbonate are more readily sanitized by the

quats than the abraded polycarbonate or mineral resin surfaces.

The quaternary ammonium compounds include alkyldimethylbenzylammonium chloride and alkyldimethylethylbenzylammonium chloride, both effective in water ranging from 500–1,100 PPM hardness without added sequestering agents. Diisobutylphenoxyethoxyethyl dimethyl benzyl ammonium chloride and methyldodecylbenzyltrimethyl ammonium chloride are compounds that require sodium tripolyphosphate to raise hardwater levels to a minimum of 500 PPM. These compounds require high dilution for germicidal or bacteriostatic action. As with other quats, these are nonconcorrosive and nonirritating to the skin and have no taste or odor in use dilutions. The concentration of quat solutions is easy to measure. The quats are low in toxicity and can be neutralized or made ineffective by using any anionic detergent. Quat sanitizers are generally more effective in the alkaline pH range. However, the effect of pH may vary with bacterial species, with gram-negative bacteria being more susceptible to quats in the acid pH range and gram-positive microbes in the alkaline range.

Quats have surfactants that cause foaming of the sanitizing solution. They can be "foamed" on, which provides a medium for them to cling to vertical and radial surfaces. When formulated with a specified detergent, they can be used as a cleaner-sanitizer. However, this application requires rinsing, although it is satisfactory for bathrooms, toilets, locker rooms, and other nonfood contact surfaces. These cleaner-sanitizers are not recommended for use in the food plant environment because there are insufficient detergent properties and pH or alkalinity levels to thoroughly clean.

The quaternary ammonium compounds should not be combined with cleaning compounds for subsequent cleaning and sanitizing because they are inactivated through detergent ingredients such as anionic wetting agents (see Chap. 9). However, an increase in alkalinity through formulation with compatible detergents may enhance the bactericidal activity of the quats.

Quats have the following major advantages:

- Colorless and odorless
- Stable against reaction with organic matter
- Resistant to corrosion of metals and not affected by water ranging from 500–1,100 PPM hardness
- Stable against temperature fluctuation with a long shelf life
- Nonirritating to the skin
- Effective at a high pH with detergency and soil penetration ability
- Effective against mold growth
- Nontoxic
- Good surfactants that provide a residual antimicrobial film

They have these disadvantages:

- Limited effectiveness (including ineffectiveness against most gram-negative microorganisms except *Salmonella* and *E. coli*) with low hard-water tolerance and low-temperature activity
- Less effectiveness against bacteriophage
- Incompatibility with inorganic phosphate- and anionic-type synthetic detergents since they are cationic molecules
- Film forming on food handling and food processing equipment
- Excessive foaming in mechanical applications and not recommended for use as cleaning-in-place (CIP) sanitizers

Acid Sanitizers

Acid sanitizers, which are considered to be toxicologically safe and biologically active, are frequently used to combine the rinsing and sanitizing steps. Organic acids, such as acetic, peroxyacetic, lactic, propionic, and formic acid, are most frequently used. Peroxyacetic acid compounds, acetic acid, octanoic acid, and water are used at such low concentrations that there is no residual vinegar flavor. The acid neutralizes excess alkalinity that remains from the cleaning compound, prevents formation of alkaline deposits, and sanitizes. Because bacteria have a positive surface charge and negatively charged surfactants react with posi-

tively charged bacteria, their cell walls are penetrated, and cellular function is disrupted. These sanitizers destroy microbes by penetrating and disrupting the cell membranes, then dissociating the acid molecule and, consequently, acidifying the cell interior. Acid treatment is dose dependent for spoilage and pathogenic microorganisms. These compounds are especially effective on stainless steel surfaces or where contact time may be extended and have a high antimicrobial activity against psychrotrophic microorganisms.

The development of automated cleaning systems in food plants, where it is desirable to combine sanitizing with the final rinse, has made the use of acid sanitizers desirable. After the final rinse, the equipment may be closed to avoid contamination and held overnight with no danger of corrosion. Although these compounds are sensitive to pH change, they are less prone to be affected by hard water than are the iodines. In the past, the disadvantage of these synthetic detergents in automated cleaning systems was foam development, which made it difficult to get good drainage of the sanitizer from the equipment. Nonfoaming acid synthetic detergent-sanitizers have become available, eliminating this problem and making these compounds even more valuable in the food industry. These sanitizers are less effective with an increase in pH or against thermoduric organisms. Acids are not as efficient as irradiation and, when applied at high concentrations, can cause slight discoloration and odor on food surfaces, such as meat. The cost-effectiveness of acid sanitizers has not been evaluated sufficiently, and experiments with acetic acid have revealed a lack of effectiveness in the reduction of *Salmonella* species contamination.

Acid sanitizers are fast acting and effective against yeasts and viruses. The pH range of below 3 is the most ideal for the performance of acid sanitizers. Acid anionic sanitizers may be incorporated as an acid rinse for equipment to leave it stainless, bright, and shiny. These sanitizers have very good wetting properties, are nonstaining and usually noncorrosive, permitting exposure to equipment overnight. Hard water and residual organic matter do not have a major effect on the effectiveness of acid anionic sanitizers and they can be applied by CIP methods or by spray, or they can be foamed on if a foam additive is incorporated. Acid sanitizers can lose all of their effectiveness by the presence of alkaline residuals or by the presence of cationic surfactants. Bacterial tolerance may increase on exposure to moderate concentrations of acids. All cleaning compounds should be rinsed from surfaces before acid sanitizers are applied.

Carboxylic acid sanitizers (also known as fatty acid sanitizers) are effective over a broad range of bactericidal activity. They are low foaming and can be used in mechanical or CIP applications. They are stable in dilutions, in the presence of organic matter, and at high temperatures. These sanitizers are noncorrosive to stainless steel, provide a good shelf life, are cost effective and act as a sanitizer and acid rinse. Carboxylic acid is less effective against yeasts and molds and not as effective above pH 3.4–4.0 as some chemical sanitizers. They are negatively affected by cationic surfactants, so thorough rinsing of detergents is essential. This sanitizer is corrosive to non-stainless steel metals, plastics, and some rubbers. Fatty acid sanitizers may be composed of free fatty acids, sulfonated fatty acids, and other organic acids. These sanitizers generally contain a mineral acid, with phosphoric being preferred. They are EPA registered as no-rinse food contact surface sanitizers and act as a sanitizer and acid rinse. These sanitizers lack effectiveness at 10 °C (50 °F) or lower.

Organic acids and bacteriocins offer potential as decontaminants. The effectiveness of organic acids in reducing populations of meatborne pathogens varies with the concentration of acid used, temperature of the acid and the carcass, contact time, spray application pressure, point at which the sanitizer is applied, tissue type, and the sensitivity of the target organisms to the specific acid. The antibacterial effects of lactic acid and acid mixtures (acetic acid with lactic or propionic acid) against gram-negative organisms are generally more extensive than their effects against gram-positive microorganisms.

Peroxy Acid Sanitizers

Peroxy acid is a strong, fast-acting sanitizer that works on the same basis as chlorine-based sanitizers, through oxidation. It is EPA registered as a no-rinse food contact surface sanitizer at the use dilution specified on the label. This sanitizer appears to be one of the most effective of those compounds available for protection against biofilms.

The low foam characteristics of the sanitizers, like chlorine, make them suitable for CIP applications. They offer a broad range of temperature activity, down to 4 °C (40 °F). As acid-type sanitizers, they combine sanitizing and acid rinse in one step. They leave no residues and are generally non-corrosive to stainless steel and aluminum in normal surface applications. Furthermore, they are relatively tolerant of organic soil.

Disadvantages of the sanitizers include loss of effectiveness in the presence of some metals contained in water. They are corrosive to some metals such as mild steel and galvanized steel and high temperatures will accelerate the corrosion rate. Full strength peroxy acid sanitizers have a strong, pungent smell. Their effectiveness against yeasts and molds vary depending on species. The newer mixed peroxy acids are more effective than the original peroxyacetic acid types on yeasts and molds.

Mixed Peroxy Acid/Organic Acid Sanitizers

This generation of mixed peroxy acid/organic acid compounds is based on the synergistic combination of organic acids and the original peroxyacetic acid. Generally, these products have the same advantages and disadvantages as the basic peroxy acid compounds. Among them are (1) effectiveness over a broad pH range, (2) effectiveness against bacteria, yeasts, and molds, (3) satisfactory activity in cold water, (4) affected minimally by water hardness, and (5) less affected by organic material than other sanitizers which operate via an oxidative mechanism, such as chlorine.

Mixed peroxy acid/organic acid sanitizers are generally more effective against various yeasts and molds than the basic peroxy acids. They may be incorporated at lower use levels than conventional basic peroxy acid compounds, producing the same efficacy, but resulting in a lower concentration. These sanitizers have higher acidity than basic peroxy acid compounds and are more effective at combining sanitizing with an acid rinse, which reduces mineral film build-up. Also, they contain a surfactant that reduces surface tension, thus, improved wetting of the treated surface. An antimicrobial in use in meat and poultry plants is a combination of sodium chlorite and citric acid.

Acid Anionic Sanitizers
These sanitizers are formulated with:

- Anionic surfactants (negatively charged)
- Acids (phosphoric acid and organic acids)

Acid anionic sanitizers act rapidly and kill a broad spectrum of bacteria and have good bacteriophage activity. They have good stability, are nonstaining, have minimal odor, are effective in a wide temperature range and are not affected by water hardness. An acidified rinse can be combined with the sanitizing step and removes and controls mineral films. These sanitizers can be corrosive to unprotected metals, and a skin irritant, inactivated by cationic surfactants, may foam too much for CIP equipment, are less effective at a higher pH, have limited and varied antimicrobial activity (including poor yeast and mold activity), and are more expensive than are the halogen sanitizers. The antimicrobial effect of acid anionic sanitizers appears to be through reaction of the surfactant, with positively charged bacteria by ionic attraction to penetrate cell walls and disrupt cellular function.

Increased use of *peroxyacetic acid* has developed in CIP sanitizing for dairy, beverage, and food processing plants. This sanitizer, which provides a rapid, broad-spectrum kill, works on the oxidation principle through the reaction with the components of cell membranes. It reduces pitting of equipment surfaces by being less corrosive than

are iodine and chlorine sanitizers. Peroxyacetic acid can be applied during an acidified rinse cycle to reduce effluent discharge, and it is biodegradable. Because this sanitizer is effective against yeasts, such as *Candida*, *Saccharomyces*, and *Hansenula*, and molds, such as *Penicillium*, *Aspergillus*, *Mucor*, and *Geotrichum*, it has gained acceptance in the soft drink and brewing industry. Peroxyacetic acid is effective for sanitizing aluminum beer kegs. Increased use of this sanitizer in dairy and food processing plants is attributable to its efficacy against various strains of *Listeria* and *Salmonella* and breaking down biofilms. The application rate of this sanitizer is 125–250 PPM. These sanitizers have the following advantages:

- Stable to heat and organic matter, have non-volatile characteristics, and can be heated to any temperature below 100 °C (212 °F) without loss of strength
- Active at a broad temperature range
- Generate low foam—suitable for CIP equipment
- Generally noncorrosive to stainless steel and aluminum
- No harmful residue
- Nonselective, permitting destruction of all vegetable cells
- Safe for use on most food handling surfaces (low toxicity—breaks down into water, oxygen, and acetic acid)
- Have rapid, broad-spectrum kill (bacteria, yeasts, and molds)
- Are pH range tolerant
- Effective against biofilms
- Relative tolerance to organic soil
- Allow sanitizing and acid rinse steps to be combined

Disadvantages include cost, odor, irritability, corrosiveness, and lower effectiveness against yeasts and molds than some sanitizers.

Acid-Quat Sanitizers
Organic acid sanitizers formulated with quaternary ammonium compounds are marketed as acid-quat sanitizers. This sanitizer is effective—especially against *L. monocytogenes*. A limitation of this type of sanitizer is that it is expensive

when compared with the halogens. These sanitizers have the following advantages:

- Aggressive against biofilm formations
- Broad spectrum of activity
- Nontoxic, odorless, colorless, temperature stable
- Formation of a residual antimicrobial film
- Stable with a long shelf life
- Mold and odor control

Disadvantages are:

- Soft metal corrosive potential
- Excessive foaming in mechanical applications
- Limited low-temperature activity
- Incompatible with anionic wetting agents
- Low hard-water tolerance
- Cost

Hydrogen Peroxide

Active chlorine is being increasingly replaced by hydrogen peroxide with or without peracetic acid because of less corrosiveness to stainless steel and potential adverse effects of chlorine. A hydrogen peroxide-based powder in 3% and 6% solutions is effective against biofilms. This antibacterial agent may be used on all types of surfaces, equipment, floors and drains, walls, steel mesh gloves, belts, and other areas where contamination exists. Also, this sanitizer is effective against *L. monocytogenes*.

Use of hydrogen peroxide for the sterilization of food packaging material is in compliance if more than 0.1 PPM can be determined in distilled water packaged under production conditions. A hydrogen peroxide solution may be used by itself or in combination with other processes to treat food contact surfaces prepared from ethylene-acrylic acid copolymers, isomeric resins, ethylene-methyl acrylate copolymer resins, ethylene-vinyl acetate copolymers, olefin polymers, polyethylene terephthalate polymers, and polystyrene and rubber-modified polystyrene polymers.

Fumigation with vapor-phase hydrogen peroxide is a potential sanitizing option. It has potent antimicrobial activity against bacteria, viruses, fungi, and bacterial spores and is a possible alternative to liquid-based disinfectants for decontamination of food contact surfaces and equipment.

Use of a foaming hydrogen peroxide/quat hybrid is considered to be the most effective sanitizer for the removal of biofilms from drains in food processing plants. This cleaning approach, known commercially as PerQuat technology, can be incorporated to clean and sanitize drains and their underground pipelines in one operation (Makdesi 2010). This technology is based the interaction of a negatively charged perhydroxide ion and a positively charged quaternary ammonium compound. When combined, the two molecules form an intimate ion pair that results in a hydrogen peroxide and quat hybrid. This combination has effective cleaning ability and antimicrobial efficacy against a wide range of microorganisms. This technology is EPA registered for the removal of biofilm drain systems.

Ozone

Because of its effectiveness against all known bacteria and their spores, ozone has emerged as a viable food safety tool (Brandt 2009). It can control ubiquitous pathogens like listeria and campylobacter and can be applied at 2–5 PPM directly to the product resulting in longer shelf life. The control of ubiquitous microbes such as those mentioned is better achieved through the continuous application of a powerful sanitizer such as ozone.

Ozone is a colorless, tasteless, odorless gas molecule comprised of three oxygen atoms that dissipates back into oxygen. This sanitizer is created by passing high-voltage electricity through air, creating a triatomic form of oxygen, or electric spark activation inside a wall-mounted unit that changes oxygen to ozone and then fed into water. It is naturally occurring in the earth's upper atmosphere. It acts as a powerful and nonselective oxidant and disinfectant (which indicates that it will attack any organic material that it contacts) and may control microbial and chemical hazards. Common by-products of ozonation are molecular oxygen, acids, aldehydes, and ketones. Ozone does not cause a harmful residue or contaminated flavor.

This sanitizer is a more powerful and faster-acting disinfectant than chlorine. It has been used safely and effectively in water treatment and is approved in the United States as generally recognized as safe (GRAS) for treatment of bottled water and has been applied in the food industry in Europe during the past. It has a broad spectrum of germicidal activity. Generally, ozone is a more effective bactericide and virucide than chlorine and chlorine dioxide and is being evaluated as a chlorine substitute. Because it oxidizes rapidly, it poses less environmental impact than some compounds.

Ozone is produced commercially through the incorporation of an ozone generator that uses electricity to generate the feed gas and ozone. The ozone is used as a gas or dissolved in water for application as an aqueous solution. A high voltage, alternating electric discharge is passed through a gas stream (dry air or oxygen). To control the electrical discharge and maintain a corona, a dielectric space or discharge gap is formed using a dielectric material such as ceramic or glass. A grounded electrode that is usually produced from stainless steel acts as a boundary to the discharge space. The most common shape for ozone generators is a cylinder, which is the most space-efficient, economic form. Care must be taken to ventilate the equipment properly as released ozone can be irritating to workers. Ozone is very unstable at a high as well as at a low pH.

This broad-spectrum germicide is effective against yeasts, molds, viruses, protozoa, as well as foodborne pathogens. The probable mode of action of ozone is through the attack on the cell membrane, rupturing and killing the cell. The most effective range for ozone is a pH of 6.0–8.5. It is effective against microorganisms in cold water, and as water temperature increases, the solubility of ozone decreases. It dissipates almost immediately at 40 °C (104 °F).

Ozone has been used to sanitize winery equipment and to disinfect water, including pools, spas, and cooling towers, and for algae control in water and wastewater treatment plants. Another application is to release gaseous ozone in cold storage rooms to control molds and eliminate ethylene, which can accelerate ripening in fruits and vegetables. Ozone is more stable in the gas phase and in an aqueous phase. This sanitizer has been used for several years as an air cleaner and deodorizer in various industries, including those that incorporate clean room technology. Storage room air can be treated to remove odors through recirculation through a separate chamber containing ozone and then through metal or charcoal oxides to remove the ozone before entry into the storage area. Further potential involves the decontamination of poultry chillers and direct decontamination of birds as they are being processed.

This sanitizer is recognized as a continuous sanitation tool because its use can eliminate chemicals and hot water, in most food plant cleaning, with the exception of isolated pieces of equipment. Furthermore, it sanitizes and dissolves grease when mixed with cold water and high pressure. When mixed in water, it can be sprayed on conveyor lines to permit processing with minimal interruption during production. Because ozone permits continuous cleaning during processing, the major portion of third-shift sanitation can be avoided. Processing plants that use ozone can potentially incorporate a third processing shift every 24-h day.

Although ozone is being used by some meat and poultry firms, it possesses limitations. Ozone is temperature sensitive, very reactive, unstable, breaks down rapidly in water, and has limited residual efficacy. It is a powerful irritant to the respiratory tract and a cellular poison that interferes with the ability of lungs to fight infectious agents. Ozone, as chlorine dioxide, has been found to produce brominated organic compounds that are alleged potential carcinogens. Furthermore, there is a high capital cost associated with the use of ozone including the need for generators at the point of use as well as energy costs to operate them. Also, ozone is corrosive to soft metals and mild steel as well as rubber and some plastics. It is less likely to be effective with heavy soil loads and has a relatively short half-life which varies depending on conditions like ambient and water temperature.

According to Brandt (2009), food processors are, by necessity, evaluating ozone systems and their ability to maintain plant cleanliness throughout the production shift. The capability of ozone as an effective degreaser and sanitizer on conveyor belts during production reduces the risk of cross-contamination. Ozone sprays can be used continuously on direct food surfaces such as conveyors, slicers, knives, and portioners to ensure their cleanliness during production. This sanitizer can be applied as interventional cleaning of indirect surfaces at breaks and shift changes, thus yielding a cleaner operation with less labor time and water and chemical usage during the sanitation operation.

Cationic Surfactants

These compounds exhibit a low killing activity against viruses, spores, and most bacteria. The low activity appears to be related to a slow kill rate and a high concentration required for microbial destruction (Stanga 2010).

Other Sanitizers

Silver has been recognized as an effective antimicrobial agent. Recently developed technology allows the "smart" (slow and steady) release, when needed, of silver ions. This technology is applicable to food and beverage applications where bacteria can thrive. Silver lined tubing has been effective against the main beverage spoiling bacteria (Kinney 2013).

The elemental ions of silver attack multiple targets in the microbe to retard growth into a destructive population. This trimodal action fights cell growth through (1) respiration prevention through inhibition of transport functions in the cell wall, (2) inhibition of cell division, and (3) disruption of cell metabolism. Depending upon the microorganism involved, this technology can reduce microbial populations within minutes to hours (Kinney 2013).

Electrolyzed oxidizing (EO) water is a novel antimicrobial agent that has been used in other countries (Koutchma 2017). Although it has advantages over conventional cleaners and sanitizers, the main disadvantage of EO water is that the solution rapidly loses its antimicrobial activity without continuous electrolysis.

Food-Grade Lubricants

Glutaraldehyde has been used to control the growth of common gram-negative and gram-positive bacteria, as well as species of yeasts and filamentous fungi found in conveyor lubricants used in the food industry. Normal wear on seals can cause gearbox and hydraulic system leaks, releasing minute levels of oil to contaminate food. When added to lubricant formulations, glutaraldehyde reduces bacterial levels by 99.99% and fungal levels by 99.9% in 30 min. Synthetic food-grade lubricants, whose polyalphaolefin's synthetic chemistry render them resistant to attack by microorganisms, are biostatic and not biodegradable.

Food-grade lubricants feature synthetic polyurea and calcium sulfate technology. Synthetic polyurea greases combat high temperatures by resisting oxidation and water and corrosion. Calcium sulfonate thickeners, although more expensive, can adapt to extreme pressures (Pellegrini 2014).

Fatty Imidazoline Compounds

These compounds form a group of synthetic lubricants. They are insensitive to cationic hardness, effective at low concentrations, and provide biocidal activity. However, these compounds do not coat surfaces effectively.

Microbicides

The microbicide, 2-methyl-5-chloro-2-methyl isothiazolone, has potential for the control of *L. monocytogenes* on product conveyors. This microbicide has been found to be effective against *L. monocytogenes* when it is incorporated in the use dilution of a conveyor lubricant at a continuous dosing rate of 10 PPM active ingredient. This biocide kills microorganisms quickly at a pH higher than 9.0, which is typical of most conveyor lubricants. Table 10.2 summarizes the important characteristics of the commonly used sanitizers. Table 10.3 matches the recommended sanitizer with the specific area or condition.

Sanitizing Applications for Pathogen Reduction of Beef Carcasses

The potential presence of *E. coli* O157:H7 on carcasses (especially beef) has necessitated the need for intervention processes to reduce microbial load, including pathogens such as *E. coli* O157:H7. Potential barriers to microbial load on carcasses are chemical and thermal sanitizers.

Table 10.2 Characteristics of commonly used sanitizers

Characteristics	Steam	Iodophors	Chlorine	Acid	Quats
Germicidal efficiency	Good	Vegetative cells	Good	Good	Somewhat selective
Yeast destruction	Good	Good	Good	Good	Good
Mold destruction	Good	Good	Good	Good	Good
Toxicity use dilution	–	Depends on wetting agent	None	Depends on wetting agent	Moderate
Shelf strength	–	Yes	Yes	Yes	Yes
Stability stock	–	Varies with temperature	Low	Excellent	Excellent
Use	–	Varies with temperature	Varies with temperature	Excellent	Excellent
Speed	Fast	Fast	Fast	Fast	Fast
Penetration	Poor	Good	Poor	Good	Excellent
Film forming	No	None to slight	None	None	Yes
Affected by organic matter	None	Moderate	High	Low	Low
Affected by other water constituents	No	High pH	Low pH and iron	High pH	Yes
Ease of measurement	Poor	Excellent	Excellent	Excellent	Excellent
Ease of use	Poor	Excellent	Excellent	High foam	High foam
Odor	None	Iodine	Chlorine	Some	None
Taste	None	Iodine	Chlorine	None	None
Effect on skin	Burns	None	Some	None	None
Corrosive	No	Not to stainless steel	Extensive on mild steel	Bad on mild steel	None
Cost	High	Moderate	Low	Moderate	Moderate

Chemical sanitizers, such as chlorine, hypobromous acid, and acetic, citric, and lactic acids, have been evaluated (see Table 10.4). These sanitizers can reduce the microbial load but do not destroy all pathogens. Past results have been inconsistent. The use of phosphates, such as trisodium phosphate and sodium tripolyphosphate, can reduce the microbial load but do not destroy all pathogens. Overall effectiveness, due to the high pH, is similar to that achieved by organic acids. More information on carcass sanitizers is included in Chap. 17.

Carcass rinse methods lack effectiveness in killing microorganisms because of the ineffective penetration of water to all of the contaminated surfaces. Hair, feathers, and scale follicles are large enough to hide bacteria but too small to admit a liquid wash or spray. An unrealistic high water pressure is needed to overcome the capillary pressure in a pore large enough to house a bacterium. The fate of *E. coli* O157:H7 cells that have been removed from carcasses by rinses

with sanitizing agents are not elucidated fully. It appears that the exposure times associated with carcass sanitizing are too short to achieve any significant direct inactivation. The primary effect of carcass rinses may be the physical removal of microorganisms.

Thermal sanitizing is the simplest form of pasteurization and may be more effective than chemical sanitizing in the destruction of pathogens on carcasses. However, hot water at or above 82 °C (180 °F) is effective, as is steam pasteurization. Hot water washes are usually designed as tunnels with conveyors that move products through hot water or steam through submersion or showering. The temperature increase reduces a number of bacteria (usually 3- or more log reduction). Steam or water is incorporated in most animal harvesting operations to reduce pathogens on the carcass surface.

Steam pasteurization involves passing carcasses through a tunnel that is approximately 12 m (40 ft) long, where large quantities of steam are applied to the carcass surface. A large percent-

Table 10.3 Specific areas or conditions for specific sanitizers

Specific area or condition	Viable sanitizer	Concentration (PPM)
Aluminum equipment	Iodophor	25
Bacteriostatic film	Quat	200
	Acid-quat	Per manufacturer recommendations
	Acid anionic	100
CIP cleaning	Acid sanitizer	130
	Active chlorine	Per manufacturer recommendations
	Iodophor	Per manufacturer recommendations
Concrete floors	Active chlorine	1,000–2,000
	Quat	500–800
Film formation, prevention of	Acid sanitizer	130
	Iodophor	Per manufacturer recommendations
Fogging, atmosphere	Active chlorine	800–1,000
Hand dip (production)	Iodophor	25
Hand sanitizer (washroom)	Iodophor	25
	Quat	Per manufacturer recommendations
Hard water	Acid sanitizer	130
	Iodophor	25
High iron water	Iodophor	25
Long shelf life	Iodophor	
	Quat	
Low cost	Hypochlorite	
Noncorrosive	Iodophor	
	Quat	
Odor control	Quat Ozone	200
Organic matter, stable in presence of	Quat	200
Plastic crates	Iodophor	25
Porous surface	Active chlorine	200
Processing equipment (aluminum)	Quat	200
	Iodophor	25
Processing equipment (stainless steel)	Acid sanitizer	130
	Acid-quat	Per manufacturer recommendations
	Active chlorine	200
	Iodophor Ozone	25
Rubber belts	Iodophor	25
Tile walls	Iodophor	25
Visual control	Iodophor	25
Walls	Active chlorine	200
	Quat	200
	Acid-quat	Per manufacturer recommendations
Water treatment	Active chlorine	20
Wood crates	Active chlorine	1,000
Conveyor lubricant	Glutaraldehyde	Per manufacturer recommendations

Table 10.4 Chemical sanitizer applications

Sanitizer	Application
Chlorine	All food contact surfaces, spray, CIP, fogging
Iodine	All food contact surfaces, approach as a hand dip
Peracetic acid	All food contact surfaces, CIP, especially cold temperature and carbon dioxide environments
Acid anionics	All food contact surfaces, spray, combines sanitizing and acid rinse into one operation
Quaternary ammonium compounds	All food contact surfaces, mostly used for environmental control; walls, drains, tiles
Ozone	Food contact and food surfaces, conveyors

age of bacteria on the carcass surface are destroyed, and the risk of enteric pathogens such as *E. coli* O157:H7 and *Salmonella* is reduced. This process involves three stages: (1) drying the washed carcass with forced filtered air, (2) immersion of the carcass in pressurized steam in a steam cabinet to envelop the entire surface area for 6–8 s to raise the temperature to approximately 82 °C (180 °F), and (3) chilling the carcass with 2–4 °C (36–40 °F) water for 6–10 s to reduce the surface temperature to 20 °C (68 °F) before storage in a chilled environment. Since meat may be contaminated during further processing, hot water and steam pasteurization can be used to decontaminate trimmings and cuts. However, meat color can change during pasteurization, reducing the desirability of the end product.

The *steam-vacuum* method was originally designed to take advantage of both hot water and steam, in combination with a physical removal of bacteria and contamination via vacuum. More recently, steam-only equipment has been designed and is used in beef processing plants for spot removal. The steam-vacuum method of pathogen reduction has resulted in a larger variation of reduction levels than other moist-heat interventions tested. This variation is attributable to repeated passes of the nozzle over the sampled surface of contaminated beef having possibly embedded bacteria, making them more difficult to remove by steam vacuumization. Plant studies have demonstrated that a commercial steam-

vacuum system can consistently outperform knife trimming for the removal of bacterial contamination of beef carcasses.

The steam-vacuum system can achieve a 5-log cycle (100,000-fold) reduction of *E. coli* O157:H7 on beef surfaces. Use of low temperature steam retards the premature warming of meat and poultry surfaces. Effectiveness is increased through air removal prior to treatment with steam since air otherwise retards the rate at which steam heats carcass surfaces.

Irradiation may be referred to as cold pressurization if treatment by radiation is noted. This is an effective method of reducing bacteria and can effectively eliminate them at a high dosage.

The effects of *cetylpyridinium chloride* (CPC) on the inhibition and reduction of *Salmonella* have been demonstrated successfully as a pathogen intervention technique for poultry carcasses. During the past, this compound has been used safely as an oral hygiene product. CPC is effective in preventing bacterial attachment and the reduction of cross-contamination. Treatment with CPC does not affect the physical appearance of poultry products. *Electrical stimulation* is another potential means of microbial load reduction on the surface of carcasses.

Activated lactoferrin (ALF) is a natural and novel antimicrobial compound that was first approved by the USDA for use on fresh beef in January 2002. Later, additional approval for ALF was granted for designation as a "processing aid" for carcass rinse treatments. Thus, this compound may be used to treat carcasses without a label declaration. The currently approved use of this carcass treatment involves a patented formulation of lactoferrin that is electrostatically applied, followed by a water rinse. This treatment physically removes bacterial contaminants from carcass surfaces, especially *E. coli* O157:H7, *Listeria monocytogenes*, and *Salmonella* spp. The growth of at least 30 species of bacteria may be retarded by ALF.

The commercial form of ALF is derived from skim milk or whey. Furthermore, the FDA has classified this compound as a GRAS substance. When lactoferrin is isolated from milk, it becomes susceptible to molecular alterations resulting from pH change, heat, proteolysis, or ionic balance, any of which can diminish antimicrobial effectiveness. "Activated" lactoferrin is the result of patented technology that provides a stabilized form of lactoferrin retaining the desired antimicrobial properties.

The ability of ALF to bind firmly to bacterial cells can cause blocking the attachment of bacteria to surfaces such as beef tissues. The physical attachment of bacterial cells, especially *E. coli* O157:H7, to carcass surfaces complicates removal and contributes to proliferation and growth of bacteria during subsequent storage. ALF can bind to tissue components such as collagen that provide anchor sites for bacterial attachment on carcasses. Because ALF has greater affinity for the anchor sites than do the bacterial cells, this substance can displace attached bacteria cells, detaching the cells and facilitating removal. The electrostatic ALF blocks bacterial cells from attachment and/or displaces cells from attachment sites. Following rinses can remove bacteria more completely. The binding of ALF to outer membrane proteins of bacterial cells disrupts the cell membrane of gram-negative cells and kills bacteria. Thus, ALF exhibits a bactericidal affect on bacterial cells that may remain on a product surface. The detrimental effect of this compound on bacterial attachment to surfaces qualifies it as a viable candidate for improved equipment sanitation as well as carcass treatments. Attachment of bacteria such as *Listeria monocytogenes* to stainless steel may be counteracted by the ability of ALF to displace attached cells. Although limited research has been conducted in this area, there appears to be potential benefits for equipment cleaning.

The binding affinity of ALF for cell surfaces may also explain the antiviral activity that has been observed for his compound. Attachment of ALF to eukaryotic cells appears to prevent the adhesion of viruses to the cell surface, which is a necessary step for the virus to infect the cell. The binding of iron is the probable exclamation of why ALF can inhibit the growth of bacterial cells. Since iron is an essential growth element for many bacteria, limiting its availability retards growth. The effectiveness of ALF as a growth inhibitor

also exists in iron-rich environments such as meat. Thus, ALF is a potential microbial inhibitor when applied to retail fresh or processed meats.

Potential Microbial Resistance

The ability of microorganisms to adapt to adverse environmental conditions presents a challenge to sanitarians. It is probable that bacteria develop resistance to sanitizing compounds, especially the quaternary ammonium compounds, similar to how they develop antibiotic resistance. Those sanitizers that kill and then rapidly disappear (oxidizers) appear to create less opportunity for resistance to develop.

Bacterial resistance to antibiotics and environmental stresses results from changes in the bacterial genome and is driven by two genetic processes and bacteria: mutation and selection known as vertical evolution. Biocides incorporated in food processing facilities provide such a powerful attack on the microorganisms that the development of resistance to the attack is retarded.

Microbial populations may not develop resistance to chlorine or quaternary ammonia because of their powerful lethal effects. Bacteria are more likely to develop resistance to organic acids than halogens. Milder organic acids treatments are safer to use and effective in some applications, but they may generate resistant strains of bacteria because they can adapt and become acid tolerant. However, a broad-spectrum biocide such as chlorine is powerful enough to prevent such change.

Sanitizer Rotation

Sanitizer rotation is a commonly employed strategy to reduce microbial resistance. The rotation of two sanitizers foils bacterial resistance to both compounds and leverages their complementary properties. The various mechanisms of biocidal attack provide logic for sanitizer rotation. As some sanitizers are more effective against gram-positive microorganisms, using them exclusively can control these microbes, but the competition for gram-negative organisms is reduced which can enhance the proliferation of these microorganisms.

If microorganisms develop resistance to one form of attack, logic would suggest a switch to a different sanitizer. Although various sanitizers may be incorporated, the most common rotation involves some form of chlorine during the week and a quaternary ammonium sanitizer to provide a residual effect over the weekend. Another viable approach is to utilize a chlorine sanitizer on Tuesday and Thursday and incorporate a quat sanitizer on Monday, Wednesday, and Friday. Chlorine and quat sanitizers should not be mixed because they can produce a dangerous reaction and a poisonous gas. Pathogens are more prone to develop tolerance to the quats and peroxy acid than oxidizers such as chlorine-based compounds (Gregerson 2009).

To complement the effectiveness of sanitizers incorporated, Gregerson (2009) suggested that alternating between alkaline-based and acid-based cleaning compounds can reduce hard-water buildup and the formation of biofilms. The rotation of sanitizers can potentially alter the environment's pH and reduce the acclamation of listeria to the particular setting.

Tests for Sanitizer Strength

Sanitarians should be trained to conduct inspection of the cleaning operation. Inspection should include the use of ATP bioluminescence technology to verify cleaning and sanitizing effectiveness. Instead of detecting the presence of microorganisms, this technology measures the presence of organic material, indicating an environment that supports microbial growth. To increase sanitizer effectiveness, a number of tests have been devised to determine the concentration of the sanitizer being tested.

Chlorine Sanitizers
The following methods can be used to determine chlorine concentration in the sanitizer being tested:

1. *Starch iodide method (iodometric)*. This is a titration test in which chlorine displaces iodine from potassium iodide in an acid solution and forms a blue color with starch. Decolorization occurs by the addition of standard sodium thiosulfate. This test is generally used to measure high residuals.
2. *o-Tolidine colorimetric comparison*. This is a test in which a colorless solution of o-tolidine is added to a chlorine solution. An orange-brown-colored compound proportional to its concentration is produced and is compared with a standardized color.
3. *Indicator paper test*. This is rapid test of limited accuracy in which test papers, usually impregnated with starch iodide, are immersed. The developed color is compared with a standard.

Iodophors
Although iodophors have a built-in color indicator that is relatively accurate, color comparative kits and other kits are available for testing.

Quaternary Compounds
There are several satisfactory tests for determining concentration of these compounds. Some reagents are available in tablets, and others use test papers by which a color comparison is made.

Study Questions

1. What are the advantages and disadvantages of hot water as a sanitizer?
2. What factors contribute to the effectiveness of a sanitizer?
3. How is chlorine dioxide produced for use in a food facility?
4. What are the advantages and disadvantages of chlorine as a sanitizer?
5. What are the advantages and disadvantages of iodine as a sanitizer?
6. What are the advantages and disadvantages of the "quats" as a sanitizer?
7. What are the advantages and disadvantages of acid sanitizers?

8. What sanitizers are frequently added to lubricants?
9. Which organic acids are applied most frequently to sanitize food contact surfaces?
10. What are the limitations of radiation sanitizing?
11. How does silver act as an antimicrobial agent?
12. What are the benefits of sanitizer rotation?
13. What are the advantages and disadvantages of ozone as a sanitizer?
14. How does a chemical sanitizer such as chlorine appear to destroy microorganisms?
15. How is silver utilized as a sanitizer?
16. What is hypobromous acid and its potential as a sanitizer?

References

Brandt J (2009). The case for ozone. *Food Qual* 15(6): 32.

Gregerson J (2009). Clean, sanitize, rinse, repeat. *Meatingplace* 05(09): S3.

Kinney G (2013). Alternative tubing eliminates bacteria in beverage delivery systems. *Food Qual* 20(2): 34.

Koutchma T (2017). Electrolyzed water as a novel cleaner and sanitizer. *Meatingplace* 10(16): 85.

Makdesi A (2010). Drain sanitation delivers peace of mind. *Food Qual* 17(3): 39.

Pellegrini M (2014). Regulating grease. *The National Provisioner* 228(6): 64.

Stanga M (2010). *Sanitation cleaning and disinfection in the food industry*, 173. Wiley-VCH: Weinheim, Germany.

Zammer C (2004). Food irradiation: Is it a matter of taste? *Food Qual* 11(3): 44.

Sanitation Equipment

11

Abstract

A major function of cleaning equipment is to dispense the cleaning compound and sanitizer to facilitate cleaning and sanitizing and reduce the microbial flora. An efficient cleaning system can reduce cleaning labor by up to 50%. High-pressure, low-volume cleaning equipment generally is among the most effective cleaning equipment in the removal of soil deposits with penetration ability in difficult-to-reach areas. Foam is easily applied, has the ability to cling to surface areas, and tends to be more effective for large surface areas. A medium similar to foam, except that less air is present and it has reduced clinging ability, is called a slurry. A gel medium is most effective for cleaning equipment with small moving parts.

A portion of the equipment used in food processing plants for fluid processing, such as for beverages and dairy products, is cleaned effectively with cleaning-in-place (CIP) units, which reduce cleaning labor. However, this equipment is expensive and is less effective where heavy soil and a variety of processing systems exist. Sophisticated CIP equipment includes a microprocessor control unit to monitor operating parameters. Parts and small utensils can be cleaned effectively with cleaning-out-of-place (COP) equipment. More sanitary lubrication of high-speed conveyors and other equipment is possible through the use of mechanized lubrication equipment.

Keywords

High-pressure • Low-volume • Low-pressure • High-temperature • Medium-pressure system • Portable cleaning units • Centralized cleaning units • Foam cleaning • Drop stations • Clean-in-place • Clean-out-of-place • Lubrication

Introduction

In Chaps. 9 and 10, we examined characteristics of cleaning compounds and sanitizers and their potential applications. This chapter provides information on cleaning and sanitizing equipment and discusses a systems approach to cleaning and sanitizing. A variety of cleaning equipment, cleaning compounds, and sanitizers are available, making selection of the optimal cleaning technique confusing. There are no cleaning compounds, sanitizers, or cleaning units available that are truly all-purpose, because such products would need to possess too many chemical and physical requirements.

Cleaning is considered to be the use of mechanical agitation and cleaning compounds to remove visible soil, biofilms, and other residual soils from the surfaces of equipment, floors, walls, and other locations in a production facility. Sanitizing is the application of chemicals or chemical treatments to remove any remaining microorganisms or debris that cannot be seen with the naked eye.

Mechanical cleaning and sanitizing equipment merits serious consideration because it can reduce cleaning time and improve efficiency. An efficient system can reduce labor costs by up to 50% and should have a "payout" period of fewer than 24 months. In addition to labor savings and increased efficiency, a mechanized cleaning unit can more effectively remove soil from surfaces than can the hand method.

Management frequently fails to recognize that there is a technology of cleaning that should be applied for effective performance. Large expenditures should not be made on effective cleaning and sanitizing equipment without hiring skilled employees to operate the equipment and qualified management to supervise the operation. Although many technical representatives of chemical companies that manufacture cleaning compounds and sanitizers are qualified to recommend cleaning equipment for various applications, people who manage the sanitation program should have enough technical expertise to understand the purpose and proper use of the equipment. It is important to approach cleaning and sanitizing problems on a technological basis. The observation of a plant during cleanup to evaluate the operation of cleaning equipment can be used to determine whether the operation is satisfactory. The majority of recent advances in equipment technology includes use of current equipment with sanitizers to reduce costs through decreasing energy use and water use.

Sanitation Costs

A typical cleaning operation has the following breakdown of costs:

Cost	%
Labor	50.0
Water/sewage	18.0
Energy	7.5
Cleaning compounds and sanitizers	6.0
Corrosion damage	1.5
Miscellaneous	17.0

The largest cost of cleaning is *labor*. Approximately 50% of the sanitation dollar is spent for cleaning, sanitizing, and quality assurance personnel and supervision. This expense, however, can be reduced more than other costs, through the use of mechanized cleaning systems. However, this further substantiates the need to have training for employees and internal knowledge within the company about sanitation to minimize costs. Redemann (2005) stated that it is important to create a culture of cleanliness in which sanitation personnel understand sanitary design limitations, sanitation Good Manufacturing Practices (GMPs), how to prevent cross-contamination, and the importance of consistently following the sanitation procedures that have been established. In addition, it is important for sanitation personnel to provide training for supporting functions including production, maintenance, and other departments, which will help ensure that sanitation procedures are followed and that there is a corporate culture of hygiene.

Water and sewage have the next highest costs. Food plants use large quantities of water for the application of cleaning compounds. In addition, this category encompasses sewage discharge costs and surcharges. Energy requirements and sewage treatment costs are major because sewage from food plants can be high in biochemical oxygen demand (BOD) and chemical oxygen demand (COD). (See Chap. 12 for more about waste treatment.) From 2006–2016, some of the greatest technology advancements have been to reduce water use for two major purposes: (1) to reduce costs and (2) to make their processes more green or environmentally friendly. A green or environmentally friendly sanitation product is a sanitizer that breaks down readily during the wastewater treatment process or one that leaves no toxic residues in the environment (Owens 2016). Some other items that can be incorporated to reduce water use include low temperature and dry sanitizers. Ecolab has an alkaline chlorine sanitizer that can sanitize metal equipment at cold water temperature (Cooper 2016).

The cost and availability of *energy* for generating hot water and steam are important factors. Most cleaning systems, cleaning compounds, and sanitizers are effective when the water temperature is below 55 °C (132 °F). A lower temperature will conserve energy, reduce protein denaturation on surfaces to be cleaned (thus increasing ease of soil removal), and decrease injury to employees.

Although *cleaning compounds* and *sanitizers* are expensive, this cost is reasonable if one considers that sanitizers destroy residual microorganisms and that these compounds contribute to more thorough cleaning with less labor. The optimal cleaning system combines the most effective cleaning compounds, sanitizers, and equipment to perform the cleaning tasks economically and effectively. Chemical costs may be reduced by the use of the correct amount of cleaning solution to perform tasks.

Improper use of cleaning compounds and sanitizers on processing equipment constructed of stainless steel, galvanized metal, and aluminum costs the industry millions of dollars through *corrosion* damage. This cost can be reduced through use of appropriate construction materials and the proper cleaning system, including noncorroding cleaning compounds and sanitizers.

An accumulation of *miscellaneous sanitation costs* includes the cost of water and sewage treatment. Miscellaneous costs encompass equipment depreciation, returned goods, general and administrative expenses, and other operating costs. The general nature of these costs makes it more difficult to identify a specific approach for their reduction. The most effective course is effective management.

Equipment Selection

Identification of the most appropriate equipment for the application of cleaning compounds and sanitizers is as important as selecting the chemicals themselves (Anon 2003). At least three sources are available to the industry to provide information related to the optimal sanitation system: a planning division (or similar group) of the food company, a consulting organization (internal or external), and/or a supplier of cleaning and sanitizing compounds and equipment. Regardless of which source is used, a basic plan should be followed to guide the selection and installation of equipment. An important factor in sanitation equipment selection is the degree of equipment disassembly, which is critical to environmental pathogen control. Other critical factors include the effectiveness of a sanitizer application, prevention of cross-contamination, and worker safety. Furthermore, it is important to clearly define and effectively communicate to employees the proper use and maintenance of sanitation equipment.

Sanitation Study

A sanitation study should start with a plant survey. A study team or individual specialist should identify cleaning procedures in use (or procedures recommended for a new operation), labor requirements, chemical requirements, and utility costs. This information is needed to determine recommended cleaning procedures, cleaning and sanitizing supplies, and cleaning equipment. The survey data should reflect required expenses and projected annual savings from the proposed sanitation system. A report of this study should be distributed to key management personnel.

Sanitation Equipment Implementation

After the appropriate equipment has been recommended and acquired, the vendor or a designated expert should supervise the installation and start-up of the new operation. Personnel training should be provided by the vendor or by the organization responsible for manufacture of the system. After start-up, regular inspections and reviews should be conducted jointly with the organization performing the sanitation study and a management team designated by the food company. In addition to daily inspections, reviews should be conducted every 6 months. Both inspections and reviews should be documented so that records are available.

Reports should contain information related to the program effectiveness, periodic inventory data, and cleaning equipment condition. Information related to labor, cleaning compounds, sanitizers, and maintenance costs provided by reports should be compared with costs projected in the sanitation study. This approach provides a way to pinpoint trouble spots and to verify that actual costs approximate projected costs. This technique will contribute to savings of up to 50% when compared with an unmonitored system.

The HACCP Approach to Cleaning

The Hazard Analysis Critical Control Point (HACCP) approach should be applied when considering the evaluation of a cleaning system. A sanitation survey will permit application of the HACCP concept. This survey should designate areas that require cleaning as highly critical, critical, or subcritical for physical and microbial contamination. These areas can be grouped according to required cleaning frequency as demanding attention:

1. Continuously
2. Every 2 h (during each break period)
3. Every 4 h (during lunch break and at the end of the shift)

4. Every 8 h (end of shift)
5. Daily
6. Weekly

Assignment of different colors to represent specific zones of a processing plant is an economical way to create convenient visual barriers between production functions. This segregation minimizes migration of materials—especially chemical and microbial contaminants from one location to another (Anon 2004).

For verification purposes, the microbial methods should be appropriate for the task. Sampling should be accomplished where the information will most accurately reflect the cleaning effectiveness. Examples are:

Flow sheet sampling is the measurement of microbial load on food samples collected after each step in the preparation sequence. When samples are collected from the first food coming into contact with cleaned equipment, the contribution of microorganisms from each piece of equipment that the food contacts can be measured.

Environmental samples taken from the processing environment are important in the control of pathogens such as species of *Salmonella* and *Listeria*. Examples are air intakes, ceilings, walls, floors, drains, air, water, and equipment.

Sanitation Hazard Analysis Work Point (SHAWP)

SHAWP is an acronym for a technique developed by Carsberg (2003). This approach requires that equipment be broken down into stages for effective cleaning. This equipment must be inspected to identify interior niches that cause microbial infestation since old and new equipment contain hidden areas that harbor microorganisms. The SHAWP is implemented and evaluated while the equipment is disassembled. A maintenance or engineering employee should either train the sanitation workers to break down the equipment or be present to break down and reinstall each piece. Training is important to send a message to the sanitation employees that management is willing to invest time and effort to improve sanitation.

Cleaning Equipment

Cleaning is generally accomplished by manual labor with basic supplies and equipment or by the use of mechanized equipment that applies the cleaning medium (usually water), cleaning compound, and sanitizer. The cleaning crew should be provided with the tools and equipment needed to accomplish the cleanup with minimal effort and time. Storage space should be provided for chemicals, tools, and portable equipment.

Mechanical Abrasives

Although abrasives such as steel wool and copper chore balls can effectively remove soil when manual labor is used, these cleaning aids should not be used on food contact surfaces. Small pieces of these scouring pads may become embedded in the construction material of the equipment and cause pit corrosion (especially on stainless steel) or may be picked up by the food, resulting in consumer complaints and even consumer lawsuits. Wiping cloths should not be used as a substitute for abrasives or for general purposes because they spread molds and bacteria. If cloths are necessary, they should be boiled and sanitized before use.

Water Hoses

Hoses should be long enough to reach all areas to be cleaned but should be no longer than required. For rapid and effective cleanup, it is important to have hoses equipped with nozzles designed to produce a spray that will cover the areas being cleaned. Nozzles with rapid-type connectors should be provided for each hose. Fan-type nozzles give better coverage for large surfaces in a minimum amount of time. Debris lodged in deep cracks or crevices is dislodged most effectively through small, straight jets. Bent-type nozzles are beneficial for cleaning around and under equipment. For a combination of washing and brushing, a spray-head brush is needed. Cleanup hoses, unless connected to steam lines, should have an automatic shutoff valve on the operator's end to conserve water, reduce splashing, and facilitate exchange of nozzles. Hoses should be removed from food production areas after cleanup, and it is necessary to clean, sanitize, and store hoses on hooks that are off of the floor. This precaution is especially important in the control of *Listeria monocytogenes*.

Brushes

Brushes used for manual or mechanical cleaning should fit the contour of the surface being cleaned. Those equipped with spray heads between the bristles are satisfactory for cleaning screens and other surfaces in small operations where a combination of water spray and brushing is necessary. Bristles should be as harsh as possible without creating surface damage. Rotary hydraulic and power-driven brushes for cleaning pipes aid in cleaning lines that transport liquids and heat exchanger tubes.

Brushes are manufactured from a variety of materials including horse hair, hog bristles, vegetable fiber, and nylon—but are usually nylon. Bassine, a coarse-textured fiber, is suitable for heavy-duty scrubbing. Palmetto fiber brushes are less coarse and are effective for scrubbing with medium soil, such as metal equipment and walls. Tampico brushes are fine fibered and well adapted for cleaning light soil that requires only gentle brushing pressure. All nylon brushes have strong and flexible fibers that are uniform in diameter, are durable, and do not absorb water. Most power-driven brushes are equipped with nylon bristles. Other synthetic brushes that are available include polyester, peek, polyethylene, polypropylene, and polyvinyl chloride. Brushes made of absorbent materials should not be used.

Scrapers, Sponges, and Squeegees

Sometimes scrapers are needed to remove tenacious deposits, especially in small operations. Sponges and squeegees are most effectively used for cleaning product storage tanks when the operation has insufficient volume to justify mechanized cleaning.

High-Pressure Water Pumps

High-pressure water pumps may be portable or stationary, depending on the volume and needs of the individual plant. Portable units are usually smaller than centralized installations. The capacity of portable units is from 40–75 L/min, (42–80 quarts/min), with operating pressures of up to 41.5 kg/cm^2 (90 lbs/0.16 in^2). Portable units may include solution tanks for mixing of cleaning compounds and sanitizers. Stationary units have capacities ranging from 55–475 L/min (60–500 quarts/min). Piston-type pumps deliver up to 300 L/min (310 quarts/min), and multistage turbines have capacities of up to 475 L/min (500 quarts/min), with operating pressures of up to 61.5 kg/cm^2 (140 lbs/0.16 in^2). The capacity and pressure of these units vary from one manufacturer to another.

In a centralized unit, the high-pressure water is piped throughout the plant, and outlets are placed for convenient access to areas to be cleaned. The pipes, fittings, and hoses must be capable of withstanding the water pressure, and all of the equipment should be made of corrosion-resistant materials. The choice of a stationary or portable unit depends on the desired volume of high-pressure water and the ease with which a portable unit can be moved close to areas being cleaned. Other uses of high-pressure water in the plant can also determine whether a stationary unit is warranted.

High-pressure, high-volume (HPHV) water pumps have been used primarily when supplementary hot, high-pressure water is desired. Because this equipment uses a large volume of water and cleaning compounds, it is frequently considered inefficient. This concept has been applied to portable and centralized high-pressure, low-volume (HPLV) equipment that blends cleaning compounds for dispensing in areas to be cleaned. With a lower volume and water temperature, it is a more efficient approach that can effectively clean areas that are difficult to reach and penetrate. Food and Drug Administration (FDA)-approved high-pressure water pumps, meaning pumps made from FDA-approved materials, should be used for all high-pressure cleaning applications in food plants.

Low-Pressure, High-Temperature Spray Units

This equipment may be portable or stationary. The portable units generally consist of a light-weight hose, adjustable nozzles, steam-heated detergent tank, and pump. Operating pressures are generally less than 35 kg/cm^2 (75 lbs/0.16 in^2). Stationary units may operate at the main hot water supply pressure or may use a pump. These units are used because no free steam or environment fogging is present, splashing during the cleaning operation is minimal, soaking operations are impractical and hand brushing is difficult and time-consuming, and the detergent stream is easily directed onto the soiled surface.

High-Pressure Hot Water Units

This equipment utilizes steam at 3.5–8.5 kg/cm^2(7.7-18.7 lbs/0.16 in^2) and unheated water at any pressure above 1 kg/cm^2 (2.2 lbs/0.16 in^2). These units convert the high-velocity energy of steam into pressure in the delivery line. The cleaning compound is simultaneously drawn from the tank and mixed in desired proportions with hot water. Pressure at the nozzle is a function of the steam pressure in the line; for example, at 40 kg (88 lbs/0.16 in^2) of steam pressure, the jet pressure is approximately 14 kg/cm^2 (30.8 lbs/0.16 in^2). This equipment is easy to operate and maintain but has the same inefficiency as the HPHV water pumps.

Steam Guns

Various brands of steam guns are available that mix steam with water and/or cleaning compounds by aspiration. The most satisfactory units are those that use sufficient water and are properly adjusted to prevent a steam fog around the nozzle. Although this equipment has applications, it is a high-energy-consuming method of cleaning. It also reduces safety through fog formation and increases moisture condensation, sometimes resulting in mold growth on walls and ceilings and increased potential for the growth of

L. monocytogenes. HPLV equipment is generally as effective as steam guns if appropriate cleaning compounds are incorporated.

High-Pressure Steam

High-pressure steam may be used to remove certain debris and to blow water off processing equipment after it has been cleaned. Generally, this is not an effective method of cleaning because of fogging and condensation, and it does not sanitize the cleaned area. Nozzles for high-pressure steam and other HPHV equipment should be quickly interchangeable and have a maximum capacity below that of the pump. An orifice of approximately 3.5 mm (0.14 in) is considered satisfactory for an operating pressure of approximately 28 kg/cm^2 (61.6 lbs/0.16 in^2).

Hot Water Wash

This technique should be considered a method instead of a kind of equipment or a cleaning system. Because only a hose, nozzle, and hot water are required, this method of cleaning is frequently used. Sugars, certain other carbohydrates, and monovalent compounds are relatively soluble in water and can be cleaned more effectively with water than can fats and proteins. Investment and maintenance costs are low, but the hot water wash is not considered a satisfactory cleaning method. Although hot water can loosen and melt fat deposits, proteins are denatured which causes the protein to cook and adhere to the surface; removal from the surface to be cleaned is complicated because these coagulated deposits are more tightly bound to the surface. Penetration of areas of poor accessibility is difficult without high pressure, and labor is increased if a cleaning compound is not applied. As with the other equipment that uses hot water, this method increases both energy costs and condensation.

Portable High-Pressure, Low-Volume Cleaning Equipment

A portable HPLV unit contains an air- or motor-driven high-pressure pump, a storage container for the cleaning compound, and a high-pressure delivery line and nozzle (Fig. 11.1). The self-contained pump provides the required pressure to the delivery line, and the nozzle regulates pressure and volume. This portable unit simultaneously measures the predetermined amount of cleaning compound from the storage container and mixes it in the desired proportion of water as the pump delivers the desired pressure. The ideal HPLV unit delivers the cleaning solution at approximately 55 °C (132°) with 20–85 kg/cm^2 (44–187 lbs/0.16 in^2) pressure and 8–12 L/min (9–13 quarts/min), depending on equipment specifications and nozzle design. However, low-pressure, medium-pressure (boosted pressure), and high-pressure equipment exists. Although high pressure is

Fig. 11.1 A portable high-pressure, low-volume unit that is used where a centralized system does not exist. This unit is equipped with racks for hoses, foamer, and cleaning compound storage and provides two rinse stations and a sanitizer unit. Two workers can simultaneously prerinse, clean, postrinse, and sanitize. This equipment can also apply foam if the spray wand is replaced with a foam wand accessory (Courtesy of Ecolab, Inc., St. Paul, Minnesota)

effective in removing heavy soils, it can create too much atomization. Therefore, the food industry has evolved primarily to medium (boosted) pressure.

The high-pressure cleaning principle is based on automation of the cleaning compound through a high-pressure spray nozzle. The high-pressure spray provides the cleaning medium for application of the cleaning compound. The velocity, or force, of the cleaning solution against the surface is the major factor that contributes to cleaning effectiveness. High-pressure, low-volume equipment is necessary to reduce water and cleaning compound consumption. This equipment conserves water and cleaning compounds, and it is less hazardous than HPHV equipment because the low volume results in reduced force as distance from the nozzle increases.

Portable HPLV equipment is relatively inexpensive and quickly connected to existing utilities. These units do require more labor than does centralized equipment because transportation throughout the cleaning operation is necessary and because less automation can be provided without a centralized system. Portable equipment is not as durable and can require an excessive amount of maintenance. High-temperature sprays tend to bake the soil to the surface being cleaned, providing the optimum temperature for microbial growth.

This hydraulic cleaning equipment is beneficial for small plants because the portable units can be moved through the facility. Portable equipment can be utilized for cleaning parts of equipment and building surfaces and is especially effective for conveyors and processing equipment where soaking operations are impractical and hand brushing is difficult and time-consuming. It appears that this method of cleaning may receive more attention in the future because it may be more effective in the removal of *L. monocytogenes* from areas that are difficult to clean with less labor-intensive equipment such as foam-dispensing units. A trend exists toward centrally installed equipment because of the potential labor savings and reduced maintenance. Since 2010, this application method has become less widely utilized due to aerosol concerns and safety hazards.

Centralized High-Pressure, Low-Volume Systems

This system, which uses the same principles as the portable HPLV equipment, is another example of mechanical energy that is harnessed and used as chemical energy. Centralized systems utilize piston-type or multistage turbine pumps to generate the desired pressure and volume. Like the portable equipment, the cleaning action of high-pressure spray units is primarily due to the impact energy of water on the soil and surface. The pump(s), hoses, valves, and nozzle parts of the ideal centralized high-pressure cleaning system should be resistant to attack by acid or alkaline cleaning products. Automatic, slow-acting shutoff valves should be provided to prevent hose jumping, indiscriminate spraying, and wasting of water. The centralized system is more flexible, efficient, safe, and convenient because there is no live steam to block vision or injure personnel.

If improperly used, this cleaning system can be counterproductive by blasting loose dirt in all directions. Therefore, a low-pressure rinse down should precede high-pressure cleaning. Most suppliers of these systems provide customers with technical assistance to match cleaning products with cleaning equipment to obtain maximum value and effectiveness.

The penetrating and cleaning action of a centralized high-pressure, (boosted pressure range) system is similar to that of a commercial dishwashing machine. The system automatically injects a cleaning compound into a water line so that the hydraulic scouring action of the spray cleans exposed surfaces and reaches into inaccessible or difficult-to-reach areas. Cracks and crevices where soil has accumulated can be flushed out to reduce bacterial contamination. Cutting and scouring action is applied to all surfaces by the jet, and chemical cleaning action is improved through the water spray, which is automatically charged with a detergent or detergent-disinfectant solution. An example of equipment components of a high-pressure cleaning system is given in Fig. 11.2, which is called a Boosted Pressure Central Foam System.

Fig. 11.2 A centralized medium-pressure (boosted pressure) system for a large cleaning operation (Courtesy of Ecolab, Inc., St. Paul, Minnesota)

The flexibility and major benefits of the centralized high-pressure cleaning system are realized if there are quick-connection outlets available in all areas requiring cleaning. Several detergents—acid, alkaline, or neutral cleaners and sanitizers—can be dispersed through the system, and mechanized spray heads can be mounted on belt conveyors with automatic washing, rinsing, and cutoff.

Centralized systems are far more expensive than portable units because they are generally custom-built. The cost varies according to facility size and system flexibility.

Factors Determining Selection of Centralized High-Pressure Equipment

Generally, two types of central equipment are most common: medium pressure (22 lbs/0.16 in^2–44 lbs/0.16 in^2) (10 kg/cm^2 boost to 20 kg/cm^2) and high pressure (40 kg/cm^2 boost to 55 kg/cm^2) (88 lbs/0.16 in^2-121 lbs/0.16 in^2). Medium-pressure systems are normally used in processing plants where heavy soils dominate. High-pressure equipment is found mostly in beverage and snack food plants where soils are light and cutting action is needed to clean processing equipment. However, several factors must be considered to determine which equipment will provide the best long-term results for each specific plant.

There is normally an inverse relationship between the rinse nozzle flow rates and pressure. Each cleaning task requires a specific impingement force to dislodge the soil and flush it off the equipment. At high pressures of 40–55 kg/cm^2 (88–121 lbs/0.16 in^2), the nozzle flow rates can average approximately 5 L/min (5.5 quarts/min); however, lower pressure requires high flow rates at the nozzles to achieve the same impingement. For example, if a plant has a medium-of pressure of 20 kg/cm^2 (44 lbs/in^2) system with 30–40 L/min (33–44 quarts/min) rinse nozzles and management wants to conserve water usage, the method to accomplish this is to increase the pressure to 40 or 50 kg/cm^2 (88 or 110 lbs/0.16 in^2) and reduce the nozzle flow rates to 10 or 15 L/min (11 or 16 quarts/min). The result is the same impingement force with a 50% reduction in rinse water usage.

Water conservation, in addition to being responsible management, carries additional benefits. Reduced nozzle flow means less sewage and less energy used to heat the water. Paybacks of less than 6 months are not unusual and often run as low as 3 months.

During the past, a trend has existed to use lower pressure in plants with a heavy soil because high pressure tends to dislodge particles with such force as to move them to another undesired location (splatter). Heavier soils require heavier impingement. Most processors with less heavy soil use medium pressure.

For long-term water conservation purposes, 40–50 kg/cm^2 (88–110 lbs/in^2) with an average rinse hose nozzle flow rate of 10–20 L/min (11–22 quarts/min) is suggested. The usual exception is if the plant has an unusually short time period with which to prepare for production. If only 4 or 5 h are available for cleanup, higher flow rates will be required. This condition is usually temporary but must be planned for, i.e., the central system must have the flow capacity.

The price of central equipment is usually the main determinant in the purchasing decision.

High-pressure equipment requires the largest investment to buy and maintain. The pumps are more expensive than medium-pressure pumps, and all of the piping, valves, and other components cost more because of the high-pressure ratings required. Usually, the benefits of low water usage outweigh the initial cost and operating expenses over the long term. Medium-pressure equipment requires less investment to purchase and operate. The pumps are mechanically less sophisticated, and none of the piping, valves, or related components requires a high-pressure rating; therefore, maintenance is usually lower than on high-pressure systems. If water usage is not a critical factor in overall plant operations, a medium-pressure system should be considered. Water conservation cannot be accomplished with medium pressure. Many processors utilize 20 kg/cm^2 (44 lbs/0.16 in^2) with 20–30 L/min (22–33 quarts/min) nozzles in most areas of the plant. Proper utilization of the equipment and training in sanitation procedures are the key elements.

Portable Foam Cleaning

Because of the ease and speed of foam application, this cleaning technique has been popular since the 1980s. With this method, foam is the medium for application of the cleaning com-

pound. The cleaning compound is mixed with water and air to form the foam. Clinging foam is readily visible and allows the worker to see where the cleaning compound has been applied, thus reducing the chance of job duplication.

Foam cleaning is beneficial for cleaning large surface areas because of its ability to cling, increasing the contact time of the cleaning compound. This technique can clean the interior and exterior of transportation equipment, ceilings, walls, piping, belts, and storage containers. It is similar in size and cost to portable high-pressure units. Portable cleaning units that may be used to apply cleaning compounds by foam are illustrated in Fig. 11.3a, b. This equipment requires a foam-charging operation to blend the cleaning compound, water, and air prior to use. This system is able to conserve water use when compared to high- and medium-pressure cleaning system.

Centralized Foam Cleaning

This equipment applies cleaning compounds with the same technique used by portable foam equipment except that drop stations for quick connection to a foam gun are strategically located throughout the plant. Centralized equipment provides desirable features similar to the centralized high-pressure system. As with portable foam

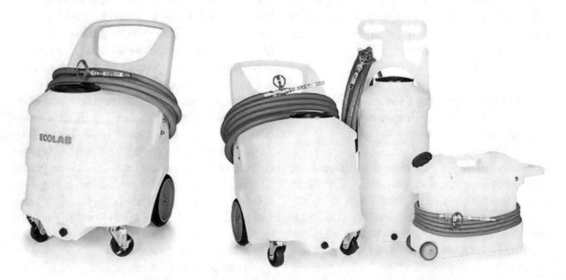

Fig. 11.3 Portable air-operated foam units that apply the cleaning compound as a blanket of foam (Courtesy of Ecolab, Inc., St. Paul, Minnesota)

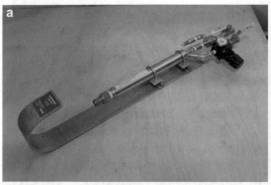

Fig. 11.4 (**a** and **b**) Scheme of a wall-mounted foam and rinse station that can provide foam application at convenient locations in a food plant through automatic metering and mixing of the cleaning compound (Courtesy of Ecolab, Inc., St. Paul, Minnesota)

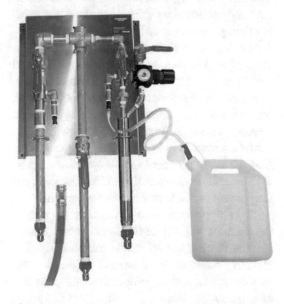

Fig. 11.5 A triple drop wall sanitize (S)/rinse (R)/foam (F) station that can be used to provide a quick connection for foam detergent, rinse, and sanitizer applications (Courtesy of Ecolab, Inc., St. Paul, Minnesota)

can be located in convenient areas where cleaning is concentrated. The equipment shown in Fig. 11.4a, b can blend and dispense cleaning compounds through an adjustable air regulator and water-metering valve. The easily accessible chemical-metering pump and other controls are in the latching stainless steel cabinet. This equipment contains a built-in vacuum breaker and check valves in the air and water lines.

Figure 11.5 features a triple drop wall sanitize (S)/rinse (R)/foam (F) station that can be used to dispense foam, high-pressure water for rinsing, and a sanitizer. The S, R, and F represent sanitize, rinse, and foam and have quick disconnect hose attachments. The foam station provides adjustable air and detergent regulators to create proper foam consistency. The rinse unit can provide up to 69 kg/cm^2 (150 lbs/0.16 in^2) of pressure.

Portable Gel Cleaning

cleaning, the cleaning compound is automatically mixed with water and air to form foam. This equipment does not require the foam-charging operation that portable foam units require. Equipment components of a centralized foam cleaning system are illustrated in Figs. 11.4 and 11.5.

The compact wall-mounted foam generation unit shown in Fig. 11.4a, b is designed to blend and dispense cleaning compounds from reservoirs or original shipping containers. Wall-mounted units

This system is similar to portable high-pressure units except that the cleaning compound is applied as a gel (due to air restriction) rather than as foam slurry or high-pressure spray. This medium is especially effective for cleaning food packaging equipment because the gel clings to the moving parts for subsequent soil removal. Equipment costs and arrangement are comparable to those for portable foam and high-pressure units.

Centralized or Portable Slurry Cleaning

This method is identical to foam cleaning except that less air is mixed with the cleaning compounds. A slurry is formed that is more fluid than foam and penetrates uneven surfaces more effectively. The exposure time of the application of slurry cleaning is less than with foam, since the foam has superior clinging ability.

Combination Centralized High-Pressure and Foam Cleaning

This arrangement is the same as centralized high-pressure cleaning, except that foam can also be applied. This method offers more flexibility than most cleaning equipment because foam can be used on large surface areas with high pressure applied to belts, stainless steel conveyors, and hard-to-reach areas. A system with these capabilities is expensive because most must be custom designed and built.

Cleaning-in-Place

As labor rates continue to increase and hygienic standards are raised, cleaning-in-place (CIP) systems become more valuable. CIP is defined as the cleaning of vessels and pipelines with minimal disassembling of equipment components (Cooper 2016). The solution is jetted of sprayed through spray devices on a tank or vessel surface through controlled velocity. Dairies and breweries have incorporated CIP for many years. It has been adapted sparingly in other plants because of equipment and installation costs and the difficulty of cleaning certain processing equipment. Because of these limitations, CIP is considered a solution for specific cleaning applications and is custom designed. CIP equipment is best used for cleaning pipelines, vats, heat exchangers, centrifugal machines, and homogenizers.

CIP systems are those in which the equipment is cleaned and sanitized using an automated and enclosed cleaning system. They are used extensively in the beverage industry, dairy industry, aseptic processing operations, and in operations where fluids are handled and processed. There are some CIP operations that require some manual operation prior to start-up. In some operations, start-up requirements may mandate that the crew manually make proper connections to the unit operations that are to be cleaned using the CIP system.

Custom-designed CIP equipment can vary in the amount of automation, according to cleaning requirements: from simple cam timers to fully automated computer-controlled systems. The choice depends on capital availability, labor costs, and type of soil. It should be designed by a reliable consulting firm and/or a reputable equipment and detergent supplier. These organizations can provide site surveys and confidential reports on the hygienic status of existing equipment and cleaning techniques.

Small-volume plants cannot always justify full automation. With reduced automation, the required circuits can be set manually by means of a flow-selector plate. Pipelines can be brought to a back plate with required connections made by a U-bend inserted in the appropriate parts. Microswitch logic can be interlocked to the CIP set. With full automation, the entire process and CIP operations can be automatically controlled. Electric interlocks negate the possibility of an error in valve operation.

The CIP principle is to combine the benefits of the chemical activity of the cleaning compounds and the mechanical effects of soil removal. The cleaning solution is dispensed to contact the soiled surface, and the proper time, temperature, detergency, and force are applied. For this system to be effective, a relatively high volume of solution has to be applied to soiled surfaces for at least 5 min and up to 1 h. Therefore, recirculation of the cleaning solution is necessary for repeated exposure and to conserve water, energy, and cleaning compounds.

For optimal use of water and reduced effluent discharge, CIP systems are designed to permit the final rinse to be utilized as makeup water for the next cleaning cycle. The dairy industry has attempted to recover a spent cleaning solution for further use by concentration through ultrafiltration or through use of an evaporator. Various installations have incorporated systems that integrate the advantages of single-use systems of known reliability and flexibility with water and solution recovery procedures that aid in reducing the total amount of water required for a specific cleaning operation. These installations combine the spent cleaning solution and past rinses for temporary storage and use as a prerinse for the next cleaning cycle. Thus, the requirements of water, cleaning compounds, and required energy are reduced.

Properly designed CIP systems are capable of cleaning certain equipment in food plants as effectively as dismantling and cleaning by hand. In many food plants, CIP equipment has completely or partially replaced hand cleaning.

The simplified flow chart in Fig. 11.6 illustrates how a CIP system operates. The arrangement illustrates how to provide mixing and detergent tank(s), pipelines, heat exchanger(s), and storage tank(s). These features permit cleaning of storage tanks, vats, and other storage containers by use of spray balls. Pipelines and various plant items can be cleaned by a high-velocity cleaning solution of water and designated cleaning compounds, which are recirculated. A typical cleaning cycle for the CIP system is outlined in Table 11.1.

The plant layout for a CIP system is important because dismantling of equipment is unneces-

sary. The use of crevice-free joints of pipe work and the design of all tanks should be considered so that they can be enclosed with smooth walls that can be cleaned by liquid spray. Spray balls, whether fixed or rotating, should produce a high-velocity jet of liquid in a 360° pattern to cover the interior of the tank and thoroughly remove residual soil or other contamination.

Development of circuits is important. The circuits must be flexible. The location of every pipe should be permanent and based on its possible function during cleaning. Large processing operations may be separated into several major circuits for separate cleaning. Circuit design should be based on soil characteristics. Circuit development can permit a limited cleanup force to proceed through the plant in an orderly sequence as process operations are completed.

Use of a drain selector valve facilitates the direction of flush water, cleaning compounds, and rinse water directly to a sewer instead of discharging onto the floor, with subsequent splashing and chemical damage. The selector valves and auxiliary tank in the spray cleaning circuit permit flushing with clear water from the supply tank, discharge to the sewer, recirculation of the

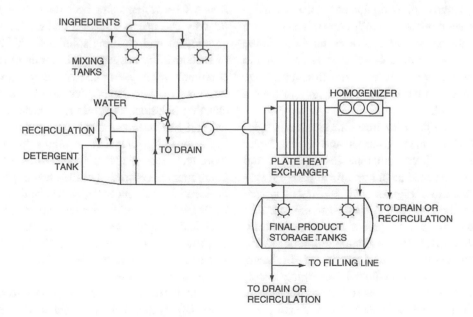

Fig. 11.6 Flow arrangement of a COP (*Source*: From Jowitt (1980))

Table 11.1 Typical cycle for CIP system

Operation	Function
Preliminary rinse (hot or cold water)	Remove gross soil
Detergent wash	Remove residual soil
Rinse	Remove cleaning compounds
Sanitization	Destroy residual microorganisms
Final rinse (optional, according to sanitizer use)	Remove CIP solutions and sanitizers

cleaning solution, and rinsing with clear water metered continuously from the supply tank with subsequent discharge to the sewer.

There are two basic CIP designs: single use and reuse systems. Another approach has been to incorporate combined systems, which provides the best characteristics of the single use and reuse equipment. This type of unit is referred to as a *multiuse system*.

Single-Use Systems

Single-use systems use the cleaning solution only once. They are generally small units, frequently located adjacent to the equipment to be cleaned and sanitized. Because the units are located in the area where cleaning is accomplished, the quantity of chemicals and rinse water can be relatively small. Heavily soiled equipment makes a single-use system more desirable than the others because reuse of the solution is less feasible. Some single-use systems are designed to recover the cleaning solution and rinse water from a previous cycle for use as a prerinse cycle in the subsequent cleaning cycle.

When compared to other CIP systems, single-use units are more compact and have a lower capital cost. These units are less complex and may be purchased as preassembled parts for easier installation. Figure 11.7a illustrates typical single-use equipment. A single-use unit consists of a tank with level probes and pneumatically controlled valves to inject steam, introduce water, and regulate the circuit, inclusive of discharge, overflow, and through-flow. Discharging is normally accomplished at the end of the rinse cycle. A single-use system includes a centrifugal pump and control panel and a program cabinet with

temperature controller, solenoid valves, and pressure and temperature instrumentation.

A typical sequence for cleaning equipment such as storage tanks or other storage containers takes about 20 min, with the following procedures:

1. Three prerinses of 20 s with intervals of 40 s each to remove the gross soil deposits are initially applied. Water is subsequently pumped with a CIP return pump for discharge to the drain.
2. The cleaning medium is mixed with injected steam (if used) to provide the pre-adjusted temperature directly into the circuit. This status is maintained for 10–12 min prior to discharge of the spent chemicals to the drain or recovery tank.
3. Two intermediate rinses with cold water at an interval of 40 s each are followed through transfer to water recovery or to the drain.
4. Another rinse and recirculation are established and may include the injection of acid to lower the pH value to 4.5. Cold circulation is continued for about 3 min, with subsequent drainage.

Reuse Systems

Reuse CIP systems are important to the food industry because they recover and reuse cleaning compounds and cleaning solutions. This benefit of sustainability has led to it becoming the most commonly used type of CIP at present time. It is important to understand that contamination of cleaning solutions is minimal because most of the soil has been removed during the prerinse cycle, enabling cleaning solutions to be used more than once. For this system to be effective, the proper concentration of the cleaning solution is essential. The concentration can be determined by following the guidelines recommended by the chemical supplier and the equipment vendor. Sequencing versatility permits the timing and sequencing of operations (acid/alkaline or alkaline/acid) to be varied.

A tank for each chemical is provided with reuse CIP systems. A hot water tank or bypass loop is necessary to save energy and water if a hot

a b

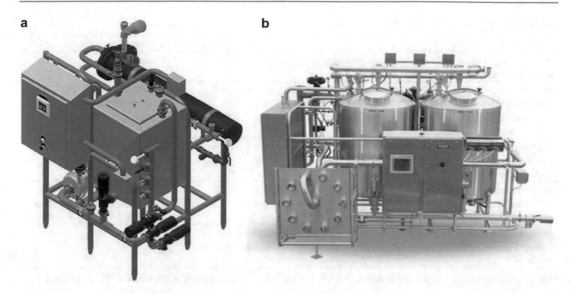

Fig. 11.7 (**a** and **b**) A CIP (**a**) single-use solution unit and (**b**) a reuse unit that are parts of systems containing a water supply tank and CIP circulating unit (Courtesy of Ecolab, Inc., St. Paul, Minnesota)

water rinse is used. The cleaning solution is frequently heated with a coil.

The basic parts of a CIP reuse system are an acid tank, alkaline tank, freshwater tank, return-water tank, heating system, and CIP feed and return pumps (Fig. 11.7b). Remote-controlled valves and measuring devices are provided with the piping layout of this cleaning system. The predetermined cleaning operations have automatic sequencing through a program control unit. With this system, the cleaning solution is transported from the CIP unit through the production plant and the equipment to be cleaned.

Two-tank systems for reuse of the wash water consist of one tank for rinse water and another for reclaiming the cleaning solution. CIP equipment with three tanks includes one tank for the cleaning solution, one for reclaiming the prerinse solution, and one for a freshwater final rinse. Both single-use and reuse systems require careful design and monitoring to avoid the danger of unwanted mixing of food products with cleaning solutions (Giese 1991).

Two tanks for alkaline cleaning compounds are frequently provided for solutions of differing concentrations. The less concentrated solution can be used for cleaning tanks, other storage facilities, and pipelines. The stronger solution is available for cleaning the plate heat exchanger. Pumps that feed the cleaning compounds into the tanks are used to automatically adjust the use of neutralization tanks with automatically adjusted acid concentrations.

Two CIP circuits can be cleaned simultaneously by the addition of extra CIP feed pumps. The tank capacity is determined by circuit volumes, temperature requirements, and desired cleaning programs. In mechanized plants, a central control console uses remote-controlled valves to switch the cleaning circuits on and off. Through use of a return-water tank, water consumption of a reuse system can be optimized. Recirculation of the cleaning solution is usually necessary for best results; thus, reuse equipment has a higher initial cost but permits operational expense savings.

The ideal CIP reuse system has the ability to fill, empty, recirculate, heat, and dispense contents automatically. A typical operation of this system with a program for storage tank and pipeline cleaning with recovery of the cleaning solution is described in Table 11.2.

Table 11.2 Operation of an ideal CIP reuse system

Operation[a]	Time (min)	Temperature
Prerinse: application of cold water from the recovery tank with subsequent draining	5	Ambient
Detergent wash: a 1% alkaline-based cleaning compound purges the remaining rinse water to the drain with subsequent diversion by a conductivity probe to the cleaning compound tank for circulation and recovery	10	Ambient to 85 °C (185 °F), depending on equipment to be cleaned and type of soil
Intermediate water rinse: softened cold water from the rinse forces out the remaining cleaning solution to the cleaning solution tank; water is then diverted to the water recovery tank	3	Ambient
Acid wash: an acid solution 0.5–1.0% forces out residual water to the drain; then this solution is diverted through a conductivity probe to the acid tank for recirculation and recovery	10	Ambient to 85 °C (185 °F), depending on equipment to be cleaned and type of soil
Final water rinse: cold water purges out the residual acid solution with collection in the water recovery tank; overflow is diverted to the drain	3	Ambient

[a]Pasteurizing equipment tanks and pipelines may also be subjected to a final flush of hot water at 85 °C (185 °F).

Fig. 11.8 In this single-use CIP system with limited recovery, an additional tank with a high-level probe is mounted so that the wash and rinse water can be collected for the next prewash cycle (*Source*: From Jowitt (1980))

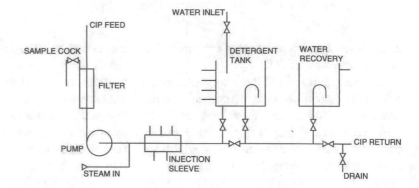

Multiuse Systems

These units, which combine the features of both single-use and reuse systems, are designed for cleaning pipelines, tanks, and other storage equipment that can be cleaned effectively by the CIP principle. These systems function through automatically controlled programs that entail various combinations of cleaning sequences involving circulation of water, alkaline cleaners, acid cleaners, and acidified rinses through the cleaning circuits for differing time periods at varying temperatures. The benefits of multiuse systems include the incorporation of the best operations from single-use and reuse CIP systems. Multiuse systems have the ability to cir-

culate sanitizer, hot water sanitize, and optimize water, energy, and chemical use (Fig. 11.8).

An example of a multiuse CIP system is presented in Fig. 11.9a, b. This versatile unit accommodates differing CIP sequencing, chemical strength, and thermal performance. The multiuse CIP system may contain tanks for chemical and water recovery, with an associated single pump, recirculating pipe work, and heat exchanger. The plate heat exchanger heats the incoming water and cleaning liquid to the required temperature. Flexibility of temperature control, optimal utilization of tank capacity, and flexibility in heating of water or cleaning solutions can be realized through the use of a heat exchanger.

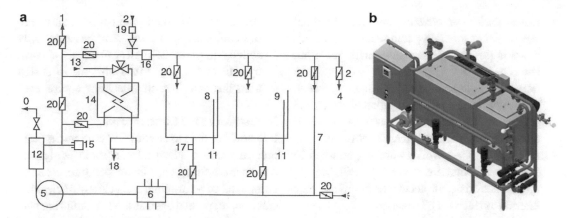

Fig. 11.9 (**a**) Typical multiuse CIP system, simplified. *1* CIP feed, *2* CIP return, *3* water inlet, *4* drain, *5* puma pump, *6* injection sleeve, *7* recirculating loop, *8* detergent tank, *9* water recovery, *10* sample cock, *11* overflow, *12* filter, *13* steam in, *14* "paraflow" heat exchanger, *15* temperature probe, *16* soluvisor, *17* conductivity probe, *18* condensate, *19* no-flow probe, *20* butterfly valve. (**b**) Typical diagram of a commercial multiuse CIP system

An automatic multiuse CIP system follows the following operation sequence:

1. *Prerinse.* This step occurs from water recovery or the water supply provided at the desired temperature. The solution from this operation can be directed to the drain or diverted by a re-circulatory loop for a timed period, then transferred to the drain.
2. *Cleaning solution recirculation.* The recirculation step occurs by the cleaning compound vessel or the bypass loop. A desired combination of cleaning chemicals can be used for variable recirculation times, and the chemical injection can boost the strength or use of the solution. The plate heat exchanger or its bypass loop can contribute to the cleaning solution recirculation. With a bypass loop, variable-temperature programming permits total detergent-tank heating. Cleaning solutions can be recovered or drained.
3. *Intermediate rinse.* This operation is similar to the prerinse, except that it is important to remove residual cleaning chemicals from the previous operation.
4. *Acid recirculation.* This optional operation, which is similar to the cleaning recirculation operation, may occur with or without an acid tank. With an acid tank, the re-circulatory loop is established on water, either through the plate heat exchanger or via the plate heat

exchanger bypass loop. The acid is injected to a preset strength based on timing for a specific circuit volume.

5. *Sanitizer recirculation.* This operation, designed to reduce microbial contamination, is similar to the acid injection operation except that heating is not normally required.
6. *Hot water sterilization.* Variable times and temperatures are available for this operation, which involves use of a recirculation loop on freshwater via the plate heat exchanger. The spent water can be either returned to water recovery or drained.
7. *Final water rinse.* Water is pumped via the CIP route and sent to water recovery. Water rinse times and temperatures are variable.

The desirable features of CIP equipment are:

- *Reduced labor.* Manual cleaning is reduced because the CIP system automatically cleans equipment and storage utensils. This feature becomes increasingly important as wages increase, and it becomes more difficult to locate dependable workers.
- *Improved sanitation.* A automated operation cleans and sanitizes more effectively and consistently. Through timed or computer-controlled equipment, cleaning and sanitizing operations are more precisely controlled.

- *Conservation of cleaning solution.* Optimal use of water, cleaning compounds, and sanitizers is possible through automatic metering and reuse.
- *Improved equipment and storage utilization.* With automated cleaning, equipment, tanks, and pipelines can be cleaned as soon as use is discontinued so that immediate reuse is possible.
- *Improved safety.* Workers are not required to enter vessels that are cleaned with CIP equipment. The risk of accidents from slippery internal surfaces is eliminated.

The disadvantages of CIP systems are:

- *Cost.* Because most CIP systems are custom designed, design and installation costs add to the high price of the equipment.
- *Maintenance.* More sophisticated equipment and systems tend to require more maintenance.
- *Inflexibility.* These cleaning systems can effectively clean in only those areas where

equipment is installed, whereas portable cleaning equipment can cover more area. Heavily soiled equipment is not as effectively cleaned by CIP systems, and it is difficult to design units that can clean all processing equipment.

Microprocessor Control Unit

It is now possible to control CIP equipment more precisely. More advanced electronic equipment documents a CIP cycle with temperature, cleaning compound concentration, and velocity of cleaning solution, all plotted against time. Monitoring systems, interfaced with CIP controllers, permit tracking of cleaning parameters to provide for troubleshooting and process control. This makes it possible to improve product protection, reduce labor and costs, and improve efficiency.

Sophisticated equipment that documents a CIP cycle is illustrated in Fig. 11.10a, b. Programming flexibility enables this unit to be used for the operation of a wide variety of CIP systems. The principle involved with this

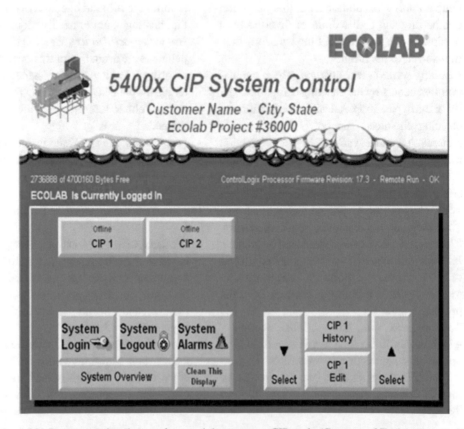

Fig. 11.10 (**a** and **b**) Instrumentation that regulates and documents a CIP cycle (Courtesy of Ecolab, Inc., St. Paul, Minnesota)

innovation is that a graphic recording of the CIP unit is provided to monitor operating parameters, including supply and return temperature, pressure, flow, pH, and conductivity. This equipment can spread the data out in detailed chronological graphic form for temperature, return velocity, and solution concentration. Items such as valve pulsings, pump cavitation, and program stepping can be charted and documented, which are not clearly visible in normally chronological CIP recording charts. Therefore, this feature is a valuable tool for CIP monitoring, documentation, and maintenance.

Computer-based CIP monitoring equipment contains a control panel equipped with a display for the operator. However, primary monitoring is performed through printed records of CIP performance. The printer can produce a series of strip charts. On each chart, a graph of time is plotted against a corresponding CIP parameter. The unit is programmed to account for variation of cleaning parameters for different cleaning circuits. The same printer can record CIP cycles for various equipment, storage vessels, tank trucks, or transfer lines. Some equipment contains alarms that warn of performance outside the limits of programmed set points.

Figure 11.11 illustrates a microprocessor and distribution system that includes a pump-and-fill station for dispensing cleaning compounds into rugged, capped allocation containers. The service station provides air, water, electrical, and thermostat control connections. Designated employees can access the system using a swipe card, thus simplifying production selection by application. Management personnel can access the unit remotely via modem to track chemical usage by application. This system can lower chemical costs by 15–20% and reduce the cleaning cycle time by 10% as a result of more efficient chemical allocation.

The microprocessor control unit enhances cleaning effectiveness and reduces cleaning costs through precise control of the variables associated with mechanized cleaning. One of these units can be designed with the capacity for as many as 200 separate and variable programs that can provide product recovery, rinse and/or cleaning compound recovery, manual rinsing, sanitizing cycle, concentration of chemical strength, extended wash duration, and many other options. The microprocessor control unit can be designed with self-contained, online programming while running via an integral keypad or an offline

Fig. 11.11 Use of a microprocessor for programmed distribution of sanitation compounds (Courtesy of Ecolab, Inc., St. Paul, Minnesota)

programming package available for use on personal computers. Since approximately 2012, it has been possible to be control CIP systems remotely through tablets and smartphones.

Cleaning-Out-of-Place

Systems designed for cleaning-out-of-place (COP) require cleaning by disassembly and/or removal from the normal location. The parts are then placed into COP tanks and cleaned using water movement, which removes soil from the components. Fluid flow is utilized in the application of force for cleaning. Regulatory agencies have previously used velocity as a means of measuring the fluid flow force, employing the rule of thumb of 1.5 M/sec (5 feet/sec). This guideline should no longer be emphasized because COP equipment can effectively clean with less velocity. Velocity and turbulence, the actual cleaning force, are not equally related under all conditions of flow.

Many small parts of equipment and utensils, as well as small containers, can be washed effectively in a recirculating parts washer, also called a *COP unit*. These units, like sanitary pipe washers, contain a recirculating pump and distribution headers that agitate the cleaning solution. Also, a COP unit can serve as the recirculating unit for CIP operation. The normal wash recirculation time is approximately 30–40 min, with an additional 5–10 min for a cold acid or sanitizing rinse.

A COP unit is frequently constructed with a double-compartment stainless steel sink equipped with motor-driven brushes. The same motor also pumps a cleaning solution through a preformatted pipe onto the brushes. Desired temperature of the cleaning solution (45–55 °C) (113–130 °F)is maintained through a thermostatically controlled heater. The first compartment is allocated to use of the cleaning solution. The cleaned parts or utensils are rinsed with a spray nozzle in the second compartment. Drying is normally accomplished by air within the COP unit or on a suitable drain board or rack.

Equipment that functions as a COP unit contains a brush assembly and a rinse assembly. A tank is included for the cleaning solution. Many COP units contain rotary brushes for cleaning both the inside and outside of parts and utensils, with the cleaning solution being introduced through the brushes that clean the inside.

The major appeal of COP equipment is that it can effectively clean parts that are disassembled as well as small equipment and utensils. This equipment can also reduce labor requirements and improve hygiene. COP units are considered reasonable in cost to buy and maintain. Their major limitations for small-volume operations are initial cost, maintenance, and labor requirements for loading and unloading these washers.

The COP concept is frequently used to clean equipment and utensils for the food preparation and foodservice industry. Stainless steel bins may be cleaned and sanitized in an enclosed stainless steel cabinet washer through the use of a computer-controlled cycle. A programmable logic controller governs the timing of each sequential step in the cleaning operation. Further discussion of COP equipment in the dairy and foodservice industries is given in Chaps. 16 and 21.

Sanitizing Equipment

Equipment for the application of sanitizing compounds can vary from hand sprayers, such as units used to apply insecticides and herbicides, to wall-mounted units and headers mounted on processing equipment. Many mechanized cleaning units may contain sanitizing features as part of the system.

Centralized HPLV cleaning and foam cleaning equipment include sanitizing lines with stations for application of the sanitizer by hose and wand or by spray headers on processing equipment, especially moving belts or conveyors. A benefit of the latter feature is that sanitizing is mechanized and can be uniformly administered through the use of timer switches. Metering of sanitizing compounds provides a more accurate and precise application of the sanitizer. The high-pressure rinse water passes through flow control with the orifice size necessary to achieve a prespecified flow rate. To sanitize, the high-pressure water passes through the sanitizer injector, which meters a specific amount of sanitizer into the water

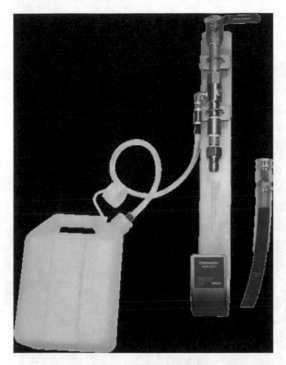

Fig. 11.12 Flood sanitizing unit (Courtesy of Ecolab, Inc., St. Paul, Minnesota)

stream. A flood sanitizing nozzle may be incorporated to spread the solution effectively without automation (Fig. 11.12).

An onboard tire sanitizer is available to reduce contamination of tire treads and wheels of delivery and service vehicles and other related equipment that may need sanitizing before facility entry. This equipment combines spray nozzles positioned above vehicle wheels with a storage tank and pump to apply disinfectants to tires while the vehicle is in motion.

Sanitation Application Methods

The application methods that are available provide acceptable methods for transport of the sanitizer to the desired area. The optimal method depends upon individual operations.

Chemical sanitizers are normally applied by one of the following methods.

- *Spray sanitizing*. This method involves use a sanitizer dissolved in water and a spray device

to transport the sanitizer to the area to be sanitized.

- *Fogging*. Fogging involves application of the sanitizer as a fine mist to sanitize the air and surfaces in a room.
- *Flood sanitizing*. This method involves the application of a sanitizer dissolved in water and applied in a large quantity to ensure extensive exposure. The use of flood sanitizing has been increased to combat the proliferation of *L. monocytogenes*. The disadvantages of this method are the cost of the sanitizer and water and the wet condition created.
- *Immersion/COP sanitizing*. This technique involves the submersion of equipment, utensils, and parts in a tank that contains a sanitizing solution.
- *CIP sanitizing*. CIP sanitizing involves sanitizing by circulation of the sanitizer inside pipes, lines, and equipment.
- *Sanitizing belt treatment*. Acid liquid belt treatment for meat and poultry, fruits and vegetables, and cheese processing conveyors can be incorporated to apply a sanitizer such as a peroxyacid solution for continuous or intermittent belt treatment during production.
- *Doorway sanitation options*. For regular or intermittent high traffic areas and doorways, infrared sensors detect motion and automatically dispense a sanitizing spray or foam to worker boots and the wheels of plant vehicles and equipment. For low traffic door and hallways where foam is not desired, a sanitizer spray can be set for 10 s of sanitizing every 10–15 min.

Lubrication Equipment

Figure 11.13a–d illustrates typical equipment for effective maintenance of high-speed bottling and canning conveyors used in the beverage industry and on shackle chains and conveyors, smokehouse chain drives, and other applications requiring precise continuous conveyor and/or chain lubrication. This principle involves water under pressure during a reciprocating piston in the chemical pump. This piston subsequently drives a chemical-concentrate-metering piston that draws

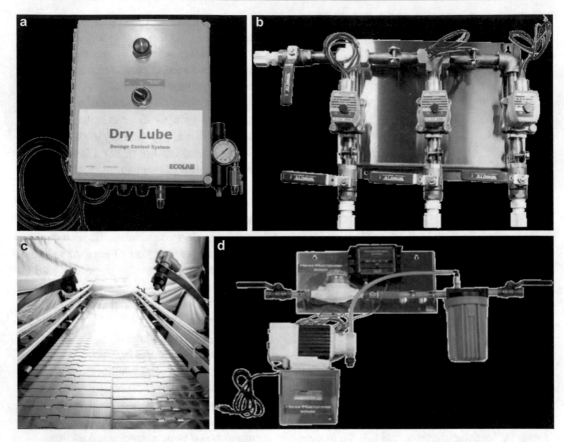

Fig. 11.13 (**a**, **b**, **c**, and **d**) Lubrication equipment for high-speed conveyors, drives, and shackle chains (Courtesy of Ecolab, Inc., St. Paul, Minnesota)

the lubricant from the drum and injects it into the water from the cylinder. Dry silicone-based lubricants have become available for use over the last few years that can be incorporated in some equipment, especially in high-speed bottling that reduces water use and water treatment.

Note: The authors wish to thank Chase Cooper with Ecolab, Inc. for his input on recent innovations in sanitation equipment.

Study Questions

1. What is CIP equipment and how does it function?
2. Why should a microprocessor control unit be incorporated in CIP equipment?
3. How does high-pressure, low-volume cleaning equipment function?
4. What are the advantages and disadvantages of high-pressure, low-volume cleaning equipment?
5. What is the difference between centralized and portable cleaning equipment?
6. Why is foam cleaning a popular and accepted method of cleaning?
7. What is the difference between the medium used for slurry and gel cleaning?
8. Which type of nozzle provides the most effective coverage for large surfaces in a minimum amount of time?
9. What is COP equipment and how does it function?
10. What is a CIP reuse system?
11. What is a CIP multiuse system

12. What are the advantages and disadvantages of CIP equipment?
13. What is the typical cycle for a CIP system?
14. What are some ways to make a sanitation program more sustainable and save costs?
15. What is a sustainable chemical?
16. What is the benefit of using a dry lubricant?

References

Anon (2003). Getting the most from your sanitation equipment. *Food Saf Mag* 9(1): 34.
Anon (2004). Keeping clean with color. *Meat Process* 43(4): 52.
Carsberg HC (2003). Why document SSOPs. *Food Qual* 10(6): 48.
Cooper CT (2016). Personal communication. Ecolab: Minneapolis, MN.
Giese JH (1991). Sanitation: The key to food safety and public health. *Food Technol* 15(12): 74.
Jowitt R (1980). *Hygienic design and operation of food plant*. AVI Publishing Co.: Westport, CT.
Owens E (2014). Defining sustainable sanitation products. Birko. Available at http://www.birkocorp.com/general/defining-sustainable-sanitation-products/. Accessed August 1, 2016.
Redemann R (2005). Basic elements of effective food plant cleaning and sanitizing. *Food Saf Mag* 11(2) Available at https://www.foodsafetymagazine.com/magazine-archive1/aprilmay-2005/. Accessed August 1, 2016.

Waste Product Handling

12

Abstract

The optimal waste treatment system is most effectively identified through a survey to ascertain waste volume and characteristics and water consumption records. Wastewater pollution is measured through biochemical oxygen demand (BOD), chemical oxygen demand (COD), dissolved oxygen (DO), total organic carbon (TOC), settleable solids (SS), total suspended solids (TSS), total dissolved solids (TDS), and fats, oil, and grease (FOG). Two important aspects to consider when treating industrial waste loads are (1) the hydraulic load and (2) the pollution strength of the wastewater. More composting is being practiced as a method for solid waste disposal.

Wastewater can be salvaged through recycling and reuse and the recovery of solids. The basic phases of wastewater treatment are pretreatment by flow equalization, screening, and skimming; primary treatment by sedimentation and flotation; secondary treatment by anaerobic lagoons, aerobic lagoons, trickling filters, activated sludge, oxidation ditch processes, land application, and rotating biological contactors (RBCs); and tertiary treatment by physical separation, tertiary lagoons, and chemical oxidations. Disinfection of treated wastewater should follow other treatment phases to reduce the reaction of organic matter with the disinfectant. Odors from wastewater may be treated through the incorporation of a biofilter.

Keywords

Biochemical oxygen demand • Chemical oxygen demand • Dissolved oxygen • Pretreatment • Primary treatment • Secondary treatment • Tertiary treatment • Total organic carbon

Introduction

Waste materials are produced by the food industry as a by-product of food processing and preparation. Many food processors consume large quantities of water. Water serves several functions in food processing including cleaning, conveying, steam generation, heat exchange, and as an ingredient. Thus, the industry should accept the challenge of handling residues and wastes as part of the production process and apply techniques to improve productivity, quality, and efficiency.

Waste materials generated from food processing and foodservice facilities present a challenge because they contain large amounts of carbohydrates, proteins, fats, and mineral salts. For example, the wastes from dairy plants, food freezing and dehydration plants, and processing plants for red meats, poultry, and seafood can produce distinct odors and heavy pollution of water if the discharge is not properly treated. Organic matter from waste materials should be treated through biological stabilization processes before it is discharged into a body of water. Improper waste disposal is a hazard to humans and to aquatic forms of life. This treatment incorporates biological processes for treatment of the effluent to meet Environmental Protection Agency (EPA) discharge limits, which is critical to the treatment plant's operation.

Federal, state, and local regulatory agencies, as well as the public, are demanding improved waste management by the industry. Processors and regulatory agencies are responsible for the disposal of waste materials promptly and completely. Accumulation of wastes can attract insects and rodents, produce odors, and become a public nuisance or an unsightly condition inside or outside the plant. The integration among controlled production processes, with low level of losses, and the treatment system and handling of residues (solids, liquids, and gases) is fundamental to the administration of waste product handling with an acceptable cost.

The major problem with these wastes is that the organic matter provides a food source for microbial growth. With an abundant food supply, microorganisms multiply rapidly, reducing the dissolved oxygen in the water. Water normally contains approximately 8 parts per million (PPM) of dissolved oxygen. A minimum standard for fish life is 5 PPM of dissolved oxygen. If values are below this level, fish can suffocate. Furthermore, if dissolved oxygen is eliminated from water through high organic matter content, a septic condition with foul odors and darkening of water occurs. Septic conditions with sulfur-containing proteins or water with a high natural content of sulfates can produce hydrogen sulfide, which has a foul odor and can blacken buildings.

Waste disposal from food processing and foodservice facilities presents a hazard if proper handling is not exercised, because of the high content of organic matter which is measured as biochemical oxygen demand (BOD). Most facilities that discharge a large quantity of effluent with a high BOD into a municipal treatment system have to pay a surcharge because of the increased wastewater treatment load. Because of this burden on small municipal treatment facilities, many large firms elect to treat effluent discharge partially or completely. The large volume of wastewater produced in food plants contains vast quantities of organic residues. The intermittent production schedule of many plants places greater demands on wastewater treatment systems. During processing, water is essential to help cleanse the product and to serve as a cleaning medium and conveying unwanted materials to the sewage system. This water becomes a challenge during wastewater treatment because it contains suspended and dissolved organic matter.

Strategy for Waste Disposal

A waste disposal survey is needed to identify the quantity of waste materials and the waste characteristics that will be discussed in this chapter.

Planning the Survey

The first step in a waste disposal survey is an operations study, which identifies sources of wastes. Construction drawings showing the plot with piping plans and equipment layouts should be studied to determine all sources of incoming and outgoing water. The piping plans should show water lines, storm sewer lines, sanitary sewer lines, and processing waste drains and lines. The pipe sizes, the locations and types of connections to processing equipment, and the flow direction should be included on the drawings.

The operating schedule of the food plant—the number of shifts and types and volumes of products produced in a single day and over a week, a month, a season, and the entire year—is important to this survey. Production records for several preceding years can provide this information. Water consumption records should also be examined.

An initial waste survey is conducted to ensure that a plant can comply with federal, state, and local effluent requirements. The EPA periodically monitors waste discharges to check the accuracy of reports submitted by applicants and permit holders. This survey is also beneficial to determine locations and types of required monitoring equipment to establish a continuous monitoring program. Another advantage of an initial survey is to determine whether waste treatment is needed to meet discharge regulations and, if needed, the most ideal waste treatment approach.

Conducting the Survey

Information obtained from the operations study should determine what to include in the survey. It may be necessary to conduct individual surveys in each season if the types and volumes of products processed in the plant vary widely with the season of year, as is typical in many food plants, especially fruit and vegetable processing plants. These steps must be part of the survey: determination of the water balance, sampling of wastewater, and determination of extent of pollution.

Determination of Water Balance
Wastewater volume and flow rates from all sources should be measured through meters placed on all incoming water lines. Suitable measuring devices are Parshall flumes, rectangular and triangular weirs, and Venturi tubes and orifices. Through calculation of the water balance for an entire plant, the quantity of water in the waste effluent, together with the quantities lost through steam leaks, evaporation, and other losses, and the amounts used in the products of the plant in a given period of time can be determined. All of these quantities should equal the amount of water supplied to the plant during a given period. This calculation can be used to identify hidden water losses or major leaks, which can affect the sanitation program and cause increased waste, additional effluent discharge, and reduced profits.

Sampling of Wastewater
Samples of the wastes should be obtained in proportion to flow rates. Random "grab" samples—taken by catching a given quantity of the effluent discharge in a container without consideration of variations in volumes and flow rates and changes in plant operations—are of limited value for determining the true characteristics of wastes and can provide misleading results. Statistical sampling at planned times during the operating and nonoperating periods, and in proportion to flow, can provide valid data related to the characteristics of the wastewater effluent from a plant.

The sampling device should be located in the wastewater discharge system to obtain a representative sample. Samples should be collected where wastes are homogeneous—perhaps below a weir or flume. Caution should be exercised to avoid sampling errors resulting from a deposition of solids upstream from a weir or from accumulation of grease immediately downstream. The sample should be collected near the center of the channel and at 20–30% of the depth below the surface, where the velocity is sufficient to prevent deposition of solids. Sewers and deep, narrow channels should be sampled at 33% of the water depth from the bottom to the surface, with the collection point rotated across wide channels. During sample collection and handling, agitation should be avoided for dissolved oxygen determination. Food plant wastes readily decompose at room temperature; thus, it is important to chill samples promptly to 0–5 °C (32–40 °F) if they are not analyzed immediately after sampling.

Determination of Extent of Pollution

A large percentage of the waste discharged in fruit and vegetable waters, wash water from animal harvesting, and cleanup water discharge are product pieces (larger pieces can be removed by screening). Finer solids (which pass through a screen) and organic matter in colloidal and true solution usually have an oxygen demand in excess of the dissolved oxygen content of the water. Discussion of measurements for water pollution determination follows.

Biochemical Oxygen Demand A frequently used method of measuring pollution strength is the 5-day BOD test. The BOD of sewage, sewage effluents, and waters of industrial wastes is the oxygen (in PPM) required during stabilization of the decomposable organic matter by action of aerobic microorganisms. The sample is stored in an airtight container for a specified period of time and temperature. Complete stabilization can require more than 100 days at 20 °C (68 °F). Because such long periods of incubation are impractical for routine determinations, the procedure recommended and adopted by the Association of Official Analytical Chemists (AOAC) is a 5-day incubation period and is referred to as the 5-day *BOD*, or *BOD₅*. This value is only an index of the amount of biodegradable organic matter, not an actual measure of organic waste.

Domestic sewage that contains no industrial waste has a BOD of approximately 200 PPM. Food processing wastes are normally higher and frequently exceed 1,000 PPM. Table 12.1 gives

Table 12.1 Typical composition of wastes from food and related industries

Type of waste	BOD₅ (PPM)	Suspended solids (PPM)
Dairy and milk products	670	390
Food products	790	500
Glue and gelatin	430	300
Meat products	1,140	820
Packing house and stockyard waste	590	600
Rendered products	1,180	630
Vegetable oils	530	475

the typical amount of BOD₅ and suspended solids from food and related industries. Note that the BOD data and values for suspended solids generally show a parallel relationship. Although BOD is a common measurement of pollution of water and the test is relatively easy to conduct, it is time-consuming and lacks reproducibility. Tests such as chemical oxygen demand (COD) and total organic carbon (TOC) are quicker, more reliable, and more reproducible.

Chemical Oxygen Demand The COD test for measuring pollution strength oxidizes compounds chemically rather than biologically by a dichromate ($K_2Cr_2O_7$) acid reflux method. Because it is a chemical analysis, this test also measures non-degradable materials, which are not measured by BOD testing. When a plant monitors effluent to be discharged for municipal treatment, daily COD measurements can serve as a guide to determine whether and when a biological or chemical effluent could create a treatment problem at the wastewater treatment plant. This test, however, gives no indication as to whether the organic matter can be degraded biologically and, if so, at what rate. Some molecules are not oxidized as a result of this type of treatment. Although overlapping occurs, this test does not duplicate the BOD₅. Data from the COD test closely relate to dissolved organic solids. Unless a ratio has been established for COD/BOD, regulatory agencies have not accepted COD data as a substitute for BOD data in the past.

Dissolved Oxygen Dissolved oxygen (DO) concentration is of major concern for both wastewater and receiving water because it affects aquatic life and is important in treatment systems such as aerated lagoons. Determination of dissolved oxygen can be accomplished by an iodometric titration procedure using the azide and permanganate procedures to remove interfering nitrite and ferrous ions, even though this method is not considered to be very reliable. Alternatively, electrode probes can be used for this measurement. They are faster and more convenient than the iodometric titrimetric method and more adaptable for use in most industrial wastewaters.

However, certain metal ions, gaseous oxidants stronger than molecular oxygen, and high concentrations of cleaning compounds will interfere with the electrode probes used to measure dissolved oxygen.

Total Organic Carbon Total organic carbon determines all materials that are organic. It measures the amount of CO_2 produced from the catalytic oxidation of solid matter in wastewater. This method of pollution measurement is rapid and reproducible and correlates highly with standard BOD_5 and COD tests, but it is difficult to conduct and requires sophisticated laboratory equipment. This test can be effectively conducted where total solid matter is mostly organic and if the operation has a large volume of effluent. However, the cost of performing TOC analysis is frequently prohibitive for smaller and/or seasonal processing plants.

Residue in Wastewater Residue can be considered pollution because it affects the measurements that have been discussed previously. Residues of evaporation (total solids) and the volatile (organic) and fixed (ash) fractions are routinely recognized.

Settleable solids (SS) settle to the bottom in 1 h. They are usually measured in a graduated Imhoff cone and reported as mL/L SS. Settleable solids are an indication of the amount of waste solids that will settle out in clarifiers and settling ponds. This examination technique is easy to perform and can be conducted at field sites.

Total suspended solids, sometimes referred to as *nonfilterable residue*, are determined by filtration of a measured volume of wastewater through a tared membrane filter (or glass fiber mat) in a Gooch crucible. The dry weight of the total suspended solids (TSS) is obtained after 1 h at 103–105 °C (218–222 °F).

Total dissolved solids (TDS), or filterable residue, are determined by the weight of the evaporated filtered sample or as the difference between the weight of the residue on evaporation and the weight of TSS. These pollutants are difficult to remove from wastewater, so knowledge of them is essential.

Treatment requires microorganisms, which are normally present, for conversion to particulate matter, i.e., microbial cells.

Fats, oil, and grease (FOG) are detrimental to biota and are unaesthetic. Interchange of air and water is reduced through the thin film created by FOG, which is detrimental to fish and other marine life. Water fowl are also affected by heavy oil films. These compounds increase oxygen demand for complete oxidation.

Although *turbidity* is not a pollutant, it is caused by the presence of suspended matter (organic matter, microorganisms, and other soil particles). Turbidity is an optical property of the sample, which causes light to be scattered and/or absorbed, rather than transmitted. It is measured by a candle turbidimeter. This measurement is not an accurate indication of suspended matter that has been determined gravimetrically because the latter method involves particle weight and the former relates to optical properties.

In waste material, *nitrogen* can exist in forms ranging from reduced ammonium to oxidized nitrate compounds. High concentrations of the nitrogen forms can be toxic to certain plant life. The most common forms of nitrogen found in wastewater are ammonia, proteins, nitrites, and nitrates. The reduced forms, i.e., organic nitrogen and ammonia, can be measured by the total Kjeldahl nitrogen (TKN) method. Other tests are necessary to measure the oxidized forms, i.e., nitrate and nitrite.

Phosphorus occurs in wastewater as phosphate in the forms of orthophosphate and polyphosphate. This element is present as either mineral or organic compounds. Although trace amounts of soluble phosphates occur in natural waters, too much is detrimental to marine life. Routine analyses measure only soluble orthophosphate. Analyses for total phosphates, polyphosphates, and precipitated phosphates are accomplished by converting the polyphosphates and precipitated phosphates to orthophosphate by acid hydrolysis, with subsequent testing for orthophosphate by colorimetric methods. With the required chemical reagents and a colorimeter or spectrophotometer, these tests may be performed through a trained technician.

The use of sulfur dioxide in pretreatment of fruits or sodium bisulfide in processing may cause the *sulfur* content of wastewater to be high enough to cause pollution problems. These pollutants exist primarily as sulfite and sulfate ions or precipitates. Also, sulfides require more available oxygen if present in water. Sulfide ions combine with various multivalent metal ions to form insoluble precipitates, which can settle out and be removed with the sludge. Sulfate and sulfide determinations are possible with a trained technician and minimal equipment. Sulfides contribute to an undesirable odor and taste in drinking water. Thus, it is important to test for these compounds if the wastewater is discharged into a stream that supplies drinking water.

Solid Waste Disposal

Solid waste residues are composed of process discards, residues of the process of wastewater treatment, and organic and inorganic garbage. The process of biotransformation of these residues in fertilizer is an alternative that should be evaluated, because of the possibility of biogas generation in one of the parts of a fermentation process.

Disposal of solid wastes is a major challenge for the food industry. In food industries such as canneries, up to 65% of the raw materials received must be disposed of as solid waste. The most common method for disposal has been to truck the wastes to municipal garbage dumps. If a dump is not nearby and the wastes are disposed on the plant site, odor and insect problems will be created. The waste materials should be hauled to a disposal site at regular intervals and the containers washed immediately after dumping.

Some processing firms handle solid wastes by composting, and the finished compost can then be applied to the soil as fertilizer. A typical analysis of composted material is 1.25% nitrogen, 0.4% phosphates, and 0.3% potash. Some municipal waste treatment facilities manufacture and sell solid waste materials for agricultural application. Burying wet food residues in trenches or earthen cells should be avoided because the organic matter undergoes anaerobic decomposition. This bacte-

rial activity rapidly lowers the pH and a "pickled" condition materializes. Wet food residues should be buried by covering thin layers of the waste material with sufficient soil to absorb the existing moisture and covering the site with a thick layer of compacted earth.

If composting is used, the organic matter in waste material must be stabilized through microbial action. Humus, which results from stabilization of waste material, improves fertility and tillage properties. The basic composting procedures have four steps:

1. Solid waste material should be comminuted (pulverized) to expose the organic matter to microbial attack.
2. The comminuted waste should be stacked in windrows approximately 2 m (2.2 yards) high and 3 m (3.3 yards) wide.
3. Aeration should be provided.
4. After extensive aeration, the compost should be comminuted again.

Addition of an inoculum will accelerate the composting process. This process is produced through those aerobic thermophilic microorganisms present in the waste material in 10–20 days, depending on temperature and waste composition.

In addition to compost, various food product wastes can be dehydrated and ground for feed use. An example is the press liquors of tomato processing wastes. The residue from alcohol manufacture can be dried and fed to livestock. Citrus wastes, including activated sludge from treatment, can also be dried and used as animal feed because they contain B vitamins and protein. Processed whey and rendered animal by-products are also valuable foods for animal consumption.

Liquid Waste Disposal

Wherever food is handled, processed, packaged, and stored, wastewater is generated. Quantity, pollutant strength, and nature of constituents of processing wastewater have both economic and environmental consequences concerning treatability and disposal. Economics of treatment are affected by the amount of product loss from the processing operations and by the treatment costs of this waste material. Significant characteristics that determine the cost for wastewater treatment are the relative strength of the wastewater and the daily volume of discharge.

These residues are normally destined to the biological system after screening and decantation for the removal of solids. Many food processors do not use decantation before sending these wastes to municipal sewer treatment resulting in an increase of dissolved solids, hindering purification, and increasing equipment maintenance and energy consumption for treatment because of larger density residues.

The largest volume of solid residues in wastewater is composed of residues from sieves, flotation sludge, and waste from the biological processes involved. Sieve residues and flotation sledge have value as a by-product because they can be converted to animal foods and fertilizer. However, sludge through the use of chemical flotation that incorporates metallic coagulants should not be used for animal food.

Wastewater can be salvaged through recycling, reuse, and the recovery of solids. The degree of conservation and salvage value of wastewater is based on factors such as wastewater treatment facilities for recoverable materials, operating costs of independent treatment, market value of the recoverable materials, local regulations regarding effluent quality, surcharge cost for plants discharging into public sewers, and anticipated discharge volume in the future. The economics of disposal of solids, concentrates, blood, and concentrated stick (in wet rendering) determine how much of these polluting solids are kept out of the sewer. A wastewater control plan must be able to remove and convey organic solids using "dry" methods, without discharging those solids to the sewer and by using a minimal amount of water in the cleaning operation.

Spent cleaning compounds and sanitizers are discharged into waste treatment facilities. The toxicity of these materials causes concern because sanitizers, which destroy microorganisms, are toxic by definition. However, they meet requirements of the Food and Drug Administration (FDA) as an indirect food additive because these organic compounds are diluted in water and their toxic properties become reduced to a safe level. Many of the ingredients used in cleaning compounds and lubricants are generally recognized as safe as food additives. It appears that the major concerns for wastewater treatment of this effluent are pH fluctuation and possible long-term exposure to trace heavy metals. However, these effects can be controlled and waste minimized through appropriate plant design and optimal concentration use of cleaning compounds and sanitizers.

Cleaning compounds and sanitizers increase BOD/COD because they utilize surfactants, chelators, and polymers in addition to organic acids and alkalis. Conveyor lubricants utilize similar materials that increase the BOD/COD of the effluent. However, these compounds account for less than 10% of the BOD/COD contributions from a food processing plant. Water volumes associated with sanitation from a food processing plant can account for up to 30% of the total water discharge. Because of the low BOD/COD contributions, pH of wastewater is a major concern.

A eutrophic condition can develop from the discharge of biodegradable, oxygen-consuming compounds if inadequately treated wastewater is discharged to a stream or other body of water. If this condition continues, the ecological balance of the receiving body of water will be harmed.

Coagulants and polymers derived of cellulose, starch, and sugar destined to the removal of oils and greases by flotation generating a material that can be processed in digesters are available. This approach is possible mainly due to the absence of metallic ions (iron and aluminum). The challenge, in the use of these products, is to find the ideal balance between the costs and benefits.

It is usually more economical to invest in waste prevention techniques and utilization of waste products than in waste treatment facilities. Yet many food plants generate waste effluents that pollute. Insufficient treatment capacity of many municipal waste treatment plants necessitates special waste facilities for a large percentage of food plants. Wastewater treatment is still a developing technology and one that is going to need the cooperation of the EPA, suppliers, and processors.

Pretreatment

The pretreatment of food processing wastewater is frequently required prior to discharge into a municipal waste treatment system. A sewer use ordinance defines specified municipal discharge limitations that determine the degree of pretreatment required. The EPA has previously concluded that many wastewaters from processing plants are compatible and biodegradable.

Municipal sewage plants normally place certain restrictions on wastewater discharge from food processing plants. Although toxic substances are not frequently associated with food process waste streams, certain wastes that are present cannot be treated and can cause obstruction and required additional maintenance. Troublesome wastes include oils and fats, plant and animal tissues, and waste materials. Therefore, some form of isolation and pretreatment of the waste stream is essential prior to discharge in a municipal waste treatment facility.

If increased waste load reduces the ability of the municipal waste treatment system to treat the additional waste adequately, the food processor usually has to accept more responsibility related to pretreatment or support of a municipal waste treatment plant modification or expansion program. The processor should calculate the cost of the added sewage treatment load and determine that the projected cost should be handled by pretreatment or by paying a surcharge to a municipal expansion program keyed to specific wastewater parameters.

Surcharge calculations start with a flow base rate and utilize multipliers for concentrations of such ingredients as BOD_5, suspended solids, and grease. An example would be to charge the flow base rate to all sewer users as 50% of the water bill, including flow from private supplies. Treatment costs chargeable to BOD and suspended solids frequently include surcharges for concentrated wastes when above an established minimum based on normal load criteria.

Small plants frequently determine that it is advantageous to provide only enough pretreatment of wastewater to ensure compliance with municipal regulations. In contrast, larger proces-

sors have discovered that providing pretreatment beyond the level required by the ordinance can be advantageous. Some plan to provide enough pretreatment to reduce the surcharge for discharging untreated wastewater. Many large-volume processors treat all of their wastewater to avoid high surcharges or because the municipal plant lacks the capacity to handle the addition effluent.

The following advantages of pretreatment of wastewater beyond the level required by the local ordinance should be considered:

- Grease and solid materials from plant and animal products frequently have a good market value. Demand from soap plants, feed plants, and other industries can make a recovery of waste solids a profitable operation. Such operations also reduce the amount of wastewater treatment.
- If municipal charges and surcharges are high, additional pretreatment can be economically advantageous because better pretreatment will reduce these charges.
- Municipality complaints can be reduced through additional treatment responsibilities assumed by the food processor.

The following disadvantages can discourage pretreatment of wastewater:

- Pretreatment facilities are expensive and increase the complexity of the processing operation.
- Maintenance costs, monitoring costs, and record keeping of a wastewater treatment operation can be expensive.
- Pretreatment facilities are placed on the property tax roll unless state regulations permit tax-free waste treatment.

If pretreatment is conducted, this process should be based on facts revealed from the waste disposal survey. Results from the plant survey and review of viable waste conservation and water reuse systems are essential for identification, design, and cost estimates of a pretreatment system. Cost estimates should include those parts of the pretreatment attributable to flow, such as dissolved air flotation and grease basins. Thus, major in-plant expenses for waste conservation and water recycling can be determined based on the estimated reduction in flow, BOD, suspended solids, and grease.

The most common pretreatment processes include flow equalization and the separation of floatable matter and SS. Separation is frequently increased by the addition of lime and alum, ferric chloride (FeCl₃), or a selected polymer. Paddle flocculation may follow alum and lime and lime or ferric chloride additions to assist in coagulation of the suspended solids. Separation is usually accomplished by gravity or by air flotation. Screening by vibrating, rotary, or static-type screens is a step that precedes the separation process and concentrates the separated floatables and settled solids.

Flow Equalization

Flow equalization and neutralization are used to reduce hydraulic loading in the waste stream. Facilities required are a holding device and pumping equipment designed to reduce the fluctuation of effluent discharge. This operation can be economically advantageous, whether processing firms treat their own wastewater or discharge into a municipal sewage treatment facility after pretreatment. An equalizing tank has the capacity to store wastewater for recycling or reuse, or to feed the flow uniformly to the treatment facility day and night. This unit is characterized by a varying flow into and a constant flow from the tank. Equalizing tanks can be lagoons, steel construction tanks, or concrete tanks, often without a cover. It is important to integrate the discard flow of the process to the normal capacity of the treatment equipment that has been installed.

Screening

The most frequently used process for pretreatment is screening, which normally employs vibrating screens, static screens, or rotary screens. In fact, food processing wastewater should be screened prior to treatment or disposal. Vibrating and rotary screens are more frequently used because they can permit pretreatment of a larger quantity of wastewater that contains more organic matter. These screening devices are well adapted to a flow-away

(water in forward flow and passing through with solids constantly removed from the screen) mode of operation and can vary widely in mechanical action and in mesh size. Mesh sizes used in pretreatment range from approximately 12.5 mm (0.5″) in diameter for a static screen to approximately 0.15 mm (0.006″) diameter for high-speed circular vibratory polishing screens. Screens are sometimes used in combination (e.g., prescreen polish screen) to attain the desired efficiency of solids removal.

Skimming

This process is frequently incorporated if large, floatable solids are present. These solids are collected and transferred into some disposal unit or preceding equipment. In meat operations, floatable oils, fats and greases, and settleable solids are collected and conveyed to a rendering operation for further processing. Lime and FeCl₃ or a selected polymer may be added to enhance separation of solids, and paddle flocculation may follow to assist with the coagulation of these solids.

One approach to wastewater treatment involves pH adjustment to 5.5–6.5 with ferric chloride and a cationic polymer added to the wastewater as a coagulant, thus providing optimal polymer flocculation. This technique involves pumping pH-adjusted wastewater and injecting an anionic acrylamide copolymer to flocculate the coagulated suspended solids. The stream flows through a flocculating mixer and is injected with dissolved air and fed into a dissolved air flotation unit. Then, floating material is skimmed off as solids sludge and pumped to a holding tank. The solids can age for a day and are then pumped from the holding tank and injected with another cationic acrylamide copolymer through flocculating tubes and belt-pressed to produce a 26% solids cake. This cake is mixed with lime and dried in a rotary drum drier.

Primary Treatment

The principal purpose of primary treatment is to remove particles from the wastewater. Sedimentation and flotation techniques are used.

Sedimentation

Sedimentation is the most common primary treatment technique used to remove solids from wastewater influent because most sewage contains a substantial amount of readily settleable solid material. As much as 40–60% of the solids, or approximately 25–35% of the BOD_5 load, can be removed by pretreatment screening and primary sedimentation. Some of the solids removed are refractory (inert) and are not measured by the BOD test.

A rectangular settling tank or a circular tank clarifier is most frequently used in primary treatment. Many settling tanks incorporate slowly rotating collectors with attached flights (paddles) that scrape settled sludge from the bottom of the tank and skim floating scum from the surface.

Design of a sedimentation system should incorporate sizing of the detention vessel and the attainment of a quiescent state for the raw wastewater. Temperature variation of the wastewater also affects sedimentation because of the development of heat convection currents and the potential interference with marginal setting participles. Grease removal is accomplished during this pretreatment process through elimination of the surface scum.

Flotation

In this treatment process, oil, grease, and other suspended matter are removed from wastewater. A primary reason that flotation is used in the food industry is that it is effective in removing oil from wastewater.

Dissolved air flotation (DAF) removes suspended matter from wastewater by using small air bubbles. Flocculants and polymers are added to the wastewater to separate grease, oils, and fats from the water. When discrete particles attach to tiny air bubbles, the specific gravity of the aggregate particle becomes less than that of water. The particle separates from the carrying liquid in an upward movement by attaching to the air bubble. These particles are then floated for removal from the wastewater. This pretreatment process involves contact of the raw wastewater with a recycled, clarified effluent that has been pressurized through air injection in a pressure tank. The combined flow stream enters the clarification vessel, and the release of pressure causes tiny air bubbles to form, which move up to the surface of the water, carrying the suspended particles with them.

Air bubbles, which incorporate the flotation principle by removal of oil and suspended particles, can be created in the wastewater by (1) the use of rotating impellers or air diffusers to form air bubbles at atmospheric pressure, (2) saturation of the liquid medium with air and subsequent combination of the mixture to a vacuum to create bubbles, and (3) saturation of air with liquid under high pressure and subsequent release to form bubbles.

One approach to wastewater treatment involves pH adjustment to 5.5–6.5 with ferric chloride and a cationic polymer added to the wastewater as a coagulant, thus providing optimal polymer flocculation. This technique involves pumping pH-adjusted wastewater from a holding tank and injecting an anionic acrylamide copolymer to flocculate the coagulated suspended solids. The stream flows through a flocculating mixer and is injected with dissolved air and fed into a dissolved air flotation unit.

Flocculating agents are commonly used to pretreat wastewater prior to treatment by a DAF unit. Treatment by DAF is widespread because of the relatively fast passage and because solids of nearly the same as or lower density than water can be removed. This treatment technique requires high investment and operating costs, especially for chemical additives and sludge handling.

DAF systems maintain a concentration of bacteria that are kept alive within the system to biodegrade pollutants in the effluent. A dewatering device, such as a belt filter, can be incorporated with DAF. After floatable oils and grease are captured, they can be chemically treated and the material conditioned, similar to a liquid-solid separation process.

Flotation technology has also been adapted to sludge handling and to secondary and tertiary treatments. Food processors with substantial quantities of grease and oil in their wastewater use this technique as part of their waste treatment systems. A problem of flotation has been the presence of a turbulent flow; however, commercial high-rate flotation devices that eliminate turbulent flow are now available. The installation of lamellas (vertical baffles) can prevent unfavorable currents and short-circuiting and, with a properly designed feed well, can improve solid-liquid separation, producing higher underflow solid concentration in gravity thickeners and better effluent quality in gravity clarifiers.

Collected sludge from primary treatment contains approximately 2–6% solids, which should be concentrated before final disposal. Sludge treatment and disposal costs are the major expenses of sewage treatment if this product is not used as a fertilizer or for some other practical application. Some treatment systems biodegrade most of the organic matter and create little sludge. These systems can reduce treatment and disposal costs. If sludge is recovered as a by-product, disposal costs can be reduced, and the value of the salvaged material can provide enough profit to defray other treatment costs. Recovered solids (sludge) can also be treated by biological oxidation methods as a means of ultimate disposal.

A method incorporated in some processing operations utilizes a series of coagulants formed from cornstarch to separate oil, grease, and suspended solids from wastewater prior to its discharge. The resultant grease and solids recovered from the DAF can be rendered. These starch-based coagulants are normally added to an equalization tank prior to the DAF system, where they can reduce the surface charge on the solids and grease, allowing the materials to coalesce and be removed by DAF.

Wastewater treatment has normally involved the removal of solids from liquids. Equipment exists that utilizes a water loop principle to filter water from behind a chiller and flow it through a series of filters before returning it to the chiller. In this process, organic matter is filtered out so that the water can be recycled. Furthermore, water concentrates of as little as 3% of organic matter can be recycled through rendering equipment such as a disc dryer to concentrate a product into dry powder, with the vapors directed back into the evaporative system to be used as an energy source. The evaporative system provides free energy.

Sludge originating from the ponds of stabilization presents another problem because of the amount of chemicals, both organic and inorganic, from the lingering decomposition of wastewater residues. So, the cleaning and emptying of ponds should be preceded by a detailed evaluation of the wastewater with a forecast of a phase of controlled microbial stabilization of the sludge, before its deposition in agricultural areas.

Adjustment of the quotation stage can be difficult because of the lack of integration between production personnel and the operators of the flotation equipment. One of the most frequent problems is the discard of the tanks and hot water in short intervals of time, causing an abrupt elevation of the volume, temperature, and pollutants going to the flotation phase and causing problems for the biological process, hindering the purification, and elevating the presence of pollutants.

Secondary Treatment

Treatment through biological (or bacterial) degradation of dissolved organic matter through biological oxidation is the most common technique for secondary treatment. However, secondary treatment can range from the use of lagoons to sophisticated activated sludge processes and may also include chemical treatment to remove phosphorus and nitrogen or to aid in the flocculation of solids.

Most lagoons are earthen basins that contain a mixture of water and waste. The mixture in the lagoon is removed continuously without emptying the lagoon. The design of most lagoons is similar. A dike or berm usually surrounds a lagoon as a lip of the basin that prevents spills and overflows. The depth of an impoundment (lagoon) depends on the volume of waste to be handled, with increased depth necessary to contain unforeseeable events, such as weather.

To accommodate such unforeseeable environmental changes, there is usually a storm event space left free of water. This is usually the amount of precipitation determined to have accumulated in 24 h during the worst storm in the previous 100 years or the amount of precipitation from the wettest month in 25 years. Additional space reserved for safety measures includes wind setup and wave run-up spaces to prevent overflows.

Circular or square lagoons enhance mixing and are usually less expensive to construct. If rectangular lagoons are used, a length/width ratio of 3:1 or less is recommended. Narrow areas isolated from the main body of water should be avoided because they may encourage mosquito proliferation. Although most lagoons are approximately 3 m (10 ft) deep, a greater depth requires less land, enhances mixing, and minimizes odors.

Lagoons must be sealed to prevent seepage that causes groundwater contamination. A lagoon can be sealed with hard-packed clay soil or with an industrial liner. A minimum of 30 cm (12″) of clay seal on the bottom and sides is required for most locations, but local ordinances may vary in their regulations. As lagoon depth increases, a thicker seal is required. Soil type, depth to water table, and depth to bedrock should be considered when locating a lagoon.

Although primary treatment removes screenable and readily settleable solid material, dissolved solids remain. The primary purpose of secondary treatment is to continue the removal of organic matter and to produce an effluent low in BOD and suspended solids. Microorganisms most frequently involved in biological oxidation of existing solids are those that naturally occur in water and soil environments. Microbial flora involved in biological oxidations can assimilate some of the dissolved solids and convert them into terminal oxidation products, such as carbon dioxide and water, or into cellular material that can be removed as particulate matter. Microbial cellular matter and assimilated organic matter continue to undergo aerobic degradation via the following endogenous respiratory reaction:

$$C_2H_9O_3N + 4O_2 \rightarrow 0.2C_3H_9O_3N + 4CO_2$$
$$+ 0.8NH_3 + 2.4H_2O$$

Oxygen is required for these reactions. After treatment, the microbial suspended solids are separated from the water by gravity sedimentation. Some of the dissolved solids and small-suspended solid matter in the form of colloidal and supra-colloidal particles escape secondary clarification. If the effluent concentration is too high, the flow should be filtered before discharge, or clarification can be improved by the addition of flocculating chemicals.

Anaerobic Lagoons

Anaerobic lagoons can be designed with either a single stage or multiple stages. The disadvantages of multiple-stage lagoons are increased construction and land costs. Advantages are:

• There is less floating debris on the second and third stages, with a reduction in clogging of the flushing system or irrigation pump.
• The first lagoon, containing a higher concentration of waste, will not overflow.
• An adequate amount of bacteria will be available for waste treatment.
• The resulting effluent will be treated more thoroughly.

Lagoon start-up should be planned to minimize the amount of biological stress. Time is required for the appropriate bacteria to become established. Because anaerobic bacteria are slow growers, it may require a year or more for a lagoon to become fully mature. Lagoons should be started up in late spring or summer to permit bacterial establishment during the warmer weather. The amount of waste added should be increased gradually over 2–3 months.

Lagoons will accumulate fluid over time, due to precipitation, and should have fluid removed periodically. Typically, 40–50% of the active lagoon volume should remain, and fluid removal should be done only during warmer months to ensure that the bacteria can replenish themselves and will not decline below an effective level. In multiple-stage lagoons, the effluent should be removed from the last stage.

After 10–20 years, a lagoon will build up sludge that should be removed to prevent biological overloading. Three techniques are used for sludge removal. The first technique involves agitation equipment to resuspend the sludge and pump it out while the contents are thoroughly mixed. The remaining sludge will resettle once the agitation is stopped. The second technique involves the use of a floating dredge to move across the lagoon, while a pump located on the dredge pumps the sludge over to another pump located on the shore. The second pump either sends the sludge to a holding tank or applies it to the land. The third technique is to pump the liquid to a lagoon and permit the remaining sludge to dry naturally. This long process may require several months.

Waste sludge may be produced by both primary and secondary treatment. Sludge typically requires further stabilization before final disposal. Anaerobic and aerobic lagoons are frequently referred to as *stabilization ponds*. They have been used for wastewater treatment and sludge stabilization. The use of this treatment technique has increased because of the relatively low capital investment, low operating costs, and ease of operation. Anaerobic and aerobic lagoons are not well suited where land costs are extremely high or for extremely large waste loads.

The treatment principle underlying lagoons is biological oxidation and solids sedimentation. Dissolved, suspended, and settled solids are converted to volatile gases, such as oxygen, carbon dioxide, and nitrogen, water, and biomass, such as microflora, macroflora, and fauna. Anaerobic and other lagoons equalize the discharge flow to further treatment facilities or receiving waters.

The depth of anaerobic lagoons varies from 2.5–3.0 m (2.8–3.3 yards). Surface area-volume ratios should be minimal. Anaerobic conditions are created throughout the entire lagoon, through heavy organic loads. Under anaerobic conditions, anaerobes digest the organic matter. Operating temperatures of 22 °C (72 °F) or higher are needed, with 4–20 days of detention. BOD reduction efficiency is typically 60–80% but is a fraction of the influent BOD and the determination time. Anaerobic lagoons are used as primary or secondary treatment of primary effluents containing high organic loads or as sludge treatment systems. Anaerobic lagoons are normally followed by aerobic lagoons or by trickling filters because their effluents remain high in organic matter.

Some treatment processes incorporate a combination of anaerobic and aerobic treatment. A completely mixed anaerobic tank reactor provides an environment for breaking down complex organic compounds into CO_2, CH_4, and simple organic compounds. The anaerobic tank reduces BOD_5 by 85–95%. The gases separate from the water and contain approximately 65–70% CH_4. The effluent flows on to an aerobic reactor for further treatment.

The previously described process involves the flow of anaerobically treated water to a degasification and flocculation tank, followed by a lamella clarifier, where the anaerobic microorganisms are separated and returned to the anaerobic tank. The supernatant flows by gravity to an aeration basin, where oxygen is supplied through mechanical aerators. Because the aeration step of the process has to remove only 5–15% of the original BOD_5, aerobic energy requirements are reduced. This process further involves settling out of aerobic sludge in the final clarifier, with a return to the aeration basin. Surplus sludge is recirculated into the anaerobic tank, where it enhances the bacterial activity and undergoes decomposition.

A combination of anaerobic and aerobic treatment can handle wide effluent variations. Anaerobic treatment responds slowly to flow variation because of the slow growth rate of the anaerobic microorganisms, but the faster-growing aerobic microorganisms can generally treat the higher loads in the anaerobic effluent. (Note: It is then no longer anaerobic.)

Aerobic Lagoons

Aerobic lagoons use mechanical aerators to supply atmospheric oxygen for aiding biological oxidation. Mechanical agitators pull air underwater and circulate it horizontally. Because oxygen transfer occurs underwater, neither freezing nor clogging occurs. Aerated lagoons are classified as either aerated facultative lagoons (which have enough mixing to dispense dissolved oxygen but not enough to keep all the solids suspended) or as completely mixed aerated lagoons, which are mixed enough to keep all solids suspended. Approximately 20% of the BOD sent to an aerobic lagoon is converted to sludge solids, and the BOD influent is reduced by 70–90%. The solids produced will partially decompose in anaerobic sludge banks in facultative lagoons, but the completely mixed effluents usually require additional treatment, such as clarification or polishing pond treatment.

Trickling Filters

Trickling filters reduce BOD and SS by bacterial action and biological oxidation as wastewater passes in a thin layer over stationary media (usually rocks) arranged above an overdrain. Biological degradation occurs almost exactly as in the activated sludge process, except that the filter is a three-phase system in which the biofilm is fixed on the solid medium (stones or plastic). Aeration is accomplished by exposing large surface areas of wastewater to the atmosphere. Layers of zoogloea (filter sludge) grow on and attach to the medium surface.

The efficiency of trickling filters is affected through temperature, waste characteristics, hydraulic loading rate, characteristics of the filter media, and depth of the filter. Media characteristics such as size, void space, and surface area, as well as hydraulic loading rates, tend to affect the performance of trickling filters more than other factors do. Removal efficiency is relatively independent of surface organic loading rate within broad ranges. Incorporation of plastic media with more surface area and void space than rock filter media has permitted improvements in design and efficiency. This treatment method is considered more rugged in operation and easier to maintain than activated sludge plants.

Activated Sludge

The activated sludge process is widely used for wastewater treatment. It requires a reactor that is an aeration tank or basin, a clarifier, and a pumping arrangement for returning a portion of the settled sludge to the reactor and discharging the balance to waste disposal. Primary treatment is optional. A portion of the clarifier-settled sludge is returned and mixed with wastewater entering the reactor. The resulting biological solids concentration is much higher than what could be

maintained without the recycle. The term *"activated sludge"* applies because this returned sludge has microorganisms that actively decompose the waste being treated. This mixture of influent wastewater and returned biological suspended solids is termed the *mixed liquor*. The activated sludge process is frequently called the *fluid-bed* biological oxidation system, whereas the trickling filter is referred to as a *fixed-bed* system.

The conventional activated sludge system has been designed for continuous secondary treatment of domestic sewage. It is not effective in treating inorganic dissolved solids but is very effective for the removal of all organic matter in the wastewater. This process may incorporate either surface aerators or air diffusers to achieve mixing. The influent organics are mixed with the activated sludge and undergo biological decomposition as they pass from the influent end of the reactor to the discharge end. The detention time in the reactor can vary from 6 h to 3 days or more, depending on the strength of the wastewater and the method of operation selected. When the activated sludge contacts the influent waste, there is a short period (less than 30 min) when influent particulate matter is rapidly absorbed onto the gelatinous matrix of the returned sludge. Absorption removes a large portion of the influent BOD. The aeration mechanical and electrical equipment components of an activated sludge system are relatively expensive, and the energy costs are relatively high. This process can be operated at high efficiency (95–98%) and can be modified to remove nitrogen and phosphorus without the use of chemicals.

The *extended aeration process* is a modification of the activated sludge plant. A typical application is the pasveer and carousel type of oxidation ditches used in Europe and in other countries that serve a large population. The term *extended aeration* was given to this process because it is operated to minimize waste sludge production. This results in a lengthening of the aeration time to maintain the mixed liquor suspended solids at a concentration that will still settle efficiently in the clarifier. This sludge is sufficiently mineralized, and the excess quantity does not require any further treatment in a digester before dewatering. However, more power is consumed in extended aeration systems because all organics are stabilized aerobically. The major advantage of this process is that it is generally capable of giving high BOD removal efficiency (95–98%) while minimizing waste sludge handling. This process is operated without primary treatment.

The aerobic digestion of sludge achieves volatile solids stabilization similar to that of aerobic digestion if mechanical or pneumatic aeration is provided. This approach is sometimes used to stabilize surplus biological sludge generated in the activated sludge process and in its modifications or in trickling filtration. It can also be used to stabilize primary sludge generated by settling prior to biological treatment.

The *contact stabilization process* is another modification of the activated sludge process, where advantage is taken of the fact that substrate removal occurs in two stages. The first stage, which lasts 0.5–1.0 h, involves rapid adsorption of the colloidal, finely suspended and dissolved organic compounds in the sewage by the activated sludge solids. In the second phase, the adsorbed organic material is separated by gravity sedimentation, and the concentrated mixed liquor is oxidized in 3–6 h. The first step occurs in the contact tank and the second in the stabilization tank. Therefore, the adsorption phase is separated from the oxidation decay phases.

In systems based on the use of stabilization ponds, a solid layer is formed in the surface of the ponds, increasing the sludge formation in the bottom, reducing the detention time (reduction of useful volume), and hindering the operation of the system.

Oxidation Ditch

This treatment technique has been developed as an efficient, easy-to-operate, and economical process for treating wastewater. The process maintains waste materials in contact with the sludge biomass for 20–30 h under constant mixing and aeration. After the biological reactor step, the stabilized suspended solids enter a clarification step, which removes them from the water by settling.

An oxidation ditch can accommodate BOD loadings of from 200–500 g (7–17.5 oz.)/day applied for each cubic meter (1.3 cubic yards) of available aeration space. Sludge solids should have a 16–20-day turnover (i.e., solids retention time or sludge age). For each kg (2.2 lbs.) of BOD applied, approximately 200–300 g (7–10.5 oz.) of new sludge solids can be produced, with an expected BOD reduction of 90-95%. Temperature can have a significant influence on the waste removal performance of the oxidation ditch. Pinpoint biological flocs may develop and be discharged with the clarifier effluent, decreasing the performance efficiency under cold-weather operating conditions.

The typical oxidation ditch aeration basin design is either a single closed-loop channel or multiple closed-loop channels with serial flow. An attractive feature of oxidation ditches is that a minimum of operation attention is required once a proper operation is established. Several food processors use oxidation ditches for wastewater treatment.

There is a current interest in the use of the total barrier oxidation ditch (TBOD) design for treating municipal food processing and industrial wastewater. TBOD biologically purifies water as it mixes oxygen with waste particles and permits the bacteria to feed on these pollutants. The system can achieve high oxygen transfer efficiency at a single point along the ditch, which allows for effective process control and design flexibility. A constant, powerful flow of wastewater is then maintained, preventing settling of the biomass at the bottom of the ditch reactor. The aeration and pumping unit consists of submerged, turbine draft tube aerators that transfer oxygen into the mixed liquid.

Land Application

Land disposal methods are designed to eliminate direct discharges into receiving waters. Disposal is accomplished by evaporation and percolation of wastewater through the soil. The two types of land application techniques that are the most efficient are *infiltration* and *overland flow*. With land application techniques, the pollutants can harm vegetation, soil, and surface and groundwaters if not properly operated. However, both of these treatment techniques can effectively remove organic carbon from high-strength wastewater. Pollutant removal efficiencies of approximately 98% for the infiltration flow system and 84% for the overland flow system can be attained. The advantage of higher efficiency obtained with an infiltration system is offset by its more expensive and complex distribution system. Less pollution of potable groundwater supplies is usually experienced with the overland flow system.

Wastewater has been spread on the land by the ridge-and-furrow irrigation technique. The land is prepared by forming rows of ditches (furrows) which are intermittently flooded with wastewater. Organic pollutants in the wastewater are retained in the soil and oxidized by indigenous bacteria. Crops are planted on the ridges. Also, the spray irrigation technique has been adapted as another method of land application. A cover crop is normally planted to enhance percolation and increase evaporative losses through transpiration. The organic matter is oxidized by soil organisms and partially utilized by the cover crop.

Although land application has been a standby in the past for discharge of some food processing wastes, this approach is now limited. Hydraulic loads that are high may necessitate an unreasonably large amount of land. Runoff and proper utilization of nutrients can restrict the vegetation. Buildup of minerals and other materials in the soil has the potential for long-term liability for residues possibly as yet undiscovered.

Rotating Biological Contactor

The rotating biological contactor (RBC) is an attached growth type of biological treatment system similar in concept to the trickling filters. Initial costs of this equipment are high, but operating costs and space requirements are moderate. This system consists of a number of large-diameter (approximately 3 m or 10 ft) and light-weight discs that are mounted 2–3 cm (0.6–1.2″) apart (to prevent bridging between the growths) on a horizontal shaft (in groups or packs, with baffles between each group to minimize surging or short-circuiting) to form an RBC unit (Fig. 12.1). The discs are partially (30–40%)

immersed and rotate slowly (0.5–10 RPM) as wastewater passes through a horizontal open tank, which usually has a semicircular bottom to fit the contour of the discs.

The RBC unit functions by attachment of microorganisms to the surface of the discs and grows by assimilating nutrients from the wastewater. Aeration is achieved through direct exposure of microorganisms to air when the surface of the disc is rotated above the water and by a thin film of water, which is aerated as it adheres to the disc's surface and rises out of the water. Acceleration in rotational speed increases the dissolved oxygen in the tank. The biofilm undergoes sloughing as in trickling filters, and these solids must be settled and removed. Although this treatment process is considered to be secondary treatment, primary sedimentation may be eliminated if the wastewater suspended solids are not unusually high (greater than 240 mg/L or 0.00085 oz./1.05 quarts).

Magnetic Separation

This secondary physical treatment method has applications for tertiary use. The organic waste solids in suspension are chemically treated with magnetite (Fe_3O_4). Alum or ferric chloride coagulation flocculation is performed, and the coagulated particles subsequently contain magnetite. This process consists of a chamber containing a stainless steel wool matrix located in a magnetic field. The magnetized coagulated particles in wastewater suspension are passed through the chamber and adhere to the stainless steel wool in

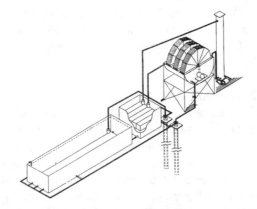

Fig. 12.1 Rotating biological contactors used to remove ammonia from water of an aquaculture production operation

the magnetic field. The collected organic waste is removed through reducing the magnetic field to zero and washing out the waste solids. This process was developed in Australia and has seen limited applications in North America.

Tertiary Treatment

Tertiary treatment processes for wastewater, which are collectively known as *advanced wastewater treatment*, are incorporated to improve the quality of waste treatment effluents. Tertiary waste treatment is applied to food processing wastewaters to remove pollutants of food processing, such as colors, odors, brines, and flavoring compounds. Some of the processes for tertiary treatment of municipal waste treatment are frequently used as a primary waste treatment for certain food processors. Water quality of treated wastewater should be equivalent to freshwater before becoming waste material.

Physical Separation

Sand filters and microstrainers have been developed for tertiary wastewater treatment and purification. Both of these physical separation methods remove suspended solids down to the micrometer particle range.

The *microstrainer* is a rotating cylinder covered by a screening material (usually fine mesh nylon or metal fabric) housed in a horizontal position in an open tank. Wastewater enters the inside of the cylinder and is filtered through the screen. As the cylinder rotates slowly, an exposed section above the wastewater surface is backwashed to clean the screen and to collect the solids into a separate channel. Particle removal by microstraining is a function of screen pore size, which normally ranges from 20–65 µm (0.0008–0.003 in). This is a relatively low-cost method of tertiary treatment because the screens are self-cleaning and operating and maintenance costs are low. The effectiveness of this treatment method is limited by partial screen clogging, with a resultant decrease in the life of the screen. Also, microorganisms can grow in secondary water inside the cylinder, causing slime formation on the screen. Ultraviolet light or chlorination treatment has been used to reduce slime formation.

The *rapid sand filter* and *mixed media and continuous countercurrent filtration* are frequently used in tertiary wastewater treatment. This treatment method requires underdrains for removal of clarified liquids and a system for recovering collected solids. Automatic backwash mechanisms are available to enable self-cleaning of filters.

Physical-Chemical Separation

Food processing wastewaters contain a substantial amount of dissolved solids that can be removed effectively by various physical-chemical separation methods. One of the least costly tertiary treatments for removing refractory organics is *activated carbon adsorption*. The affinity of the organic solute for the carbon depends on the type of carbon and the solubility coefficient of the solute to water.

Ion exchange processes remove minerals, either cations or anions, by replacing them with other ions through the medium of charged resins. Multivalent ions are usually replaced by monovalents, such as Na^+ or H^+, and the anions are replaced with OH^- or Cl^-. The principal purpose of this technique is to remove minerals considered harmful to the water supply or to recover valuable minerals from industrial processing wastewater.

The ion exchange resin usually consists of a network of cross-linked organic molecules known as *polymers*, which contain reactive functional groups that are usually strongly acidic, weakly acidic, or strongly basic. The resin is charged with ions such as H^+ or Na^+, which are replaced by multivalent ions from the wastewater passing through. Periodic recharging of the resin is necessary. This can be accomplished using strong acid or base solution. Ion exchange is especially beneficial for demineralizing water and whey. With the development of pulse-type ion exchange units, this method of treatment is becoming economically feasible.

Electrodialysis is used to remove minerals from brines and to demineralize whey. This process functions through the principle of alternately located cation- and anion-selective membranes placed in a current path. As ionic solutions pass through as a function of electric current, cations are transported through the cation-selective membrane and anions through the anion-selective membrane. Portions of the solution within the electrodialysis unit become concentrated with ions, while the remainder is demineralized. Because of problems related to precipitation of salts and membrane clogging by organic components in the water, electrodialysis as a tertiary treatment method has limited utility.

Tertiary Lagoons

These maturation lagoons, which are known as *polishing ponds*, are used for tertiary treatment of secondary effluents from activated sludge or trickling filter systems. This type of lagoon is usually from 0.3-1.5 m (1–5 ft) deep. Natural aeration, mechanical aeration, or photosynthesis provides the oxygen source. The waste removal efficiency of this system is influenced by temperature. This simple method of treatment requires practically no equipment or power, and minimal attention is required for the day-to-day operation. However, the land requirement of this process is the highest of the treatment methods.

Chemical Oxidations

Chemical oxidations through various chemicals are used for further oxidizing wastewater components in the tertiary treatment process. *Ozone* is a viable chemical oxidation treatment process. Ozone-generation equipment has made the process economically feasible. Ozone is a strong oxidant that breaks down in water to form oxygen and nascent oxygen, which rapidly reacts with organic matter. This process also disinfects, removes taste and odor, and bleaches. Other chemicals used in chemical oxidations are chlorine, chlorine dioxide, oxygen, and permanganate.

Disinfection

The major purpose of disinfection is to reduce the total bacterial concentration and eliminate the pathogenic bacteria in water. A potable water supply requires zero or very low bacterial concentration to avoid disease transmission. The total number of coliforms, instead of the presence of specific pathogens, is often used as an indicator for sanitary quality and the efficiency of disinfection. There are many chemical disinfectants and physical methods that can be incorporated for disinfection. Disinfection or physical removal of microorganisms can be accomplished through the use of a membrane filter. A sterile bacteria-retaining membrane has a nominal porosity of 0.22 μm, whereas a 0.45 μm membrane is small enough for applications where specific microorganisms are the target (Stanga 2010).

For public health reasons, treated wastewaters should be disinfected before final discharge. Addition of a chemical disinfectant to water provides maximal time of contact between the chemical and organisms, assuring efficient bactericidal action. Less disinfection is required as a result of the removal of microbes by primary and secondary wastewater treatment and by death of pathogenic microorganisms from extended exposure to natural environments. A variety of chemical disinfectants are available for use in water treatment. Examples are chlorine, iodine, bromine, quaternary ammonium, and ozone. Chlorine, as gaseous chlorine, or solid component such as calcium or sodium hypochlorite is the most common chemical used for disinfection due to low cost, high efficiency, and ease of application. Pre-chlorination, or source water chlorination, is designed to minimize operational problems associated with biological slime formation on filters, pipes, and tanks and to lessen potential taste and older problems. Post-chlorination, or terminal disinfection, is a primary exercise for microbial reduction in water. The addition of chlorine either immediately before the clear well or immediately before the sand filter is most common. Because of the potential reaction of disinfectants with organic matter, it is more practical to disinfect

Table 12.2 Microbial characteristics of domestic wastewater

Microorganism	Quantity per 100 mL (0.21 pints) wastewater
Total bacteria	10^9–10^{10}
Coliforms	10^6–10^9
Fecal streptococci	10^5–10^6
Salmonella typhosa	10^1–10^4
Viruses (plaque-forming units)	10^2–10^4

at the end of wastewater treatment. Table 12.2, which relates the typical microbial population and load in domestic wastewater, illustrates the amount of contamination that can occur from wastewater of food processing operations.

Chemical oxidants; ultraviolet, gamma, and microwave irradiation; and physical methods, such as ultrasonic disruption and thermal application, are used as disinfectants. Chlorination has received less emphasis in recent years because of potentially carcinogenic organohalides in chlorinated waters. In addition, over-chlorination of wastewater effluents can be toxic to fish. Chlorination and other chemical treatments do not kill all microorganisms. Certain algae, spore formers, and viruses (including pathogenic viruses) survive chlorination treatment. Surfactants used in the treatment process must be biodegradable.

Antimicrobial agents, such as sanitizers, incorporated in a food plant's sanitation program can present a challenge since they may destroy microorganisms involved in wastewater treatment. Sanitizers, pH, flow, BOD loading, solids, temperature, and other toxic materials, under certain conditions, can adversely affect the operation of a waste treatment plant. However, when sanitizers are properly used according to label directions, they do not normally interfere with the delicate microbial process in most treatment plants. Yet, accidental or mass discharge due to a spill of any sanitizer or chemical may complicate the treatment process. The sanitizer used by food processors that causes the most concern is the quaternary ammonium chlorides (QACs). These sanitizers are stable and effective over a broad pH range. However, there are four factors that may counteract the microbial activity of the sanitizers that may reach a treatment plant. They are inactivation, adsorption, biodegradation, and acclimation. In most treatment plants, there are enough cationic chemicals entering the waste treatment system to inactivate the QACs. Furthermore, chlorine and iodine rapidly lose activity in the waste stream and rarely enter the treatment plant. The dilution of acid sanitizers or carboxylic acid sanitizers normally raises the pH above 4.0 reducing their antimicrobial activity. Peroxyacetic acid sanitizers are like chlorine, very unstable

when mixed with general plant effluent, and are not expected to reach the treatment plant.

It is practical to disinfect moderate volumes of effluent with ultraviolet irradiation equipment, an effective method with no residual effects that harm flora or fauna in receiving water. Thermal treatment is effective but is impractical for large volumes of effluent. Membrane technology in the application of water treatment for the beverage industry will be discussed in Chap. 20.

Odor Removal

Treated water may be safe to drink yet have an unpleasant taste and odor because of the activity of some microscopic organisms such as algae, especially during the summer months. Thus, deodorization is essential to remove the taste and odor in treated water. Air stripping and aeration is a treatment to bring water into contact with air to expedite the transfer of a gas between the two phases. Applications include the removal of hydrogen sulfide that causes an unpleasant odor, carbon dioxide to reduce the demand of lime in the subsequent softening treatment, and trace volatile organic contaminants. Involved equipment includes diffused aeration, spray nozzles, and tray aerators. Odors produced during wastewater treatment and the drying of sludge may be removed through the incorporation of a biofilter. A biofilter that has been incorporated in odor removal utilizes recycled wood to grow bacteria which consume odor compounds. This technique can remove up to 95% of odor compounds such as ammonia.

Study Questions

1. What is biochemical oxygen demand?
2. What is chemical oxygen demand?
3. What are the advantages and disadvantages of pretreatment of wastewater?
4. What are the three methods of wastewater pretreatment?
5. Briefly describe two methods of primary treatment of wastewater.

6. Why are anaerobic lagoons used as a method of secondary treatment of wastewater?
7. How do aerobic lagoons function?
8. What is activated sludge?
9. What is the function of sand filters and microstrainers?
10. What is the most ideal method for disposing of solid wastes from a food processing operation?

11. What is a viable technique for odor removal of wastewater?
12. Why is chlorination of wastewater receiving less emphasis?

References

Stanga M (2010). *Sanitation cleaning and disinfection in the food industry*, 525. Wiley-VCH: Weinheim, Germany.

Pest Control

13

Abstract

Pests of major significance to the food industry include the German cockroach, American cockroach, Oriental cockroach, housefly, fruit fly, Norway rat, house mouse, pigeon, sparrow, and starling. Control of pests can be most effective through prevention of entry into food establishments and the elimination of shelter areas and food sources for subsistence and reproduction. If pests become established, pesticides, traps, and other control techniques are essential. These eradication devices should be considered a supplement to, rather than a replacement for, effective sanitation practices. Because pesticides are toxic, these compounds should be selected and handled carefully. Precautions during use, storage, and disposal are essential. Although a trained employee can handle pesticides, a professional exterminator should be employed for complex and hazardous applications.

Keywords

Integrated pest management • Dirty 22 • Pesticides • Sanitation • Cockroach • Housefly • Norway rat

Introduction

The intent of this chapter is to provide the sanitarian additional understanding of the impact of insects, rodents, and birds on the contamination of food supplies. The purpose of discussing pest control is to acquaint readers with the major pests that can contaminate the food supply and how to control the presence of these unwanted guests. The intent of this chapter is not to train pest control experts. There are relatively few species of insects, rodents, and birds that the food sanitarian has to contend with, but those encountered can cost the food industry billions of dollars every year. An effective program against pests begins with a basic understanding of the characteristics of pest contamination sources and a comprehensive knowledge of safe and effective extermination and control procedures. If a pest control operator is not incorporated to control pests, one or more employees (depending on the size of the organization) should be trained and held responsible for maintaining effective pest control. Even

© Springer International Publishing AG, part of Springer Nature 2018
N.G. Marriott et al., *Principles of Food Sanitation*, Food Science Text Series,
https://doi.org/10.1007/978-3-319-67166-6_13

if a pest control operator is used, the sanitarian and other food safety personnel in the plant should have training in pest control. The US Food and Drug Administration (FDA) has created a list of 22 species that are specifically linked to food-borne illness, which include four species of rodents, four cockroach species, two species of ants, and 12 species of flies (Higgins 2015).

Thorough housekeeping is an effective practice in ridding the premises of pests. A clean operation facilitates the extermination of pests within the building(s) and controls the entry of pests from the outside. In addition, pests have more difficulty finding suitable shelter where they can thrive, reproduce, and infest the facilities. Elimination of shelters, rubbish, decaying material, discarded supplies, and equipment will discourage the presence of insects and rodents. Pests may be found in enclosed areas under shelves, platforms, chutes, and ducts, especially if debris is allowed to accumulate in these areas. The same is true for breaks in walls and insulation. Rentfrow et al. (2008) suggested a list of good manufacturing practices to country ham processors for preventing pest infestations, including:

1. A gravel dead zone (no grass or shrubs that can harbor pests) at least 0.6 m (2 feet) immediately around the aging house.
2. Keep the space outside the plant free of garbage and debris and keep the outdoor trash away from the aging house.
3. Keep areas inside the plant clean and sanitized.
4. Regularly clean the walls and floors of the aging house so that the moisture and fat accumulated on those surfaces will not harbor mites.
5. Personnel who discovers mite infestation should change clothes and shower to avoid cross-contamination.
6. Clean and sanitize the aging house and racks thoroughly at the end of each aging period prior to hanging a new batch of hams.

Other non-fumigant alternatives to control pests in commodities and structures include integrated pest management (IPM), cold treatments, heat treatments, contact insecticides, controlled/modified atmosphere, and inert dusts. The concept of IPM is defined by the Food and Agriculture Organization (FAO) Panel of Experts as "a pest management system that, in the context of the associated environment and the population dynamics of the pest species, utilizes all suitable techniques and methods in as compatible a manner as possible and maintains the pest population at levels below those causing economic injury" (Kogan 1998). The main focus of an IPM program is prevention, which includes using physical barriers to prevent insects from entering, frequent cleaning and sanitation, controlling the environmental temperature and humidity. Another key focus of IPM is to monitor pest populations and to ensure that the preventive actions are effective through inspections. Most ham producers in the United States fumigate when mite infestations are visually observed. However, a bait trap was recently developed to detect and monitor mite populations in dry-cured ham processing plants, which may assist future decisions on when to apply a treatment to prevent product infestation (Amoah et al. 2016). Cold treatments can be applied as part of an IPM program for stored products. The population of most pests of durable products starts to decline slowly at 10 °C (50 °F) with ceased reproduction. At −15 °C (50 °F), most pest species in durable commodities will be controlled after a few days (UNEP 2007).

Insect Infestation

Arthropod pests are projected to cause postharvest losses between 8% and 25% in developed countries and 70–75% in developing countries. These losses are attributable to pest consumption and contamination.

Cockroaches

Cockroaches are the most common pests among food processing plants and foodservice facilities throughout the world. Control of these pests is essential because they carry and spread various disease organisms. Many carry approximately 50 different microorganisms (such as *Salmonella* and *Shigella*), poliomyelitis, and *Vibrio cholerae*, the causative agent of cholera.

Cockroaches spread disease organisms through contact with food, predominantly through biting and chewing. Cockroaches prefer foods that contain a large amount of carbohydrates but will feed on any substance that humans will consume, human waste, decaying materials, dead insects (including other cockroaches), shoe linings, and paper and wood materials. Cockroaches are most active at night and in dark areas, due to less human activity. Cockroaches are most commonly brought into buildings on product and/or supply shipments. Employees also may carry cockroaches to work with them on their personal belongings.

Some cockroaches will travel between and enter buildings through drain/sewer pipes, underground utility lines, and steam tunnels. The sewers and floor drains can provide a never-ending route for cockroaches to enter the building unless the sewer connection is addressed. Large cockroaches, such as American, Oriental, Brown, and Turkestan species, can survive outdoors in southern regions. Therefore, conducive conditions around plants need to be evaluated and monitored to eliminate or reduce potential harborage.

Cockroaches multiply rapidly through the production of small egg cases on a monthly basis that typically contain 15–40 eggs. The egg case is deposited in a hiding place for added protection. Young cockroaches feed on the same material as the adults shortly after they hatch. Immature cockroaches look like adults except that they are smaller and do not have wings. Young cockroaches develop wings after growing larger and shedding their skin several times. Cockroaches live up to over a year and mate several times. Identification of the specific kind of cockroach infesting an establishment can aid in the determination of the control technique. There are four cockroach species that are on the FDA's Dirty 22 list. These species include the German, American, Oriental, and brownbanded cockroaches.

Species

German Cockroach (*Blatella germanica*)

The German cockroach is 13–20 mm long and pale brown, with two dark brown stripes behind the head. Adults of both sexes have well-developed wings. The female carries the egg case protruding from the tip of the abdomen until hatching occurs. During the approximate lifetime of nine months, an adult female produces approximately 130 offspring.

In food establishments, German cockroaches can infest the main processing or preparation rooms in addition to storage areas, offices, and welfare facilities. German cockroaches are effective "hitchhikers" and may hide in the incoming supplies, packing materials, cardboard boxes, pallets, etc. All incoming materials should be routinely inspected for a potential pest infestation. They prefer to inhabit warm crevices near heat sources and are not usually found in storage areas below ground level. German cockroaches are especially common in restaurants and may be found from floor to ceiling levels in rooms.

American Cockroach (*Periplaneta americana*)

This species is approximately 40–60 mm long and is the largest cockroach in the United States. Adults are reddish brown to brown, and the young are pale brown. The female hides egg cases as soon as they are produced. This species produces more offspring than the German cockroach because the adult female lives for 12–18 months, lays as many as 33 egg cases, and produces approximately 430 offspring.

American cockroaches tend to inhabit open, wet areas, such as basements, sewers, drainage areas, and garbage areas, although this species may be found in storage rooms. They tend to stay in places that are slightly cooler and have larger cracks and crevices than does the German cockroach. This species is most frequently found in large storage areas below ground level, on loading docks, or in basements of food processing plants.

Oriental Cockroach (*Blatta orientalis*)

The Oriental cockroach is shiny, dark brown to black, and approximately 25 mm (1 in) long. The wings are short in the male and absent in the female. Young cockroaches of this species are pale brown. Egg cases from the females are hidden soon after their formation. Females live 5–6 months and can produce one egg case per month for an approximate production of 80 offspring. This species prefers a habitat similar to the American cockroach. In food plants, they normally inhabit below ground storage areas or those areas with a moist environment.

Brownbanded Cockroach (*Supella longipalpa* (Serville))

Brownbanded roaches obtain their name from the two lighter bands they have throughout their dark brownish bodies. Male brownbanded cockroaches have full wings that reach past the pointer of their pointed abdominal areas; however, females have underdeveloped wings that prohibit them from flying. The brownbanded roach could live for about 206 days. Brownbanded cockroaches prefer warm and dry locations, such as near refrigerator motor housings, on the upper walls of cabinets, and inside pantries, closets, dressers, and furniture in general. They can also be found behind picture frames and beneath tables and chairs and inside clocks, radios, light switch plates, doorframes, and dressers. It is common to find them hiding nearer the ceiling than the floor and away from water sources. Accurate identification is paramount to controlling brownbanded cockroaches. Control strategies for other cockroaches will not be efficacious for brownbanded cockroaches.

Field Cockroach (*Blatella vaga*)

The field (vaga) roach was introduced from Asia to California, Arizona, New Mexico, and Texas. Unlike the German roach, the field (vaga) cockroach normally lives outdoors where it feeds on decaying vegetation. This species is not repelled by light and is often active during the daytime. Although primarily a field insect, in many cases, it is actually attracted to lights on buildings or streetlights and then enters the structure through cracks and crevices during drier months in search of water.

Detection

Cockroaches may be found in any location where food is being processed, stored, prepared, or served. These insects tend to hide and lay eggs in dark, warm, hard-to-clean areas. Their favorite harborages are small spaces in and between equipment and shelves and under shelf liners. When cockroaches need food that is not in these areas or when they are forced out by other cockroaches, they will come out into the light.

One of the easiest methods of checking for cockroach infestation is to enter a darkened production or storage area and turn on the lights. Also, a strong, oily odor that arises from a substance given off by certain glands of this insect can indicate the presence of cockroaches. Cockroaches deposit their feces almost everywhere they have visited. These droppings are small, black or brown, and almost spherical.

Control

Control of this pest in food establishments should be a continuous operation through effective sanitation and use of chemicals. The most important form of control is effective sanitation. These pests require food, water, and a sheltered hiding place. Exterior lighting, including parking lot lights, should consist of sodium-vapor bulbs yellow lights) that attract fewer insects than the standard incandescent type (Eicher 2004). Because these insects will eat almost anything, elimination of debris and maintenance of a tidy operation, including welfare facilities, through an ongoing sanitation program is the foundation for cockroach control. Integrated pest management (to be discussed later) is more effective than insecticides (DeSorbo 2004).

Infestation is reduced through filling cracks in floors and walls with caulking or other sealants. It is especially important to seal spaces where large pieces of equipment are improperly fitted to their bases or to the floor. These spaces provide an ideal habitat for these pests. Airflow in a facility should be positive to reduce insect entry. Physical exclusion is a key management process to keep these species out of the facility. Eicher (2004) suggests a rule of thumb for protection against insect entry is the elimination of cracks that permit the sight of light. Infestation is reduced by deprivation of easy access via other sources. These hitchhikers can enter food establishments as cockroaches or as eggs in boxes, bags, raw foodstuffs, or other supplies. Incoming materials should be thoroughly examined and any insects or eggs removed. Cartons and boxes should be removed from the premises as soon as the supplies have been unpacked.

The use of chemical control should follow sanitary practices. Chemical control can be handled through a pest control operator, but integrated chemical control and sanitary practices can be more effective and more economical. Because insects such as cockroaches become inactive at approximately 5 °C (40 °F), refrigerated storage and refrigeration of other areas will reduce infestation. Cockroach control is usually based on the use of baits and bait stations, fungi, and possible nematodes.

Diazinon offers potential for the control of cockroaches. A residual insecticide such as diazinon sprayed in hiding places is considered effective if these pests have not developed a resistance to this compound. This compound is sometimes supplemented with a pyrethrin-based nonresidual insecticide to force the insects from the hidden areas to the sprayed region, where improved contact with the insecticide can occur. Other compounds, such as flowable microencapsulated diazinon, are available for the control of cockroaches and other insects through spot, crack, or crevice treatment but not for application in food handling areas. The liquid pesticide, cyfluthrin, a parathyroid-class chemical, is a nerve toxin that kills insects. This chemical, which has a very low toxicity to humans and pets, may be found in commercial insecticides such as *Raid*. The powder, disodium octaborate tetrahydrate, is a boric acid formulation with low toxicity for humans and pets but causes insects to dehydrate and die (DeSorbo 2004). Any compound applied as an insecticide for the control of cockroaches or other pests should be used according to the directions on the label.

Baiting is an effective method to control or eliminate brownbanded cockroaches. Baits containing hydramethylnon, fipronil, sulfluramid, boric acid, or abamectin can provide a high level of control when applied to those areas where cockroaches harbor. Insecticidal dusts like boric acid, silica aerogel, and diatomaceous earth can provide additional control. The use of residual insecticidal sprays or aerosol foggers within a structure is of little value in controlling brownbanded cockroaches. In fact, these applications may disperse the cockroaches making control difficult and lengthy.

Other Insects

The most common of the seasonal insects in foodservice and food processing plants are flies. The most populous varieties of flies associated with these establishments are the housefly and the fruit fly. There are 11 other flies species that are included in the FDA's Dirty 22 list, which include the little housefly, latrine fly, Cosmopolitan blue bottle fly, Holarctic blue bottle fly, Oriental latrine fly, blue bottle fly, secondary screwworm, stable fly, green bottle fly, black blow fly, and red-tailed flesh fly (Higgins 2015).

The *housefly (Musca domestica)*, which is found throughout the world, is an even greater pest than the cockroach. It is a pest to all segments of the community, transmitting a variety of pathogenic organisms to humans and their food. Examples are human disease such as typhoid fever, dysentery, infantile diarrhea, and streptococcal and staphylococcal infections.

Flies transmit diseases primarily because they feed on animal and human wastes and collect these pathogenic microorganisms on the feet, mouth, wings, and gut. These pathogens are deposited when the fly crawls on food or in the fly excrement. Because flies must take nourishment in liquid form, they secrete saliva on solid food and let the food dissolve before consumption. Fly spittle, or vomitus, is loaded with bacteria that contaminate food, equipment, supplies, and utensils.

Control of flies can be a challenge because these pests may enter a building that has openings only slightly larger than the head of a pin. Flies normally remain close to the area where they emerged as adults, even though they are lured to locations with odors and decaying materials. Air currents frequently carry flies a much greater distance than they normally travel. Flies are most likely to reside in warm locations protected from the wind, such as electric wires and garbage can rims. Houseflies lay an average of 120 eggs within a week of mating and can produce thousands of offspring during a single breeding season. Warm, moist, decaying material that is protected from the sunlight provides an ideal environment for housefly eggs to hatch, with subsequent growth of fly larvae or maggots.

Houseflies are more abundant in the late summer and fall because the population has been building rapidly during the warm weather. When adult flies enter buildings for food and shelter, these pests generally remain. Flies are most active in a 12–35 °C (54–95 °F) environment. Below 6 °C (42 °F) they are inactive, and below −5 °C (24 °F) death can occur within a few hours. Heat paralysis sets in at approximately 40 °C (104 °F), and death can occur at 49 °C (120 °F). It is difficult to control the size of a housefly population because they frequently breed in areas away from food establishments where decaying material exists. Therefore, the most effective means of controlling the fly population is to prevent them from entering processing, storage, preparation, and serving areas and reducing their population size within these areas.

Prevention of entry into food establishments can be accomplished by prompt and thorough removal of waste materials from food areas. Air screens, mesh screens (at least 16 mesh, recommended by the US Public Health Service), and double doors discourage fly entry. Doors should be opened for receiving and/or shipping for a minimal amount of time, and air screens should be operational. Self-closing doors should remain open for a minimal amount of time. A stealth trap has been developed by Ecolab for use outside of plants that has a dark reflective surface, is laced with a pesticide, and is more effective when lighting is included (Higgins 2015). To reduce attraction of flies around a food establishment, outdoor garbage storage should be as far away from doors as possible. If garbage is stored inside, this area should be separated by a wall from other locations and refrigerated to reduce decay and fly

activity. Garbage should be stored in closed containers.

If flies have entered a facility, they can be controlled by the use of an electric flytrap or by other commercial traps, which attract adult flies to blue lights, killing them in electric grids. Electric flytraps should be used all day, and the catch basin should be cleaned daily. Chemical control through aerosols, sprays, or fogs, using chemicals such as pyrethrins, can aid in fly control. The limited results are temporary, and the use of chemicals is restricted in food facilities. Therefore, one should try control by exclusion and through the use of flytraps. At this writing, flytraps that contain the insecticide nithiazine appear to be effective against fly control outside of buildings. BASF has developed Alpine Pressurized Fly Control for indoor use with dinotefuran as the active ingredient. In addition, Syngenta has introduced a granular bate named Zyrox, with the active ingredient cyantraniliprole; Maxforce Fly Spot with the active ingredient imidacloprid is also available for use inside food processing facilities.

Fruit flies (Drosophila melanogaster), which are smaller than the housefly, are also considered seasonal and are most abundant in late summer and fall. Adult fruit flies are approximately 2–3 mm long, with red eyes and light brown bodies. They are attracted to fruit, especially decaying fruit. These pests are not attracted to sewage or animal waste; thus, they carry less harmful bacteria. The life cycle and feeding habits of fruit flies are similar to those of houseflies, except that these insects are attracted specifically to fruits. These pests proliferate most rapidly in late summer and early fall, when rotting plants and fruits are more abundant. The life span of a fruit fly is approximately one month.

Total eradication of the fruit fly is difficult. The use of mesh screens and air screens will decrease entry into food establishments. When entry occurs, electric traps are somewhat effective. One of the most effective methods of controlling these pests is to avoid accumulation of rotting fruits and fermenting foods.

The *cigarette beetle*, one of the most common stored product insects, infests tobacco and dried plant materials such as herbs, spices, grain, processed plants, and dried flowers. Although this insect is frequently mistaken for the drugstore beetle, it is distinguished through its serrated antennae versus the clubbed antennae of the drugstore beetle. When viewed from the horizontal position, the cigarette beetle's head points downward, giving it a "humped" appearance. This insect generally lives 30–90 days, and the larvae feast on the surrounding food supply. This pest is attracted to subdued lights, insect light traps, and pheromone traps. Monitoring grid patterns and trend analysis reports can determine infestation points. These insects can be controlled through identification and removal of infested stored materials and product storage. Methoprene is currently the treatment of choice to control cigarette beetles. A potential technology to control cigarette beetles is heat treatments. Thermal treatments in which ambient air temperature is increased to 48 °C (118 °F) and held for 24 h are lethal for most insects (Hirsch 2004).

Miscellaneous insect pests that plague food processing and foodservice operations are ants, mites, beetles, and moths. The last two are generally found in dry storage areas. These pests can be identified through their webbing and holes in food and packaging materials. These insects can be kept in check through a tidy environment, good ventilation, cool and dry storage areas, and stock rotation.

Ants frequently nest in walls, especially around heat sources, such as hot water pipes. If infestation is suspected, sponges saturated with syrup should be placed in a number of locations to serve as bait in determining where the insecticides should be applied. Because ants, beetles, and moths can thrive on very small amounts of food, good housekeeping and proper storage of food and supplies are essential safeguards against these pests.

Mites are ubiquitous in the environment. However, they are generally not present at levels such that they are visible to the human eye or infesting the food product. Mites reproduce under favorable conditions but are able to reproduce and cause infestations when dried and fermented products are produced such as dry-cured meats, aged artisan cheeses, and fermented food products. Mites on aged meats can still be controlled

through methyl bromide fumigation in the United States, but integrated pest management should be used to control infestations.

Silverfish and firebrats can reside in cracks, baseboards, window and doorframes, and between layers of pipe insulation. Because these pests thrive in undisturbed areas, their presence suggests inadequate and/or infrequent cleaning. Silverfish prefer a moist environment, e.g., basements and drains. The firebrat is more likely to be found in warmer environments, such as around steam pipes and furnaces.

Insect Destruction

Pesticides

Pests should be destroyed without chemicals, if possible, because of the controversy and potential danger of pesticides. However, if these techniques are ineffective, it is necessary to use pesticides or fumigants. To ensure proper and effective application of pesticides, the use of a professional pest control firm should be considered. Restricted pesticides should be applied by a commercial applicator. Even if an exterminating firm is contracted, supervisory personnel from the food establishment should have a basic knowledge of these pests, insecticides, and regulations pertaining to the use of these chemicals.

Residual insecticides are applied to obtain insecticidal effects for an extended period of time. In residual treatment, the chemicals are normally applied in spots or cracks and crevices. Some residual insecticides cannot be legally used in food areas. Therefore, extreme caution should be taken to avoid contamination of food, equipment, utensils, supplies, and other objects that come in contact with workers. People who use these chemicals should be familiar with the terms on the product labels, which describe the types of authorized applications and potential effects.

Another method of residual application of insecticides is *crack and crevice treatment*. Small amounts of insecticides are applied to cracks and crevices where insects hide or in areas where these pests may enter buildings—for example, expansion joints between the various elements of construction and between equipment and floors. Treatment at these locations is critical because these openings frequently lead to voids, such as hollow walls or equipment legs and bases. Other important areas where treatment is essential are conduits, junction or switch boxes, and motor housings.

For example, several pyrethroids have been identified as effective control agents against stored product pests. Populations of *Acarus siro* and *Lepidoglyphus destructor* were inhibited by over 99% when treated with 2 parts per million (PPM) bifenthrin applied to grain (Collins 2006). Moreover, deltamethrin and bifenthrin inhibited the population growth of *T. putrescentiae* by approximately 80% at 4 PPM applied on grain (Collins 2006). Chlorfenapyr is a broad-spectrum insecticide—acaricide that can regulate all developmental stages of phytophagous mites and various insect pests. It has moderate mammalian toxicity (LD_{50} 441 mg/kg) and persists in the environment for a long period of time ($DT_{50} > 365$ days) (Dekeyser 2005). Athanassiou et al. (2014) assessed the residual effect of chlorfenapyr for residual control of stored product psocid species, *Liposcelis entomophila* (Enderlein), and *Liposcelis paeta* Pearman (Psocoptera, Liposcelididae). Chlorfenapyr caused 99–100% and 92–100% mortality of *Liposcelis bostrychophila* and *L. paeta*, respectively, at rates of 13.8 mg/m² or higher after 3 days of exposure. Bifenazate is mainly effective at controlling mobile stages of spider mites but is also active against eggs of particular mite species like *Tetranychus urticae* Koch. The combination of chlorpyrifos-methyl and deltamethrin has been registered in the United States (US) for direct application to protect wheat and small grains, with activity against several species of stored grain insects. Abbar et al. (2016) reported that chlorfenapyr could potentially be used as part of a mold mite (*T. putrescentiae*) integrated pest management program. This insecticide is a proacaricide that is effective at controlling all stages of spider mites and eriophyoid mites. It was demonstrated to control mites on wood, metal, and concrete for up to 8 weeks after its application.

Nonresidual insecticides are applied for the control of insects only during the time of treatment and are applied either as contact or as space treatments. *Contact treatment* is the application of a liquid spray for an immediate insecticidal effect. *Contact* refers to actual touching of the pests. This treatment method should be used only when there is a high probability that the spray will touch the pests. In *space treatment*, foggers, vapor dispensers, or aerosol devices are used to disperse insecticides into the air. This technique can control flying insects and crawling insects in the exposed area. Space spraying should be done to control the insect population.

Nonresidual insecticides may be dispensed through fogging as aerosols in food production areas when food is not exposed. This technique is used to apply pyrethrins, which are usually synergized with piperonyl butoxide. Other common insecticides are pyrethroids. Aerosol applications, which effectively kill flying and exposed insects, are frequently dispensed on a timed-release basis at a prearranged convenient time when food production and contact do not occur.

Fumigants are used in the food industry primarily to control insects that attack stored products. Their primary feature is the ability to reach hidden pests. These compounds are normally used for space treatment, typically on weekends, when processing operations are ceased for safety precautions. To ensure adequate dispersion, fumigants are often applied with air-moving equipment, such as ventilation machinery or fans. The major mode of fumigant action is through the activation of respiratory enzymes within the pest. Oxygen assimilation is blocked or delayed by most fumigants. The following chemicals are common fumigants for insects:

• *Methyl bromide*: This nonflammable fumigant was widely used prior to 2000 since it is the most effective fumigant at controlling the majority of food pests including insects, mites, rodents, microflora, and nematodes. Methyl bromide penetrates food products and structures and acts as a respiratory toxin, apparently absorbed through the insect's cuticle. Prior to 2006, the United States used approximately

21,000 tons of methyl bromide annually, primarily for soil fumigation (85%), for commodity and quarantine treatment (10%), and for structural fumigation (5%) (EPA 2006). However, methyl bromide also depletes the stratospheric ozone layer. Methyl bromide is classified as a Class I ozone-depleting substance (ODS) with an ozone-depleting potential (ODP) of more than 0.2 (EPA 2007). According to an international agreement (the Montreal Protocol) that was ratified by more than 180 countries, methyl bromide would be phased out of all industries by 2015. As of 2015, the United States does not produce any methyl bromide and is only allowed to use methyl bromide for quarantine, imports, and dry-cured pork. The phaseout was scheduled to reduce methyl bromide use by 25% by 1999, 50% by 2001, 70% by 2003, and 100% by 2005 (with the exception of critical use exemptions). Only yearly requests for critical use exemptions, quarantine and pre-shipment applications, and emergency uses are exempt from the phaseout (Johnson et al. 2012). In 2014, preplant strawberries (415 MT), postharvest commodities (0.74 MT), postharvest structures (22.8 MT), and dry-cured ham (3.7 MT) were nominated for critical use exemptions. For preplant strawberries, black root rot or crown rot disease, yellow/purple nutsedge infestation, and/or nematode infestation are the three major conditions that require the critical use (EPA 2012). Postharvest structures of rice millers and pet food manufacturers were exempted in 2014 due to beetle, weevil, moth, or cockroach infestation and the presence of sensitive electronic equipment that are susceptible to corrosion (EPA 2012). Postharvest commodities such as walnuts, figs, dried plums, and dates were exempted in 2014 since rapid fumigation is required for these commodities to meet critical marketing seasons (EPA 2012). Dry-cured pork products have received a critical use exemption due to the red-legged beetle, ham skipper, dermestid beetle, and/or ham mite infestations.

- *Phosphine*: This is a flammable fumigant that has been approved for use in many food

storage situations and is typically applied to cereals, nuts, dried fruits, pulses, oilseeds, and dried animal products. Phosphine rapidly diffuses in air as it has a similar density to that of air. Although PH_3 is extremely volatile and diffuses rapidly, minimal residues persist in food commodities following fumigation (Longobardi and Pascale 2008). The legal limit of phosphine in processed food products is 0.01 PPM. Phosphine is highly toxic to organisms undergoing oxidative respiration but is nontoxic to organisms that can survive in low-oxygen environments (<1%) or that can anaerobically respire. The toxicity of phosphine to all stages of 13 species of stored product beetles was determined. Several PH_3-based pest control procedures generate PH_3 by decomposition of metal phosphides. The most widely used phosphides are aluminum phosphide or magnesium phosphide (Longobardi and Pascale 2008). Phosphine is a strong reducing agent of biological redox systems, especially the components of the mitochondrial electron transport chain that is probably the site of its action in insects. Phosphine is believed to disrupt normal oxygen metabolism in insects, which causes the production of highly deleterious "oxyradicals" and other intermediates. Studies on isolated rat liver have shown that PH_3 inhibits the mitochondrial oxygen uptake due to its reaction with cytochrome C and cytochrome C oxidase. PH_3 also inhibits insect catalase activity. There are multiple factors that influence metabolism, and mitochondrial function has a direct influence on phosphine toxicity. Mitochondrial membrane potential, rate of electron flow through the mitochondrial respiratory chain, ATP levels, metabolic supply versus demand, and mitochondrial generated oxidative stress are all metabolic factors that contribute to the effectiveness of phosphine at eradicating pests (Zuryn et al. 2008). Damaging corrosion to electrical systems and devices is why phosphine fumigation is rarely used in buildings such as food plants. Extreme caution is needed during fumigation to protect electronic devices from corrosion. If the susceptible electronic apparatus cannot be removed from the fumigation area, protection action such as lubricating oil spray or a layer of paraffin may be applied to the copper materials.

- *Ethylene oxide*: This nonresidual fumigant is normally mixed with carbon dioxide in a ratio of 1:9 (by weight) to reduce flammability and explosiveness. This insecticide, most frequently used for stored commodities, should be applied through a professional pest control operator. Ethylene oxide has also been used to sterilize equipment, treat spices, and control arthropod infestations and microbial growth.
- *Carbonyl sulfide*: Carbonyl sulfide is toxic to a large number of species of stored product insects. It has been patented as a fumigant for control of insects and mites in postharvest commodities. It is versatile, being toxic in short exposure periods or for a longer exposure time. This fumigant shows no adverse effects on seed germination and is an effective fumigant for other commodities. It has shown *promise* for controlling the navel orangeworm, cigarette beetle, dried fruit beetle, sawtoothed grain beetle, codling moth in walnuts, and Mediterranean fruit fly in lemon. One potential downside of fumigation with carbonyl sulfide is off-odors, such as in bread that was made from fumigated wheat.
- *Sulfuryl fluoride*: Sulfuryl fluoride (SF) is produced by DOW AgroSciences under the trade name ProFume. Sulfuryl fluoride is as effective as methyl bromide at controlling grain weevils, flour beetles, and other pests associated with some food products. In general, it is effective against the adult, pupal, and larval life stages of insects, but it is not as active as methyl bromide against the egg stage. SF reduces the amount of oxygen taken up by insect eggs. Eggs are less susceptible than adults primarily because the eggshell limits the passage of sulfuryl fluoride. It has been reported that SF fumigation of food products can lead to the absorption of SF by oils and the binding of sulfate and fluoride ions to proteins, resulting in the possibility of the presence of unsafe residues in the product. The mode of action for SF is that it breaks down to fluoride and sulfate inside the insect body. Fluoride is the primary toxin which interferes

with the metabolism of stored fats and carbohydrates that the insect needs to maintain a sufficient source of energy by disrupting glycolysis and the citric acid cycle.

Other Chemical Methods of Insect Control

Other potential methods of insect control include the use of *baits*. Baits are a combination of insect-attracting foods, such as sugar, and an insecticide. Although baits are not always as convenient to use as other methods, they can be effective in controlling inaccessible areas of ant and cockroach infestations and in reducing outside fly populations. Because baits are a poisonous food, special precautions should be exercised in their use and storage. Commercial dry granular baits should be scattered thinly over feeding surfaces daily, or as needed, to provide initial knockdown and control of populations. Granular fly baits are satisfactory for outdoor use only. Liquid baits consist of an insecticide in water with an attractant such as sugar, corn syrup, or molasses. They may be applied using a sprayer or sprinkling can to walls, ceilings, or floors frequented by flies. Fly bait should be used regularly during the summer months to control population growth.

Ozone has been deemed "generally recognized as safe" (GRAS) as a disinfectant for foods. The strength of the case for using ozone may rest with its versatility and environmental benefits over some existing food sanitizing methods. Ozonated water can be used on food products as a disinfectant wash or spray. When dispersed into water, ozone can kill bacteria like *E. coli* faster than traditionally used disinfectants, such as chlorine, since it is a more powerful oxidant than chlorine. Ozone also acts as a disinfectant in its gaseous state and can be applied to sanitize food storage rooms and packaging materials, which may help to control insects during the storage of foods and help prevent the spoilage of produce during transport. Ozone can eliminate insects in grain storage facilities without harming food quality or the environment. Ozone gas is unstable and decays naturally into diatomic oxygen, thus leaving no residues. The rate of decay of ozone is generally dependent on temperature and the surfaces with which it comes in contact with. Ozone is usually produced industrially by subjecting oxygen to an electric arc although it can be chemically generated with UV light. Several studies have been reported regarding ozone fumigation on fruits, vegetables, and stored grains. In addition, ozone toxicity has been tested on a variety of insect pests. Bonjour et al. (2008) evaluated ozone for its effectiveness against six species of stored grain insect pests. The rice weevil, *Sitophilus oryzae*, adults were the most susceptible species with 100% mortality reached after 2 days exposure at 50 ppmv and after four days exposure at 25 ppmv ozone. *Tribolium castaneum* adults had 100% mortality after four days exposure at the 50 ppmv concentration.

Use of Food-Grade Coatings

Edible coatings have been applied for many different purposes on a large variety of food products such as fresh fruits and vegetables, confections, and meat products. Edible films can be made by (1) spraying/brushing the solution with film ingredients or dipping the product into a solution and (2) making stand-alone film first and then cover it to the surface of food. A plasticizer, such as glycerol, is usually added to the coating solution to keep the film from being brittle. It is important that the coating materials applied on food products are generally recognized as safe (GRAS) by the Food and Drug Administration and the US Department of Agriculture. For example, edible coatings can be used as part of an integrated pest management program to control ham mite infestations on hams. A previous study on dipping ham slices/cubes directly into mineral oil, propylene glycol, 10% potassium sorbate, glycerin, and hot lard indicated that both lard and propylene glycol were effective at controlling mite reproduction under laboratory conditions. Xanthan gum, agar, propylene glycol, alginate, and carrageenan +propylene glycol alginate were tested by using proprietary formulations with water and propylene glycol. Coatings were applied by dipping the cubes into the gel solutions. Twenty large (mostly adult female) mites were placed on each cube of ham, and the cube was placed in a mite-proof, ventilated glass container and incubated for 2 weeks. All treatments with 50% PG were effective at controlling mite reproduction (Zhao et al. 2016).

Mechanical Methods

None of the conventional devices to control insects mechanically is especially effective. Fly swatters are contaminated and spread insect carcasses and parts when being used, so they should not be permitted in food processing, storage, preparation, or sales areas. A viable mechanical device for the control of insects is the air curtain, which not only reduces cold air loss in a refrigerated facility but also protects against insect, rodent, and dust entry into food establishments. Air curtains can be used for personnel doors and entrances large enough for loading trucks or for the passage of large equipment. An air curtain supplies a downward-directed fan that sweeps air across the door opening at rates of up to 125 m³/min (4,500 ft³/min). Air curtains are most effective if the area being protected is under positive air pressure. This equipment is normally mounted outside and above the opening to be protected.

Insect Light Traps

One of the safest and most effective methods of fly control is the use of insect light traps. This technique does not have the potential hazard of toxic sprays. Insect light traps use a high-voltage, low-amperage current on a conducting grid (Fig. 13.1) placed in front of a quasi-ultraviolet (UV) irradiation source. This equipment attracts the flies toward the light source, where they are electrocuted. Some light traps contain a "black light," which is effective at night, and a "blue light," which is effective in the daytime.

Insect light traps in food processing plants and warehouses should be installed in stages, as follows:

- *Stage 1, interior perimeter:* These units should be placed near shipping and receiving doors, employee entrances, and personnel doors that provide access to the outside or anywhere else that flying insects may enter. Units should be placed 3–8 m (10–25 ft) inside the doors, away from strong air currents, and out of traffic areas, where forklifts or other equipment may damage the units.
- *Stage 2, interior:* These units should be placed along the path that insects may follow to the processing areas. Within the processing areas,

Fig. 13.1 An insect light trap that attracts flies to the light source, subsequently electrocuting them (Courtesy Gilbert Industries, Inc. Jonesboro, Arkansas)

units with wings should be used to prevent dead insects from falling on the floor or on processing equipment.

- *Stage 3, exterior perimeter:* Covered docks, especially if refuse is being staged, should be protected. The units should be installed between the insects and the entrances but not directly at the entrances.

Although this method of control can be effective, some precautions should be considered. The UV light source should be replaced in the spring to attain optimal effectiveness. The trap should be strategically located to obtain optimal exposure and not to attract insects from the outside. The pan that collects the electrocuted insects should be emptied regularly to prevent infestation by dermestid beetles and pests that feed on dead insects.

Sticky Traps

These traps can consist of sticky flypaper, pieces of waterproofed cord, or flat pieces of plastic covered with a slow-drying adhesive. Yellow plastic strips with a sticky covering will catch a wide variety of flying insects. Some sticky traps contain pheromones so that a specific insect species can be caught. Light trap models use a low-voltage electric pulse to stun the insects, which then fall down onto the glue board. This approach reduces the production of insect fragments and

does not create the bug zapping sound generated by the electrocution traps.

Biological Control

The use of biological control is frequently incorporated into integrated pest management (IPM) programs (discussed near the end of this chapter). One of the most widely used biological control schemes for the control of phytophagous insects is the development and incorporation of host-plant resistance. Resistance is attained through the use of plant species that are known to be refractory to attack. One of the promising techniques is the incorporation of gene splicing and recombinant DNA manipulation, which is being investigated universally. Other possibilities are the use of viruses, fungi, and bacteria to produce diseases in specific pests and of growth regulators, hormones, and pheromones that can influence sexual activity, primarily those that sterilize male pests. Equally important are growth regulators that interrupt the life cycle of insects and prevent their reproduction, usually in the pupal stage of development. Growth regulators have been evaluated experimentally to control mosquitoes, fleas, and other insects. Insects can be potentially controlled by the use of milled diatomaceous earth. The milling process fragments the diatom shell into sharp microscopic particles, which penetrate the insects' wax coating whenever contact is made, causing moisture depletion and death. If particles of the shell enter the body cavity, they interfere with digestion, reproduction, and respiration.

Pheromone Traps

Pheromones are chemical substances emitted by insects to communicate with others of the same species. Types of pheromones include sex attractant, aggregation, fear, and territorial boundary markers. Natural and synthetic sex attractant pheromones lure male insects into sticky traps where they become permanently trapped and die. Some of these traps are based on the use of a specific sex pheromone and have a trapping chamber where the insects are caught. Some are constructed with a plastic funnel leading into the reception chamber, which contains an insecticide strip. Recently developed products containing microencapsulated pheromones provide a slow chemical release over a long period of time. Chemical attractants are now being used to control fruit flies.

Common pheromones for pest control are aggregation (usually produced by long-lived adult insects) and, more frequently, sex pheromones (usually produced by short-lived adults). Aggregation pheromones, usually produced by the male, can cause a response from both sexes.

Pheromone traps can be used in pest management for:

1. Detection and monitoring. Information such as presence, location, and amount of a species can determine when appropriate action should be taken (i.e., pesticide application).
2. Mass trapping. Larger traps with a larger quantity of pheromone can be incorporated to catch insects.
3. Confusion. Sex pheromones can confuse mating instincts of male insects to prevent their location of females.

The use of pheromones in pest management offers the following advantages.

1. Economy. A small amount is required and traps are easy to use.
2. Species specific. A pheromone used to attract a specific species does not attract or harm beneficial species.
3. Nonpoisonous. No known safety hazards exist to humans or other animals.
4. No insect resistance. Sex attractants are fatal to the insects being trapped.

Hydroprene, a nonpesticide insect growth regulator (IGR), is appropriate for cockroach control in sensitive environments because of its margin of safety and toxicity. It has been approved by the Environmental Protection Agency for use in areas where food is present. An IGR can be destructive through disruption of the normal growth and development of immature cockroaches. Growth and development abnormalities include deformed wings and the inability to reproduce.

Trap Placement

Trap placement affects the success of the pest control program. Traps for houseflies and other filth flies should be placed a maximum of 1.5 m (5 ft) above the floor (Mason 2003). Ceiling-mounted traps in a location that permits inspection and cleaning should be installed for night fliers. If light traps are needed near bay doors, they should be placed above the top of the doorway and perpendicular to the door, so that the light is not directed outside. Electric flytraps should not be installed outside near the loading dock because they will attract more flies than can be caught. If a food facility is located near a large body of water, light traps can be placed 9 m (30 ft) or more away from the building with the back of the trap toward the water. Insects that are attracted to the lighted building will be attracted toward the water and away from the food facility. Light traps should not be installed at ceiling level directly over or next to exposed food or within 4 m (13 ft) of a door because of potential attraction of insects to the site, risk of insect fragment contamination, and the possibility of trap interception failure. Traps should not be placed where damage from forklifts, other equipment, or strong air currents can be avoided.

Monitoring of Infestations

A systematic inspection or surveillance and the recording of the species of pests present, their quantity, and origin should be established. Monitoring should include raw materials, adjuncts, and production and storage premises. Laboratory testing of samples should be performed using a filth test method. These methods can be found in the *Official Methods of Analysis* published by the Association of Official Analytical Chemists or in other specialized analytical publications. Insects, insect fragments, eggs, larvae, and chrysalises should be identified, counted, and recorded to permit immediate pinpointing of dangerous infections or the appearance of abnormal variations. The same should be done for rodent hairs and excrement.

Rodents

Rodents such as rats and mice are difficult to control because they have highly developed senses of hearing, touch, and smell. These pests can also effectively identify new or unfamiliar objects in their environment and protect themselves against these changes in the surroundings. The National Pest Management Association (NPMA) established rodent management and other pest management standards for food plants in 2013 (NPMA 2013). The Food Safety Modernization Act (FSMA) has stated that rodent management programs need to be risk-based programs based on the specific pressures of an individual facility (Black 2015). Global Food Safety Initiative (GFSI) benchmarked audits also require a proactive approach to pest management. There are four rodents on the FDAs Dirty 22 list and include the Norway rat (*Rattus norvegicus*), the roof rat (*Rattus rattus*), the Polynesian rat (*Rattus exulans*), and the house mouse (*Mus musculus*).

Rats

Rats can force their entry through openings as small as a quarter, can climb vertical brick walls, and can jump up to a meter (40 in) vertically and 1.2 m (48 in) horizontally. These rodents are strong swimmers and are known for their ability to swim up through toilet bowl traps and floor drains. Rats are dangerous and destructive. The National Restaurant Association has estimated that the loss from rodent damage could be as high as $10 billion per year. This includes consumption and contamination of food and structural damage to property, including damage from fires caused by rats' gnawing on electrical wiring. Of greater importance than economic losses from rat infestation is the serious health hazard from contamination of food, equipment, and utensils. Rats directly or indirectly transmit diseases such as leptospirosis, murine typhus, and salmonellosis. Several million harmful microorganisms can be found in one rat dropping. When droppings dry and fall apart or are crushed, the particles can be carried into food by air movement within a room.

The most abundant rat in the United States is the Norway rat, a red-brown to gray-brown rodent, sometimes known as the *sewer rat*, *barn rat*, *brown rat*, or *wharf rat*. Norway rats are 18–25 cm (7–10 in) long, excluding the tail, weigh 280–480 g (10–18 oz), have a rather blunt nose and a thick set body, and tend to live in burrows. The roof rat is generally found in the south and along the Pacific coast and Hawaii. This rat, which seeks an elevated location for its habitat, has more coordination than the Norway rat. It is black to slate gray; 16.5–20 cm (6.5–8 in) long, excluding the tail; and weighs 220–340 g (8–12 oz). Roof rats will burrow or create nests in trees, vines, and other locations above the ground. As of 2015, this rat has made its way to the center of the United States and Canada through transportation routes and throughout the Mississippi Valley (Black 2015).

The female rat becomes fertile within 6–8 weeks after birth and can produce 6–8 young per litter, 4–7 times per year, if conditions are optimal for reproduction and survival. The typical female weans an average of 20 offspring per year. Rats that receive an adequate amount of food will usually not move more than 50 m (170 ft) from their nest if mates are available. However, rat populations will adjust as food becomes scarce in one location or as a portion of the population starts to die from eradication methods. Rats and mice instinctively avoid uninterrupted expanses, especially if this potential barrier is lightly colored. Therefore, a potential rodent deterrent can be created by the construction of a 1.5-m (5 ft)-wide band of white gravel or granite chips around the outside perimeter of a building.

Mice

Mice, found frequently as the *Mus musculus domesticus* and *M. musculus brevirostris* varieties, are almost as cunning as rats. They are known to enter a building through a hole as small as a nickel. They are skilled swimmers that can swim through floor drains and toilet bowl traps, and they have an excellent sense of balance. Like rats, mice are filthy rodents and can spread diseases similar to those spread by rats. The house mouse, which is found everywhere in the United States, has a body length of 6–9 cm (2.5–3.5 in) and weighs approximately 14–21 g (0.5–7.5 oz). It has a small head and feet and large prominent ears. Mice attain sexual maturity in approximately 1.5 months. Female mice produce 5–6 offspring per litter, up to 8 times per year. The typical female weans 30–35 young per year. Mice do not need a source of water because they can survive on water that they metabolize from food sources. However, they will drink liquids if available.

Mice are easily carried into food premises in crates and cartons. They are easier to trap than rats because they are less wary. Metal and wood-based snap traps are normally effective. Several traps may be spaced about 1 m (40 in) apart. Mice usually accept a new object, such as a trap, often after about 10 min. Black (2015) stated that pest management providers in Great Britain have experienced trap aversion by house mice, indicating that mice are evolving or learning to avoid traps. Sodium fluorosilicate and the anticoagulant chlorophacinone are poisonous tracking powders that are effective in mice control. Except for red squill, mice are destroyed with the same poisons as rats.

Determination of Infestation

Rats and mice are nocturnal animals. Because they tend to be inactive during daylight hours, their presence is not always immediately detected. The presence of fecal droppings is one of the obvious signs of rodent infestation. Rat droppings range from 13–19 mm (0.5–0.75 in) in length and up to 6 mm (0.25 in) in diameter. Fecal material from the house mouse is approximately 3 mm (0.12 in) long and 1 mm (0.04 in) in diameter. Fresh droppings are black and shiny, with a pasty consistency. Older fecal material is brown and falls apart when touched.

Rats and mice generally follow the same path or runway between their nests and sources of food. In time, grease and dirt from their bodies form visible streaks on floors and other surfaces. Because rodents tend to keep in contact with surfaces when they travel, runways along walls, rafters, steps, and inner sides of pipes are frequently

visible. Rat and mouse tracks can be seen on dusty surfaces with light shining from an acute angle. Rodent tracks are identified through spreading talc in areas with suspected rodent activity. Urine stains may be detected through the use of long-wavelength UV light, which will cause a yellow fluorescence on burlap bags and a pale, blue-white fluorescence on kraft paper.

The incisor teeth of rats are strong enough to gnaw through metal pipes, unhardened concrete, sacks, wood, and corrugated materials to reach food. Teeth marks can be observed if gnawings are recent. A bumping noise at night, accompanied by shrill squeaks, fight noises, or gnawing sounds are clues that rodents may be present.

Control

Control of rodents, especially rats, is difficult because of their ability to adapt to the environment. The most effective method of rodent control is proper sanitation. Without an entrance to shelter and the presence of debris, which can nourish rodents, these pests cannot survive and will migrate to other locations. Without effective sanitation practices, poisons and traps will provide only a temporary reduction in a rodent population.

Prevention of Entry

Protection against rats is accomplished most effectively through the elimination of all possible entrances. Poorly fitting doors and improper masonry around external pipes can be flashed or covered with metal or filled with concrete to block entry of rodents. Vents, drains, and windows should be covered with screens. Because decay in building foundations will permit rats to burrow into buildings, masonry should be repaired, and fan openings and other potential entrances should be blocked. Recent innovations include bug and rodent blocking loading dock doors and bug and rodent blocking mesh seal doors.

Rodent control is enhanced through depriving them of a location to reside (harborage). Shapton and Shapton (1991) have suggested that outside equipment must be raised 23–30 cm (9–12 in) clear of the surface to prevent rodent harborage. Shrubbery should be at least 10 m (33 ft) away from food facilities. Katsuyama and Strachan (1980) recommend that a grass-free strip 0.6–0.9 m (2–3 ft) in size be covered with a layer of gravel or stones 2.5–3.8 cm (1–1.5 in) deep around food processing buildings. This feature helps to control weeds and rodents and is convenient for the sanitation inspection rodent bait stations or traps placed against the building. Shapton and Shapton (1991) suggested that employees not eat on the plant grounds because dropped food attracts rodents, birds, and insects.

Elimination of Rodent Shelters

Crowded storage rooms with poor housekeeping provide sheltered areas for rodents to build nests and reproduce. Rodents thrive in areas where garbage and other refuse are placed. These sheltered areas are less attractive to rodents if garbage is stored 0.5 m (20 in) above the floor or ground. If waste containers are stored on concrete blocks, hiding places beneath them are eliminated. Waste containers should be constructed of heavy-duty plastic or galvanized metal with tight-fitting lids. Housekeeping can be improved, with concomitant protection against rodent infestation, by storing foodstuffs on racks at least 15 cm (6 in) above the floor or away from the walls. A white strip painted around the edge of the floor of storage areas reminds workers to stack products away from the walls and aids in the identification of rodent infestation through the presence of tracks, droppings, and hair.

Elimination of Rodent Food Sources

Proper storage of food and supplies combined with effective cleaning can aid in the elimination of food sources for rodents. Prompt cleaning of spills, regular sweeping of floors, and frequent removal of waste materials from the premises also reduce available food for rodents. Food ingredients and supplies should be stored in properly constructed containers that are tightly sealed.

Eradication

The more effective methods of eradicating rodents are poisoning, gassing, trapping, and ultrasonic devices.

Poisoning

Poisoning is an effective method of eradication; however, precautions are necessary because poison baits are hazardous if consumed by humans. Examples of rodenticides are the anticoagulants, such as 3-(α acetonylfurfuryl)-4-hydroxycoumarin (fumarin), 3-(α acetonylbenzyl)-4-hydroxycoumarin (warfarin), 2-pivaloyl-1, 3-indandione (pival), brodifacoum, bromadiolone, chlorophacinone, and difethialone. These multidose poisons must be consumed several times before death occurs, and accidental consumption of poisoned bait does cause danger.

The multiple-dose anticoagulants (chronic poisons), although safer than most other poisons, should be prepared and applied according to directions. The ideal locations for application are along rodent runways and near feeding sites. Fresh bait should be put out daily for at least two weeks to ensure that the poison is effective. Brodifacoum, difethialone, and bromadiolone are considered second-generation rodenticides that are single-dose treatments.

Anticoagulant rodenticides are commercially available in several forms. They are sold as ready-to-use baits that can be placed in plastic or corrugated containers near rodent runways; in pellet form, mixed with grain for use in rodent burrows and dead spaces between walls; in small plastic packages for placement in rodent hiding places; in bait blocks; and as salts that are mixed with water. The sanitarian or pest control operator should record the location of all bait containers for easy inspection and replacement. If bait is not consumed after two or more inspections, it should be relocated.

Anticoagulants have been extensively used to eradicate rats. One unfortunate result is that rats have become increasingly resistant to them. Consequently, new control strategies are being studied that utilize alternative cycles of anticoagulant and acute (fast-acting) rodenticides.

Bromethalin, a nonanticoagulant, has been reformulated and remarketed by two manufacturers since its introduction in the early 1980s. This rodenticide produces death in rodents 1–3 days compared to 5–7 days for anticoagulants but is approximately twice as expensive as anticoagulant baits.

If immediate death of rodents is required, single-dose (acute) poisons, such as red squill and zinc phosphide, are available. These poisons can be mixed with fresh bait material, such as meat, cornmeal, and peanut butter. These baits should be prepared and administered according to directions provided by the manufacturer. Unfortunately, some of the single-dose poisons are effective against only Norway rats.

Baits should be deposited in several locations because rodents frequently travel only a limited distance from their shelter. If sufficient food and shelter are available, rats tend to stay within a radius of 50 m (165 ft). Mice tend to journey about 10 m (33 ft) under similar conditions. If baits are dispersed too sparsely or are not strategically located, rodents may not locate the poison. Where signs of rodent activity are recent and numerous, baits should be dispersed liberally and replaced frequently. Rodents that are killed by single-dose poisons may die in their nests. Dead rodents should be removed and burned or buried. Most mice are destroyed from the same compounds as rats.

Although the use of bait is one of the most effective methods of eradication, rats that have suffered a toxic response by ingesting a poison, such as discomfort and pain but not death, may avoid the bait. They also become cautious if dead or dying rats are near bait. Therefore, the most acceptable bait is the type with which the rat is most familiar. Bait shyness and avoidance may be countered by the use of prebait, nonpoisoned bait introduced for approximately 1 week. Then the prebait is replaced with the same bait containing a rodenticide. Prebaiting is especially important if single-dose poisons are used but is not recommended when anticoagulants are incorporated. Because mice have weaker avoidance instincts than rats, prebaiting for mice is not necessary.

Tracking Powder

These compounds kill rats or, in the case of nontoxic powders, identify their presence and number. These powders may contain an anticoagulant or a single-dose poison. This poison kills rodents when they groom themselves after running through the powder. Such powders are effective if the food supply is abundant. It is best to use self-contained bait boxes placed inside the buildings where the food products are processed, prepared, or stored to restrict the spread of these poisoned baits. Tracking powders are less effective against rats than mice, but sodium fluorosilicate is an effective rodenticide.

Gassing

This technique should be used only if other eradication methods are not effective. If this approach is necessary, rodent burrows should be gassed with a compound such as methyl bromide only by a professional exterminator or a thoroughly trained employee. Rodent burrows should not be gassed if they are less than 6 m (20 ft) from a building because burrows can extend beneath a closely located building.

Trapping

This is a slow but generally safe method of rodent eradication. Traps and bait stations should be tamper resistant so that nontarget animals cannot get into them and placed at right angles to rodent runways, with the baited or trigger end toward the wall. Food that appeals to rodents can be used as bait. Traps should be checked daily, with trapped rodents removed and bait replaced as needed. Trapping should be considered a supplement to other methods of eradication, and an abundance of traps should be used. The sanitarian should be aware of the rat's innate shyness and adaptability. Rats can avoid traps as effectively as they can bait. An effective mousetrap is the glue board, which physically prevents a mouse from escaping by the glue sticking to its feet. After use, the pest control operator should discard the disposable tray and mouse and place a new tray in the most strategic location.

Ultrasonic Devices

This eradication method uses sound waves that are supposed to repel the entry of rodents into areas where the device is installed. The most appropriate time to hit rodents with noise is when they first arrive. Although this method can reduce the presence of rodents, with prolonged hunger, rodents ignore the sound barriers. Furthermore, ultrasound does not provide randomly and continually varying frequencies, which may be more effective. Machines are available that emit a combination of three or four different sounds, not any one of which are totally effective but in combination provide enough stress that rodents will leave the area. If infestation is established, it may require 6–9 days for riddance, but the induced stress makes the rodents more vulnerable to being caught through trapping.

Rodent Management Self-Assessment

Black (2015) recommended exploring the following questions to determine if changes need to be made to a company's rodent management program:

1. Have there been any changes, additions, or subtractions to rodent control devices in the program in the last 5 years?
2. Is the rodent management program focused on the plant's area of biggest risk?
3. Is it possible to quickly access pest management data without calling the vendor?
4. Are the recommendations of the pest management company clear, concise, and actionable?
5. Is the pest management company's service report clear?
6. What actions need to be taken by plant personnel and the pest management company after each service by the pest management company?

If the answer to all of these questions is not yes, changes need to be made to the rodent management program such that the answer to all of these questions become yes.

Birds

Birds such as *Columba livia* (pigeons), *Passer domesticus* (sparrows), and *Sturnus vulgaris* (starlings) may present problems for the food facility. Their droppings are unsightly and can carry microorganisms detrimental to humans. Birds are potential carriers of mites, mycosis, ornithosis, pseudotuberculosis, toxoplasmosis, salmonellosis, and organisms that cause encephalitis, psittacosis, and other diseases. Insect infestations may also occur from those brought into the plant by birds. The close association of birds, such as European starlings, with people in urban areas presents a threat because of their propensity for transmittal of fungal and bacterial diseases directly and also to serve as reservoirs for viral encephalitis (Gingrich and Oysterberg 2003).

A bird population can be reduced through proper management and sanitation. *Exclusion* is an effective and less objectionable method to control bird infestation. Holes and gaps can be eliminated through sealing with hardware cloth, mortar patching, netting, expandable foam, and sheet metal. If sanitary practices are followed to remove food from the site, birds will not be attracted. Entry into buildings can be reduced through the installation of screens on doors, windows, and ventilation openings.

Trapping is generally considered an acceptable method of bird control. Traps should be prebated for 1–2 days to permit acclimation. Wires that administer a mild electric shock and pastes that repel birds are also effective in preventing them from roosting near food establishments. However, electric wires are expensive and require frequent inspection and maintenance. Flashing lights and noisemaking devices have a limited effect on birds, which soon become accustomed to this equipment. Other techniques that may be effective if conducted repeatedly are removal of bird nests and spraying of birds with water as a form of harassment. The most effective procedure for eradication is employment of an exterminator who specializes in bird control. A professional exterminator provides expertise and equipment required for the safe use of chemicals to combat birds.

Bird density can be reduced through the use of commercially available chemical poisons, although these compounds should not be used inside a food establishment. Strychnine has been used in the past; however, its incorporation is restricted by some local regulations. Strychnine alkaloid is used at a concentration of 0.6% to coat baits such as cereal grains. Dead birds should be removed so that dogs and cats will not eat them and suffer from secondary poisoning. Another compound that controls bird density is 4-aminopyridine. In addition to killing birds, it causes the affected birds to make distress sounds and to behave abnormally, thus frightening away those that remain. Azacosterol is a temporary sterilant approved only for the control of pigeons. A biological control method such as this offers potential with less risk than other compounds but provides only a short-term solution, especially in a long-lived species such as pigeons. Minimal intermediate value from this compound is provided to the sanitarian that must rid a bird population immediately.

Birds can be controlled through trapping. Live decoys are required for maximal efficiency. Starlings have been trapped effectively through decoys and an Australian crow trap. Tunnel traps and sparrow traps can also be effective. Pigeons can be trapped with a device containing bars that swing inward into a trap baited with grain. A major limitation of trapping is the cost of labor and materials. There are a variety of bird control measures that can be used without resorting to bird poisons. These controls include bird spikes and gutter spikes that birds cannot land on, electronic shock tracks, bird netting, misting systems, and sonic devices.

Although frequently used except at airports and large military bases, the employment of a falconer and trained peregrine falcons can be effective (Gingrich and Osterberg 2003). When falcons are observed, other birds leave quickly. This biological control method is expensive and may require a falconer to be present for up to a week to prevent new flocks of birds from occupying territories occupied by the departed flocks.

Use of Pesticides

Insecticides should not be sprayed in food areas during hours of operation. They should be applied only after the shift, over the weekend, or at other times when the food establishment is closed. Precautions should be taken to ensure against spattering or drift of the insecticide out of the treatment area to adjacent surfaces or onto food. Insecticidal dusts, which generally contain in dry form the same toxic compounds present in sprays, are also available. They require more skill in application than do sprays and should be administered only by professional pest control operators.

Prior to the use of insecticides approved for edible food products or supply storage areas, all exposed food and supply items should be covered or removed from the area to be treated. The equipment used in spraying inevitably will become contaminated and must be thoroughly cleansed before reuse. This is best accomplished by scrubbing with a cleaning compound and hot water and then rinsing. Products containing residual-type insecticides should not be used on any surfaces that come into contact with food. A fumigation procedure is not recommended unless it appears to be the only effective method and even then only when it is carried out by a professional fumigator. Under no circumstances should regular plant personnel or supervisors attempt this type of work unless they are thoroughly trained. Even when professional fumigators are used, the plant managers should ensure themselves that all precautions have been taken in accordance with accepted safety practices.

The following precautions, suggested by the National Restaurant Association Education Foundation (1992), should be considered when applying pesticides:

1. Pesticide containers should be properly identified and labeled.
2. Exterminators employed should have insurance on their work to protect the establishment, employees, and customers.
3. Instructions should be followed when using pesticides. These chemicals should be used

for only the designated purposes. An insecticide effective against one type of insect may not destroy other pests.

4. The weakest poison that will destroy the pests should be used with the recommended concentration.
5. Oil-based and water-based sprays should be used in appropriate locations. Oil-based sprays should be applied where water can cause an electrical short circuit, shrink fabric, or cause mildew. Water-based sprays should be applied in locations where oil may cause fire, damage to rubber or asphalt, or an objectionable odor.
6. Prolonged exposure to sprays should be avoided. Protective clothing should be worn during application, and hands should be washed after the application of pesticides.
7. Food, equipment, and utensils should not be contaminated with pesticides.
8. If accidental poisoning occurs, a physician should be called. If a physician is unavailable, a fire department, rescue squad, or poison control center should be contacted. If immediate assistance cannot be obtained, treatment should include induction of vomiting by inserting a finger down the throat, with a follow-up of two tablespoons of Epsom salts or milk of magnesia in water, followed by one or more glasses of milk and/or water. If the poison does not present immediate danger, no action should be taken until a physician arrives. Poisoning from heavy metals should be treated with the administration of a half-teaspoon of bicarbonate of soda in a glass of water, one tablespoon of salt in a glass of warm water (until vomit is clear), two tablespoons of Epsom salts in a glass of water, and two or more glasses of water. If strychnine poisoning occurs, administer one tablespoon of salt in a glass of water within 10 min to induce vomiting, followed by one teaspoon of activated charcoal in half a glass of water. The victim should then be laid down and kept warm.

Chemical pesticides are not considered to be a substitute for effective sanitation. Rigid sanitary practices are more effective and more economical than are pesticides. Even with effective pesticides, pests will return when unsanitary conditions prevail.

To minimize possible contamination, a food facility should store on the premises only pesticides essential to control pests that present a problem to the establishment. Pesticide supplies should be checked periodically to verify inventories and to inspect product condition. The following storage precautions should be observed:

1. Pesticides should be stored in a dry area and at a temperature that does not exceed 35 °C (96 °F).
2. The area where pesticides are stored should be located away from food handling and food storage areas and should be locked. These compounds should be stored separately from other hazardous materials, such as cleaning compounds, petroleum products, and other chemicals.
3. Pesticides should not be transferred from their labeled package to any other storage container. Storage of pesticides in empty food containers can cause pesticide poisoning.
4. Empty pesticide containers should be placed in plastic receptacles marked for disposal of hazardous wastes. Even empty containers are a potential hazard because residual toxic materials may be present. Paper and cardboard may be incinerated, but empty aerosol cans should not be destroyed through burning. Local regulatory requirements related to restricted pesticides and general use and disposal should be followed.

Integrated Pest Management

Because of limitations of chemical pesticides, integrated pest control programs based on predicted ecological and economic consequences have been developed. Most single insect control methods have not been successful, and insect resistance to pesticides has become extensive. Thus, a variety of methods have been selected and integrated into a control program for the target pest. This program is called *integrated pest management* (IPM). Its major objective is to control pests economically through environmentally sound techniques, many of which use biological control. The goals of IPM are to use pesticides wisely and to seek alternatives to commonly used pesticides.

IPM implies that pests are "managed" and not necessarily eliminated. However, the ultimate objective of pest management in food processing is to prevent or control pest infestations. Several food processing and preparation firms have discovered the benefits of IPM as a means for pest control, due to the progress accomplished in the development and implementation of these methods since the early 1970s (Brunner 1994). Economic, social/psychological, and environmental advantages may be attained through IPM. Outlook for the acceptance of IPM methods is encouraging and should continue to improve over time with continued exposure. The apparent benefits are realized through lower costs, increased pest control, and reduced pesticide usage. Pest control practices are classified as inspection, housekeeping, and physical, mechanical, and chemical methods. The integrated use of these practices in a complementary manner is essential for economical, effective, and safe pest management. A brief discussion of control practices follows.

Components of a rodent integrated pest management program for food plants include (1) exclusion and (2) sanitation. These components must comprise the main thrust of the rodent control program. Entry prevention is paramount to rodent control. In addition to the implementation of sanitation and rodent proofing efforts, the use of rodent baits and traps provide a preventive and remedial role in a rodent IPM program. Baits and traps are typically incorporated in a "perimeter defense" program. IPM emphasizes sex pheromones because they are environmentally friendly, species specific, and effective at low doses.

An integrated pest management program for a dry-cured ham plant is specific for the mold mite (*Tyrophagus putrescentiae*) since it is the primary pest of dry-cured ham. The IPM program includes (1) sanitation; (2) monitoring of mites, including trapping; (3) crack and crevice spray; (4) the use of food-grade coating; and (5) fumigation as a

last resort. All of these items can be included in a HACCP-based preventive IPM program where fumigation only occurs if the target pest is not under control.

Inspection

Inspection is a preventive, monitoring control measure that is time-consuming but important and cost-effective. Increased practice of IPM to replace chemical control practices has made inspection a more critical function. This function can identify existing problems and detect potential problems and can monitor an ongoing sanitation problem. Both formal and informal inspections should be conducted periodically (e.g., monthly). Formal inspections should be conducted with a predetermined frequency. These inspections should be thorough and should evaluate the overall progress and effectiveness of pest management. If well-qualified inspectors can be obtained from outside the plant (e.g., corporate staff inspector, consultant, or contracting inspection service representative), this resource should be used.

Informal inspections should be conducted periodically through plant personnel assigned to specific work areas. Supervisory personnel should encourage and expect awareness of sanitation problems that may reduce pest control effectiveness among plant personnel as they conduct their normal tasks. Inspections should include raw materials, manufactured or prepared products, site, facilities, and equipment. Inspectors should be equipped with a flashlight, equipment-opening tools, and sample containers. An inspection form should be devised as a guide and for recoding results. These forms provide written identification of potential problems and identification of problem areas.

Housekeeping

Mills and Pedersen (1990) suggested that standards of cleanliness and cleaning schedules must be established with direct accountability for cleaning activity. These authors suggested that, in many areas, cleaning must be continuous, as even small amounts of undisturbed product residues can attract infestation and provide adequate pest harborage. Furthermore, this residual material contains allergens and is the major cause of asthma in inner-city children (Desorbo 2004).

Physical and Mechanical Methods

Because many pesticides once commonly used are no longer allowed in the control of pests, physical and mechanical methods have become more important. Examples are rodent traps, glue boards, and electric flytraps. Generally, these methods are noncontaminating and can fill some of the gaps in an IPM program left by reduced or restricted pesticide use. One of the effective methods is temperature manipulation, which is sometimes combined with forced air movement. Because the optimal temperature for most insect species is 24–34 °C (75–92 °F), variation above or below this range can reduce pest proliferation.

Insects depend as much on suitable moisture levels as on acceptable temperatures; thus, moisture content is critical in determining whether proliferation occurs. Lower moisture content (especially below 12%) of foods discourages insect growth. Several forms of radiation, such as radio frequencies, microwaves, infrared and ultraviolet light, gamma rays, X-rays, and accelerated electrons can effectively disinfect food products, but not all of these methods are effective and practical. Gamma rays, X-rays, and accelerated electrons have commercial applications for insect disinfection.

Chemical Methods

Pesticides and other chemicals, such as repellents, pheromones, and sticky materials for traps, barriers, or repellency, are incorporated when needed. Whoever applies pesticides must be trained to know the safe, approved, and effective use of each chemical. Application of restricted-use pesticides requires state certification of the

application. IPM-targeted establishments have been treated with nonvolatile, low-toxic methods such as gel bait formulation, hydramethylnon, which is safe for commercial food handling areas. When applied among cockroach populations, these insects consume the bait and return to harborage, where they excrete feces containing fipronil, another active ingredient in the formulations. Consumption of the contaminated feces by other cockroaches gives them a lethal dose. When the cockroaches die, others may consume the carcasses and die as well (DeSorbo 2004).

The Environmental Protection Agency (EPA) classifies pesticides as being either for general use or restricted use. Those classified as restricted use are more likely to adversely affect the environment or to injure the applicator. Thus, these pesticides can be purchased and used by only certified applicators or by persons directly under a certified applicator's supervision. Through an EPA-approved program, states train and certify applicators.

The pesticide storage area should be large enough to store normal supplies of pesticide materials adequately and neatly. This should be in a separate building, if possible, or stored in isolated areas from food. The area should be equipped with power ventilation exhausting to the outside and should never be cross-ventilated with food processing or food container storage areas. This storage area should be totally enclosed by walls, and the door should be locked to prevent unauthorized entry. The storage environment should be dry, with the temperature controlled sufficiently to protect the pesticides. Pesticide containers should be stored with the label plainly visible and a current inventory maintained. Pesticide handling and application equipment should include rubber gloves, protective outer garments, and respirators such as dust masks or self-contained breathing apparatus (SCBA)equipment.

Chemosterilants offer potential for the control of rodents. A single oral dose of alpha-chlorohydrin (which is effective in sexually mature male rats) high enough to cause sterility is effective within 4 h. As an acute toxicant, it compares favorably with similar rodenticides. After ingestion, rats and mice rapidly degrade alpha-chlorohydrin. Thus, there is no danger to nontarget species that may eat rats or mice killed by this compound. Since there is no secondary or cumulative toxicity, alpha-chlorohydrin is biodegradable and poses no known long-term danger to the environment.

Although more costly than conventional methods, IPM principles will be applied to future pest control programs because of the success of this program and increased environmental concerns associated with the indiscriminant use of chemical insecticides. The control of insects in commodities by the IPM technique influences the overall infestation levels in plants processing these materials in foods.

Insect-Resistant Packaging

Insect-resistant packaging is a control strategy that may not always be incorporated when considering nonchemical control or exclusion techniques. Stored product insects vary in their ability to contest packages (Arthur and Phillips 2003). These pests may be penetrators, capable of boring through packaging materials or invaders that can enter through seams or openings. Insects may vary in their ability to enter packages at different life stages (Mullen 1997). Packaging films may vary in their ability to prevent insect entry. For example, polypropylene films are more resistant to insect entry than those manufactured from a polyvinyl chloride polymer.

Study Questions

1. What adverse effects do cockroaches have on a food facility?
2. How are cockroaches best controlled?
3. Why are flies so unsanitary?
4. How are flies destroyed most effectively?
5. What is the difference between a residual and a nonresidual insecticide?
6. How does an insect light trap destroy flies?
7. What are insect pheromones?
8. How are rats and mice controlled most effectively?

9. How are birds controlled most effectively?
10. What is integrated pest management?
11. What are the merits of integrated pest management?
12. What are pheromones?
13. Why was it decided that the fumigant methyl bromide would be phased out of the industry
14. What is the Dirty 22 list?

References

Abbar S, Schilling MW, Phillips TW (2016). Efficacy of selected pesticides against Tyrophagus putrescentiae (Schrank): Influence of applied concentration, application substrate, and residual activity over time. *J Pest Sci* 90: 379.

Amoah B, Schilling MW, Phillips TW (2016). Monitoring Tyrophagus putrescentiae (Schrank) (Acari: Acaridae) with traps in dry-cured ham aging rooms. *Environ Entomol*. https://doi.org/10.1093/ee/nvw059

Arthur F, Phillips TW (2003). *Stored-product insect pest management and control*, eds. Hui YH, et al., 341. Marcel Dekker, Inc.: New York.

Athanassiou CG, Kavallieratos NG, Arthur FH, Throne JE (2014). Residual efficacy of chlorfenapyr for control of stored-product psocids (Psocoptera). *J Econ Entomol* 107: 854.

Black J (2015).Three reasons rodent management is important this winter season. *In*, Get ready for fall pests. *Food Proc* (10): 1.

Bonjour EL, Jones CL, Noyes RT, Hardin JA, Beeby RL, Eltiste DA, Decker S (2008). Efficacy of ozone against insect pests in wheat stored in steel grain bins. In *Proceedings of the 8th International Conference on Controlled Atmosphere and Fumigation in Stored Products*. p. 522.

Brunner JF (1994). IPM in fruit tree crops. *Food Rev Int* 10: 135.

Collins D (2006). A review of alternatives to organophosphorus compounds for the control of storage mites. *J Stored Prod Res* 42: 395.

Dekeyser MA (2005). Acaricide mode of action. *Pest Manag Sci* 61: 103.

DeSorbo MA (2004). Combating cockroaches. *Food Qual* 11(5): 24.

Eicher E (2004). Environmentally responsible pest management. *Food Qual* 11(5): 29.

EPA (2006). Environmental Protection Agency. Final rulemaking: The 2006 critical use exemption from the phase-out of methyl bromide. Retrieved from http://www.epa.gov/spdpublic/mbr/

EPA (2007). Environmental Protection Agency. U.S. nomination for methyl bromide critical.

EPA (2012). Methyl bromide critical use renomination for post-harvest commodities. U.S. Environmental Protection Agency.

Gingrich JB, Oysterberg TE (2003). *Pest birds: Biology and management at food processing facilities*, eds. Hui YH, et al. Marcel Dekker, Inc.: New York. p. 317.

Higgins KT (2015). *In*, Get ready for fall pests. *Food Proc* (10): 1.

Hirsch H (2004). Pest of the month: Cigarette beetle. *Food Saf Mag* 10(1): 59.

Johnson JA, Walse SS, Gerik J (2012). Status of alternatives for methyl bromide in the United States. *Outlooks Pest Manag* 23: 53.

Katsuyama AM, Strachan JP (1980). *Principles of food processing sanitation*. The Food Processors Institute: Washington, DC.

Kogan M (1998). Integrated pest management: Historical perspectives and contemporary developments. *Ann Rev Ent* 43: 243.

Longobardi F, Pascale M (2008). Rapid method for determination of phosphine residues in wheat. *Food Anal Methods* 1: 220.

Mason L (2003). Insects and mites. In *Food plant sanitation*, eds. Hui YH, et al. Marcel Dekker, Inc.: New York. p. 293.

Mills R, Pedersen J (1990). *A flour mill sanitation manual*. Eagan Press: St. Paul, MN. p. 55.

Mullen MA (1997). Keeping bugs at bay. *Feed Manag* 48(3): 29.

National Restaurant Association Education Foundation (1992). *Applied foodservice sanitation*. 4th ed. John Wiley & Sons: New York, NY, in cooperation with the Education Foundation of the Restaurant Association.

NPMA, National Pest Management Association (2013). Pest management standards for the food plant. 2013. Pages. http://www.npmatesting.com/PDFs/Foodplantstandards2012.pdf

Rentfrow G, Hanson DJ, Schilling MW, Mikel WB (2008). The use of methyl bromide to control insects in country hams in the Southeastern United States. Extension Publication. University of Kentucky Extension/National Country Ham Association. Publication # ASC-171.

Shapton DA, Shapton NF (1991). Buildings. In *Principles and practices for the safe processing of foods*. Butterworth-Heinemann: Oxford. p. 37.

UNEP (United Nations Environment Programme) (2007). Montreal protocol on substances that deplete ozone layer. *Progress report on technology and economic assessment panel*. http://ozone.unep.org/teap/Reports/TEAP_Reports/Teap_progress_report_April2007.pdf

Use Exemptions from the 2007 Phase-out of Methyl Bromide (2007). October 4. http://www.epa.gov/ozone/science/ods/index.html

Zhao Y, Abbar S, Phillips TW, Schilling MW (2016). Development of food-grade coatings for dry-cured ham. *Meat Sci* 113: 73.

Zuryn S, Kuang J, Ebert P (2008). Mitochondrial modulation of phosphine toxicity and resistance in *Caenorhabditis elegans*. *Toxicol Sci* 102(1): 179.

Sanitary Design and Construction for Food Processing

14

Abstract

Sanitary design and construction or renovation of food facilities is essential to maintain a sanitary operation. Hygienic design begins with a site free of environmental contamination such as polluted air, pests, and pathogenic microorganisms. Site preparation is necessary to attain proper drainage and the reduction of contamination from the environment. All portions of a food facility should contain smooth, impervious surfaces that discourage pest entry. Loading dock design should include dock seals.

Facility and equipment design enhances cleaning effectiveness and avoids microbial growth niches. Process design should incorporate a flow that prevents finished items from making contact with raw materials and unprocessed products. During construction (especially renovations), the suppression of dust particles can be reduced by creating a negative pressure in the construction area and the erection of temporary walls to separate the construction area from food production. Although expensive, stainless steel should be considered for food contact surfaces.

Keywords

Construction • Design • Equipment • Materials • Renovation

Introduction

A food establishment should follow a sanitary design strategy to ensure that the facility can be cleaned to protect against spoilage and pathogenic microorganisms. The major sources of cross-contamination from the physical facility and the food processed, prepared, or stored are product flow design and personnel contamination.

Higher hygienic standards in food processing operations are being adopted. New and renovated food processing and foodservice facilities should be planned to enhance a hygienic operation and effective cleaning. Depending on location, different building codes, permits, and other regulations will determine some of the design processes. Because most equipment and facilities are designed to feature functionality, hygienic design and construction principles should be emphasized to

© Springer International Publishing AG, part of Springer Nature 2018
N.G. Marriott et al., *Principles of Food Sanitation*, Food Science Text Series,
https://doi.org/10.1007/978-3-319-67166-6_14

ensure a sanitary operation. Hygienically designed facilities enhance the wholesomeness of all foods and improve the effectiveness and efficiency of a sanitation program.

A facility should be as sustainable as possible. The facility should come as close as possible to a net zero on utility consumption and disposal. Since poorly designed facilities do not provide sustainability, the structure of the facility and its environment is critical to sustainability.

Schug (2015) suggested that designers working on a new site from scratch can take advantage of building information modeling (BIM) tools in 3D formats that illustrate processors hygienic interior elements such as sanitary workflow, construction materials, and optimal equipment replacement. Designers can use BIM data to determine how to best access equipment for cleaning, sanitizing, and maintenance.

Site Selection

Site selection plays an important role in the development of a hygienic operation. Food establishments should not be constructed near chemical plants that emit noxious odors nor close to salvage or water disposal operations. Food products that are relatively high in fat will readily pick up bad odors and flavors, and pathogenic microorganisms can be picked up by the wind and blown on the manufactured products unless special filters are added to the intake air systems. Drainage is important, as sites located close to standing water with poor drainage are more likely to have *Listeria monocytogenes* in the facility and on manufactured products. Also, drainage will benefit the foundation. Large bodies of water will attract scavenger birds that carry *Salmonella*. Standing water in an environment conducive to insects provides water to sustain the lives of rodents and other pests. For protection against pathogenic microorganisms, a food manufacturing or distribution facility should not be located near existing pest harborages.

To facilitate a sanitary environment, the location of a food plant near small streams and drainage ditches should be avoided, as should locations near refuse dumps, landfills, and equipment storage yards. Land reclaimed from swampy ground or disposal areas for refuse should not receive serious consideration.

The selected site should permit future expansion. Overcrowded facilities are inefficient and pose a sanitation-related liability. Water availability and adequate waste disposal facilities should be considered. Trees and foliage provide food and/or harborage for birds and should not be planted close to the buildings; furthermore, existing growth should be removed. Parking lots should be paved to prevent dust and be well drained to facilitate prompt removal of rainwater. A perimeter chain-link-type fence that surrounds the property should be considered.

Site Preparation

If present at the site, toxic materials should be removed to reduce potential contamination. The site should be graded to prevent standing water, which provides breeding sites for insects (especially mosquitoes). Storm sewers should be provided. If the local municipality requires landscaping for aesthetic reasons, shrubbery should be at least 10 m (11 yards) from buildings to reduce protection for pests such as birds, rodents, and insects. Grass should not be present within 1 m (3.25 ft) of building walls so that a pea gravel strip of 7.5–10 cm (3″) deep can be laid over polyethylene or the equivalent to discourage rodent entry. Parking lots should be physically separated from the interior perimeter of the facility to address biosecurity, congestion, and security. Incorporation of radio frequency identification (RFID) and barcode ID badges for security scanning of people entering and exiting the facility can improve those issues mentioned.

Building Construction Considerations

Generally, the most ideal shape for a cold storage building is a cube. Shorter distances are the logistic prerequisite for more efficient product movement and uniform temperature distribution to reduce microbial proliferation. Design layout should provide for a single, one-way flow of raw materials from receipt and storage to finished products to minimize contamination of processed and/or semi-processed products and increase handling efficiency. Entrance from non-production to production areas should be restricted to passage through clothes changing rooms where personnel are required to wash and change into appropriate clothing.

Ease of cleaning of facilities and equipment is paramount to the attainment of a sanitary operation. Equipment with less parts, easy disassembly and assembly, and accessibility can be more efficiently and effectively cleaned. Furthermore, quality assurance technicians and/or inspection personnel and maintenance personnel can perform their responsibilities more effectively.

Johnson (2014) suggested that to emphasize sanitation, operations efficiency, and maintenance, a dedicated passage for plumbing, electrical, pneumatics, hydraulics, and data lines should be incorporated. By segmenting these utilities and dropping down from the supply lines, a barrier is provided to enhance sanitation, safety, and accessibility. Also, he indicated that airflow should be dispersed with socks or diffusion chambers to enhance air circulation. Airflow should be designed to go from the cleanest room to the least clean. Welfare areas should be located to permit a barrier for hygienic purposes but to permit workers to go to and from breaks and meals in a reasonable time.

Temperature and moisture control in processing areas is essential to attain hygienically designed food facilities. Microbial harborage and proliferation can be reduced if the optimal temperature is attained. The mixing of dry environments with wet processing makes cleaning difficult and enhances microbial growth.

Walls

The foundation and walls of a food processing or foodservice facility should be impervious to moisture, easily cleaned, and constructed to prevent rodent entry. Accepted practices during the past have included slab floors that contain footers constructed with a rodent flange 60 cm (2 ft) below grade, extending 30 cm (1 ft) out at right angles to the foundation to prevent rats from burrowing under the floor slab and gnawing their way into the building. If a basement is planned, the floor should be tied directly to the solid wall foundation to create a solid box as a pest barrier.

The most appropriate walls are poured concrete, trowelled smooth to a maximum of nine holes per square meter (approximately 1 square yard), none of which exceeds 3 mm (0.12"). Poured concrete is more expensive and requires on-site construction of forms and finishing, but it does not have seams that require caulking that is needed for precast or tilt-up construction.

If concrete block wall construction is incorporated, it must be a high-density type. Less porous material reduces moisture absorption and microbial growth. An effective sealer can close pores to improve hygienic design. Past experience has indicated that when concrete blocks are laid, the first course should have the center core filled with mortar to provide an effective seal against insects entering through the joint created at the junction with the foundation. Walls should be covered at the floor, to a minimum radius of 2.5 cm (1"). Concrete blocks should be capped off to prevent access by rodents and insects. Caulk has limited application for either permanent construction or temporary repairs. Although caulking can be functional when sealing to prevent water from entering seam joints, it will eventually dry and shrink with resultant loosening and need for replacement.

Corrugated metal siding is not recommended because it is not reliable in stopping the entry of insects and rodents and this material is damaged easily. If corrugated metal is incorporated, the outside corrugation should be blocked and caulked at the top and at the foundation to discourage pest entry. To reduce pest invasion, wall penetration for utility access should be sealed the same day that this operation is performed.

Wet processing areas should have glazed ceramic tile or baked-on enamel-insulated metal paneling to enhance the ability to clean inside walls. This material is resistant to food, blood, acid, alkali, cleaning compounds, and sanitizers.

Tile walls are expensive to install but inexpensive and easy to maintain. Epoxy paints over a compatible sealer provide additional protection.

Loading Dock

Loading docks and platforms should be constructed at least 1 m (3.3 ft) above the ground. The underside of the dock opening should be lined with a smooth, impervious material, such as plastic or galvanized metal, to prevent rodents from climbing into the building. Rodent access should be denied through a dock or platform overhang of 30 cm (1 ft) that will not permit a roosting location for birds. Pest entry is discouraged through truck door seals and air curtains.

The truck dock area should be equipped with dock seals. This design prevents the entrance of insects, and if the plant is under positive pressure with air flowing out of the openings that do occur around the seal, dust contamination is reduced. Dock seals can replace overhead canopies that require constant monitoring to prevent pest entry, especially birds.

Roof Construction

A logical roof type for precast concrete wall panels is a precast double tee. This design is attractive and hygienic. Pitch and gravel roofs should not be installed over food processing or preparation areas, as they are difficult to clean. Low-moisture materials, such as grain, starch, and flour, can be carried out through vents and will attract birds and insects and encourage the growth of weeds, bacteria, molds, and yeasts. The roof can be improved with the addition of a lighter colored membrane or solar panels for facilities located in sunny regions. The barrier between the exterior and interior is one of the most important areas of a facility. Securing the interior envelope will benefit the infrastructure. Smooth membrane-type roofs should be considered because they can be swept, hosed, and kept clean more

effectively than other roofs. Roof openings for air handling or other uses should be screened, flashed, or sealed to prevent the entry of contaminants such as insects, water, and dust. Roof opening caps and mounted air-handling units should be insulated with sandwich panel insulation, as open insulation is difficult to clean and can become infested with insects.

Windows

Effective environmental control and adequate lighting negate the need for windows, which can present a sanitation hazard, due to breakage and contamination from pests, dust, and other sources. Windows increase maintenance through required repair, cleaning, and caulking. If windows are installed, it is best if they cannot be opened and construction of unbreakable polycarbonate material should be considered. Furthermore, the sill on the outside should be sloped at a 60° angle to prevent bird roosting and debris accumulation. The next best design for windows is to place them flush with the outside wall and to use the same slope for the inside sill. Some municipalities require windows to conform to local fire codes.

Doors

Doors provide an entry for pests and airborne contaminants. A double-door entry reduces airborne and pest contamination. The exterior of the doors should be equipped with air curtains. Air curtains should have enough air velocity (minimum of 500 m/min or 1650 ft/min) to prevent the entry of insects and air contaminants and should extend completely across the opening with a down-and-out sweep. Air curtains should be wired directly into the door opening switch to permit air movement simultaneously with the door opening and closing.

Romakowski (2015) suggested that sandwich panels provide an energy-efficient and more hygienic operation because of their ideal physical properties. They are durable and enhance cold insulation for reduced microbial proliferation.

Ceilings

False ceilings are discouraged because the area above can become infested with insects and other contamination. If a dropped ceiling is installed, it should be constructed as if it is another floor sealed off from the processing area below and should contain utility runs, air-handling ducts, and fans. Construction usually includes catwalks so that the maintenance crew can service the equipment or lines passing through the area. This area should be kept pressurized to avoid dust infiltration. The exposed side of a suspended ceiling is attractive and easy to clean. Day-to-day operations beneath the ceiling can continue in a sanitary and efficient manner independent of what occurs above the ceiling. Isolation of pipes, electrical, and other services improves hygiene. Walk-on ceilings have merit because installation work can be completed above and below the ceiling simultaneously. Where lighting is not recessed into a walk-on ceiling loft, the fixtures should be sloped instead of flat to prevent dust collection.

Ceiling construction should be a smooth concrete slab of exposed double tees with caulked joints. If exposed structural steel is used over processing areas, it should be enclosed in concrete, granite, or the equivalent to avoid overhead areas that collect dust and debris or provide rodent runways or insect harborage. Metal panels should not be installed because their high heat transfer rate can cause moisture condensation. Furthermore, the metal expansion and contraction complicates the maintenance of seals at the joints, resulting in harborages for insects. Fiberglass batting should not be installed, as rodents live and thrive in it. Preferred insulation is Styrofoam and other insert materials. The hazards of asbestos prohibit its use. Of major importance is heat differentiation between the exterior roof barrier and the internal ceiling panel to reduce the development of condensation.

Floors and Drainage

The floor surface needs to be cleanable and slip-resistant. Johnson (2014) indicated that the use of curbing to provide a continuum from the floor to the wall has been successful. He suggested that the curbing be 1 m (1.2 yards) high. Floors may range from plain, sealed concrete in warehouses to acid brick in high-impact, high-temperature, and high-chemical-exposure areas. However, plain concrete floors may spall, and the exposed aggregate creates protection for microorganisms. Monolithic floors should be considered because they are seamless, easier to apply, and less expensive than brick or tile. These floors are both epoxy- and polyurethane-based and are either rolled or trowelled on by hand. Floors in food facilities should be impervious to water, free of cracks and crevices, and resistant to chemicals. Although tile floors provide an acceptable surface, with heavy wear, grouting loss can occur, which results in the penetration of water. Plastic or asphalt membranes may be laid between the underlying concrete surface and the tile or brick. Acid brick floors deserve consideration because of their durability and ease of replacement in case of breakage and their reduced moisture accumulation under cracks and holes.

The functional layout, including equipment location, should be developed before the floor is designed to ensure that any possible discharges are routed directly to drains, which should be located at the lowest point to eliminate water pooling. However, equipment should not be located directly over drainage channels because this arrangement may restrict access for cleaning. Clean-in-place equipment discharge should connect directly to a drain. The type of drain incorporated depends upon the processing operation. If an operation involves an extensive amount of water and solids, channel drains may be the most suitable. Aperture channel drains are more favorable for operations generating large volumes of water with little solids. They recommended round bottoms no deeper than 150 mm (6″) with easily removable gratings to enhance safety and more rapid and effective cleaning.

Processing and Design Considerations

Whether building a new building, expanding or remodeling, or attempting to solve a moisture problem in an existing plant, an understanding of airflow dynamics and heat and humidity conditions is critical to controlling moisture and condensation. Condensation control is a complex challenge with the need to address components such as temperature, moisture, pressure, and filtration. Therefore, knowledge of psychrometrics (the science of moist air) is essential. Pehanich (2006) indicated that in order to contend with moisture challenges, an air balance study should be conducted. This study typically starts with analysis of how air is moving in and out of a plant. This analysis is followed by the introduction of a device to record the temperature, air pressure, and humidity within the plant over 48 h.

Condensation control is important because beads of moisture can become contaminated with the condensation carrying dirt, microorganisms, and other contaminants. Thus, contamination can contribute to reduced product quality and potentially lead to product recall and/or foodborne illness. Furthermore, moisture accumulation can damage the facility structure, inviting pest infestation and other sources of contamination. Cold pipes should be insulated with a polyvinyl chloride (PVC) covering and a sealed vapor barrier. This precaution is important for protection during wash-down. If warm moist air gets to these pipes, leakage through the insulation may occur.

Neither standard industrial refrigeration nor fans will consistently control humidity. Hot water, moist products, workers, and process heat contribute to moisture accumulation. The moisture that air cannot absorb results in condensation. With air potentially changing every 2–3 min, contamination can spread. Although fans have been used in some food processing plants to draw warm air in or pull cold air out, this creation of negative air pressure can cause condensation. However, air can be scrubbed with microbe filtration if an air filtration system is installed.

Mechanical dehumidifiers are available for humidity control that cannot be accomplished by refrigeration and fans. These dehumidifiers set two coils in series, passing cold air first through a cooling coil and then through a reheat coil. The principle is to overcool the air so that the cool air is colder than needed to meet the room space and is at 100% relative humidity. When the air passes through the second coil, it is reheated 3 °C (38 °F). It comes off at 85% humidity (because warm air has higher moisture carrying capacity) and can now absorb more moisture. A desiccant dehumidifier moves plant air and passes it over a desiccant wheel which absorbs moisture from the air and can reduce relative humidity to as low as 15%. However, a desiccant wheel has limitations because as it passes through a hot section (ovens and/or fryers), it collects heat. When the hot wheel reenters the air stream, it passes heat to other areas of the plant, elevating ambient air temperature. Bakeries require vapor barriers to prevent condensation caused mainly by oven exhaust and moisture emitted by baked products. A desiccant wheel costs more than a mechanical system but does more dehumidifying. A mechanical dehumidifier is more cost-effective than desiccant systems which cost more. However, desiccant systems absorb more moisture.

Appropriate facility design incorporates a product flow that permits finished items from making contact with raw materials or unprocessed products. The ideal flow provides for raw materials and adjuncts to enter the process near the receiving dock, flowing sequentially into the preparation area, process area, packaging area, and to storage. This design flow permits proper air pressure conditions to the overall plant efficiency. Some personnel doors support this concept because they are designed so that workers must pass from a "clean" to "less clean" area. Return to the cleaner area may require a uniform change and a sanitizing step, followed by entrance through an air lock or pressurized vestibule.

Processing equipment should have 1 m (3.3 ft) of clear space around it to facilitate maintenance and cleaning. A minimum of 0.5 m (20″) of clearance over each piece of equipment should be provided to permit effective cleaning. Floor-mounted equipment should be either sealed directly to the floor or mounted at least 15 cm (6″) from the floor. The processing layout should permit the location of equipment for accessibility to maintenance, sanitation, and inspection. Areas that are difficult to reach and clean are less likely to be cleaned frequently and thoroughly.

Equipment openings and covers should be designed to protect stored or prepared food from contaminants and foreign matter that may fall into the food. If an opening is flanged upward and the cover overlaps the opening, contaminants, especially liquids, are prevented from entering the food contact area. Failure to provide devices that extend into the food contact areas with a watertight joint at the point of entry into the food contact area may cause liquids to contaminate the food by adhering to shafts or other parts and running or dripping into the food. An apron on parts extending into the food contact area is an acceptable alternative to a watertight seal. If the apron is not properly designed and installed, condensation, drips, and dust may gain access to the food. Equipment containing bearings and gears that require lubricants should be designed and constructed to prevent lubricant leaks, drips, or entry into food or onto food contact surfaces. Condenser units that are an integral component of equipment should be separated from the food and food storage space by a dust-proof barrier. A dust-proof barrier between the condenser and food storage areas of equipment protects food and food contact areas from dust contamination that is accumulated and blown about during the condenser's operation.

Airborne contamination is attributable to the cause of some pathogenic contamination. Unfiltered air and negative air pressure in areas where the product is exposed contribute to microbial contamination in the plant environment. Thus, airflow design is as important to hygiene as is the design and construction of floors, walls, and ceilings. The zone with the highest pressure should be the area where the product is last exposed to the open air and packaged. The airflow from this zone is outward to the processing/preparation area and on to the storage zone. Dust collection is more effective if conducted under a positive pressure.

If an air-handling system is currently designed, the opening of an outside door provides an air stream exiting the building; whereas, in a negative air pressure situation, an opened door causes an incoming breeze containing outside contamination. The continual influx of unfiltered air complicates the overall cleaning of a plant, equipment, overhead pipelines, and other structural features. An air filtration system with a nitrogen generation unit is being installed in high-moisture food plants to improve hygienic conditions. Sterile filters can remove up to 100% of all visible particles. Membrane nitrogen generators convert air into nearly pure nitrogen that is injected into packages to eliminate oxygen that can reduce storage life.

Appropriate design is essential to prevent growth niches. There are many possible mechanisms: aerosols, stress cracks (caused by fluctuating pressures) in walls covered with other materials such as stainless steel or glass board, jacketed vessels, and heat exchangers. These mechanisms result in microbial transfer to growth niches. Furthermore, microbial biofilms are involved in this transfer.

Design Practices to Prevent Pest Infestation

The topography near a food facility should be sloped to permit water flow away from the building without the formation of puddles. Puddles provide available water for pests and attract them close to the facility. A rodent lip installed 60 cm (2 ft) down on the foundation and extending out 30 cm (1 ft) prevents rats from burrowing under the slab and entering the plant by chewing through expansion joints or through drains inside the building.

Cavities within walls should be avoided because they become nests for rodents and insects. All parts of the structure should allow easy cleaning of ledges, scale pits, and elevator pits. Proper installation of electric lines, cables, conduit, and electrical motors should be conducted to eliminate harborage sites. Motor housings provide ideal nesting sites for mice. Ventilation stacks should be equipped with adequate screening to prevent pest entry.

Locker rooms and eating areas are vulnerable to pest entry because of traffic, food particles, and moisture. These facilities should be designed and constructed with interiors that can be cleaned, covered wall/floor junctions, and smooth, water-impermeable walls and washable floors. Drinking fountains, vending machines, and other fixtures should be mounted far enough away from the walls for access to routine cleaning or mounted on casters for moving during cleaning. Locker tops should contain a 60° slope to avoid debris accumulation. These facilities should not open directly into a processing room or any area with exposed food. The toilet facilities should have a negative air pressure, and the internal air should be exhausted directly to the outside.

The best opportunity to keep birds out of a food processing facility is through the proper design. Since birds will utilize small gaps and cracks, or protected sites for entry, nesting, or resting sites, spaces under corrugated roofs should be blocked to preclude such activity. The materials for this purpose may include hardware cloth, expandable foam, sheet metal, and bird netting. Signs from the side of buildings should be removed or placed tightly against the side of buildings to prevent nest building. If sign removal is not possible or placement is not flush, the gaps between the building and sign should be blocked with an appropriate netting or screening material.

When designing new dock areas and protected overhangs, the use of tubular supports (square or oval) should be considered instead of I-beams. This practice deserves serious consideration because I-beams provide abundant nesting and roosting areas. The ends of the tube members should be completely sealed to prevent pest entry into the interior area. Potential exclusion materials are hardware cloth, expandable foam, and sheet metal. Overhangs in loading/receiving dock areas should be constructed using a cantilever design that limits the number of open supports. If horizontal supports are required, they should be tubular instead of I-beams. Window ledges and other similar structures should be eliminated if possible

to avoid roosting and nesting. Openings into the building and areas under corrugated roofs should be sealed.

Lights should be erected on poles distanced from the building and directed toward the area to be illuminated to eliminate roosting and nesting sites for birds and attraction of light to flying insects. Since insects are attracted to the area of greatest light intensity, they will gravitate toward the light itself located several meters (yards) from the building. Birds may be repelled from lights through the installation of metal or plastic "bird spikes" affixed to the light with a high-quality weather-resistant adhesive. Building lights should be sodium vapor lamps instead of mercury vapor lamps since the former are generally contracted to insects, while the latter are highly attractive.

Equipment Design for Ready-to-Eat Processing Operations

Adequate space should be provided to accommodate required equipment. As equipment becomes larger and more specialized, a proper allocation of space and functional layout becomes more critical.

In 2014, the American Meat Institute Foundation (AMIF) released its new sanitary equipment design principles as a follow-up to efforts extended during 2002. This effort has been a major factor in food safety advances in meat and poultry plants, especially the reduction of *Listeria monocytogenes* in ready-to-eat (RTE) meat and poultry products. An equipment design task force developed these principles to meet the expectations of the meat and poultry industries. Ten design principles have provided guidelines for equipment suppliers and users to identify collectively sanitary issues of common concern before equipment manufacture while creating a standardized food safety focus. Also, the principles included a checklist, glossary, and new photo examples for plants to incorporate when evaluating their equipment.

The following guidelines for equipment design in RTE processing operations have been adapted from a sanitary design checklist developed by the AMIF:

1. Food processing and handling equipment should be designed and constructed to ensure that it can be effectively and efficiently cleaned.
2. Construction materials should be completely compatible with the product, environment, cleaning and sanitizing compounds, and cleaning and sanitizing methods. Equipment construction materials should be inert, corrosion-resistant, nonporous, and nonabsorbent. All equipment and/or component surfaces should be paint-free. Through elimination of incompatible materials in the construction of processing equipment, the processor reduces the likelihood of creating an environment conducive to microbial proliferation.
3. All parts of the equipment are to be accessible for inspection, maintenance, cleaning, and/or sanitation. Disassembly and assembly should be facilitated by the equipment design to optimize sanitary conditions.
4. Elimination of product or liquid collection through self-draining equipment that will assure that debris, water, or product liquid does not accumulate, pool, or condense to increase contamination of the equipment or product zone areas is essential.
5. Hollow areas of equipment (e.g., frames and rollers) must be eliminated where possible or permanently sealed. Bolts, studs, mounting plates, brackets, nameplates, junction boxes, end caps, sleeves, and other such items must be continuously welded to the surface of equipment and not attached by grilled or packed holes. Open, inverted angle supports should be incorporated for equipment legs and bracing. Open supports should be mounted with the internal angle facing downward or out to the side to eliminate locations that are difficult to reach for cleaning. The legs should be designed to support the equipment off of the floor at least 30 cm (12″).

6. All parts of the equipment must be free of niches such as pits, cracks, corrosion, recesses, open seams, gaps, lap seams, recessed fasteners, protruding ledges, nuts and bolts and other fasteners, inside threads, bolt rivets, and dead ends. Control boxes should have sloped, cleanable tops. They should be mounted on support posts or framework with a minimum of 4 cm (1.6″) clearance from the nearest surface to permit sufficient cleaning behind the control boxes. All nuts (cap, wing, or others) should be mounted on the equipment exterior, and exposed threads in product zones should be covered with sealed cap nuts. All welds must be continuous and fully penetrating.
7. During normal operations, the equipment should perform so that it does not contribute to unsanitary conditions or the harborage and growth of bacteria. During processing, moisture and product buildup should be minimal in different product zones. Belt construction for food conveyance should incorporate a nonabsorbent, nonporous material of modular plastic belting.
8. Maintenance enclosures (e.g., electrical control panels, chain guards, belt guards, gear enclosures, junction boxes, pneumatic/hydraulic enclosures) and human-machine interfaces (e.g., push buttons, valve handles, switches, touch screens) must be designed, constructed, and maintainable to ensure that the product, water, or product liquid does not penetrate into or accumulate in or on the enclosure and interface. Equipment with bearings and gears that require lubricants should be located out of the product zone and designed and constructed so that the lubricant cannot leak, drip, or be forced into food or onto food contact surfaces. All bearings need to withstand cleaning and sanitizing. Synthetic rubbers or elastomers are potential materials for seals which need to withstand

temperature and moisture variations. The physical design of the enclosures and rolled edges should be sloped, rolled, or pitched to avoid creating flat areas that are difficult to access and clean.

9. Design of equipment must ensure hygienic compatibility with other equipment and systems (e.g., electrical, hydraulics, steam, air, water). All faceplates on gauges/sensors/sight glasses or other surfaces should be made of shatterproof, easily cleanable material such as polycarbonate or other material that is cleaned easily. The hygienic compatibility to the equipment with other systems is both a processor and equipment manufacturer responsibility.

10. Procedures for cleaning and sanitizing must be written clear and validated. Compounds recommended for cleaning and sanitizing must be compatible with equipment and the manufacturing environment.

Ten sanitary operation practices adapted from those provided by Seward (2004) are the following:

Principle 1: **Identify distinct hygienic zones established in a facility**. A distinct separation should be maintained to reduce the transfer of contamination throughout the plant.

Principle 2: **Control personnel and material flow to reduce hazards**. Traffic and process flow should be established to control the movement of employees, visitors, supplies, product, and rework, to reduce food safety risks.

Principle 3: **Control water accumulation**. To reduce microbial growth, design and construction should reduce water accumulation through effective floor drainage and the absence of pockets, ledges, and nooks.

Principle 4: **Control temperature and humidity**. Heating/ventilation and air-conditioning (HVAC)/refrigeration systems serving processing areas should maintain specified room tempera-

tures and control the room's dew point and prevent condensation.

Principle 5: **Control air quality and flow**. Air movement should be from cleaner to less clean areas. Incoming air should be filtered. Outdoor makeup air should be provided to maintain specified airflow, and pressurized and source capture exhaust should be provided to manage high concentrations of heat, moisture, or particulates generated.

Principle 6: **Provide site accommodations**. Access control is essential to rigid sanitation. Adequate lighting and water management systems are necessary to facilitate sanitary conditions.

Principle 7: **Provide a building envelope for sanitary conditions**. The building envelope (skin or shell) should be constructed to prevent pest entry and facilitate easy cleaning and ongoing inspection.

Principle 8: **Provide interior space conducive to rigid sanitation**. The area should facilitate cleaning and maintenance of building components and processing equipment.

Principle 9: **Incorporate "sanitation friendly" construction materials and utility systems**. Construction and renovation materials should be designed to prevent contamination, impervious, easily cleaned, and resistant to corrosion and wear.

Principle 10: **Incorporate an integrated sanitation system**. Food facilities should have an integrated sanitation such as hand sinks, sanitizers, doorway foamers and/or footbaths, hose stations, cleaning-out-of-place (COP) equipment, and equipment washers to enhance hazard control.

The "clean room" design has become prominent. Increased emphasis on sanitation has resulted in more interest in surfaces (including wall panels) made from stainless steel as a construction material, even though this material is very expensive. Additional concepts that are being promoted include the integration of entry and exit vestibules with garment changing facilities for traffic into and out of exposed ready-to-eat product areas. Also, there is a trend toward the removal of all refrigeration coils from RTE areas and the utilization of more roof-mounted refrigeration air units and to duct the air into the necessary spaces. This practice is being conducted to reduce dirt or dust accumulation.

Additional construction trends include expanded polystyrene (EPS) panels and doors for walk-in coolers and freezers, food processing areas, and low-temperature distribution warehouses. EPS insulation manufactured from small, uniform polystyrene beads contains only stabilized air, to ensure stable and consistent settings. In addition to stainless steel construction, fiberglass-reinforced plastic finishes for the packaging area and vestibule are being incorporated. Plastic materials should be nonporous and corrosion-resistant.

The following belt conveyor design options which have been adopted (Anon 2004) merit consideration:

1. Hinges that open wide around the sprockets to maximize cleaning access to the hinge area but close on the conveyor bed to prevent debris from clogging the belt offer improved sanitation.
2. Hinge openings large enough to permit spray to reach the top and bottom surfaces.
3. Allowance of catenary sag to enable more effective cleaning because the extra space enhances water spray penetration to loosen soil and scraps in the hinge area.
4. Drive bars underneath to channel water and debris to the side away from the production line to reduce moisture absorption and microbial growth.
5. Compatible with the belt lifters. When lifting the belt, a belt-lifting device, whether portable or frame mounted, should lift the belt evenly across its width without causing damage.
6. Designs should be tested to validate or improve hygienic features.

Renovation Considerations

If an establishment cannot produce safe food in a facility without excessive modifications, new construction may be the most viable option. If renovation is practical, modifications such as wall treatments, gap-filling materials, and resurfacing materials may be viable applications. Structural rehabilitation, surface preparation, retrofit design modifications, and food-grade protective polyurea coating are possible considerations.

Preparation for renovation should involve a plan for the reduction of the spread of particles from the contaminated construction site to the processing and/or storage area. Thus, the new site should be sealed off before construction through building false walls, either taped-down sheet plastic or a temporary wall out of plywood on the renovation side. An ideal arrangement is the erection of stud walls with insulation. Fiberglass-reinforced panels on the production/storage side with caulked joints provide an impervious barrier to construction debris and other contamination.

A plant-wide air balance study to determine how to maintain positive pressure in the processing area should be considered. Positive pressure may be obtained through a ventilation system that pumps a higher volume of air into the production side. Further hygienic considerations involve ventilation of the construction area to the outside without location of the exhaust too close to the plant's fresh air. Although pathogens such as *Listeria monocytogenes* are not typically airborne, they can be carried during construction of an expansion or renovation. Thus, suppression of dust is essential to prevent contamination. Reduction of dust particles can be accomplished by creating negative pressure in the construction area and an erection of temporary walls to separate the construction area from food production. Cramer (2006) suggested that a quaternary ammonium spray sanitizer at 800–1,000 parts per million (PPM) be applied to the walls and floor areas during the demolition process.

Construction Materials

Stainless steel, although expensive, should be considered for food contact surfaces. This inert material resists corrosion, abrasion, and thermal shock, is cleaned easily, and is resistant to sanitizers. The high chromium content (12% or more of the steel) provides corrosion resistance. The most commonly used stainless steel is type 304 of the 300 series. Type 316 contains approximately 10% nickel instead of the usual 8% and is used more frequently for corrosive products such as fruit juices and drinks. Type 316b offers more resistance to high-salt-content products. Corrosion resistance is enhanced through passivation—a cleaning and corrosion protection treatment for stainless steel and other metals accomplished with an acid solution that removes contaminants from the metal surface and coats the surface in a protective film.

Epoxy-coated and sealed floors should be considered. Painted surfaces should be avoided because they can chip and rust due to contact with high-caustic/chlorine combinations.

Study Questions

1. Why is site selection important when building a food facility?
2. What site selection considerations should be adopted when building a food facility?
3. What site preparation should be conducted before building a food facility?
4. Why is it recommended that parking facilities be separated from the interior perimeter of a food facility?
5. What are the desired characteristics for the walls of a food facility?
6. Why is corrugated metal siding not recommended for food facilities?
7. What roof construction is preferred for food facilities?
8. Why are windows not recommended for a food facility?
9. Why should air curtains be installed?
10. Why are false ceilings not recommended in food facilities?
11. What is the best flow design for food products?
12. What is psychrometrics?
13. What is the importance of positive air pressure in a food plant?
14. How can the welfare facilities of food facilities be designed to reduce pest entry?
15. Why is stainless steel superior to other materials for food facilities?
16. What are monolithic floors and why are they incorporated in food plants?
17. Why should pea gravel be located within 1 m (3.3 ft) of the walls of food processing plants?
18. What is passivation?

References

Anon (2004). Selecting easy-to-clean conveyor belts for superior sanitation. *Food Saf Mag* 10(1): 46.

Cramer M (2006). *Food plant sanitation*, 91. CRC Press: Boca Raton, FL.

Johnson J (2014). 5 pillars of plant design. *The National Provisioner* 228(5): 56.

Pehanich M (2006). Controlling moisture in the plant. *Food Process* 67(06): 51.

Romakowski D (2015). Revolving doors with antimicrobial properties. *Fleischwirtschaft International* 30(4): 35.

Schug D (2015). Eliminating hygienic hazards. *Food Eng* November: 69.

Seward S (2004). How to build a food-safe plant. *Meat Process* 43(4): 22.

Low-Moisture Food Manufacturing and Storage Sanitation

15

Abstract

Rigid sanitation practices are essential in low-moisture food manufacturing and storage facilities to maintain product acceptability and to comply with regulatory requirements. A sanitary operation should be complemented with appropriate facility site selection and hygienic design of the building and equipment. Unprocessed materials should be sampled during the receiving operation to verify that they are not infested with insects, molds, rodents, or other unacceptable contaminants. Insect-resistant packaging should be considered for nonperishable items.

Separate storage areas should be provided for raw materials, supplies, cleaning compounds and sanitizers, lubricants, and pesticides. Toxic materials should be stored in separate, locked rooms where access is limited to authorized personnel. During storage, unprocessed and manufactured products should be protected from contamination through effective housekeeping practices. Storage areas require routine inspection to observe for microbial and pest infestation. Inspection and cleaning frequency of storage areas depends on temperature and humidity. Prompt disposal of defective products is essential. Cleaning in the manufacturing area should be done daily. Cleaning equipment consists of basic cleaning tools for low-moisture product areas, including vacuum equipment, powered floor sweepers and scrubbers, and compressed air for certain applications. Periodic deep cleaning should be considered for areas and equipment that are difficult to clean.

Keywords

Cleaning • Design • Equipment • Insects • Storage

© Springer International Publishing AG, part of Springer Nature 2018
N.G. Marriott et al., *Principles of Food Sanitation*, Food Science Text Series,
https://doi.org/10.1007/978-3-319-67166-6_15

Introduction

An effective and practical sanitation program is essential for low-moisture food manufacturing plants. It is necessary to ensure that the operation complies with the US Food and Drug Administration (FDA), state, and local requirements. Furthermore, rigid sanitation in low-moisture food manufacturing operations is necessary to ensure that consumers are provided with safe and wholesome foodstuffs. Effective sanitation in low-moisture food manufacturing is essential to maintain an acceptable operation. A tidy operation can be more efficient, assist in the promotion of branded products and company image, and determine whether an operation remains profitable or even stays in business. Failure to exercise proper sanitation can lead to customer dissatisfaction, decreased sales, and damage to a firm's reputation.

The Office of the Inspector General of the Department of Health and Human Services has indicated that low-risk food operations, such as bakeries, bottlers, and food warehouses, are becoming riskier because of ineffective inspection. Although firms engaged in interstate commerce are regulated by the FDA and subject to inspection by state and local authorities, where inspections are made, the surveillance is often cursory, with primary emphasis on birds, rodents, and insects. Firms operating under unsanitary conditions put the population's health at risk.

Sanitary Construction Considerations

The following discussion about sanitary construction considerations will relate specifically to low-moisture food manufacturing and storage facilities and supplement what is mentioned in Chap. 14 about building construction considerations.

Site Selection

Sites should exhibit the following hygienic characteristics:

- Nearly level to a slight slope with adequate drainage
- Free of springs or water accumulation
- Accessible to municipal services (sewage, police, and fire)
- Remote from incinerators, sewage treatment plants, and other sources of noxious odors or pests
- Located within an air quality district tolerant of emissions from thermal processing
- Located away from areas prone to flooding, earthquakes, or other natural disasters

Exterior Design

The exterior should incorporate smooth, tight, impervious walls free of ledges and overhangs that could harbor birds. They should also contain sanitary seals against rodents and insects. Driveways should be paved and free of vegetation, trash, and water accumulation areas. Regular sweeping should be conducted to keep dust from blowing into storage areas.

Interior Design

Interior design considerations discussed in this chapter relate to low-moisture production and storage facilities. Further discussion of hygienic interior design, exterior design, and site selection and preparation is presented in Chap. 14.

Walls and Framing

Exposed structural members may be satisfactory in non-product areas, as long as they can be kept clean and dust-free. Reinforced concrete construction is preferred for product areas, and interior columns should be kept to a minimum. Personnel doors should be fitted using self-closing devices (hydraulic or spring hinges) and screened. Gaps at door bases should not exceed 0.6 cm (0.24″), and 20-mesh (minimum) screening should be incorporated.

Walls should be free of cracks and crevices and impervious to water and other liquids to permit easy and effective cleaning. Wall finishes should consist of appropriate food-approved materials, as dictated by the function of each area. Glazed tile for surface finishes on processing area walls should be considered, with fiberglass-reinforced plastic panels painted with epoxy or coated with other materials meeting the company and regulatory standards. Alternatives to painting in food areas should be considered. Although paint is inexpensive, it tends to crack, flake, and chip with age and requires more frequent maintenance.

Insulation should be installed carefully in bakery facilities because it constitutes a potential dust and insect harborage. Even though inert, it should be applied to the outside of the building.

Ceilings

The use of suspended ceilings is satisfactory in nonfood areas if the space above the ceiling can be inspected and kept free of pests, dust, and other debris. Ceiling panels must be sealed into the grid but be easily removable. This feature is difficult to accomplish with most designs. However, suspended ceilings can provide a shelter for pests and may become moldy if wet, thus providing a source of contamination. In flour-handling areas of bakeries, dust may accumulate above the ceiling very rapidly, leading to insect, microbial, fire, and even explosion hazards.

Overhead structural elements, such as bar joists and support members, should be avoided whenever possible. Precast concrete roof panels provide a clean, unobstructed ceiling. Precast panels can be fabricated with a smooth interior surface, coated to resist dust accumulation, and easily cleaned. Overhead equipment supports; gas piping; water, steam, and air lines; and electrical conduits should be designed to avoid passing over exposed food areas, cluttering the ceiling, and dripping dust or moisture onto people, equipment, and product. A mechanical mezzanine to house utility equipment above can result in an easily cleaned ceiling, free from horizontal pipe runs and ductwork (Fig. 15.1).

Floors

Floors in wet-washed areas must be impervious to water, free of cracks and crevices, and resistant to chemicals and acids. Floor joints must be sealed, and wall junctions must be covered and sealed. Expansive concrete should be used whenever possible to minimize the number of joints. Floors should be sloped to drains with a pitch of 21 mm/m (0.85″/40″ or a 2% grade) for proper

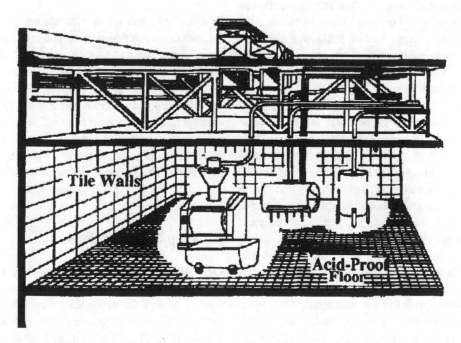

Fig. 15.1 Mechanical mezzanine separates ductwork and utility support equipment from the bakery mixing room. This arrangement reduces the need for overhead cleaning, improves access for equipment maintenance, and enhances product safety

drainage and wet cleaning. Process equipment should be connected to drain lines, and drip pans should be used to control floor spillage.

A perimeter setback of 0.5 m (20″) should be provided around all storage areas. Floor striping of setback spaces should be provided. Stored food must be segregated from nonfood items. Examples of products that must be segregated to avoid cross-contamination are bulk or palletized agricultural commodities and biologically active materials (i.e., pesticides, petroleum products, paints, cleaning compounds, and aromatic hydrocarbons).

The ideal floor material depends on the operation and type of traffic. For packaging and oven areas of bakeries, reinforced concrete, coated or hardened to prevent dust, may be adequate. However, areas such as those for liquid fermentation and dough handling that are often wet-cleaned and exposed to hot water, steam, acids, sugar, and other ingredients or sanitizing chemicals should have a surface composition tailored to the use and abuse the floor is expected to receive.

Chemical-resistant floors are most appropriate for wet areas. Monolithic materials, such as epoxies or polyester, and tile or brick should be considered because they are often less expensive. Toppings bonded directly to a substrate, such as concrete, should be used. They function as a resistant, watertight barrier protecting the concrete. However, they can crack, allowing liquids to enter. Only materials with proven success should be considered.

"Dairy" tile or pavers should be considered for areas with heavy traffic and those that come in contact with product or cleaning solutions. This material, when properly installed with acid-resistant bonds, is very durable and has minimal sanitary problems. It is cleaned easily and can be manufactured with a nonslip finish. This material is an expensive option, but it can be the most economical.

Floors in specialized areas, such as coolers and freezers, must be constructed with appropriate materials designed for their intended uses and be properly insulated and ventilated. An uninsulated freezer floor will eventually permit the ground beneath to freeze deep and hard enough to cause cracking or buckling of the freezer floors, with resultant jamming of the doors.

Ventilation and Dust Control

Dust control is very important. Although organisms from unprocessed low-moisture materials are usually harmless, they have been found to contain *Salmonella*, pathogenic mold spores, and other undesirable organisms. The manufacturing process, by heating the product above the pasteurization temperature, usually kills vegetative organisms, but spores may survive in the interior, especially in relatively soft, high-moisture baked foods. Furthermore, finished food can become contaminated from raw material dust within the plant, especially in coolers and packaging equipment.

To maintain acceptability, facilities must be designed so that finished foods are not contaminated. This practice requires a superb sanitation design and follow-through procedures, proper equipment arrangement, and proper ventilation and dust control. The proper selection of temperature/humidity controls will minimize the opportunity for bacterial growth.

Equipment Considerations

Equipment features that will enhance productivity include separating heating and cooling equipment from the processing areas by using a mechanical mezzanine, high-efficiency motors and electrical equipment (see Fig. 15.1), the latest technology in controls and automation to the maximum degree that is cost-effective, and flexible modular design for responding to changing markets and business demands. Furthermore, all equipment should meet the latest requirements from regulatory or advisory agencies.

Sanitary Considerations

Dry cleaning methods are incorporated where products are hygroscopic and if water can react to form hard deposits which are difficult to remove. Failure to control moisture can enhance the growth of pathogens such as *Salmonella* spp. in the processing or storage environment with subsequent contamination. Facilities that are usually dry-cleaned include the production and/or storage of flour, peanuts or peanut butter, dry milk products, candy, snack mixes, and dry infant formulae. Vacuuming with a suitable exhaust filter is frequently the desired method since it does not spread dirt and dust. Disinfection is frequently accomplished by the application of 70% alcohol with subsequent drying before equipment reassembly.

Because an operation such as bread making is a fermentation process, it is necessary for facilities such as bakeries to be maintained in a sanitary condition. Naturally occurring organisms must be prevented from fermenting the dough in competition with the desired yeast inoculum. An ineffective sanitary facilities design can result in the growth of wild microbial strains such as *Bacillus subtilis* or *mesentericus* "rope" formers, which can degrade product acceptability. Once established in the facility, these organisms are very difficult to remove totally and to control.

Single and multistory functional design considerations include:

1. Service areas, including truck docks, rail sidings, parking lots, boiler rooms, and trash collection, should be located away from processing areas to minimize food contamination. These areas should be well drained through grading or a drainage system.
2. Overhead piping and ductwork should be minimized in food processing areas.
3. The number of interior walls should be minimized to improve air circulation and to simplify cleaning.
4. Electric motor control centers, instrument panels, and other plant control functions should be grouped or centralized for more efficient and effective cleaning.

Sanitation features that are integrated into plant design were given increased emphasis by the FDA's promulgation of good manufacturing practices (GMPs). For low-moisture food products, current GMPs (CGMPs) as they relate to the design and construction should provide:

1. Adequate space for equipment installation and storage of materials

2. Separation of operations that might contaminate food
3. Adequate lighting
4. Adequate ventilation
5. Protection against pests

The best way to achieve these objectives is to keep the plant interior spaces simple and uncluttered. This characteristic facilitates sanitation, cleaning, and inspection. Plant sanitation criteria and engineering specifications should be thoroughly integrated into the layout by the design team, including the plant's technical and engineering staff, and a contracted design and engineering firm. A representative from the production staff should be consulted.

The following suggestions should be considered for a viable sanitation program:

1. A full or part-time experienced sanitarian schedules should be incorporated.
2. Cleaning records should be maintained and kept current.
3. Employees should be trained in and practice GMPs.
4. The sanitation program should be periodically evaluated to verify program effectiveness.

Other Considerations

The following design considerations complement Chap. 14, which contains more general location and design information:

- Locate the plant management office and the laboratory centrally for proper supervision and quality control.
- Locate ingredients storage near the mixing and use areas.
- Locate secondary equipment, such as boilers and refrigeration equipment, to minimize pipe and utility runs.
- Arrange manufacturing equipment for convenient cleaning-in-place (CIP).
- Use proven equipment or allow time for testing any equipment or process that does not have a known track record.

- Apply state-of-the-art controls and automation to the greatest extent that is cost-effective.
- Check plant design for compliance with federal, state, and local regulations.

Receipt and Storage of Raw Materials

Sampling for Acceptability

It is imperative that food production and storage facilities inspect incoming materials for possible contamination. Because it is impractical to sample all of the raw materials being received, a sampling protocol should be devised to determine whether products should be accepted or rejected. A statistically valid sample is necessary to determine acceptance or rejection with reasonable confidence. All deliveries should be checked for evidence of contamination and adulteration by soil, water, insects, rodents, birds, and foreign substances such as debris, chemicals, oil, and grease. Also, shipping container integrity for packaged ingredients and supplies should be examined. Damaged and contaminated supplies and/or products should be destroyed or removed promptly to reduce contamination. More information about statistical sampling and statistical quality control is provided in Chap. 8.

Transport Vehicle Inspection

Inspection of low-moisture raw materials should begin with an examination of the transport vehicle before, during, and after unloading. The overall condition of the vehicle should be appraised, and it should be checked for dead areas where product and dust can collect and harbor insects, whether the containers are full or empty. Areas adjacent to doors or hatches should be observed for insects. This inspection is accomplished by examination for crawling or flying insects and their tracks. It is important to check for nesting materials, odor, and fecal material. Pellets and odors may indicate rodents, and feathers or droppings may reveal contamination by birds.

Product Evaluation

An effective food warehouse sanitation program requires that the materials received, including foods and their packaging materials, must not be exposed to contamination from insects, birds, rodents, or other vermin or through the introduction of filth or other contaminants. To reduce contamination from raw materials being received, product evaluation is essential. Although moisture content may be determined objectively through the analysis for percentage of moisture, a subjective evaluation should also be conducted. A sour or musty odor can result from mold growth, which indicates high moisture content in products such as cereal grains. Such a discovery indicates that additional inspection should be conducted, with sampling to identify the specific characteristics of the problem. Cereal grains above 15.5% moisture should not be put in long-term storage because of potential insect development and mold growth. Evaluation of products being received should also include checking for pesticide odors that may be associated with the presence of insects. The inspection process should also determine whether the pesticide has made the product unacceptable.

Samples taken when materials are received should be evaluated to determine the amount of individual kernels that are damaged by insects. Further examination should be conducted to determine amounts of dust and other foreign material, webbing, evidence of molds and odors, live and dead insects, rodent droppings, and rodent-damaged kernels. These defects can be determined through visual inspection. Internal infestation in the form of immature insects inside of the kernels can be determined with X-ray equipment or by cracking-flotation methods. Samples should also be examined for rodent filth, such as droppings and hair.

The inspection of inbound goods is an appropriate prevention measure to reduce pest damage because incoming items can contaminate the end product. Because pests or their contamination can enter buildings as "hitchhikers," incoming ingredients, packaging materials, pallets, and machinery should be inspected. A food processor has the right to reject any materials coming into the plant or to hold any questionable shipment for further evaluation. Decisions related to rejection should be made by qualified personnel.

Product Storage and Stock Rotation

Foods and other materials should be received into a processing plant or warehouse for handling or storage in a way that will facilitate cleaning and the implementation of insect, rodent, and other sanitary controls. Effective procedures for stock rotation appropriate to the specific food should be adopted and implemented. Damaged foods should be promptly detected, identified, and separated from other products for additional inspection, sorting, and disposition. If any product is determined to present a contamination hazard to other foods, it should be removed from the facility promptly.

Many of the low-moisture food processing plants store material such as grain for processing. Unfortunately, when grain is stored, it initially contains mold spores and insect eggs in enough quantity to infest and damage the product if specific environmental conditions occur. Physical damage to the kernel itself can allow entry of infesting or infecting agents. Biological damage from insects through penetration of the kernel permits fungal entry through inoculation of the inner tissues.

Grain to be placed in storage for more than one month should receive special treatment. In addition to being inspected for verification that infestation and infection have not occurred, it is necessary to maintain a maximum of 13.5% moisture content. These authors suggest that cleaning the grain before storage using aspiration or other methods can remove dockage, external insects, weed seeds, and foreign materials and can improve its storability. Furthermore, as grain is being stored, chemical grain protectants can be applied to provide residual protection against insects.

A modified atmosphere, such as carbon dioxide and nitrogen, may be incorporated to fumigate grains. Fumigation by inert gases is receiving more attention because of increased restrictions on the use of chemicals. Although inert atmospheres do not represent a residual hazard, the environment in a storage bin with an inert gas can be as deadly to humans as if it contained a lethal concentration of a chemical fumigant. Insect feeding and reproduction can be reduced in temperate regions if storage bins are equipped with aeration systems.

Control of dust in handling and storage of low-moisture foods can improve housekeeping and pest control. The containment of dust production reduces deposits on floors, walls, ledges, overhead objects, and equipment, with a resultant decrease in cleaning time. Dust control is enhanced through suction (reduced pressure) on grain handling equipment such as conveyors, receiving hoppers, bucket elevators, and bins, as well as at points in the handling system where product is transferred from one piece of equipment to another (e.g., from spout to conveyor belt, conveyor to bin, and bin to conveyor).

The application of highly refined oils to grain as it goes to storage is an effective way to reduce dust when handling grain. Oil, which may be added to levels of up to 200 parts per million (PPM), should be applied to the grain as closely as possible to the point of discharge from the transport vehicle to reduce dust formation and to provide a grain-protection treatment.

Although the sanitation of root crops, such as potatoes, during storage is not as critical as for other foods, storage conditions must be controlled to prevent *Fusarium* tuber rot and bacterial soft rots. Well-ventilated storage rooms with concrete floors have enabled the potato storage industry to exert adequate control over its product.

Bulk storage of oils and shortenings normally occurs in large carbon steel or stainless steel tanks. Thus, appropriate sanitation can be attained by proper cleaning of these containers through washing with a strong alkaline solution or alkali and detergents before use. Hygiene conditions can be enhanced further through the nitrogen blanketing of process and deodorized oils. However, precautions are essential during the bottling and emptying of unprocessed, processed, and deodorized oils to prevent excessive splashing and agitation, which can potentially promote oxidative deterioration. Cleaned bulk tanks (especially carbon steel tanks) should be recoated with oil to seal them for rust prevention.

Pest Control

Since past management and control is presented in Chap. 13, only stored product insect pests will be discussed here. Stored product pests are classified into two groups, based on characteristics of their life cycle. Internal pests spend most of their life cycle within a whole seed or kernel of grain and rarely feed in processed foods. External pests normally feed on processed foods and spend most of their lives on milled grains and grain-based food products. Adults of some species can utilize nonfood products, especially pollens and molds.

Internal Feeders

Weevils Adults of rice, maize, and granary weevils range from 0.3 to 0.6 cm (0.12–0.24″) long. Adult weevils are commonly called snout beetles because the head is elongated into a "snout" that contains the mouthparts. The larvae are small, white, legless grubs that spend the entire larval stage inside whole kernels of grain. Rice/maize weevils are capable of flight, whereas granary weevils cannot fly.

Lesser Grain Borer The adult lesser grain borer is a cylinder-shaped, dark brown beetle approximately 0.3-cm (0.12″) long. Its head is tucked so far under the prothorax that it is not visible from above. It specializes in consuming grain and grain products and is most commonly found in wheat and wheat-based products, but its eggs can infest corn, rice, and barley. Adults live 4–5 months and are strong fliers.

Angoumois Grain Moth The adult Angoumois moth is a small, buff-colored insect with a wingspan of approximately 0.125 cm (0.05″). The most distinctive identification of this moth is the long hairs on the fore and rear wings that give them a fringed appearance. Larvae bore into the kernels where they feed and develop. Corn, bar-

ley, rice, rye, and oats are their preferred foods. Adults do not feed on grain or other food products and do not cause damage. Larvae may be found developing in caked material. Pheromone traps are very effective for monitoring adult male populations.

External Feeders

Indian Meal Moth The adult Indian meal moth has a wingspan of approximately 0.125 cm (0.05″). The copper-colored band of scales on the forewings identifies this species. The larvae feed on most grain-based products but also on chocolate, beans, spices, cocoa, nuts, and dried fruit. Larvae leave webbing behind as they feed, frequently causing particles of dry food to clump. The webbing may contain frass (feces). Their tendency is to crawl up vertical surfaces, making observation of this insect easier than many other pests. The presence of larvae can identify an emerging pest population or locate one in existence. Pheromone traps are effective in monitoring adult male populations.

A characteristic of this insect and many other stored product moths is the ability to diapause. This period of slowed or suspended growth or dormancy can be initiated in response to cold temperatures, large population levels, or short photoperiod. An unheated warehouse that cools during the winter may give the appearance that control has been achieved, when in fact the larval population has diapaused and will resume activity, typically in the spring, when environmental conditions favor growth.

Mediterranean Flour Moth The wingspan of the adult Mediterranean flour moth is approximately 2.5 cm (1″). The forewings are pale gray with transverse black lines and flecks, whereas the hind wings are gray to dirty white. Other characteristics, including diapause, are similar to the Indian meal moth. Pheromone trapping is very effective for this insect.

Flour Beetles Adult red flour beetles and confused flour beetles are approximately 0.3–0.5 cm (0.12–0.20″) long. Each antenna of the red flour beetle ends abruptly in a three-segmented club, while the antennae of the confused flour beetle gradually enlarge. The sides of the red flour beetle's thorax are curved, whereas sides of the confused flour beetles' thorax are nearly straight. Red flour beetles are not strong flyers but have that capability, while confused flour beetles do not fly. These beetles are major pests of flour. They rely on other insects or rodents to first damage the kernels, since they cannot feed on whole grains. Flour beetles may be found in grain fines, peas, beans, other vegetables, dried fruit, chocolate, spices, rodent baits, botanical drugs, dried milk, peanuts, and forest products. Both species are capable of breeding year-round in heated buildings. In unheated buildings, only the adults are likely to be observed during cold weather. The confused flour beetle is more common in the cooler areas of the world, while the red flour beetle is more prevalent in warmer climates. However, both species have become widely distributed and can sustain populations in any geographic location. When large beetle populations exist, both species give flour and other processed foods a grayish tint. Both species produce secretions that impart foul odors to food products. Since the flour beetles live for over 3 years, their persistence is important in pest management.

Drugstore Beetle The adult drugstore beetle is 0.15–0.35 cm (0.06–0.14″) long, light brown to red-brown, and hump-backed with an invisible head when viewed from above. The wing covers have pits arranged in longitudinal rows or grooves. The antennae have a three-segmented club. In wooden pallet storage areas where food residues are absent, identification should focus on a wood-boring beetle. The larva is capable of feeding on a whole kernel of grain but is more likely to consume processed grain products. These larvae also feed on leather, wool and other textiles, spices, tobacco, and botanical drugs. Since these larvae can perforate tinfoil and sheet lead, many kinds of packaging are readily penetrated. These beetles fly and are attracted to light.

Cigarette Beetle The adult cigarette beetle is 0.15–0.35 cm (0.06–0.14″) long, light brown with a humped shape, and similar in appearance

to the drugstore beetle except for smooth wing covers and sawlike antennae. The larvae avoid light through harborage within the food source. Although this is a pest that attacks tobacco, it also feeds on grain products, vegetables, dried fruits, textiles, spices, botanical drugs, dried flowers, and books. This insect is known for its ability to penetrate packages and is a strong flyer with peak flight activity in the late afternoon and early evening.

Grain Beetles The sawtoothed grain beetle and the merchant grain beetle are similar in appearance but can be distinguished from other food pest insects by the six sawlike projections on each side of the prothorax. The sawtoothed grain beetle is distinguished from the merchant grain beetle by its smaller eyes, and the area behind the eyes is larger. The merchant grain beetle is a weak flyer, and the sawtoothed grain beetle cannot fly. Since these beetles are not attracted to light, light traps are not effective monitoring tools. If a large population develops within a food ingredient, the resulting food product will have an off-flavor that is objectionable to humans.

Other grain beetles include the flat grainPest control:stored product insects:external feeders beetle and the rusty grain beetle. Both of these are approximately 0.15 cm (0.06″) long and are among the smallest grain-infesting beetles. The antennae of the male flat grain beetle are about the same length as the body, whereas the female flat grain beetles and both sexes of the rusty grain beetles have short antennae. The geographic range of the flat grain beetle is restricted by low temperature and low humidity, and the rusty grain beetle is a more abundant in the wet tropics. Although these insects cannot feed on intact grain kernels, those with very small cracks or defects are vulnerable to attack. The larvae also feed on dead insects.

Spider Beetles There are various kinds of spider beetles that are so named because of their very small head and prothorax and large abdomen, causing a resemblance to spiders. They are 0.075–0.475 (0.03–0.18″) cm long with voluntary legs that also make them look similar to spiders. They are scavengers that are found feeding on milled or processed grains, dried fruits, dried meats, animal droppings, textiles, dead insects, and vertebrates. They remain active during freezing temperatures and pose a problem in unheated facilities during the entire year.

Mealworms These are among the largest beetles closely associated with the food industry. Adult mealworms are oval shaped with 11-segmented antennae. Dark mealworms are black and yellow mealworms are shiny dark brown to black. They thrive on old, moldy, off-condition grains or grain products but will feed on cereals, crackers, and meat. The ingestion of mealworm eggs can cause severe gastrointestinal illness. They can fly and are attracted to lights.

Structure-Infesting Pests

The structure-infesting pests, cockroaches, and flies are discussed in Chap. 13. Other structure-infesting pests include psocids, commonly called booklice. Psocids are 0.75–6.25 mm (0.0003–0.0026″) long, colorless to gray or light brown insects with scale-like wings (usually nonfunctional). Adults survive 1–3 months and feed primarily on molds. They can also feed on starches, starchy glues used in bookbinding, and dead insects. Raw grains and finished food products are vulnerable to this pest if they become moldy or are stored under humid conditions. Many species reproduce by parthenogenesis. Dry conditions or low humidity stops or retards development or causes desiccation or death. During hot humid weather, the psocid population increases on composite fiber "slip sheets" used to separate palletized stacks of recently manufactured metal cans. Without the use of plastic slip sheets or sterilization of these cans before use, some of these insects can be canned with the product. An effective way to eliminate psocid infestation is to reduce the relative humidity to less than 50%, increase air movement to increase moisture evaporation, and disinfect to reduce mold growth. Psocids contaminate food products by their presence but usually cause minimal direct damage to bulk grains.

Insect-Resistant Packaging

Insect-resistant packaging is a strategy that should be considered when nonchemical control or exclusion techniques are addressed. It is possible to evaluate effectiveness of packaging materials to determine which will protect the product during storage and shipment. The potential exists for the incorporation of natural chemical repellents into packaging material and new glues and ceiling methods to improve the structural integrity of insect-resistant packaging. Packaging films differ in their ability to prevent insect entry. Potential exists for the incorporation of natural chemical repellents into packaging material and new glues and sealing methods to improve the structural integrity of insect-resistant packaging.

Product Storage Housekeeping

Unsanitary conditions may be prevented through effective maintenance and housekeeping. Bulk storage areas (especially interior areas) should be maintained so that they are free of cracks or ledges that collect dust and other debris, which provide an environment conducive to insect growth, whether full or empty. Empty bins or other storage containers should be inspected for residues of product stored previously and for overall condition. Residual material that can support insect growth should be removed before products are stored.

Tunnels, gallery floors, and associated areas of storage bins or similar facilities should be maintained in a sanitary condition. Periodic inspections are an essential part of effective sanitation for stored products. As in other areas, inspectors should examine the dust on the floors and walls for insect tracks and for resting or flying moths. Inspections include examination for damp lower areas that collect dust and provide conditions conducive to the proliferation of molds, mites, and fungus-feeding insects. Further inspection should include checking for any unusual odors that could indicate mold, insects, or chemicals. It is especially important to inspect handling equipment, such as elevators and conveyors, that may harbor residual product. Unused equipment may retain residual material that will encourage insect growth and subsequent migration to storage areas or contamination of new product.

Storage areas require regular inspection to observe for live insects on product surfaces, floors, and walls. Thermocouple cables should be used for grain in extended storage so that temperature can be monitored. Increased temperature during storage should be investigated. Samples should be taken with probes or as the product is transferred to another location to determine whether the temperature rise could cause developing populations of insects or molds. Mold growth can normally be controlled through drying or blending with other dry products. If insects are present, treatment or fumigation should be conducted. Heating from insect infestation can also cause moisture to spread, with resultant mold development. Inspection should be accompanied through a complete record of inspections, cleaning and fumigation, or other corrections administered.

Sanitation requirements for product storage are similar to those for bulk storage. An orderly storage arrangement is essential to ease inspection and cleaning and to reduce the potential for sanitation problems. Records of regular inspections and housekeeping are essential. Inspectors and other employees should be aware of the presence of pests and of eradication methods.

To ensure effective sanitation, bags and cartons be stacked on pallets and spaced away from the walls and from each other for inspection and that the surrounding area be cleaned. Stock should be rotated to reduce insect infestation and rodent entrance. Inspections should include visual observation, using a light for looking in dark corners, under pallets, and between stacks. Insects may be detected while flying; crawling on walls, ceilings, and floors; and while hovering over bags and cartons. Product spoilage should be sifted to detect insects. Additional information on insect and rodent control is provided in Chap. 13.

Cleaning and inspection frequency depends on temperature and moisture conditions. Under ambient temperature (25–30 °C or 78–86 °F) conditions, the life cycle of many insects that infest low-moisture grains and foods is approxi-mately 30–35 days. Insect reproduction normally ceases when the storage temperature is below 10 °C (50 °F). When storage temperature increases, the cleaning and inspection interval should be decreased. Raw material or product temperature has more influence on insect growth than ambient temperature. Areas where high-moisture (humid) conditions exist will require more frequent inspections and cleaning. High-moisture conditions should be reduced through proper ventilation. Moist materials that remain static at room temperature or above will increase immediate development of molds, yeasts, and/or bacteria. Suction can be used to remove moist air.

Ledges and other locations that can accumulate static material should be eliminated. External supports, braces, and other construction features and/or equipment should be designed to prevent material accumulation. Dust can adhere to moist surfaces and provide an excellent habitat for molds.

Heat treatment (superheating) can combat pests in dry storage and production areas where unprocessed materials are stored. However, this practice is energy-intensive because of the amount of heat required to kill insects, especially during cold weather. Portable heating units may be used to superheat an individual piece of equipment that may be infested with insects. When designing new or renovated facilities, consideration should be given to the potential for heat treatment. Maintenance of a cold environment is less practical because of refrigeration costs and possible equipment or facility damage from freezing. It may be impractical to maintain a moderately low temperature that will retard insect inactivity.

Inspection of Raw Materials and Product Storage Area

Inspections should be conducted and reported. An inspection report format should be developed with a numerical scoring system. Scoring and rating values should be defined, with a description for each value.

Inspection in the processing and storage areas should emphasize the identification of potential product contaminants and prompt and proper corrective action to prevent contamination. The lower minimum water activity (A_W) of low-moisture foods reduces the chance of microbial spoilage; thus, more emphasis should be placed on other forms of contamination, which are discussed in the following paragraphs.

Overhead areas should be examined for flaking paint, obstructions to cleaning, dust accumulation, and condensation. Ground-level, basement, and above-ground-level inspection should focus on broken windowpanes and absence of or damage to screens. Open windows or other entry avenues for pests are potential sources of contamination and should be reported and/or corrected on a continual basis. Evidence of pests, such as insect trails in dust, rodent droppings, and bird droppings or feathers should be identified through periodic inspection and inspection by employees on a continual basis. Evidence of pests should be reported so that appropriate action can be taken to identify the problem source and correct it. All employees should be alert for evidence of pest activity.

Inspection of equipment exteriors is accomplished on a continual basis through operations personnel. Overhead equipment should be inspected regularly. Equipment interiors should also be examined periodically for sanitation-related problems during maintenance inspection. Some equipment contains dead spots where product can accumulate. Therefore, inspection of equipment should be performed routinely when the equipment is not in operation. Equipment, especially conveyors, should be constructed, if possible, so that interiors are accessible through clean-out openings or by easy disassembly. This design also facilitates equipment cleaning during routine housekeeping. If feasible, equipment not in use should be removed from the facility. Equipment that is used infrequently should be left "open" so that any product filtering into it will pass through or will be observed easily. The openness feature will also enable easy and effective cleaning.

The condition of the facility itself should exclude contaminating factors, such as insects, rodents, and birds. Any defects discovered should be reported and corrected immediately.

Cleaning of Low-Moisture Food Manufacturing Plants

Cleaning in the manufacturing area of low-moisture food plants should be accomplished daily. Some of the cleaning should be done while the plant is operating to ensure that the facility remains tidy, but most of the equipment cleaning (especially equipment interiors) should be done while the manufacturing portion is not in operation. Some of the required cleaning can be combined with routine maintenance operations. Easily stored, conveniently located equipment encourages employees to accomplish the cleaning necessary to control infestation.

Dry cleaning is the most viable method for low-moisture food processing plants. When water is introduced, some material is not removed from cracks and crevices. Most of the dry cleaning equipment is easy to use. Hand brooms, push brooms, and dust and wet mops provide the basic equipment used for cleaning. Brushes, brooms, and dustpans remove the heaviest debris accumulations and function well on semi-smooth surfaced floors. Dust mops provide a more rapid means of cleaning on smooth floor surfaces with low levels of dust accumulation. In many production areas, vacuuming provides the most acceptable means of equipment cleaning. Vacuum cleaning provides one of the most thorough methods of cleaning because it removes light and moderate accumulations of debris from both smooth and irregular surfaces. Dust is contained and does not require a secondary means of collection. Smaller operations can more effectively utilize portable vacuum equipment, whereas larger facilities can benefit from an installed vacuum system. Centralized debris collection and disposal is more convenient with additional access to difficult-to-reach areas. In large storage areas with nonporous floors, a mechanical scrubber or sweeper should be used to more efficiently and effectively maintain a clean environment. Cleaning tools such as brooms, scrubbing brushes, squeegees, and floor scrubbers should be accomplished at a level of 600 PPM of a sanitizer such as a quaternary ammonium compound.

The suggested procedures for dry cleaning of food equipment are:

1. Shut off power to equipment being cleaned.
2. All equipment or operating systems should be emptied, purged, and disassembled as needed for cleaning—including the removal of belts, dividers, guards, hose, molds, and lids.
3. Materials and tools for cleaning should be prepared and assembled.
4. Dry-clean by picking up large debris and using approved vacuum to reach confined areas.
5. Use rags or appropriate materials to wipe and clean grease from fittings, linings, and similar parts.
6. Use approved brushes to brush down equipment.
7. Inspect equipment to verify that all soils are dislodged, and ascertain the absence of residue or soils.
8. Reassemble equipment.
9. Sanitize equipment.
10. Document that all procedures were followed.

Wet cleaning has been incorporated in some peanut processing operations even though dry cleaning is generally the preferred method. According to Hui (2015), the application of dry and wet cleaning is one of the most debated issues for the sanitation and safety in peanut processing. If wet cleaning is incorporated, the equipment utensils involved in processing should be handled in a dedicated area separate from the processing operation with one or more cleaning compounds and sanitizers, pressurized water, and cleaning tools. Wet cleaning procedures for higher moisture food establishments are mentioned in Chaps. 16, 17, 18, 19, 20, and 21.

Pre-cleaning of utensils and food equipment permits the removal of debris to facilitate further cleaning. Some heavily soiled surfaces should be presoaked to facilitate cleaning. Pre-cleaning should involve the scraping of debris on equipment and utensils over a waste disposal unit, scupper, or garbage receptacle, or the debris should be removed in a ware-washing machine with a prewash cycle. With heavy soil, utensils and equipment should be pre-flushed, scrubbed with abrasives, or presoaked. *Wet cleaning* is conducted to remove completely loosened organic soils through manual or mechanical operations. Standing water should be removed from the area being cleaned with the application of 800 PPM of a sanitizer such as a quaternary ammonium compound (Cramer 2006).

A compressed air line is widely used to remove debris from equipment and other difficult-to-reach areas. Furthermore, the use of compressed air is safer than depending on employees to work from a ladder with a brush. However, compressed air disperses dust from a specific location to a less confined area and may spread an infestation if it exists. Compressed air should be incorporated at low volume and low pressure, depending on the cleaning operation, to minimize dust dispersal. Employees who use compressed air should wear safety equipment, such as dust respirators and safety goggles.

Specialized tools are required for certain equipment cleaning. Cylindrical brushes are used for spouts. They can be either dragged through spouts by rope or cords or operated on flexible motorized shafts. Dough mixers need to be flushed and then cleaned with an alkaline cleaner that contains an emulsifier to ensure that the fat will be removed.

The maintenance of a tidy operation depends on proper organization and installation of equipment and on cleaning of individual pieces of equipment and of the surrounding area. Ingredients and supplies should be properly stacked in a designated storage area. Receptacles should be conveniently located for the disposal of bags, film, paper, and waste products from manufacturing, packaging, and shipping.

Suggestions about how to respond to difficult-to-clean equipment include the following:

1. Provide maintenance and cleaning personnel training in the design of equipment being cleaned.
2. Assign cleaning accountability to one or more workers.

3. Require knowledge or training of the English language.
4. Use video tapes or other visuals to train workers.
5. Review and update training programs quarterly.
6. Require that all equipment be certified as acceptable for use in food plants.
7. Secure knowledge of equipment harborage sites.
8. Disassemble equipment before cleaning.
9. Identify more hygienic equipment designs for future purchases.
10. Conduct microbial testing of cleaned equipment.
11. Inspect to ensure that equipment is clean and require a signature for verification.
12. Provide rewards for sanitation employees to recognize their superior performance.

The following suggestions apply to maintaining a sanitary environment:

1. Water that contacts equipment, raw materials, and finished products must be clean.
2. All equipment (including belts and other parts), containers, and hand utensils must be clean at the start of production.
3. All employees should be free of communicable diseases, open sores, or any signs of infection of the hands and arms.
4. All production employees should wear clean, light-colored clothing and head covering.
5. All personnel must wash their hands and arms before entering their work area.
6. An adequate number of toilet facilities should be provided for employees.
7. The welfare facilities should be ventilated to the outside and kept free of odors.
8. The doors of all toilet rooms should be solid, tight, and self-closing and should not open directly into production rooms.
9. Employees should not store their clothing in any area except in locker rooms that are clean and in good repair, closets, or protected space provided for this purpose.
10. Employees should wash their hands in a toilet facility instead of a sink provided for washing equipment and utensils.

11. After cleaning, multiservice containers, utensils, and disassembled piping and equipment should be transported and stored in a method to facilitate drainage and reduce contamination.
12. Bottled drinks should be filled, capped, sealed, and packaged in a hygienic environment to reduce contamination.

Deep cleaning is a concept that involves the disassembly of complex equipment to clean all components for effective soil removal followed by dry steam heating metal surfaces to 82 °C (180 °F). Steam penetrates cracks, crevices, and pores to facilitate cleaning effectiveness. Although steam is not always the most practical approach, under pressure it is drier and penetrates more effectively than water and larger-molecule chemicals. It is imperative that equipment and other surfaces be free of debris before steam application to avoid baking of soil to the equipment or area being cleaned and that the surfaces being cleaned attain the temperature indicated previously to ensure microbial destruction.

Shipping Precautions

Prior to loading, the truck, trailer, or rail car interiors should be inspected for general cleanliness and freedom from moisture and foreign materials that may cause product contamination or damage packaged products or their containers. If necessary, the transportation equipment should be cleaned, repaired, or rejected before loading is accomplished. Care should be exercised during loading to avoid product spillage or damage. The staging area and loading dock should be free from accumulations of debris and spillage.

Other Checkpoints

The plant and site should be kept free of liquid or solid emissions that could be sources of contamination. Materials that are stored in the open should be stacked neatly and away from buildings on racks above ground level. Activities that may cause contamination of stored foods with chemicals, filth, or other harmful material should be separated from the storage and processing operations. To conduct appropriate inspection, it is essential to know how to inspect and what is needed to make a good inspection.

Study Questions

1. What percentage slope should exist in wet-washed areas of low-moisture food plants?
2. What chemical-resistant floors are recommended in wet-washed areas?
3. What is the maximum percentage moisture for cereal grains placed in long-term storage to be protected against insects and molds?
4. How can dust be reduced in low-moisture food plants?
5. How can compressed air be used to clean in low-moisture food plants?

6. What precautions are necessary if suspended ceilings exist?
7. What is unacceptable ceiling in low-moisture food establishments?
8. What can happen if grain kernels are physically damaged?
9. How often should cleaning in the manufacturing area of low-moisture food plants be conducted?
10. What is one of the most thorough methods of cleaning in low-moisture food plants?
11. How does insect-resistant packaging prevent infestation of these pests?
12. Which beetle is similar in appearance to the drugstore beetle?

References

Cramer M (2006). *Food plant sanitation* 145. CRC Press: Boca Raton, FL.
Hui YH (2015). *Plant sanitation for food processing and food service* 12. CRC Press: Boca Raton, FL.

Dairy Processing Plant Sanitation

16

Abstract

Plant layout and construction affect microbial contamination and overall wholesomeness of the product. It is especially important to ensure that clean air and water are available and that surfaces in contact with dairy foods do not react with the products. Soils that are found in dairy plants include minerals, proteins, lipids, carbohydrates, water, dust, lubricants, cleaning compounds, sanitizers, and microorganisms. Effective sanitation practices can reduce soil deposition and effectively remove soil and microorganisms through the optimal combination of chemical and mechanical energy and sanitizers. This condition is accomplished through the appropriate selection of clean water, cleaning compounds, cleaning and sanitizing equipment, and sanitizers for each cleaning application. A current trend has been toward modification of cleaning-in-place (CIP) systems to permit final rinses to be utilized as makeup water for the cleaning solution of the following cleaning cycle and to segregate and recover initial product-water rinses to minimize waste discharges. Every processing facility should verify the effectiveness of its cleaning and sanitation program through daily microbial analyses of both product and various equipment and areas. Recent advances in technology have allowed tracking of cleaners and sanitizers in the CIP system. In addition, current computer, sensor, and traced chemistry technology allows real-time understanding of concentration in CIP systems so that system variation can be monitored and corrective actions can be taken.

Keywords

Milk • Cheese • Soft cheese • *Listeria monocytogenes* • Cleaning-in-place (CIP) • Tracer chemistry • Sensor technology

Introduction

The dairy industry has the reputation for being a food industry leader in hygienic design and practices, as well as in the implementation of sanitation standards. Also instrumental in the leadership role has been recognition by the industry of the primary need for good sanitation practices to ensure improved stability and high quality of dairy products that require refrigeration.

The physical and chemical properties of dairy products, especially fluid items, have made the automated cleaning of processing facilities possible. Some of the following components have been developed that contribute to automation:

- Permanent piping of nearly all "welded" construction has been installed to reduce the amount of manual cleaning of tubing and fittings.
- Control systems based on relay logic, dedicated solid-state controllers, small computers, and programmable logic controllers wired or programmed to control complex cleaning sequences have been developed.
- Automatically controlled cleaning-in-place (CIP) systems have provided a method to ensure uniformly thorough cleaning of tanks, valves, and pipes on a daily basis.
- Air-operated, CIP-cleaned sanitary valves have eliminated the manual cleaning of plug-type valves and provided for remote and/or automatic control of CIP solution flow.
- Silo-type storage tanks and dome-top processors have been designed for effective cleaning by CIP equipment.
- Processing equipment has been designed for CIP cleaning (homogenizers, plate heat exchangers, certain fillers, and the self-desludging centrifugal machine).
- Sensor and Trace Chemistry Technology have been developed to optimize CIP systems through detecting system variation and determining corrective actions.

These components function most effectively when properly integrated into a complete cleaning system that is designed and installed for auto-mated control of all cleaning and sanitizing operations.

The source of the milk supply is of major concern. Even the most effective pasteurization process cannot upgrade quality or eliminate the problems created by undesirable bacteria in the raw supply. Although pasteurization is an effective weapon against pathogenic and spoilage microorganisms, it is only a safeguard measure and should never be used to cover up an unsanitary raw supply or improper sanitation.

Polyphosphates and synthetic surface-active agents have been responsible for changes in cleaning operations to keep pace with new materials and cleaning and sanitizing equipment. These advances have enabled the formulation of specific cleaning compounds that adapt to water conditions, types of metals, and soil characteristics. They also have the buffering ability to minimize corrosion. They have opened up a new avenue of close union and intimate association between cleaning compounds and sanitizing agents to enhance the value of both phases of sanitization. Other recent advances include:

1. The use of alcohol as a carrier to decrease water use and allow better cleaning and sanitizing in areas where minimal water can be used
2. The use of mixed peracetic acid-based sanitizers to save water, energy, and time and lengthen fluid milk code dates

Role of Pathogens

From 1998 through 2011, 148 outbreaks due to the consumption of raw milk or raw milk products were reported to the Centers for Disease Control and Prevention (CDC). These resulted in 2384 illnesses, 284 hospitalizations, and two deaths. Most of these illnesses were caused by *Escherichia coli, Campylobacter, Salmonella, or Listeria* (CDC 2016). Out of the 104 outbreaks reported with substantial data, 82% involved at least one person younger than 20 years old.

Despite the industry's reputation for hygienic design and practices, pathogens have continued to invade dairy products. During 1985, a large outbreak of salmonellosis occurred in pasteurized milk. Other recent foodborne illness outbreaks from the ingestion of dairy products have included staphylococcal food poisoning caused by ice cream, the implication of campylobacteriosis that has occurred sporadically without a finite determination of the mode of transmission, and listeriosis from contaminated cheese. The latter outbreak was responsible for several deaths. As a result, the dairy industry has recalled a large number of food products at great expense. These events have brought the full force of the regulatory agencies upon the industry and motivated several dairy processors to invest heavily in the improvement in sanitation of their production facilities. These experiences have underscored the importance and urgency of effective sanitation programs. Because pathogens are discussed in Chap. 3, only *Listeria monocytogenes* and *Escherichia coli* O157:H7, the pathogens of greatest concern in dairy products, will be discussed here.

Listeria monocytogenes

The discovery of *L. monocytogenes* in fermented and unfermented dairy products has prompted food manufacturers to renew their concern about plant hygiene and product safety. *Listeria monocytogenes* is widely distributed in nature and often carried in the intestinal tract of cattle. Approximately 5% of normal, healthy humans are fecal excretors of this microorganism. Approximately 5–10% of raw bovine milk is contaminated with *L. monocytogenes*. This microbe has been isolated from improperly fermented silage, leafy plants, and the soil, with the latter being a reservoir of *Listeria* organisms.

Listeria recalls of ice cream and cheese products have precipitated major processing and sanitation operation changes in dairy processing plants. Many processors are voluntarily adopting Grade A standards required for the production of pasteurized milk. The importance of an effective sanitation program to combat *L. monocytogenes* has contributed to major increases in training,

supervision, total employee count, and salaries of sanitation workers in dairy processing plants.

The epidemiologic implication of pasteurized milk in the Massachusetts listeriosis outbreak in 1983 and in the outbreak in Los Angeles in 1985, attributable to a Mexican-style soft cheese, led to the establishment of the US Food and Drug Administration (FDA) standard methodology for detection of this pathogen. These events also contributed to a decision to conduct a large survey for pathogenic microorganisms in the dairy industry. This survey revealed that, in nearly all instances, post-processing contamination was responsible for contamination of *L. monocytogenes*. Buchanan et al. (2016) indicated that listeriosis outbreaks from dairy products have not decreased in frequency over the last 10 years. Illness outbreaks have been reported for unpasteurized milk, queso fresco cheese, and ice cream from 2012 to 2015 (CDC 2016), highlighting the need for continuous improvement in *Listeria* control through use of the guidelines provided by FDA (2008).

Specific guidelines have been developed for controlling *L. monocytogenes* in dairy processing facilities in the United States. These guidelines stress the need to (a) decrease the possibility that raw products will contain *Listeria* organisms, (b) minimize environmental contamination in food processing facilities, and (c) use processing methods and sanitation techniques that will reduce the probability that this pathogen will occur in food.

Properly constructed and maintained facilities and equipment are fundamental to an effective cleaning and sanitation program for the control of *L. monocytogenes*. Construction characteristics that will be described in this chapter and in Chaps. 17, 18, and 19 should be considered when planning a program for the control of this pathogen.

Listeria monocytogenes is sensitive to sanitizing agents commonly employed in the food industry. Chlorine-based, iodine-based, acid anionic, and/or quaternary ammonium-type sanitizers are effective against this pathogen when used at concentrations of 100 parts per million (PPM), 25–45 PPM, 200 PPM, and 200 PPM, respectively. Although these concentrations may require adjustment to compensate for in-plant use (as may oxidation-reduction factors relating to water quality and hardness), recommended concentrations should not be markedly exceeded, as use of extremely concentrated sanitizing solutions heightens the danger to employees, increases the risk of chemical contamination of food, and, in some instances, causes corrosion of equipment.

Quaternary ammonium-based sanitizers are not recommended for food contact surfaces and should not be used in cheese factories, as lactic acid starter culture bacteria are inactivated rapidly by small residues of these sanitizers. *Listeria monocytogenes* also has the ability to adapt and become tolerant to quaternary ammonium-based disinfectants (Buchanan et al. 2016). In contrast, acid anionic and iodine-type sanitizers are best suited for equipment surfaces, with the former readily neutralizing excess alkalinity from cleaning compounds and preventing the formation of alkaline mineral deposits. The use of steam should be discouraged (due to energy costs) and, if used, should be confined to closed systems because of potential hazards associated with aerosol formation. Sanitizing with hot water is not recommended because of the energy costs of heating water and because high temperatures cannot be maintained easily.

Effectiveness of a *Listeria* control program can be measured by conventional and routine, preoperative microbial monitoring, such as aerobic plate count and coliform count (see Chaps. 3 and 8). However, industry experience has suggested that the most accurate measurement relies on specific testing for *Listeria* organisms in the plant environment. Environmental sampling should be organized to guide preoperative sanitation practices and direct management toward a *Listeria*-controlled operation.

Escherichia coli O157:H7

Outbreaks of this pathogen associated with raw milk have challenged investigators to further research this microorganism in dairy products. This pathogen can grow in cottage cheese and cheddar cheese but is inactivated by the pasteurization of milk. It has been associated with unpasteurized milk and was associated with an outbreak in cheese as recently as 2010 (CDC 2015).

Buchanan and Doyle (1997) suggested that alternative technologies to thermal processing control *E. coli* O157:H7 while maintaining the acceptability of dairy products. A viable alternative technology for dairy, meat, and poultry products is ionizing radiation. This pathogen is relatively radiation-sensitive, and radiation pasteurization doses of 1.5–3.0 KGy appear to be destructive at the levels that are most likely to occur in ground beef (Clavero et al. 1994).

Salmonella

Milk and milk products have been identified as a vehicle for transmission in approximately 5% of salmonellosis cases, although the sources of infection in Maine were identified in most cases (CDC 2000). Salmonellosis is commonly diagnosed in dairy animals (Wells et al. 2001), and there is evidence that it is shed from the mammary gland (Radke et al. 2002). Fecal contamination is another alleged major source of contamination in raw milk. Van Kessel et al. (2003) evaluated the efficacy of a portable real-time polymerase chain reaction (PCR) system for the detection of salmonella in raw milk. They found that the portable real-time PCR techniques yields results in 24 h compared with the 48–72 h required for a traditional culture. Non-pasteurized milk can be a vehicle for *Salmonella*. It is important to use pasteurized milk in the production of dairy products. The popularity of farmer markets and the local fresh food movement brings new concerns for the dairy industry since unpasteurized milk is sometimes sold and can be contaminated with pathogenic bacteria.

Sanitary Construction Considerations

The considerations most important to dairy plant sanitation are drainage and waste disposal. Storm and sanitary sewers must be adequate and readily available. In rural areas and municipalities with limited treatment facilities, dairy processors frequently must provide their own waste disposal facilities. An adequate supply of potable water and acceptable drainage and waste disposal are essential. Other considerations are mentioned in Chap. 14.

Floor Plan and Type of Building

The layout and construction of a dairy plant are subject to the approval of one or more regulatory agencies. All equipment and utensils should be purchased subject to the approval of the various regulatory authorities.

Ventilation is important, especially in areas where excess heat that is produced during processing must be removed. The ventilation should be tailored to the different types of rooms and should have the flexibility to meet the needs of any future alterations in production. It is frequently necessary to filter incoming air, especially if the plant is located in a heavy industrial area. Also, the control of humidity, condensation, dust, and spores should be considered.

Construction Guidelines

Unless construction is carefully planned, the structure and equipment can contribute to contamination. This problem can be helped by reducing overhead equipment to a minimum, which reduces contamination due to maintenance of this equipment. Overhead equipment is also difficult to clean. A separate service floor that will accommodate a major portion of the ducts, pipe works, compressors, and other equipment should be provided. This arrangement results in a clear ceiling, which is easy to clean and keep sanitary.

Some other design and construction characteristics that are conducive to effective sanitation are:

- All metal construction should be treated to withstand corrosion.
- Pipe insulation should be of a material that is resistant to damage and corrosion and will endure frequent cleaning.

- Chronic condensation points should be protected by the installation of a drainage collection system.
- All openings should be equipped with air or mesh screens and tight-fitting windows.

Structural finishes should be of materials that require minimal maintenance. Walls, floors, and ceilings should be impervious to moisture. Floor materials should be resistant to milk, milk acids, grease, cleaning compounds, steam, and impact damage. Epoxy, tile, and brick are good choices. Paint should not be used if suitable alternatives exist. If paint is applied, it should be of a grade that is acceptable for food plants. Floor drains should be designed to control insect infestation and odors. A slope of approximately 2.1 cm/m (0.25 in/ft) is recommended to reduce accumulation of water and waste on the floor, which could hamper the sanitation and lead to growth of *L. monocytogenes*.

Floor drains and ventilation systems contribute to contamination from airborne microorganisms instead of acting as a sanitation barrier. A properly designed ventilation system with air filtration can improve air quality. Inexpensive filters remove dust and other contaminants that would normally be drawn into these spaces or rooms.

Equipment should be designed and oriented for easy cleaning and reduction of contamination. Traditionally, equipment layout has been important to operational efficiency, with the effects on the sanitation operation of only secondary importance. The most critical considerations related to equipment sanitation include a location to permit sanitary operations between equipment and walls or partitions, an exterior with an easy-to-clean surface, and a design to permit effective sanitation between the equipment and floor. All equipment should be accessible, easy to clean, and designed for draining and sanitizing.

Soil Characteristics in Dairy Plants

In the dairy industry, soil consists primarily of constituents of minerals, lipids, carbohydrates, proteins, and water. Other soil constituents may be dust, lubricants, microorganisms, cleaning compounds, and sanitizers.

White or grayish films that form on dairy equipment are usually milkstone and waterstone. These films usually accumulate slowly on unheated surfaces because of poor cleaning or use of hard water or both. Calcium and magnesium salts precipitate when sodium carbonates are added to hard water. During cleaning, some of this precipitate may adhere to equipment, leaving a film of waterstone. When proteins denatured by heat adhere to surfaces and other components absorb them, milkstone may form quickly on heated surfaces. Because they become less soluble at high temperatures, calcium phosphates from milk are present in large quantities. The nature of soil on heated and unheated surfaces usually differs in composition. Thus, each type of soil requires a different cleaning procedure. Milkstone is usually a porous deposit that will harbor microbial contaminants and eventually defy sanitizing methods. It can be removed through an acid cleaner to dissolve the alkaline minerals and remove the film. Heavy soil deposits require a stronger cleaning compound than lighter soils. Also, freshly deposited soil on an unheated surface is more readily dissolved than the same soil that has dried or has baked on a heated surface.

Soil deposition can be reduced and subsequent removal eased by application of the following principles:

- Generally, product surfaces should be cooled before and immediately after emptying of heated processing vats.
- Foams and other products should be rinsed after the production shift and before they dry.
- Where possible and practical, the soil deposits should be kept moist until the cleaning operation starts.
- Rinsing should be accomplished with warm (not hot) water.

Soil deposition is increased in ultrahigh-temperature heaters if milk contains high acidity and is complicated by low-velocity movement and poor agitation during the operation. Preheating and holding at a high temperature reduce film deposition.

The nature of the surface determines the ease or difficulty of soil removal. Pits in corroded surfaces, cracks of rubber parts, and crevices in insufficiently polished surfaces protect soil and microorganisms from the effects of cleaning compounds and sanitizers. The soil to be removed determines the cleaning method and cleaning compound.

Biofilm Prevention

Inadequacy of cleaning/sanitation in dairy plants can lead to biofilm formation. Biofilms are a community of bacterial cells that (1) adhere to each other and surfaces such as steel, glass, and plastic, (2) are held together and protected by polysaccharides that act as a glue-like material, and (3) differ in gene expression when compared to normal planktonic cells (Sofos 2009). Bacteria are able to attach to surfaces when there are nutrient and soils on a surface due to inadequate cleaning and form biofilms. Common bacteria in biofilms include *Listeria*, *Salmonella*, *Escherichia coli*, and spoilage bacteria. Biofilm control is discussed in detail in Chap. 19. However, one additional item that should be mentioned in this chapter includes control of biofilms in dairy processing membranes. Anand et al. (2014) reported cleaning methods to inhibit bacterial biofilms in reverse osmosis membranes that are commonly used to isolate whey protein from lactose and other protein in the dairy industry.

Sanitation Principles

Cleaned and sanitized equipment and buildings are essential to the production, processing, and distribution of wholesome dairy products. The major part of the total cleaning cost is labor. Therefore, it is important to use appropriate cleaning compounds and equipment so that the sanitation program can be effectively administered in a shorter period of time and with less labor.

The sanitarian should know the time required to clean each piece of equipment with the mechanization and cleaning compounds available. Cleaning tasks should be assigned to specific employees, who should be made responsible for the equipment and area under their care. These assignments should be made through official notification or by posting the cleaning schedule or assignments on a bulletin board.

Role of Water

The major constituent of almost all cleaners, including those used by dairy plants, is water. Because most plant water is not ideal, the cleaning compounds that are selected should be tailored to the water supply or should be treated to increase the effectiveness of the cleaning compound. It is especially important to reduce suspended matter in water to avoid deposits on clean equipment surfaces. Water hardness complicates the cleaning operation. Suspended matter and soluble manganese and iron can be removed only by treatment, whereas small amounts of water hardness can be counteracted by sequestering agents in the cleaning compounds that are used in the sanitation operation. If the water is hard or very hard, it is usually more economical to pretreat the water to remove or minimize hardness.

Role of Cleaning Compounds

Like all other cleaning compounds, those used in cleaning dairy plants generally are complex mixtures of chemicals combined to achieve a specific desired purpose. The following cleaning functions are related to the role of cleaning compounds in dairy sanitation operations:

1. Prerinsing is conducted to remove as much soil as possible and to increase the effectiveness of the cleaning compound.
2. The cleaning compound is applied to the soil to facilitate subsequent removal through effective wetting and penetrating properties.
3. Solid and liquid soils are displaced through fat saponification, protein peptizing, and mineral dissolution.

4. Soil deposits are dispersed in the cleaning medium by dispersion, deflocculation, or emulsification.
5. Effective rinsing is conducted to prevent redeposition of the dispersed soil onto the cleaned surface.

The value of a cleaning compound is most accurately determined by measuring the area that can be cleaned efficiently with minimal costs. High-cost cleaning compounds are frequently the most economical because of labor, energy, and cleaning compound savings. More discussion of cleaning compounds is provided in Chap. 9.

Application of Cleaning Compounds

Identification of the optimal external energy factors and application methods is necessary to facilitate cleaning. If cleaning is done by hand, strong acids and alkalies should be avoided because they irritate human skin. Instead, emphasis should be placed on external energy, such as heat and force. Superb results depend on the use of circulation cleaning and on whether it is done in or out of place. Table 16.1 provides a guide to the most appropriate cleaning compound, cleaning procedure, and cleaning equipment for the major cleaning applications.

Role of Sanitizers

After cleaning, sanitizers should be applied to destroy microorganisms. Of the many methods for sanitizing (see Chap. 10), those most frequently used in dairy plants are steam, hot water, and chemical sanitizers.

Steam Sanitizing

Steam sanitizing is accomplished by maintaining steam in contact with the product contact surfaces for a designated time. The effective procedures have been found to be 15 min of exposure when the condensate leaving the assembled equipment is at 80 °C (176 °F). This method of sanitizing has limited utility because it is difficult to maintain a constant required temperature and because the energy costs are excessive. Steam

Table 16.1   Optimal cleaning guides for dairy processing equipment

Cleaning applications	Cleaning compound	Cleaning medium	Cleaning equipment
Plant floors	Most types of self-foaming or foam boosters added to most moderate to heavy-duty cleaners	Foam—pressure control centralized or portable system cleaners should be used with heavy fat or protein deposits	Portable or centralized foam cleaning equipment with foam guns for air injection into the cleaning solution
Plant walls and ceilings	Same as above	Foam	Same as above
Processing equipment and conveyors[a]	Moderate to heavy-duty alkalies that may be chlorinated or nonalkaline	Pressure-controlled system	Portable or centralized system cleaners with variable pressure; low-volume equipment; sprays should be rotary hydraulic
Closed equipment	Low-foam, moderate to heavy-duty chlorinated alkalies with periodic use of acid cleaners as follow-up brighteners and neutralizers	CIP	Pumps, fan or ball sprays, and CIP tanks; sensor technology with tracer chemistry

[a]Packaging equipment can be effectively cleaned with gel cleaning equipment

application can also be more dangerous than other sanitizing methods and is not usually recommended.

Hot Water Sanitizing

The Pasteurized Milk Ordinance requires that for hot water sanitizing, equipment surfaces must be exposed to a minimum of 77 °C (170 °F) water for 5 min. The International Dairy Federation recommends 85 °C (185 °F) for 15 min. FDA regulation, 21 CFR 129.80, establishes that hot water sanitizing of enclosed systems must be at a minimum of 77 °C (170 °F) for at least 15 min or at 94 °C (200 °F) for at least 5 min. A proper time and temperature combination is essential.

Hot water is pumped through the assembled equipment to bring the product surfaces in contact with water at a given temperature for a specified time. Water temperature maintained at 80 °C (176 °F) at the equipment outlet for 5 min serves as the sanitizer. This technique is expensive because of the required energy costs.

Hot water is relatively inexpensive, easily available, and effective in microbial destruction as well as having a broad antimicrobial activity. It is generally noncorrosive and provides sufficient heat penetration into difficult-to-reach areas such as behind gaskets and in threads, pores, and cracks.

The use of hot water has limitations since it is comparatively slow and requires a lengthy process involving heat and cooldown, compared to chemical sanitizing. Furthermore, it can close film and scale formation or heat fixing of any remaining soils, making future cleanup more difficult. Hot water can shorten equipment life because of thermal expansion and contraction stress and cause premature failure. Equipment must be designed to withstand a temperature in excess of 82 °C (180 F), and hot water in the system creates condensation within the plant production environment, and water heated above 77 °C (172 °F) is hot enough to cause serious burns.

Chemical Sanitizing

This method is accomplished by pumping an acceptable sanitizer such as the halogens (usually chlorine or iodine compounds) through the assembly for at least 1 min. This technique requires contact of the sanitizer with all of the possible product surfaces. Because contact of the sanitizer with the surface is essential, the application method in dairy operations is important.

For large-volume, mechanized operations, the sanitizer can be applied through sanitary pipelines by *circulation* or pumping of a sanitizing solution through the system. The appropriate amount of sanitizing solution is prepared in a container and pumped. A slight backpressure should be built up in the system to ensure contact with the upper inner surface of the pipeline.

Small operations that cannot justify mechanization can sanitize by the *submersion* of equipment, utensils, and parts in the sanitizer solution. This process normally involves submersion for approximately 2 min and then draining and air-drying on a clean surface.

Closed containers, such as tanks and vats, are easily and effectively sanitized through *fogging*. The strength of the sanitizing solution should be twice that of the ordinary use solution, and it should be given at least 5 min of exposure.

If a sanitizer is applied through *spraying*, all surfaces should be contacted and completely wetted. As with fogging, the sanitizing solution strength should be twice that of the ordinary use solution.

If mechanized sanitizing equipment is unavailable, large open containers, such as cheese vats, can be sanitized by *brush application*. All areas should be touched with the brush. This method has high labor costs.

Sanitized surfaces should not be rinsed with water; otherwise, equipment and utensils can be recontaminated with aerobic microorganisms that reduce product stability. Furthermore, other recontaminations of the sanitized surfaces should be avoided.

Cheese ripening rooms possess an environment that encourages mold growth. *Ozone* is effective in the inactivation of airborne molds in this environment but not surface molds. Serra et al. (2003) indicated that it was necessary to wipe the surfaces with a commercial sanitizer to decrease the viable mold load on these surfaces. Air control with high efficiency particle filtration systems is essential to mold and yeast control (Roginski 2014).

Cleaning Steps

Dairy operations require multiple cleaning procedures that are separated into eight steps in this section:

1. *Cover electrical equipment*. Covering material should be polyethylene or equivalent.
2. *Remove large debris*. This task should be accomplished during the production shift and/or prior to prerinsing.
3. *Disassemble equipment as required*.

4. *Prerinse*. Prerinsing can effectively remove up to 90% of the soluble materials. This operation also loosens tightly bound soils and facilitates penetration of the cleaning compound in the next cleaning step.
5. *Apply cleaning compound*. This step can be simplified through proper selection and use of processing equipment and cleaning equipment, proper location of equipment, and reduction of soil accumulation. Further reduction of soil buildup is possible through use of the minimum required temperature for heating products a minimum amount of time; cooling product heating surfaces, when practical, before and after emptying of processing vats; and keeping soil films moist by immediate rinsing of foam and other products with 40–45 °C (104–113 °F) water and leaving it in the processing vats until cleaning.
6. *Postrinse*. This step solubilizes and carries away soil. Rinsing also removes residual soil and cleaning compounds and prevents redeposition of the soil on the cleaned surface.
7. *Inspect*. This step is essential to verify that the area and equipment are clean and to correct any deficiencies.
8. *Sanitize*. A sanitizer is added to destroy any residual microorganisms. By destruction of microorganisms, the area and equipment contribute to less contamination of the processed products.

The important components that are part of every cleaning process include prerinsing, cleaning, rinsing, and sanitizing steps. In addition, the effectiveness of the procedure is based on cleaners and sanitizers that are chosen, cleaner and sanitizer concentrations, and time of application.

Other Cleaning Applications

When mechanized cleaning is not practical, hand cleaning should be done, following these guidelines:

• Cleaning applications should involve a prerinse of water at 37–38 °C (99–101 °F).
• The cleaning compound used should have a pH of less than 10 to minimize skin irritation.

Table 16.2 Special considerations for hand cleaning dairy plant equipment

Equipment	Recommended cleaning procedures
Weigh tanks (can receiving and/or in-plant can transfer)	Rinse immediately after milk has been removed; disconnect and disassemble all valves and other fittings; wash weigh tank, rinse tank, and fittings; sanitize prior no next use.
Tank trucks, storage tanks, processing tanks	Remove outlet valve, drain, rinse several times with small volumes of tempered (38 °C, 100 °F) water, remove other fittings and agitator; brush or pressure-clean vats, tanks, and fittings; rinse and reassemble after sanitizing fittings just before reuse. Thoroughly clean manhole covers, valve outlets, slight glass recesses, and any airlines. High-pressure sprays are preferable to keep the cleanup personnel out of the tanks or vats and to minimize damage to surface and contamination of cleaned surfaces.
Batch pasteurizers and heated produce surfaces	Lower temperature to below 49 °C (120 °F) after emptying product; immediately rinse, with brushing to loosen burned-on products. If the vat cannot be rinsed, fill with warm (32–38 °C, 90–100 °F) water until cleaning. Clean the same as for other processing vats.
Coil vats	Although not in general use, they are difficult to clean because of inaccessibility of some surfaces of the coil. After prerinsing, fill with hot water. Add cleaning compounds and rotate, while all exposed coil surfaces are brushed.
Homogenizers	Prerinse while the unit is assembled; dismantle and clean each piece; place clean parts on a parts cart to dry. Sanitize and reassemble prior to use.
Sanitary pumps	After use, remove head of pump and flush thoroughly with tempered (38 °C, 100 °F) water; remove impellers and place them in the bucket containing a cleaning solution of 49–50 °C (120–122 °F). Wash intake and discharge parts and chamber. Brush impellers and place them in a basket on a parts table to dry.
Centrifugal machines	Non-CIP types must be cleaned by hand. Rinse with 38 °C (100 °F) until discharge is clear. Dismantle, remove bowel and discs, and rinse each part before placing in the wash vat. A separate wash vat is desirable for separator and clarifier parts. Each disc should be washed separately, rinsed, and drained thoroughly. If a separator is used intermittently during the day, it should be rinsed after each use, with at least 100 L (26.4 gallons) of tempered water. Use of a mild alkaline wetting agent can improve rinsing efficiency.

The temperature of the cleaning solution should be maintained at 45 °C (113 °F). Solution-fed brushes can be used effectively with hand cleaning operations. Filler parts and other parts that are difficult to clean should be cleaned with cleaning-out-of-place (COP) equipment to move the surface lubricant and other deposits more effectively.

- The postrinse operation should use water tempered to 37–38 °C (99–101 °F), with subsequent air-drying.
- The sanitizing operation should include a chlorine sanitizer applied by a spray or dip.

Table 16.2 classifies and summarizes special considerations for various types of dairy plant hand cleaning equipment.

Cleaning Equipment

Cleaning of dairy facilities involves physical removal of soil from all product contact surfaces after each period of use, with subsequent application of a sanitizer. Although surfaces that contact non-products are less critical, they must be cleaned. The techniques for cleaning dairy plants vary depending on the plant size. The major portion of a large-volume plant is cleaned by some CIP system. This cleaning technique is the recognized standard for cleaning pipelines, milking machines, bulk storage tanks, and most equipment used throughout the processing operation. Because the normal period of use for dairy processing plant equipment is less than 24 h, this equipment and the area are cleaned daily. Longer and continued use of piping and storage systems can reduce the cleaning frequency to once every 3 days.

CIP and Recirculating Equipment

Effectiveness of the CIP approach depends on the process variables, time, temperature, concentration, and force. Rinse and wash time should be minimized to conserve water and cleaning compounds but should be long enough to remove soil and to clean effectively and efficiently. Time is affected by temperature, concentration, and force. An energy-efficient CIP system can reduce cleaning costs by over 35% with approximately 40% less energy.

A salmonellosis outbreak in pasteurized milk during the 1980s that was allegedly caused by a CIP cross-connection between raw and pasteurized products has been responsible for the installation in many dairies of a completely separate CIP system for the receiving area of the plant.

Temperature of the cleaning solution for CIP equipment should be as low as possible and still permit effective cleaning with minimal use of the cleaning compound. Rinse temperature should be low enough to avoid deposits from hard water.

Force or physical action determines how effectively the cleaning compound is introduced to the areas to be cleaned and how it is controlled by the system design. Adequate force (or physical action) can be ensured by the selection and utilization of appropriate high-pressure pumps to provide sufficient turbulence in and through pipelines and storage tanks, achieving maximum efficiency.

CIP operations in dairy plants are normally divided into two major categories: spray cleaning and line cleaning. Other closed circuits, such as high-temperature short-time (HTST) units, are frequently used. Although many types of spray devices are utilized in the dairy processing industry, permanently installed fixed-spray units are more durable than are portable units and rotating or oscillating units. Other advantages include no moving parts, stainless steel construction, and less performance difficulty, due to minor variations in supply pressure.

The line cleaning principle can involve product piping CIP circuits with readily available points from which a circuit can be fed and to which it can be returned. Return lines from storage tanks to a return pump should have an approximate 2% pitch continuously toward the return pump inlet. Control of pressure and flow should be provided for each spray device.

Shell and tube heat exchangers that are equipped with return-bend connections of CIP design can be incorporated into CIP piping circuits or may be cleaned independently as a separate operation. Triple tube-type tubular heat exchangers can be installed so that they will be self-draining. Plate-type heat exchangers are more widely used than are tubular units because of ease of inspection, flexibility of design, and ease of adaptation to new applications.

In CIP, the cleaning compound must be applied forcefully enough to provide intimate association with the soiled surfaces, and it must be continuously replenished. Various forms of CIP equipment systems are available. (The basic forms are discussed in Chap. 11.) Some CIP systems have been modified to permit use of final rinses as the cleaning solution for makeup water of the following cleaning cycle and to segregate and recover initial rinses to minimize waste discharges.

Installations since the mid-1970s have incorporated CIP systems that combine the advantages of the flexibility and reliability of single-use systems with water and solution recovery techniques that aid in reducing the amount of water required for a cleaning cycle. The intent of these systems is to recover the spent cleaning solution and the postrinse water from one cleaning cycle for temporary storage and reuse of the detergent-rinse water mixture as a prerinse for the subsequent cleaning cycle. This approach reduces the total water requirement of spray-cleaning systems by 25–30%, as compared with alternative approaches. Through this technique, steam consumption is reduced by 12–15% and cleaning compound consumption by 10–12%, because a prerinse of the spent solution adds heat to the vessel as it removes the soil. If a CIP recirculating unit is used to clean equipment with a large quantity of insoluble soil, a powered strainer, centrifuge, or settling basin may be incorporated in the return system to prevent this material from recirculating and impairing the spray action. Proper operation of the entire CIP system should be verified from data collected on recording charts, which can be stored for future reference.

CIP systems can be optimized using 3D TRASAR™ technology, which combines CIP methodology, advanced analytics, sensor technology, and tracer chemistry to optimize cleaning and sanitizing while minimizing costs. This technology minimizes temperature, which saves energy, saves water, optimizes CIP processes, allows for multiple CIP programs at variable temperatures, detects system variability, determines corrective actions, tracks where the cleaner has contacted the equipment surfaces, and monitors both cleaner and sanitizer concentration. For fluid milk, use of this technology increases production, extends shelf life, and decreases cleaning times.

COP Equipment

The following steps are recommended when COP equipment is used in dairy plants:

1. A prerinse with tempered water at 37–38 °C (99–101 °F) to remove gross soil
2. A wash phase through circulation of a chlorinated alkali cleaning solution for approximately 10–12 mins at 30–65 °C (86–150 °F) for loosening and eradicating soil not removed during the prerinse phase
3. A postrinse with water tempered to 37–38 °C (99–101 °F) to remove any residual soil or cleaning compound.

Cleaning of Storage Equipment

Appropriately designed storage tanks with properly installed spray devices are essential for effective spray cleaning. The fixed-based spray that is permanently installed has become more prominent in the industry than the rotating and oscillating spray devices. It requires less maintenance, is constructed of stainless steel without moving parts, and endures. Performance of this unit is not affected by minor variations in supply pressure, and spray is continuously applied to all of the surfaces. Cylindrical and rectangular tanks can be properly cleaned when sprayed with 4–10 L/min/m² (4.2–11 quarts/min/11 ft²) of internal surface, with patterns designed to spray the upper one-third of the storage container. Because the equipment contains heating or cooling coils with complex agitators, a special spray pattern is normally required, as is a subsequent increase in pressure and volume to cover all of the surfaces.

The vertical silo-type tank requires flow rates of 27–36 L/linear meter (6–8 gallon/yard) of tank circumference. Because of the difficulty in reaching the spray devices for occasional inspection and cleaning, non-clogging disc sprays are normally used in this type of storage vessel. Although most spray cleaning is conducted with standard sprays, special devices such as disc sprays, ball sprays, and ring sprays are available for use with vacuum chambers, dryers, evaporators, and complex vessels with special processing features.

Cleaning of large tanks that use spray devices differs from line cleaning applications because prerinsing and postrinsing are generally accomplished through use of a burst technique in which water is discharged in three or more bursts of 15–30 s each, with complete draining of the tank between successive bursts. This procedure is more effective in removing sedimented soil and foam than is continuous rinsing, and it can be accomplished with less water consumption.

The soil deposited in storage tanks and processing vessels is more variable than that associated with piping circuits; thus, cleaning techniques for this equipment are more diverse. For lightly soiled surfaces, such as those of storage tanks for milk or low-fat milk by-products, effective cleaning can be accomplished through a three-burst prerinse of tempered water. Recirculation of a chlorinated alkaline detergent of 5–7 min at 55 °C (132 °F), application of a two-burst postrinse at tap water temperature, and recirculation of an acidified final rinse for 1–2 min at tap water temperature also contribute to effective cleaning. Recirculation time and temperature may be increased slightly for more viscous products with a higher content of fat and total solids.

Soil components from cold surfaces differ from those of burned-on deposits, which contain higher protein and mineral contents. Burned-on soil requires increased cleaning compound concentration and solution temperatures of up to 82 °C (180 °F), with an application time of up to 60 min. Excessive amounts of burned-on deposits can also be cleaned effectively with application and circulation of a hot alkaline detergent and a hot acid detergent solution.

Table 16.3 lists the typical concentration of cleaning compounds and sanitizers for various cleaning applications. Although variations can exist, the suggested concentrations should be considered.

Table 16.3 Typical concentrations for various cleaning applications

Cleaning applications	Chlorinated cleaning compounds (ppm)	Acid/acid anionic chlorine sanitizers (ppm)
Milk storage and transportation tanks	1,500–2,000	100
Cream, condensed milk, and ice cream storage tanks	2,500–3,000	100–130
Processing vessels for moderate heat treatment	4,000–5,000[a]	100–200
Heavy "burn-on"	0.75–1.0% (causticity)	Acid wash at pH 2.0–2.5

[a]An acid rinse after cleaning should be considered

Cleaning programs depend on the properties of the product passing through the system during production. In addition to cleaning applications previously discussed, the following approach is recommended for the following processing systems:

Milk, Skim Milk, and Low-Fat Product Processing Equipment Because of the mineral content of these products, the equipment can be cleaned effectively by recirculation of an acid detergent for 20–30 min, with follow-up by direct addition of a strong alkaline cleaner, which is then recirculated for approximately 45 min. An intermediate rinse of cold water may be alternated between the acid and alkaline cleaners.

Cream and Ice Cream Processing Equipment These products, which contain a higher percentage of fat and a lower percentage of minerals, can be cleaned more effectively if an alkaline cleaner is first recirculated for approximately 30 min. The concentration of the alkaline solution may range from 0.5% to 1.5% in causticity. The acid is generally added to produce a pH of 2.0–2.5. A practical rule of thumb is to use a cleaning solution temperature during the recirculating period that is adjusted to approximately 5 °C (9 °F) higher than the maximum processing temperature used during the production shift.

Cheese Making Area and Equipment

The two main types of spoilage of hard and semi-hard varieties of cheese are surface growth of microorganisms (usually molds) and gas production of microorganisms growing in the body of cheese. *Penicillium* accounts for up to 80% of spoilage cases, and other common spoilage species are *Alternaria, Aspergillus, Candida, Monilia,* and *Mucor.* Mold spoilage reduction may be accomplished through sterile filtration of air, ultraviolet disinfection of handling surfaces, ozone treatment, and antimycotic coating of packaging material. The spraying of chemical disinfectants in the air is a routine practice for mold control (Roginski 2014). *Enterobacteriaceae, Bacillus, Clostridium,* and *Candida* are some common microorganisms responsible for gas production. According to Varnam and Sutherland (1994), soft cheeses can be affected by gram-negative bacteria, such as *Pseudomonas fluorescens, P. putida,* and *Enterobacter agglomerans;* by diarrheagenic strains of *E. coli,* which come from wash water or added ingredients; and by gram-positive bacteria, such as *L. monocytogenes.*

Milk should be stored in tanks constructed with materials and designs that are easy to clean. However, silo tanks that are large and cannot be cleaned using normal cleansing methods should be equipped with CIP methods and cleaned every time that they are emptied. They should be rinsed with water to remove gross soils and washed with detergent solutions, rinsed, and sterilized. Acid solutions should be incorporated when tank materials permit their use. Chemical sterilization is the preferred method, and steam sterilization should be avoided.

As with other dairy processing plants, piping should be carefully laid out to prevent cross-contamination between pasteurized and unpasteurized milk. Separate CIP equipment should be provided for both products. Cleaning and sterilization can be achieved through circulating materials such as sodium hydroxide and nitric acid (Varnam and Sutherland 1994).

Brine tanks should be lined with a noncorrosive material, such as tiles or plastics. Brines should be maintained at the correct strength to reduce the growth of halophilic microorganisms. The walls, floors, and ceilings of ripening rooms and cheese storage areas should be washed with fungicide solutions.

Increased outbreaks of *L. monocytogenes*, *S. aureus*, and *Yersinia enterocolitica* cause concern because these organisms can attach to surfaces and cross-contaminate food products or expose workers to contamination if surfaces are not cleaned and sanitized properly. Because disinfectants affect microorganisms differently and at different concentrations, tests should be conducted to determine the appropriate disinfectants and concentrations at each step of the cheese manufacturing process.

Rapid Assessment of Cleanliness

Paez et al. (2003) evaluated a commercial ATP-bioluminescence system to evaluate cleanliness of milking machines, bulk tanks, rinse water, and milk transport tankers on an experimental dairy farm. Bioluminescence results were not reliable for rinse water, so it was suggested that surface swab evaluations were also needed for a complete hygienic assessment.

Recent Technological Advances and Green Technology

The use of 3D TRASAR™ technology, as described previously in this chapter, can be used to save water and energy, which lowers costs but also lowers the environmental impact of the product. Other methods for decreasing costs and energy and water use include:

1. Use of cleaner and sanitizer combinations. For example, a peracetic acid blend with hydrogen peroxide can be used in the CIP process for a simultaneous acid rinse and sanitizing step to reduce water consumption.
2. Use of alcohol/quaternary ammonium compounds combinations, where the alcohol is able to evaporate rapidly and minimize water use.

3. Peroxyacetic acid-based pretreatments can also be used prior to washing to remove dairy protein on pasteurizers and other dairy equipment where heating is applied.

Note: The authors want to thank Fred Sonetto with Ecolab, Inc. St. Paul, Minnesota for his input about recent innovations in dairy processing sanitation.

Study Questions

1. What construction characteristics are needed for effective sanitation in dairy plants?
2. What temperature is necessary to hot water sanitize dairy processing equipment?
3. How is chemical sanitizing of dairy processing equipment accomplished?
4. What are the two major categories of CIP operations?
5. What brushes are best for cleaning dairy processing equipment?
6. How can film deposition be decreased in ultrahigh-temperature heaters?
7. What is a preferred cleaning method for lightly soiled surfaces of storage tanks for dairy products?
8. How is the processing equipment for milk, skim milk, and low-fat dairy products cleaned?
9. How is the cream and ice cream processing equipment cleaned?
10. How do soil components from cold sources differ from those of burned-on deposits with higher protein and mineral contents?
11. How is tracer chemistry and sensor technology used to optimize CIP operations?
12. What are some methods that can be implemented to minimize water use in the dairy industry?

References

Anand S, Singh D, Avadhanula M, Marks S (2014). Development and control of bacterial biofilms on dairy processing membranes. *Comp Rev Food Sci Food Saf* 13: 18.

Buchanan RL, Doyle MP (1997). Foodborne disease: Significance of *Escherichia coli* O157:H7 and other enterohemorrhagic *E. coli*. *Food Technol* 51(10): 69.

Buchanan RL, Gorris LGM, Kayman MM, Jackson TC, Whiting RC (2016). A review of *Listeria monocytogenes*: An update on outbreaks, virulence, dose-response, ecology, and risk assessments. *Food Control.* https://doi.org/10.1016/j.foodcont.2016.12.016.

CDC. Centers for Disease Control and Prevention (2000). Surveillance for foodborne-disease outbreaks-United States, 1993–1997. *Morb Mortal Wkly Rep* (Surveillance Summary-1): 1.

CDC. Centers for Disease Control and Prevention (2015). https://www.cdc.gov/foodsafety/rawmilk/raw-milk-index.html. Accessed October, 2016.

CDC. Centers for Disease Control and Prevention (2016). https://www.cdc.gov/listeria/outbreaks/index.html. Accessed October, 2016.

Clavero MR, Monk JD, Beuchat LR, Doyle MP, Brackett RE (1994). Inactivation of *Escherichia coli* O157:H7, *Salmonellae*, and *Campylobacter jejuni* in raw ground beef by gamma irradiation. *Appl Environ Microbiol* 60: 2069.

FDA. Food and Drug Administration (2008). Guidance for industry: Control of Listeria monocytogenes in refrigerated or frozen ready-to-eat foods. http://www.fda.gov/Food/GuidanceRegulation/GuidanceDocumentsRegulatoryInformation/FoodProcessingHACCP/ucm073110.htm. Accessed October, 2016.

Paez R, Taverna M, Charlon V, Cuatrin A, Etcheverry F, Da Costa LH (2003). Application of ATP-bioluminescence technique for assessing cleanliness of milking equipment, bulk tank and milk transport tankers. *Food Prot Trends* 23: 308.

Radke BR, McFall M, Radostitis SM (2002). Salmonella muenster infection in a dairy herd. *Can Vet J* 43: 443.

Roginski H (2014). Moulds and yeasts in the dairy industry. http://www.dairyaustralia.com.au/~/media/19DE63F642E6492C89EE0673C52ECE8C.pdf. Accessed October, 2016.

Serra R, Abrunhosa L, Kozakiewicz K, Venancio A, Lima N (2003). Use of ozone to reduce molds in a cheese ripening room. *J Food Prot* 66: 2355.

Sofos J (2009). Biofilms: Our constant enemies. *Food Saf Mag* February/March: 1.

Van Kessel JS, Karns JS, Perdue ML (2003). Using a portable real-time PCR assay to detect *Salmonella* in raw milk. *J Food Prot* 66: 1762.

Varnam AH, Sutherland JP (1994). *Milk and milk products: Technology, chemistry and microbiology.* Chapman & Hall: New York.

Wells SJ, Fedorka-Cray PJ, Dargatz DA, Ferris K, Green A (2001). Fecal shedding of Salmonella spp. by dairy cows on farm and cull cow markets. *J Food Prot* 64: 3.

Meat and Poultry Plant Sanitation

17

Abstract

An efficient cleaning system can significantly reduce labor costs in meat and poultry plants. The optimal cleaning system depends on the type of soil and type of equipment present. High-pressure, low-volume cleaning equipment is normally the most effective for removing heavy organic soil, especially when deposits are located in areas that are difficult to reach and penetrate. However, foam, slurry, and gel cleaning have become more prominent because cleaning is quicker and cleaners are easier to apply using these media. Because of high equipment costs and cleaning limitations, cleaning-in-place (CIP) systems are typically limited primarily to applications that involve large storage containers.

In meat and poultry plants, acid cleaning compounds are used most frequently to remove mineral deposits. Organic soils are more effectively removed through the use of alkaline cleaning compounds. Chlorine compounds provide the most effective and least expensive sanitizer for the destruction of residual microorganisms. However, iodine compounds give less corrosion and irritation, and quaternary ammonium sanitizers have more of a residual effect. Appropriate cleaning procedures depend on the area, equipment, and type of soil.

Keywords

Contamination • Beef • Pork • Poultry • *Salmonella* • *Campylobacter* • *E.coli O157:H7* • Peracetic acid • Cleaners • Sanitizers • Cleaning procedures • Chiller water

Introduction

Meat and poultry are perishable foodstuffs, and red meat has a relatively unstable color. Poor sanitary practices increase microbial damage resulting in an undesirable color and/or flavors and reduced product safety. Effective sanitation is essential to reduce discoloration, spoilage, and pathogen growth with a resultant increase in shelf life and product safety.

© Springer International Publishing AG, part of Springer Nature 2018
N.G. Marriott et al., *Principles of Food Sanitation*, Food Science Text Series,
https://doi.org/10.1007/978-3-319-67166-6_17

Sanitation in the meat and poultry industry requires good housekeeping, beginning with the live animal or bird and continuing through serving the prepared product. The sanitation program should be thoroughly planned, actively enforced, and effectively supervised. The most successful program involves inspection by trained personnel who are directly responsible for the sanitary condition of the plant and equipment. In dry-cured and/or aged meat products, effective sanitation is essential as part of an integrated pest management program to control pest infestations in the plants.

Role of Sanitation

Meat and poultry nourish microorganisms that cause discoloration, spoilage, and foodborne illness. Methods of processing and distribution are responsible for the increased exposure of these products to microbial contamination. For example, many of today's merchandising techniques depend on appearance to sell the product. Improved sanitation is responsible for reduced contamination and increased product stability.

There are many obvious reasons for maintaining high standards of cleanliness in meat and poultry facilities. The following are a few that are important:

- These products are nutrient dense, making them vulnerable to attack by microorganisms that are present under unsanitary conditions.
- Microorganisms cause product discoloration and flavor degradation.
- Self-service merchandising of aerobically packaged fresh meat and poultry places a premium on intensive sanitation to increase shelf life.
- Improved sanitary conditions reduce waste since less discolored and spoiled product has to be discarded.
- Immaculate sanitary conditions can improve the image of a firm, whose reputation depends on the condition of the product. A sanitary product is more wholesome and superior in appearance to tainted merchandise.
- Increased emphasis on food nutrition and sanitation by regulatory agencies and consumers

suggests a need for an effective sanitation program.
- Employees deserve clean and safe working conditions. Sanitary and uncluttered surroundings improve morale and productivity and increase product turnover.
- The established trend toward increased centralized processing and packaging dictates a need for increased emphasis on sanitation. Increased processing and handling necessitate a more intensive sanitation program.
- Sanitation is good business.

Effect on Product Discoloration

Biochemical discoloration is related to the amounts of oxygen and carbon dioxide present, specifically in beef, pork, and lamb. Figure 17.1 illustrates how the partial pressure of oxygen affects the myoglobin chemical state, which ultimately influences muscle color. High carbon dioxide partial pressure can cause a gray or brownish discoloration by association of carbon dioxide with myoglobin at the free binding site, and the rate of metmyoglobin formation increases with decreasing oxygen pressure.

A major cause of discoloration is related to microorganisms. Microbes consume available oxygen at the product surface, which reduces available oxygen that is needed to maintain the muscle pigment myoglobin in the oxymyoglobin state. Oxidation can cause an abnormal brown, gray, or green discoloration of meat by oxidation of the ferrous iron of the heme compound to the ferric state and direct attack by oxygen on the porphyrin ring. The color of fresh meats becomes unacceptable when metmyoglobin reaches approximately 70% of the surface pigment. Formation of metmyoglobin is accelerated by decreased oxygen pressure as a result of oxygen consumption through the growth of aerobic microorganisms. The critical partial pressure for oxygen has been found to be 4 mm (0.15 in). Rapid oxidation to metmyoglobin occurs below this level.

Research has suggested that the primary role of bacteria in meat discoloration is the reduction of the oxygen tension in the surface tissue.

Fig. 17.1 Relationship of partial oxygen pressure to myoglobin chemical state

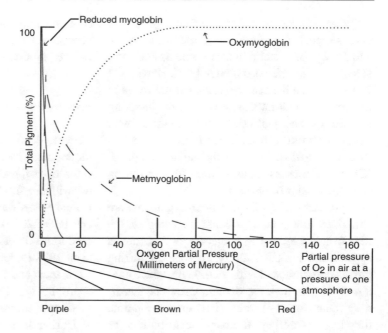

This conclusion has been based on the following observations:

1. Rate of oxygen uptake of the muscle tissue surface is related to microbial activity and color change.
2. Oxidation to metmyoglobin occurs at intermediate levels of oxygen demand of the surface tissue. With high respiration rates, reduction to myoglobin occurs, correlating with similar changes under controlled oxygen atmospheres.
3. Pigment oxidation and reduction are controlled by adjustment of oxygen level in the storage atmosphere with a light load of microorganisms.
4. Agents inhibiting high oxygen uptake rates in exposed tissues preserve color under atmospheric conditions but are ineffective under low oxygen pressures.
5. Even under high oxygen partial pressure, bacterial growth can contribute to the oxidation of Fe^{2+} to Fe^{3+} in the heme ring of myoglobin, thus further contributing to discoloration.

These observations result in the conclusion that the reduction of oxygen in muscle tissue by microbial growth or by physical effects can produce an increase in reduced myoglobin through oxidation by metabolic hydrogen peroxide produced by muscle tissue or by bacteria. With oxygen tension reduced to a low enough level, hydrogen peroxide formation is nil, and no oxidation will occur. This condition indicates that the dissociation of the oxy compound increases as oxygen tension decreases. Fresh meat pigments are more vulnerable to discoloration at oxygen tensions below that of air at atmospheric pressure.

Clearly, the growth of bacteria from poor sanitation contributes to muscle color degradation through reduced oxygen concentration and ultimate discoloration. Various genera and species of microorganisms differ in their effect on pigment alteration; however, improved cleanliness can delay the development of high numbers of microbes. Those who handle meat should strive to minimize the initial microbial load.

Meat and Poultry Contamination

During the slaughter, processing, distribution, and foodservice cycle, food items are handled frequently, often as many as 18–20 times. Because almost anything contacting meat and poultry can serve as a contamination source, the risk that this condition will occur increases each time that these products are handled.

When alive, a healthy animal possesses defense mechanisms that counteract the entrance and growth of bacteria in the muscle tissue. After slaughter, the natural defenses break down, and there is a race between humans and microbes to determine the ultimate consumer. If the handling is careless and ineffective, the microbes win. Those involved with sanitation must create a less favorable environment for the microorganisms. (Chap. 5 discusses contamination sources during slaughter and processing.)

Approximately 1 billion microorganisms are contained in a gram of soil that is attached to the hide of a live animal. A gram of manure contains approximately 220 million microbes. Sticking knives contaminated with bacteria introduce contamination through the wound. An animal's heart may beat for 2–9 min after exsanguination, thereby permitting thorough distribution of microbes. Unwashed animals have approximately 155 million microorganisms/cm^2 (0.15 in^2) of skin where the jugular vein is cut.

Although the temperature of a scalding vat is approximately 60 °C (140 °F), the microbial load is approximately 1 million bacteria per liter (1.05 quarts) of water. The dehairing operation for hogs is responsible for microorganisms being beaten into the surface skin.

Contamination during evisceration of animals is more common than at other processing steps since the stomach and intestinal contents are loaded with microorganisms. A major contamination source for meats in the abattoir is rumen fluid, which averages 1.3 billion microorganisms/mL.

Carcass surface counts of microorganisms average between 300 and 3,000/cm^2 (0.15 in^2). Beef and pork trimmings contain 10,000–500,000 bacteria per gram, depending on contamination and sanitation practices. Cutting boards on fabricating tables normally contain approximately 77,500 bacteria per square centimeter (0.15 in^2). Slicers, conveyors, and packaging equipment may increase the contamination of processed meats by 1000–50,000 bacteria per gram, depending on sanitation practices. Therefore, it is important to monitor bacterial counts on equipment and the product at every step in the production process to determine where the likelihood of

contamination most commonly occurs and intervention strategies to minimize bacterial counts on finished products.

Pathogen Control

Between 1998 and 2015, approximately 17% of foodborne illness outbreaks, and 17% of the cases of outbreak-associated foodborne disease for which a food vehicle was implicated in the United States was due to meat and poultry products (CDC 2016). This is a decrease from 23% of foodborne illness outbreak and 27% of disease from 1998 to 2008 (Goetz 2013). From 1998 to 2015, meat and poultry were also associated with approximately 17% of the deaths reported from foodborne outbreaks (CDC 2016).

During the scalding, evisceration, rinsing, and chilling phases of poultry processing, the carcasses are vulnerable to contamination by species of *Salmonella* and *Campylobacter*, *Aeromonas hydrophila*, *Listeria monocytogenes*, and other microorganisms of public health concern. *Salmonella* and *Campylobacter* have presented a serious problem for the poultry industry since these organisms are commonly present on raw poultry and are the leading cause of foodborne illness in the United States with estimated infection rates of 16.4 and 14.3 cases per 100,000 people in 2012 (MMWR 2013). In the United Kingdom, *Campylobacter* incidence on chicken was reported as 73% in retail markets, and campylobacteriosis is the leading cause of foodborne illness in the United Kingdom (The Poultry Site 2015). Poultry has been implicated in campylobacteriosis that has occurred sporadically without a finite determination of the mode of transmission. The design of poultry processing equipment, especially the plucking equipment, is such that adequate cleaning is difficult. The major risk in evisceration is the spilling of the gut content onto the carcass. *Campylobacter jejuni* will spread during the harvesting process. Regardless of the type of harvesting, heavily infected poultry flocks may result in a contamination rate of 100% for the finished product. Immersion chilling poses a contamination threat due to the entrapment of microorganisms in skin channels and the swelling of collagenous material in the neck flap area. These highly contaminated carcass parts should be trimmed to lower the microbial load. Freezing is known to reduce *Campylobacter* populations, presumably by ice crystal damage to cells and by dehydration. Current research results indicate that rinsing poultry carcasses removes a small amount of *Salmonella* organisms that may be present. Species of *Salmonella* and *Campylobacter* affix themselves to the skin and flesh of poultry so tightly that they become part of the food that is intended for human consumption.

Salmonella and *Campylobacter* Control in the Poultry Industry

Since *Salmonella* and *Campylobacter* are major contamination risks in the poultry industry, USDA-FSIS issued a compliance guideline for controlling *Salmonella* and *Campylobacter* in raw poultry (USDA 2015b) and developed performance standards for raw poultry products (USDA 2015b). These performance standards are as follows: (1) broiler carcasses, maximum of 7.5% and 10.4% incidence of contamination for *Salmonella* and *Campylobacter*; (2) turkey carcasses, maximum of 1.7% and 0.79% incidence of contamination for *Salmonella* and *Campylobacter*; (3) comminuted chicken, maximum of 25% and 1.9% incidence of contamination for *Salmonella* and *Campylobacter*; (4) comminuted turkey, maximum of 13.5% and 1.9% incidence of contamination for *Salmonella* and *Campylobacter*; (5) chicken parts, maximum of 15.4% and 7.7% incidence of contamination for *Salmonella* and *Campylobacter*. In an effort to help meet these performance standards, the compliance guidelines summarize the points in the production process where these pathogens can be prevented, eliminated, or reduced. The likely points of contaminations are discussed, and possible steps that can be taken to control contamination at each point in the process are included (USDA 2015b). Since *Salmonella* and *Campylobacter* can be prevalent in live birds prior to harvest, it should be determined 2–5 days prior to slaughter whether either of these pathogens are detectable through drag swabs, boot samples, and/or litter samples. This will allow *Salmonella*- and *Campylobacter*-positive birds to be transported, slaughtered, and processed separately from farms or houses that are negative for these pathogens. To minimize the possibility of flock contamination, a multi-hurdle pathogen reduction program should be utilized. Vaccines (*Salmonella* only), probiotics, prebiotics, competitive exclusion, and organic acids can all be used, predominantly in food and/or water, as preventative controls in live poultry. In addition to the use of these preventative methods, best practices include obtaining chicks from pathogen-free breeder flocks, use of breeding stock with innate resistance to these pathogens, and use of new or sanitized containers (USDA 2015b). Bedding, feed, water, boot dips, hygiene plans, and biosecurity measures are all items that are important to prevent contamination in the houses. In particular, bedding, feed, and water should be free of these pathogens, and prebiotics, probiotics, and/or organic acids can be used in the feed and water as well.

Processing Plant Control of *Salmonella* and *Campylobacter*

Salmonella and *Campylobacter* contamination have been linked to dirty cages and unsanitary conditions during unloading. Employee training and cleaning and sanitation of the unloading area needs to be a priority since it is a common place for fecal contamination. Feed withdrawal needs to be conducted to minimize fecal contamination, and stunning needs to be done correctly.

The scalding process can be an intervention step for *Salmonella* and *Campylobacter* if the pH is maintained at approximately 9 and a counter-current water flow is used to maintain a gradient such that carcasses flow from dirty to clean water. Cleaning birds prior to the scald tank with brushes, using organic acids in the scalder, and using a post-scald antimicrobial rinse all can help control pathogens. Picking also can lead to pathogen contamination. Besides cleaning and sanitation of picker equipment and parts, a rinse with vinegar or hydrogen peroxide can be used after picking as an antimicrobial intervention. Evisceration can lead to contamination if crops rupture. Therefore, proper sanitation processes need to be conducted. A chlorine water or peracetic acid rinse can be used as an antimicrobial intervention after evisceration, but if the sprayers are not positioned correctly, contamination can be spread instead of controlled.

Poultry plants may address the control of *Salmonella* and *Campylobacter* in their sanitation standard operating procedures if they can determine that is not likely to occur in the process, but this is not likely feasible in poultry plants. Therefore, there needs to be one or more critical control points (CCPs) within the Hazard Analysis and Critical Control Point (HACCP) plan to address the pathogen. Critical limits need to be established and evaluated for each critical control point, and monitoring must be done to determine if the CCP is in control. To prevent cross-contamination, employees on the line must sanitize their knives in either an antimicrobial solution or 180 °F (82 °C) water. If water is not available for cleaning knives, cross-contamination also occurs from knives due to multiple employees using the same knife sharpener. Since *Salmonella* and *Campylobacter* may be linked to human illness outbreaks, it can be helpful to define individual suppliers or one flock as an independent lot. This will allow less product to be recalled if an outbreak occurs.

Use of antimicrobial interventions to control *Salmonella* and *Campylobacter* is commonly used in HACCP plans for poultry products. The onus is on the company to maintain scientific support for the effectiveness of any intervention that is used. More information is available at the FSIS Compliance Guideline HACCP Systems Validation and FSIS Directive 7120.1. For any antimicrobial intervention that is used, it is imperative to measure parameters such as concentration, pH, and/or temperature when the antimicrobial is applied to the product, instead of when it is prepared.

Microbial testing should be conducted at multiple steps within a process to determine pathogen control. One way to do this is to use process and/or carcass mapping to assess the effectiveness of interventions and the entire food safety system. Process mapping could include obtaining samples after each processing step for an individual lot. Process mapping can be utilized to determine the effectiveness of the process and sanitation effectiveness through testing for indicator organisms and/or pathogens. Process mapping can be utilized to verify that both the process and HACCP system are functioning as designed. Sampling plans and statistical process control are also valuable tools that can be used to make corrective actions prior to failing a product performance standard for various poultry products. Information on how to develop these programs is provided by USDA-FSIS (USDA 2015b). The chiller tank is a primary concern for contamination since it contains chilled water that can be used as a vector to cross-contaminate poultry carcasses. Chlorine was commonly used to control pathogens in the chiller tank for many years, but is currently used less often since many countries will not import poultry products from carcasses that were chilled in water with chlorine in the chill tanks. Broiler processors in the United States commonly use peracetic acid (peroxyacetic acid, PAA) in the chiller tank. If maintained at the proper temperature and concentration, PAA can effectively control both *Salmonella* and *Campylobacter* such that performance standards can be met for carcasses. PAA concentrations should not exceed 2000 parts per million (PPM) in the chiller tank and should be buffered to a pH of 6–7 in the chill tank so that it is effective at controlling pathogens without negatively impacting meat quality. A pH greater than 7 or 8 will decrease the effectiveness of the PAA. As pH of the chill water continues to decrease further away from a pH of 6, water loss and product toughness and dryness increase. When sampling for pathogens out of the chiller tank, it is important to allow the carcass to drip for a minimum of 60 s before taking the sample so that the residual antimicrobial in the drip does not prevent live cells from being detected. In addition, additional dips or sprays are likely necessary after separating carcasses into parts since individual chicken parts can become contaminated during the process. Samples must also be collected such that they are representative of all production that occurs at the facility.

Shapton and Shapton (1991) emphasized the need for cleaning of roofs over food manufacturing areas. Process equipment and exhaust stacks may be vented through the roof. If feasible, roof-mounted process equipment should be enclosed with a floor to separate it from the processing area. Particles, especially hygroscopic powders, can deposit on the roof, especially if it is flat. When left unattended, this area may attract birds, rodents, or insects, which are known carriers of *Salmonella* organisms and of *L. monocytogenes*. Pools of water will encourage these pests. A minimum slope of 1% is recommended to ensure drainage.

Listeria monocytogenes is a challenge for meat processors because it is very difficult to eliminate this pathogen from the processing plants. It survives at cold temperatures, tolerates salt and nitrite, and can attach to stainless steel surfaces. Thus, equipment can easily provide a means of transfer of *L. monocytogenes* from one location to another, even after cleaning and sanitizing (Sebranek 2003). The incidence of *L. monocytogenes* is approximately 15–50% for poultry carcasses, 20% of dry sausage and fresh sausage, and 10% or more of ground beef samples evaluated. Growth can also occur in some cooked meat products after packaging. A significant portion of fresh meats that were incorporated as raw materials for processed products can be contaminated with this psychrotrophic pathogen, which demonstrates the importance of preventing postprocessing recontamination of ready-to-eat (RTE) products. Other viable product contamination areas include slicers, dicers, saws, lugs, tubs and other containers, hand tools, gloves, aprons, packaging materials, packaging equipment, tables, shelves, racks, and cleaning equipment. Other areas where this pathogen may be hidden include recesses, hollow rollers, motor housings, switch boxes, rusted materials, cracked or pitted hoses and door seals, walls that are cracked or pitted or covered with inadequately sealed surface panels, vacuum/air pressure pump lines or hoses, air filters, open bearings, and ice makers.

Listeria monocytogenes is often found around wet areas and cleaning aids, such as floors, drains, wash areas, ceiling condensate, mops and sponges, brine chillers, and at peeler stations. Biofilm formation is exacerbated through older and unclean equipment with exposed bolts and threads and unsealed rivets. Thus, control of *Listeria* organisms in processing plants is essential to reduce the potential of postprocessing contamination. One cannot control the growth of this pathogen through refrigeration at 4–5 °C (39–41 °F), a common storage temperature, because this microbe can survive in a 0 °C storage environment. Doyle (1987) suggested that the use of antimicrobial agents, reduced storage temperature (<2 °C, <36 °F), reformulation of products (reduced minimum water activity [A_w], pH, etc.), or postprocessing pasteurization of products may need to be incorporated for the control of such psychrotrophic pathogens in foods.

Biofilm Prevention

Inadequacy of cleaning/sanitation can lead to biofilm formation. Biofilms are a community of bacterial cells that (1) adhere to each other and surfaces such as steel, glass, and plastic, (2) are held together and protected by polysaccharides that act as a glue-like material, and (3) differ in gene expression when compared to normal planktonic cells (Sofos 2009). Techniques for preventing biofilms in food plants, including meat plants, are included in Chaps. 18 and 19. Specific information on controlling *Listeria monocytogenes* biofilms is included in this section.

Frank et al. (2003) evaluated the effectiveness of cleaning and sanitizing chemicals in the removal of *Listeria monocytogenes* biofilms coated with soil of poultry origin and applied under static conditions without heat application. Alkaline and neutral cleaning compounds were evaluated as well as sodium hypochlorite, acidified sodium chloride, peroxyacetic acid, peroxyacidic acid/octanoic acid mixture, and quaternary ammonium compound sanitizing agents. The alkaline cleaning compound removed 99% of fat and 93% of protein within 30 min. The neutral cleaning compound was equally effective at removing fat, but eliminated only 77% of protein. The alkali cleaning compound also effectively

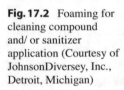

Fig. 17.2 Foaming for cleaning compound and/ or sanitizer application (Courtesy of JohnsonDiversey, Inc., Detroit, Michigan)

removed *L. monocytogenes* biofilm coated with protein. Biofilm removal is more successful if cleanup is initiated as soon as possible after the production shift ceases. More prompt cleaning after production reduces time for additional microbial growth and facilitates cleaning because of reduced drying of soil deposits. Acidified sodium chlorite and peracetic acid/octanoic acid mixture were the most effective sanitizers for the destruction of *L. monocytogenes* biofilm coated with fat and protein. Figure 17.2 illustrates how sanitizers such as those mentioned can be applied to reduce contamination from employees entering processing areas.

Pathogens such as *L. monocytogenes* can be better controlled through the reduction of cross-contamination. Employees who work in the raw and finished product areas, such as smokehouses and water- and steam-cooking areas, should change outer clothing and sanitize their hands or change gloves when moving from a raw to finished product area. Utensils and thermometers that are used for raw and finished products should be sanitized each time they are used. Frequent cleaning with floor scrubbers in essential. If ceiling condensate is present, removal should involve a vacuum unit or a sanitized sponge mop. Cleaned floors that do not dry before production start-up should be vacuumed or squeegeed. Separating the areas of the plant where raw meat is from

parts of the plant where the RTE cooked samples are located has been a crucial part of plant design and control of *Listeria monocytogenes* in RTE meat plants. If possible, employees from the RTE side of the plant should not come into personal contact with the raw side of the plant.

Although growth niches may be present in a plant, more positive sites found during environmental monitoring are not growth niches. They are transfer points (i.e., product handlers and equipment). Since the microorganisms are present in this location before the product comes to the line, transfer points are not growth niches, because the organism is eliminated during the cleaning and sanitizing process. Thus, most pathogen monitoring and control sampling occur at transfer points, not the true harborage places that are growth niches (Butts 2003), such as floors and drains.

Growth niches should be designed out of the process, but if this is not accomplished, they must be managed by minimizing their contamination potential with process control techniques. The manufacturer should consider the degree to which equipment needs to be disassembled for effective cleaning and sanitizing. The chemical sanitizer treatment being practiced, including consideration of flood sanitizing coverage and the requirements for treatment time, is another factor that will have an impact on the successful control of pathogenic growth niches. Butts (2003) suggested that the flood sanitizing step must be implemented to further assure that growth niches are managed.

The following guidelines should be considered when planning for the control of *L. monocytogenes* in meat, poultry, and other food plants.

Layout and Plant Design

Although most modern plants are much more hygienically designed than during the past, these principles to complement those mentioned in Chap. 14 should be considered.

1. Plant layout should prevent pests and vermin and should control the movement of *L. monocytogenes* between raw and cooked product areas. Examples are employee traffic patterns, support and supervisory staff movement, and food handling activities.
2. Air and refrigeration movement equipment should be designed for easy cleaning and sanitizing. Ready-to-eat areas should have a positive air pressure design.
3. All equipment and other surfaces should be smooth and nonporous so they can easily be cleaned and sanitized.
4. Floors should be surfaced with materials that are easily cleaned and will not encourage water accumulation.
5. Prevent proliferation in growth niches or other sites that lead to ready-to-eat product contamination.

Process Control

1. If the process does not contain a *L. monocytogenes* kill step, the operation should be designed to reduce contamination.
2. The kill step (if applicable) should be a critical control point in the HACCP program.
3. Implement an appropriate sampling plan to determine if the process is under control.
4. Establish appropriate corrective action.
5. Verify that the corrective action was effective.
6. Review and analyze data to ensure that the control program is effective.

Operation Practices

1. Employees should be educated about good manufacturing practices (GMPs), HACCP, and the responsibilities of each.
2. Equipment should be provided to maintain sanitary conditions such as (a) foot baths, (b) hand dips, (c) hair nets, and (d) gloves.
3. Contamination sources, especially in ready-to-eat areas, should be eliminated.
4. Management should be educated to support GMPs and HACCP.

Sanitation Practices

1. An adequate number of employees, time, and supervision should be provided for cleaning and sanitizing.
2. Written cleaning and sanitizing procedures should be developed and posted for each area in the plant.
3. Environmental sampling programs to verify the effectiveness of cleaning and sanitizing should be established.

Verification of L. monocytogenes Control

1. A microbial assay of weekly samples from plant areas, equipment, and the air supply should be conducted. It is especially important to sample points between the kill step and packaging.
2. Samples can be composited to reduce the analysis cost. If a composite sample is positive, a follow-up analysis of individual samples is necessary to determine which equipment is the contamination source.

The following important suggestions for *Listeria* control in meat plants should be considered:

1. Mechanically or manually scrub floors and drains daily. Drains should contain a "quat plug" or be rinsed with disinfectants daily.
2. Clean the exterior of all equipment, light fixtures, sills and ledges, piping, vents, and other areas in the processing and packaging areas that are not in the daily cleaning program.
3. Clean cooling and heating units and ducts weekly.

4. Caulk all cracks in walls, ceilings, and window sills.
5. Keep hallways and passageways that are common to raw and finished product clean and dry.
6. Minimize traffic in and out of processing and packaging areas and establish plant traffic patterns to reduce cross-contamination from feet, containers, pallet jacks, pallets, and fork trucks.
7. Change outer clothing and sanitize hands or gloves when moving from a "raw" to a finished product area. If possible, have separate employees in raw and cooked sections of the plant.
8. Change into clean work clothes daily. Provide some pattern of color-coding to designate various plant areas.
9. Minimize the number of visitors and require them to change into clean clothes provided at the plant.
10. Provide a plant environmental monitoring program to measure effectiveness of the *Listeria* control procedures.
11. Enclose processing and packaging rooms so that filtered air comes in, and ensure that these areas are under positive pressure.
12. Clean and sanitize all equipment and containers before their entry into processing and packaging areas.

The three alternative levels (Lazar 2004) of *Listeria* control in a plant that produces ready-to-eat (RTE) products that are post-lethality exposed include:

Alternative Level 3—basic control level addressed through effective sanitation.
Alternative Level 2—effective sanitation is combined with post-lethality treatments such as heat, antimicrobial agents, or freezing.
Alternative Level 1—effective sanitation, antimicrobial agent or process, and a post-lethality treatment combining all three strategies.

Under these three levels, specific guidance for sanitation and *Listeria* control is available in the FSIS *Listeria* Guideline: Requirements of the

Listeria Rule (USDA-FSIS 2014). In addition, there are excellent principles with respect to control of *Listeria* in RTE product provided by FDA (2008). Though not specific to meat, the principles are very helpful with respect to chemical parameters, product formulation, and sanitation.

USDA-FSIS issued a final rule in 2015 pertaining to "Control of *Listeria monocytogenes* in Ready-to-Eat Meat and Poultry Products." This rule was finalized from the interim final rule in 2003, which stated that RTE meat or poultry products must control *L. monocytogenes* in the processing environment through its HACCP plan or prevent contamination of products by the pathogen through sanitation standard operating procedures (SSOPs). The percentage positive for *Listeria monocytogenes* in FSIS testing for *L. monocytogenes* in RTE products has decreased from 0.76% in 2003 to 0.34% in 2013, which indicates that these regulations have helped improve the control of *Listeria monocytogenes*.

It has been suggested (Russell 2003b) that 28% of cattle designated for harvesting are infected with *Escherichia coli* O157:H7 and that an average of 43% of beef carcasses contain this pathogen at various stages of production. Koohmaraie et al. (2007) sampled carcasses from two plants for *Escherichia coli* O157:H7 to determine a baseline. There was 8–10% carcass incidence of *E. coli* O157:H7 in one plant and greater than 30% incidence on the carcass in the other plant.

In September 2002, the US Department of Agriculture's Food Safety and Inspection Service (USDA-FSIS) announced its plan to institute a series of additional measures to complement previous policies aimed at the prevention and control of *E. coli* in ground beef. These included:

1. All beef harvesting and grinding plants are required to acknowledge that *E. coli* is a hazard recently likely to occur in their operations unless proven otherwise.
2. All establishments producing raw beef products must reassess their mandatory HACCP plans and investigate the adequacy of existing pathogen/intervention controls. If controls are not in place or are determined to be inadequate,

a pathogen reduction step to reduce the risk of *E. coli* O157:H7 in the product must be incorporated into the production process.

3. FSIS inspectors will conduct random microbial verification testing of all beef grinding operations.

4. FSIS will increase inspections of pathogen reduction and intervention steps to verify that they are effective in reducing the incidence of *E. coli* O157:H7 under actual plant conditions.

In 2015, USDA-FSIS released their 4th revision on sampling verification activities for Shiga toxin-producing *E. coli* (STEC) in raw beef products with respect to collecting and submitting samples of raw beef products (USDA 2015a).

Temperature Control

Meat and poultry spoil when held at a high temperature. Temperature affects the rate of chemical and biochemical reactions and, especially, the lag phase of the growth pattern of microorganisms. The rates for both microbial and non-microbial spoilage increase to approximately 45 °C (115 °F). Microbial spoilage usually does not occur above 60 °C (140 °F). (Microbial growth kinetics is discussed in Chap. 3.) Microorganisms grow most rapidly between 4 and 60 °C (40 and 140 °F). This range is considered the critical zone or the danger zone. Meat and poultry must be stored out of this temperature zone and should be taken through this range as quickly as possible when a temperature change is necessary (as when cooking and chilling). Storage temperature below the critical zone does not effectively destroy bacteria but does reduce the rate of growth and multiplication of microorganisms. Below the critical zone, bacteria are less active, and some death can occur through stress.

Processing and storage at a colder temperature will reduce spoilage and microbial growth on equipment, supplies, or other areas. Under unsanitary conditions with improper temperature control, certain species of *Pseudomonas* can double in number every 20 min. Meat and poultry are generally expected to avoid spoilage at least twice as long at 0 °C (32 °F) than at 10 °C (50 °F).

Air curtains should be installed, especially when truck doors must be left open, to prevent refrigeration loss where the plant is under positive pressure. Entry of insects and dust is reduced with the use of air curtains. The air velocity should be a least 488 m/min (1,600 ft/min), measured at a distance of 910 mm (38 in) above the floor. For personnel entrances, the air stream should be continuous across the entire width of the opening, with a thickness of at least 254 mm (10 in) and a minimum velocity of 503 m/min (1,700 ft), measured 910 mm (38 in) above the floor (Shapton and Shapton 1991).

Sanitation Principles

Efficient cleaning arrangement can reduce labor costs up to 50%. Construction and equipment selection are critical for the most effective cleaning operation. It is important that the floors, walls, and ceilings be constructed of impervious material that can be easily cleaned. Floors should be sloped with a minimum of 10.5 mm/m (0.4 in/40 in).

Hot Water Wash

Hot water washing of meat and poultry soil is not effective. Hot water can loosen and melt fat deposits but tends to polymerize fats, denature proteins, and complicate removal of protein deposits by binding them more tightly to the surface to be cleaned. The main advantage of a hot water wash system is minimal investment of cleaning equipment. Limitations of this approach include increased labor requirements and water condensation on equipment, walls, and ceilings. It is difficult to remove heavy soil with this system.

High-Pressure, Low-Volume Cleaning

High-pressure, low-volume spray cleaning is a viable method in the meat and poultry industry because of the effectiveness with which it removes tenacious soils. With this equipment, the operator can more effectively clean difficult-to-reach areas with less labor, and the cleaning compound is more effective at a lower temperature. However, protein industries such as meat and dairy are moving away from using high-pressure cleaning since it can introduce contamination into the air (Cooper 2016).

This hydraulic cleaning technique may involve portable units. Portable equipment can be utilized for cleaning parts of equipment and building surfaces and is especially effective for conveyors and processing equipment when soaking operations are impractical and hand bushing is difficult and labor-intensive.

Foam Cleaning

Foam is particularly beneficial in cleaning large surface areas of meat and poultry plants and is frequently used to clean transportation equipment exteriors, ceilings, walls, piping, belts, and storage containers. Portable foam equipment is similar in size and cost to portable high-pressure units. Centralized foam cleaning applies cleaning compounds similar to how a centralized system functions.

Gel Cleaning

This equipment is similar to high-pressure units, except that the cleaning compound is applied as a gel rather than as a high-pressure spray. Gel is especially effective for cleaning packaging equipment because it clings to the surfaces for subsequent soil removal. Equipment cost is similar to that of portable high-pressure units.

Combination Centralized High-Pressure, Low-Volume, and Foam Cleaning

This system is the same as centralized high pressure except that foam can also be applied through the equipment. This method offers the most flexibility because foam can be used on large-surface areas, and high pressure can be applied to belts, conveyors, and hard-to-reach areas in a meat or poultry plant. Equipment costs for this system range from $15,000 to over $150,000, depending

on size. As of 2015, the meat industry has moved away from high-pressure cleaning unless it is absolutely necessary to remove soils and sediment since it can introduce contamination into the air (Cooper 2016). This approach has also been mainly replaced with system cleaners.

Cleaning-in-Place (CIP)

With this closed system, a recirculating cleaning solution is applied by installed nozzles, which automatically clean, rinse, and sanitize equipment. Benefits of CIP systems are discussed in Chap. 11. The use of CIP systems in the meat and poultry industry is limited. This equipment is expensive and lacks effectiveness in heavily soiled areas. CIP cleaning has some application in vacuum thawing chambers, pumping and brine circulation lines, preblend/batch silos, and edible and inedible fat rendering.

Cleaning-out-of-Place (COP)

Although some specialized applications of this cleaning technique exist in the meat and poultry industry, the use of this equipment is somewhat limited. More detailed information on this topic is presented in Chap. 11. In addition to parts washing equipment, COP units are being incorporated in the cleaning of racks and returnable containers. Typical equipment consists of a cabinet with oscillating spray bars to reach all areas to be cleaned with high-pressure volume. A complete wash and rinse cycle ranges from 5–20 min per batch, depending upon the level of soil buildup on what is being cleaned. This equipment saves water and chemical costs by recycling.

Cleaning Compounds for Meat and Poultry Plants

Many of the recent advances in cleaning have been to combine foam cleaners with sanitizing chemicals, to use cleaners that will reduce water use, and to use systems that will monitor the concentration of cleaners and sanitizers real time, instead of needing to titrate chemicals to determine concentration.

Acid Cleaners

Information about strongly and mildly acid cleaners is provided in Chap. 7.

Strongly Alkaline Cleaners

Examples of strongly alkaline compounds are sodium hydroxide (caustic soda) and silicates having high $N_2O:SiO_2$ ratios. The addition of silicates tends to reduce the corrosiveness and improve the penetration and rinsing properties of sodium hydroxide. These cleaners are used to remove heavy soils, such as those found in smokehouses.

Heavy-Duty Alkaline Cleaners

The active ingredients of these cleaners may be sodium metasilicate, sodium hexametaphosphate, sodium pyrophosphate, and trisodium phosphate. The addition of sulfites tends to reduce the corrosion attack on tin and tinned metals. These cleaners are frequently used with CIP, high-pressure, and other mechanized systems found in meat and poultry plants.

Mild Alkaline Cleaners

Mild cleaners are frequently in solution to use for hand cleaning lightly soiled areas in meat and poultry plants.

Neutral Cleaners

Information about these and other cleaning compounds is discussed in Chap. 9.

Sanitizers for Meat and Poultry Plants

To obtain maximum benefits from the use of a sanitizer, it must be applied to surfaces that are free of visible soil. Soils of special concern are fats, meat juices, blood, grease, oil, and mineral buildup. These deposits provide areas for microbial growth, both below and within the soil, and can hold food and water necessary for microbial proliferation. Chemical sanitizers cannot successfully penetrate soil deposits to destroy microorganisms.

Steam

Steam is an effective sanitizer for most applications. Many operators mistake water vapor for steam and fail to provide adequate exposure to create a sanitizing effect. Steam should not be used in refrigerated areas because of condensation and energy waste. It is also unsatisfactory for continuous sanitizing of conveyors.

Chemical Sanitizers

Chlorine is one of the halogens used for disinfecting, sterilizing, and sanitizing equipment, utensils, and water. The sanitizers most frequently used in sanitizing meat and poultry operations are the following:

- *Sodium and calcium hypochlorite*: These are more costly than elemental chlorine, but are more easily applied. Hypochlorous acid is an active germicidal agent, and the activity of hypochlorites is pH dependent. Alkalinity decreases as the germicidal activity increases.
- *Liquid chlorine*: This sanitizer is used in processing and cooling water chlorination to prevent bacterial slimes.
- *Chlorine dioxide*: This is an effective bactericide in the presence of organic matter because it does not react with nitrogenous compounds. The residual effect is also more persistent than that of chlorine. However, this sanitizer needs to be generated on site.

- *Active iodine* solutions, like active chlorine solutions, can be sanitizers. Iodophors are very stable products with much longer shelf lives than hypochlorites and are active at a low concentration. These sanitizing compounds are easily measured and dispensed, and they penetrate effectively. Their acid nature prevents film formation and spotting on equipment. Solution temperature should be below 48 °C (118 °F) because free iodine will dissipate.
- The *quaternary ammonium* compounds (quats) are widely used on floors, walls, equipment, and furnishings of meat and poultry plants. The "quats" are effective on porous surfaces because of their penetration ability. A bacteriostatic film that inhibits bacterial growth is formed when quats are applied to surfaces. Those sanitizers and compounds containing both an acid and a quat sanitizer are most effective in controlling *L. monocytogenes* and mold growth. Quats may be temporarily used when a mold buildup is detected.
- *Acid sanitizers* combine the rinsing and sanitizing steps. Acid neutralizes the excess alkalinity from the cleaning residues, prevents formation of alkaline deposits, and sanitizes. Acid sanitizers effectively kill both gram-positive and gram-negative bacteria. Sodium chlorite and citric acid are in use in meat and poultry plants as an antimicrobial (Stahl 2004). Other information about acid sanitizers may be found in Chap. 10.
- *Ozone* is incorporated to control microbial contamination in water, sprayed directly onto meat and other foods to reduce microbial contamination, and then applied onto clean food contact sources as a non-rinse sanitizer. It is an excellent biocide for chill water in slaughter plants and cooling tower operations (Stier 2002) because it breaks down to harmless compounds and will not concentrate in the system. Moisture must be present (80–90%) for ozone to be able to attack microorganisms. Although ozone can reduce pathogenic microorganisms on beef carcasses, Castillo et al. (2003) discovered that an aqueous ozone treatment provided no improvement over a hot

water wash. Too much ozone application on the meat surface will cause a pale color (Clark 2004). An additional limitation of ozone is that it does not penetrate surfaces well since it dissipates on contact with organic material and it needs to be generated on-site. This limits its effectiveness as a sanitizer

Carcass and Product Decontamination

Cleansing of cattle prior to harvesting can reduce contamination during hide pulling. Antimicrobial rinses and treatments are common in meat and poultry plants. Of the various decontamination treatments reported by Allen (2004), spray wash treatments with ethanol and 4–6% concentrations of lactic acid were the most effective in the reduction of microbial contamination. Acetic acid and lactic acid are currently the two most commonly used spray washes. Use of warm water (greater than 25 °C (80 °F)) can also be used as a decontamination treatment. Several cattle hide interventions are effective in a controlled laboratory setting, but may not be feasible for use on live animals (Allen 2004).

An application for disinfectants involves a reduction of bacteria on carcasses. Applications have focused on *acidified sodium chloride* (ASC), *hydrogen peroxide*, *trisodium phosphate*, *cetylpyridinium chloride* (CPC), and the application of an *electrochemically activated solution* (ECA). ECA is a mixture of sodium hypochlorite and peroxides to provide an electrical process that enables it to destroy a wide range of microorganisms. CPC has been successfully incorporated with lactic acid and sodium tripolyphosphate to destroy *Salmonella*.

ASC has been approved for use on meat products as well as fruits, vegetables, and seafood products. A commercial application of ASC involves 1000 PPM after prechilled carcasses are water rinsed for 10 s. Sodium chlorite acidifies in the presence of citric acid and destroys bacteria, viruses, fungi, yeast, and some protozoa by disrupting proteins in the microbial cell. It is effective in the destruction of pathogenic bacteria. This compound can be applied at room temperature through immersion or spray techniques without jeopardizing product quality. It is environmentally friendly and can be discharged into municipal and private sewage systems without additional treatment (Velazco 2003). ASC may be applied postchill to reduce *Campylobacter* spp. and *E. coli* in commercial broiler carcasses. Postchill systems may eventually be used in different applications, such as mist, spray, or bath, which could be applied closer to the final stages in processing (Oyarzabal et al. 2004).

Dipping solutions of sodium diacetate, sodium benzoate, sodium propionate, and potassium sorbate have been incorporated to inhibit the growth of *Listeria monocytogenes* in turkey frankfurters. Gombas et al. (2003) concluded that 1.8% sodium lactate combined with 0.25% sodium acetate, sodium diacetate, or glucono delta-lactone in frankfurters inhibits the growth of this pathogen and that combinations of lactate with diacetate were the most effective since this combination provided a synergistic inhibitory effect. Due to the recent trend for clean label meat products, use of buffered vinegar as an ingredient in processed products has increased. Buffered vinegar can inhibit *Listeria* growth for between 45 and 60 days in cooked, marinated chicken breast that is stored at 2–4 °C (36–40 °F) and can also inhibit *Listeria* growth on frankfurters and deli meats for up to 60 days in uncured products and as much as 105 days in cured products (Smith 2016).

The combination of acetic acid and hydrogen peroxide is effective in the destruction of listeria. Antimicrobial washes with hydrogen peroxide and organic acid reduce microorganisms on carcass surfaces more effectively than a plain water wash because of the synergistic effect between organic acids and *hydrogen peroxide*. Carcasses should be washed with hydrogen peroxide as soon as possible after hide removal for maximum effectiveness, and residues should not be left on the carcasses after treatment. *Sodium citrate* or *sodium lactate* at a concentration of 2% (wt/wt) or higher is known to inhibit *Clostridium perfringens* growth over, and 18 h cooling period (Sabah et al. 2003) and citric acid with irradiation can inhibit growth of *Listeria monocytogenes* (Sommers et al. 2003a).

An acidified calcium sulfate solution, when applied to the surface of frankfurters, reduces the growth of *Listeria monocytogenes*. Also, it prevents the regrowth of this pathogen.

During the past, treatment of frankfurters with lactic acid initially reduced the number of microorganisms, but failed to kill all of them and prevent additional growth. Lactate and diacetate additives and CPC are effective pathogen inhibitors (Petrak 2003; Sommers and Fan 2003; Sommers et al. 2003b). Post-packaging pasteurization technology, especially through heat application, has provided a means to reduce pathogen growth.

Compounds incorporated in carcass washes, such as *acidified sodium chlorite* and *ozone*, can lack effectiveness and threaten worker safety if not properly handled. Since ozone gas is a toxic respiratory irritant with limited effectiveness, it has not been further developed (Russell 2003a). Antimicrobial resistance is another potential limitation. *E. coli* O157:H7 and other pathogens may be capable of acid adaptation in processing plants.

Carcass washes lose their efficacy if microbes evolve and become resistant. To reduce this threat and increase the effectiveness of these washes, a multi-hurdle approach may be incorporated through the use of more than one rinse or other preventive measures. Some larger meat plants may have as many as five or six hurdles including *activated lactoferrin*, a nonionic surfactant, and *electrolyzed oxidizing water* (EO) (which has been effective against pathogens attached to cutting boards and as a poultry spray/dip combination).

Another carcass decontamination concept involves a wash cabinet with a water and sodium hydroxide mixture, which releases soils and contaminants from the hide. Then, the carcass is conveyed to a second cabinet, where it is rinsed with high-pressure water before being steam vacuumed with a lactic acid application (Yovich 2003). Stopforth et al. (2003) indicated that peroxyacetic acid is more effective than alkaline (quaternary ammonium) sanitizers as a decontaminant and increased destruction effectiveness is attained with the application of hot water and an acid wash as compared to washing only with water. Use of carcass washers has increased in an effort to reduce fecal contamination (Bashor et al. 2004).

Activated lactoferrin is a natural nontoxic protein that is consumer label friendly with no in-plant disposal challenges. It is FDA approved and a generally recognized as safe (GRAS). This naturally occurring protein is derived from whey and skim milk. It is the critical ingredient in mammalian mother's milk that provides suckling babies anti-pathogenic protection. Activated lactoferrin removes fimbria, which comprises the web of fibers a pathogenic bacterial cell, such as *Listeria monocytogenes*, uses to attach itself to a host. Once exposed to lactoferrin, pathogens cannot attach. It can block the attachment of *E. coli* O157:H7 and more than 30 other pathogens such as *Salmonella* and *Campylobacter*.

Phenolic compounds in wood smoke serve as antimicrobials. Liquid smoke components have been found (Sunen et al. 2003) to provide a significant inhibitory activity against *L. monocytogenes*.

Electrolyzed oxidizing water is more economical and effective than chlorine or ozone. This process relies on sodium chlorite, which is converted by an electrolyzing machine that converts the sodium chlorite, in a 12% solution in water into two antimicrobial compounds.

Barboza et al. (2002) evaluated the effectiveness of nisin, lactic acid, and a combination of lactic acid and nisin to reduce carcass contamination. They discovered that washing carcasses with water did not significantly reduce the bacterial load and that the largest reduction in bacterial contamination was accomplished with a mixture of nisin and lactic acid. A small antimicrobial peptide produced by *Lactococcus lactis* is more effective against *Listeria monocytogenes* when used in combination with lactic acid. Most of the salts of lactic acid, including potassium lactate, at up to 5%, partially inhibit the growth of this pathogen. Zinc and aluminum lactate and zinc and aluminum chloride (0.1%) work synergistically with 100 IU of nisin per milliliter to control the growth of *Listeria monocytogenes* Scott A (McEntire et al. 2003).

Mukhopadhyay and Ramaswamy (2012) published a review on the use of emerging technologies to control *Salmonella* in food products such as ozone, ultraviolet light, ultrasound, pulsed electric field, high-pressure processing, and other technology that could potentially be used as part of a hurdle technology to control *Salmonella* in foods.

Sanitation Practices

General Instructions

Approximately 50% of sanitation problems result directly from less-than-optimum sanitation procedures and chemical usage. All personnel should practice good personal hygiene, as discussed in Chap. 6. They should wear freshly laundered clothes and stay away from meat and other processing equipment if they are ill. Cleaning and sanitizing compounds should be kept in an area accessible only to a sanitation supervisor, manager, and superintendent and should be allocated only by the sanitation supervisor. Misuse of these compounds inhibits effective cleaning and may possibly result in personal injury and equipment damage. The water temperature should be locked in at 55 °C (132 °F).

Instructions provided with the portable or centralized high-pressure or foam-cleaning system should be followed. Cleaning compounds should be applied according to instructions or recommendations provided by the vendor. (Chapter 9 provides a discussion related to safety precautions when handling cleaning compounds.) The sanitation supervisor should inspect all areas nightly while the cleanup crew is on duty. All soiled areas should be recleaned prior to the morning inspection by the regulatory agency.

Chlorine papers should be used to check the sanitizing solution if automatic makeup or instructions are not available. These test papers include directions for use and are available through most cleaning compound suppliers. Other check systems for monitoring sanitation are also available and are discussed in Chap. 8. More information on these systems may be obtained from firms that sell cleaning compounds and monitoring systems.

Recommended Sanitary Work Habits

Sanitary workers should follow these general practices:

1. Store personal equipment (lunch, clothing, etc.) in a sanitary place and always keep storage lockers clean.
2. Wash and sanitize utensils frequently throughout the production shift, and store them in a sanitary container that will not be in contract with floors, clothing, lockers, or pockets.
3. Do not allow the product to contact surfaces that have not been sanitized for meat and poultry handling. If any particle contacts the floor or other unclean surface, it should be thoroughly washed.
4. Use only disposable towels to wipe hands or utensils.
5. Wear only clean clothing when entering production areas.
6. Cover the hair to prevent product contamination from falling hair.
7. Remove aprons, frocks, gloves, or other clothing items before entering toilets.
8. Always wash and sanitize hands when leaving the toilet area.
9. Stay away from production areas when a communicable disease, infected wound, cold, sore throat, or skin disease exists.
10. Do not use tobacco in any production area.

HACCP

HACCP is regulated through the Food Safety and Inspection Service of the US Department of Agriculture. (Additional discussion of HACCP is included in Chaps. 1, 7, 18, and 22.) HACCP does not necessarily include major investments or expensive microbial or other techniques. An example would be control options for the pasteurization step in pork or turkey ham processing. Design, maintenance, and process control are successful and relatively inexpensive.

An example of HACCP in a meat or poultry operation is the development of a flowchart of a meat and poultry production line. The flow pattern is a long sequence of events, with steps that are difficult or impossible to control. Many relevant factors related to hazards of each step can be identified and critical control points determined.

Livestock and Poultry Production

Animals can be produced in a specific pathogen-free (SPF) environment. Contamination can also be reduced through administration of bacterial cultures that exclude pathogens from the gut flora. The farm environment (its pastures, steams, manure, etc.) contributes to the recycling of excretion and reinfection. Sanitation practices must be established to improve hygiene in this portion of the flowchart.

Transportation

The stressful conditions of live animal transportation may cause pathogen carriers to spread these microorganisms. The challenge is to incorporate sanitary practices during transportation to reduce contamination in the processing plant.

Lairage

Stress during this phase of the flowchart can cause changes in the microbial flora composition of the intestinal tract, with the emergence and shedding of *Salmonella* organisms. Showering of animals can reduce stress and contamination.

Hide, Pelt, Hair, or Feather Removal

The protective coats of meat animals can and frequently do contain species of *Salmonella* and other detrimental microorganisms. New procedures and equipment modification are necessary to reduce contamination. A machine vision system that instantly detects trace levels of organic contamination, including ingesta and fecal material which harbors pathogens, is available and can be used in processing, distribution, and retailing environments to help workers detect organic contamination, ensuring a safer and more wholesome product.

Evisceration

Intestinal spillage and viscera rupture can occur. During poultry slaughter, a series of water or sanitizer sprays can be applied to reduce contamination. Red meat carcasses can also be decontaminated. The efficacy of spraying has not been totally resolved because this operation does not completely remove microorganisms and can spread contamination over the carcass.

Inspection

A meat inspector should use a sanitizer for the hands and knife because they can contaminate dressed carcasses.

Chilling

Controlling chilling parameters (air temperature, air movement, relative humidity, and filtering air) can reduce microbial growth. Drying of the carcass surface is important in the suppression of microorganisms (e.g., *Campylobacter* species). Trimming of the neck flap area of poultry carcasses after chilling will cause contamination. Use of up to 2,000 PPM of PAA, buffered with potassium hydroxide to a pH of 6.0–7.0, in chiller water will control *Salmonella* on broiler carcasses if the chiller water pH and PAA concentration is monitored and maintained at effective levels.

Further Processing

Chilled carcasses and cuts should not be exposed to an unchilled environment. The equipment used in this operation should be hygienically designed and sanitized before use. Safe and wholesome adjuncts should be used.

Packaging

The appropriate packaging material will protect the product from contamination. Proper storage temperatures must be maintained.

Distribution

The method of distribution must be rapid and clean. An effective temperature and sanitary environment must be maintained. The transportation environment should be monitored for sanitation and temperature control.

Sanitation Procedures

Detailed cleaning operations should be written and posted in the plant. Documentation of procedures is beneficial when supervision changes are made and for training of new employees. As mechanization increases, cleaning methods become correspondingly more detailed and complicated. Prior to adopting a cleaning procedure, it is essential to become familiar with the operation of all production and cleaning equipment. In addition to providing the necessary information, this can lead to improvements in methods that are used or should be incorporated.

The following are examples of cleaning procedures that could be used for distinct operations and areas in a plant. These examples are only guidelines. Every cleaning application should be adapted to the prevailing conditions. Although this step will not be mentioned, hoses and other equipment should be returned to their proper locations after cleanup.

Livestock and Poultry Trucks

Frequency: After each load has been hauled.

Procedure

1. Immediately after removing livestock or poultry from trucks, scrape and remove all manure that has accumulated from the premises.

2. Clean the truck beds, wheels, and frame by washing down the racks, floors, and frames with water to completely remove all manure, mud, and other debris, completely disinfecting with a quaternary ammonium sanitizer spray or by cleaning and sanitizing in one operation by spray cleaning with an alkaline detergent sanitizer.

Livestock Pens

Frequency: As soon as possible after each lot has been removed.

Procedure

1. After the livestock are taken from each pen, clean the manure from the floors and walls, and remove it from the plant premises.
2. Every four months, scrape all dried manure and loose whitewash from the gates and partitions. Sweep cobwebs from the ceilings, and whitewash the interior of the pens. Mix a cresylic acid-type sanitizer with the whitewash slurry.
3. If contagious diseases are brought into the pens, quarantine the diseased animals and destroy them separately from the healthy livestock. Remove the manure completely from the surrounding pen area (using a hose if necessary), and disinfect the pens by spraying with a quaternary ammonium sanitizer.

A general cleaning procedure for slaughter and processing areas encompasses (1) gross physical removal of debris, (2) prerinsing and wetting, (3) cleaning compound application, (4) rinsing, (5) inspection, (6) sanitizing, and (7) prevention of recontamination. The first step is essential to reduce time and water requirements and can minimize the biological load on the sewage system. Physical removal of debris also reduces splashing of large particles during the second step. The significance of the other steps has been previously alluded to and will be discussed in other chapters. The role of these cleaning procedures is illustrated in the applications to follow.

Slaughter Area

Frequency: Daily. Debris should be removed periodically during the production shift.

Procedure

1. Pick up all large pieces of extraneous material and transfer the matter to receptacles.
2. Cover all electrical connections with plastic sheeting.
3. Briefly prerinse all soiled areas with 50–55 °C (122–131 °F) water. Start working water from the ceiling and walls and the upper portion of all equipment, and continue to direct all extraneous matter down to the floor. Avoid direct contact of water with motors, outlets, and electrical cables.
4. Apply an alkaline cleaner through a centralized or portable foam system, using water that is 50–55 °C (122–131 °F). The system should be designed and operated to reach all framework, undersides, and other difficult-to-reach areas. Allow 5–20 min of exposure prior to the rinse. Although foam requires less labor, high-pressure equipment for application is more effective in penetrating hard-to-reach areas of equipment and may be more effective in the removal of *L. monocytogenes*.
5. Rinse ceilings, walls, and equipment within 20 min after application of the cleaning compound. Use the same rinse pattern as for prerinse and cleaning compound application, with 50–55 °C (122–131 °F) water.
6. Inspect all equipment and surfaces and touch up as necessary.
7. Apply an organic sanitizer to all equipment with a centralized or portable sanitizing unit. The solution should be at least 50 PPM chlorine.
8. Remove, clean, and replace drain covers.
9. Apply white edible oil to surfaces subject to rust corrosion. Any further use of oil here or for applications that follow is discouraged because the protective film contributes to microbial growth and biofilm formation.

10. Clean any specialized equipment in this area according to the manufacturing firm's recommendations.
11. Avoid contamination during maintenance and equipment setup by requiring maintenance workers to carry a sanitizer and to sanitize their work areas

Poultry Mechanical Eviscerators

Frequency: Daily. A continuous or intermittent sanitizer spray should be provided to reduce contamination.

Procedure

1. Pick up all large pieces or extraneous material, and transfer the matter to receptacles.
2. Cover electrical connections with plastic sheeting.
3. Briefly prerinse this equipment with 50–55 °C (122–131 °F) water.
4. Apply an alkaline cleaner through a centralized or portable foam system, using 50–55 °C (122–131 °F) water. Allow 10–20 min of exposure time prior to rinse down with 40–50 °C (104–122 °F) water.
5. Inspect all areas and conduct any necessary touch-ups.
6. Apply 200 PPM chlorine (or other organic sanitizer) with a centralized or portable sanitizing unit.
7. Avoid contamination during maintenance, as described previously.

Poultry Pickers

Frequency: Daily.

Procedure

1. Pick up all large debris and transfer the matter to receptacles.
2. Cover electrical connections with plastic sheeting.
3. Briefly prerinse this equipment with 50–55 °C (122–131 °F) water.

4. Apply a heavy-duty alkaline cleaner through a centralized or portable foam system on the shower cabinets. Shackles should go into the tank with the same cleaner.
5. After cleaning compound exposure for approximately 20 min, rinse down with 40–50 °C (104–122 °F) water.
6. Remove residual feathers and other debris by hand.
7. Because of the rubber fingers, apply 25 PPM iodophor as a sanitizer through a centralized or portable sanitizing unit.

Receiving and Shipping Area

Frequency: Daily.

Procedure

1. Cover all electrical connections, scales, and exposed product with plastic sheeting to prevent water and chemical damage.
2. Briefly rinse the walls and floors with 50–55 °C (122–131 °F) high-pressure water. The wall-rinse motion must be from top to bottom and side to side, with extraneous matter worked to the floor. This prerinse is designed to remove heavy soil deposits and to wet the surfaces.
3. Apply an acid cleaning detergent through a slurry or foam gun. Recommended spray temperature is 55 °C (131 °F) or lower. High-pressure output (for these cleaning operations) is 25–70 kg/cm² (350–1000 lb./in²) and 7.5–12 L/min (1–1.5 gallon/min) at the wand.
4. Apply a high-pressure rinse with 50–55 °C (122–131 °F) water, within 20 min of applying the cleaning compound.
5. Remove, clean, and replace drain covers in the proper position after rinse down.

Processed Products, Offal, and Storage Cooler

Frequency: Weekly. Processed meats, offal, and hanging meat should be rotated so that half of a section at a time can be cleaned each week.

Procedure

1. Clean each section, when empty, with a reliable floor cleaner. Apply slurry or foam.
2. Rinse thoroughly with 55 °C (131 °F) or lower temperature water within 20 min of detergent application. Do not splash water on hanging meat in the section not being cleaned. Work all debris to the floor from overhead fixtures and walls.
3. Squeegee the floor where water has accumulated to prevent it from freezing.
4. Remove, clean, and replace drain covers.

Fabricating or Further Processing

Frequency: Daily.

Procedure

1. Pick up all large pieces of lean, fat, bones, and other extraneous matter, and deposit them in a receptacle.
2. Cover all electrical connections with plastic.
3. Prerinse all soiled surfaces with 55 °C (131 °F) water. Start at the bone conveyor top and work all extraneous matter down to the floor. Avoid hosing motors, outlets, and electrical cables.
4. Following wash-down and subsequent heavy soil removal, apply an alkaline cleaner through a centralized or portable high-pressure, low-volume system, using 50–55 °C (122–131 °F) water. The system should be used such that it reaches all framework, table undersides, and other difficult-to-reach areas. Allow 5–20 min of soak time prior to rinse down. Alternative equipment for cleaning compound application is a foam unit. This unit rapidly applies the cleaner but does not penetrate as well as high-pressure, low-volume equipment and may be less effective in the removal of *L. monocytogenes*.
5. Rinse all equipment within 20 min after application of the cleaning compound. Using the same pattern as with prerinse and cleaning compound application, spray 50–55 °C (122–131 °F) water on one side of equipment at a time.
6. Thoroughly inspect all equipment surfaces and conduct any necessary touch-up.
7. Apply an organic sanitizer to all clean equipment with a centralized or portable sanitizing unit.
8. Remove, clean, and replace all drain covers.
9. Apply white edible oil to surfaces subject to rust or corrosion.
10. Avoid contamination during maintenance, as described previously.

If a bone shelter or hopper exists, it should also be cleaned, as outlined in the preceding steps. This operation should be performed twice a week during winter months and daily during the summer.

Processed Product Area

Frequency: Daily.

Procedure

1. Dismantle all equipment and place the parts on a table or rack. Disconnect all stuffing pipes.
2. Pick up all large pieces of meat and other extraneous matter and deposit in a receptacle.
3. Cover all electrical connections with plastic.
4. Prerinse all soiled surfaces with 55 °C (131 °F) water. Start at the top of all processing equipment, and direct all extraneous matter down to the floor. Avoid direct hosing of motors, outlets, and electrical cables.
5. Following wash-down and subsequent heavy soil removal, apply an alkaline cleaner through a centralized or portable high-pressure, low-volume system, using 50–55 °C (122–131 °F) water. The system should effectively reach all framework, tables, other equipment undersides, and other difficult-to-reach areas. Soak time prior to rinse down should be 5–20 min. Although foam is less effective in penetration, it is a viable cleaning medium and is easily applied.

6. Rinse all equipment within 20–25 min after cleaning compound application. Using the same prerinse pattern as with the prerinse and detergent application, spray 50–55 °C (122–131 °F) water on one side of each piece of processing equipment at a time.
7. Thoroughly inspect all equipment surfaces and touch up as necessary.
8. Apply an organic sanitizer to all clean equipment with a centralized or portable sanitizing unit.
9. Remove, clean, and replace drain covers.
10. Apply white edible oil only to surfaces subject to rust or corrosion.
11. Avoid contamination during maintenance as described previously.

Fresh Product Processing Areas

Frequency: Daily.

Procedure

1. Dismantle all equipment, and place the parts on a table or rack. Disconnect all stuffing pipes.
2. Remove large debris from equipment and floor and deposit in a receptacle.
3. Cover mixer and packaging equipment with plastic.
4. Briefly prerinse all soiled surfaces with 50–55 °C (122–131 °F) water to remove heavy debris and soak exposed surfaces. Guide hoses to force all debris toward the closest floor drain.
5. Apply an alkaline cleaner through centralized or portable high-pressure, low-volume cleaning equipment, using 50–55 °C (122–131 °F) water. Foam, gel, or slurry may be incorporated to introduce the cleaning compound. Cleaning compound application must cover the entire area, including equipment, floors, walls, and doors.
6. Rinse the area and equipment within 20–25 min after the application of the cleaning compound.
7. Inspect the area and inspect all equipment. Touch up as needed.

8. Remove, clean, and replace drain covers.
9. Sanitize all clean equipment with an organic sanitizer using a centralized or portable sanitizing unit.
10. Apply white edible oil only to surfaces subject to rust or corrosion.
11. Avoid contamination during maintenance as described previously.

Processed Product Packaging Area

Frequency: Daily.

Procedure

1. Dismantle all equipment, placing the parts on a table or rack.
2. Remove large debris from equipment and floors and place in a receptacle.
3. Cover packaging equipment, motors, outlets, scales, controls, and other equipment with plastic film.
4. Prerinse all soiled surfaces with 55 °C (131 °F) water to remove heavy debris and to soak exposed surfaces. Hoses should be guided to force all debris toward the closest floor drain.
5. Apply an alkaline cleaner through centralized or portable foam cleaning equipment, using 50–55 °C (122–131 °F) water. Cleaning compound application must cover the entire area—equipment, floors, walls, and doors.
6. Rinse the area and equipment within 20–25 min after application of the cleaning compound, using the same pattern of movement as used when applying the cleaner.
7. Inspect the area and all equipment. Touch up as needed.
8. Remove, clean, and replace drain covers.
9. Sanitize all clean equipment with an organic sanitizer using a centralized or portable sanitizing unit.
10. Apply white edible oil only to surfaces subject to rust or corrosion.
11. Avoid contamination during maintenance as described previously.

Brine Curing and Packaging Area

Frequency: Daily.

Procedure

1. Pick up all large debris and place in a receptacle.
2. Cover all electrical connections, scales, and exposed product with plastic sheeting.
3. Prerinse the area and all equipment with 55 °C (131 °F) water.
4. Place an acid cleaner in the shrink tunnel (if used), and circulate for 30 min during prerinsing.
5. Rinse shrink tunnel (if present) before detergent application.
6. Place all prerinse debris in a receptacle.
7. Apply an alkaline cleaner through a foam or slurry cleaning system, using 50–55 °C (122–131 °F) water.
8. Rinse with 55 °C (131 °F) water within 20 min after detergent application.
9. Inspect the area and equipment and touch up as needed.
10. Remove, clean, and replace drain covers.
11. Sanitize all clean equipment with an organic sanitizer that is applied through a centralized or portable system.
12. Apply white edible oil only to those parts subject to rust or corrosion.
13. Avoid contamination during maintenance as described previously.

Dry Curing Areas (Curing, Equalization, and Aging)[1]

Frequency: After product input and at the end of designated cure or equalization period.

Procedure

1. Sweep floors.

2. Remove pallets and other portable storage equipment, and rinse cure and other debris with 50 °C (122 °F) water.
3. Hose down vacated areas with 50 °C (122 °F) water.
4. Clean trolleys, trees, and other metal equipment used as outlined for wire pallets and metal containers or trolleys.
5. Sanitize cleaned areas according to manufacturer requirements with a quaternary ammonium compound for its residual effect.
6. Spray aging rooms once every 3 months with a synergized pyrethrin. Follow the directions on the label. Chlorfenapyr can be sprayed on nonfood contact surfaces instead of pyrethrin once every 2–3 months. It has the added benefit that it can be used as part of a mold mite (*T. putrescentiae*) integrated pest management program. Abbar et al. (2016) reported that it is effective for controlling all stages of spider mites and eriophyoid mites on wood, metal, and concrete for up to 8 weeks after its application.

Smokehouses

Frequency: After the end of each smoke period.

Procedure

1. Pick up large debris and place it in a receptacle.
2. Apply an alkaline cleaning compound recommended for cleaning smokehouses through a centralized or portable foam system. Figure 17.3 illustrates a unit used for cleaning smokehouses.
3. Rinse the area within 20–30 min after cleaning compound application. Start at the ceiling and walls, and work all extraneous matter down to the floor drain.
4. Inspect all areas and touch up where needed.
5. Apply a quaternary ammonium sanitizer with a sanitizing unit at the entry area to reduce air contamination.

Smokehouse Blower

Frequency: After each use cycle.

---

[1]To reduce mold growth, filtered air or air conditioning with a filter is recommended for aging rooms.

20 GALLON FOAM UNIT

Fig. 17.3 A portable air operated foaming unit for the application of a cleaning compound as blanket of foam. (Courtesy of Ecolab, Inc., St. Paul Minnesota)

Procedure

Blades

1. Remove the blower housing access panel and drain plugs; soak with an alkaline solution.
2. Start the blower and flush with steam.
3. Stop the blower and flush again with water. Repeat the operation until the equipment is clean.

Housing

1. Soak the inside of the plenum well, and wash the blower evolute wall with the alkaline cleaning solution.
2. Flush the housing with steam, then with water. Repeat until the housing is clean.
3. Replace drain plugs and access panel.

Smokehouse Steam Coils

Frequency: Depends on amount of use.

Procedure

Coils

1. Open the coil chamber access door and soak with an alkaline cleaning solution, brushing vigorously.
2. Flush the coils with steam, then with water. Repeat until the metal is shining.

Chamber Around Coils

1. Brush the cleaning solution on the inside of the chamber walls.
2. Use 55 °C (130 °F) water to flush the chamber wall clean.
3. Close the coil chamber access door.

Smokehouse Ducts and Nozzles

Frequency: Depends on amount of use.

Procedure

Outside Ducts

1. Remove the ductwork at the back of the house and remove carbon deposits. Disassembly is not necessary if the ducts have access panels.
2. Spray the inside surface with an alkaline cleaning solution.
3. Flush the outside ducts clean with 90 °C (194 °F) water or steam, followed by hot water until the metal is exposed.

Inside Return Ducts

1. Mark the positions of the slide panels over the return ports for setting back to their original openings.
2. Open the ports all the way and use as access doors for applying an alkaline cleaning solution to the ducts.
3. Use 90 °C (194 °F) water for flushing the return ducts. Repeat until the metal shows.
4. Reset the slide panels to the originally marked positions.

Inside Jet Ducts

1. Open side access panels (or drop hinged panel, depending on type of house).
2. Soak the inside ducts and nozzles with a cleaning solution.
3. Use water at 90 °C (194 °F) to flush these ducts clean. Repeat until the metal is exposed.
4. Close the access panel (or hinged panel).

Exhaust Stack

1. Disassemble the stack (or open access panels).

2. Soak the stack interior with an alkaline cleaning solution.
3. Flush the stack with 90 °C (194 °F) water or steam, followed by hot water. Repeat until the metal shows.
4. Reassemble the stack (or close the access panels).

Smoke Generator

Frequency: Depends on amount of use.

Procedure

Filter

1. Soak the filter in an alkaline cleaning solution.
2. If mineralization has occurred, cut the frame apart, and clean the leaves individually. Reweld the frame after cleaning. Avoid warping.

Baffle and Cascade Chamber

1. Mechanically or hand brush the baffles (especially the edges) with a wire brush.
2. Scrape the edges of the cascade water outlet.

Wash Chamber

1. Disassemble the duct that connects the smoke generator to the smokehouse.
2. Remove soot and ash from the chamber below the filter.
3. Clean the duct and chamber surface until the metal shows.

Spiral Freezer

Frequency: After use.
 Procedure: See instructions for specific equipment to be cleaned.

Precautions

1. To minimize friction, regularly wash the spiral with a foaming cleanser.
2. When the track is warm, wipe with a cloth that has been dampened with a detergent solution. If the track is cold, a dry cloth may be used. Tie the cloth to the underside of the conveyor belt and let it be drawn through the spiral.
3. Defrosting the evaporator coil alone is insufficient for cleaning. Coils may appear clean, but grease, oils, salts, food adjuncts, and organic materials often remain hidden on internal surfaces. Therefore, it is necessary to clean and sanitize contaminated sites with warm water and a pH-balanced detergent. Cleansing solutions typically include an etching agent, a degreaser, inhibitors, metal protectors, stabilizers, and water. A mildly alkaline cleanser is often recommended for cleaning the evaporator coil.
4. If the freezer has been supplied with a recirculating CIP system, use a low-foaming detergent. Otherwise, a high-foaming detergent is best. A chemical supplier should be consulted to determine the best cleaner.

Wash Areas

Frequency: Daily.
 Procedure: See instructions for specific equipment to be cleaned.

Precautions

1. Use a separate wash area for raw and cooked product equipment to reduce the spread of *Listeria* and spoilage microorganisms.
2. Provide this operation in an area where clean equipment does not cross fresh product areas of the plant.

Packaged Meat Storage Area

Frequency: At least once per week and more often in a high-volume operation.

Procedure

1. Pick up large debris and place in a receptacle.
2. Sweep and/or scrub with a mechanical sweeper or scrubber, if available. Use cleaning

compounds provided for mechanical scrubbers, according to directions provided by the vendor.

3. Use a portable or centralized foam or slurry cleaning system with 50–55 °C (120–130 °F) water to clean areas heavily soiled by unpackaged products or other debris. The cleaning through rinsing down process should be conducted as previously described for production and processing areas.

4. Remove, clean, and replace drain covers, if present.

Low-Temperature Rendering (Edible)

Frequency: Daily.

Procedure

1. Remove all large pieces of fat and tissue from the grinding equipment and store in a cool area.

2. Drain the system so that no lard, tallow, or melted fat remains.

3. The entire system should be flushed with 55–60 °C (130–140 °F) water to remove heavy accumulations of deposits from the equipment and piping.

4. Disconnect the system where possible to allow the water and scrap to drain from each piece of equipment. Dismantle dead ends and T-joints in the piping to allow scrap accumulations to be removed from these sections.

5. Open the equipment and dismantle where possible to allow cleaning of all surfaces that come in contact with the product. Place parts, pipe sections, and other sections in a sink or truck to soak in an alkaline cleaning solution. Follow specific instructions from the manufacturer for dismantling and cleaning the equipment.

6. Remove large scraps of product from the interior of the equipment.

7. Spray clean all exposed surfaces of the equipment throughout the system with an alkaline detergent sanitizer. Take special care to remove all possible products from the interiors of augers, pump screws, cutters, grinders, centrifuge chambers, and tanks. Spray clean the cooling rollers where they are operating without refrigeration. Clean parts and pipe sections in a truck with a scrub brush and an alkaline cleaning solution.

8. Clean the centrifugal equipment and piping that cannot be dismantled to allow the interior surfaces to be spray cleaned by circulating a solution of a heavy-duty alkaline cleaner through the equipment and piping. While circulating the cleaning solution, operate the centrifuges at reduced speeds to provide a scrubbing action in the system. Although CIP equipment is expensive, this system can be utilized effectively in this cleaning application, due to the potential savings in labor.

9. Circulate the cleaning solution for at least 30 min.

10. Drain the system and flush with 55–60 °C (130–140 °F) water until the effluent is free of scraps.

11. Transfer all scraps that are flushed out of the equipment to the inedible department.

Wire Pallets and Metal Containers

Frequency: Prior to use.

Procedure

1. Use high-pressure water at 55 °C (130 °F) or lower as a prerinse.

2. Preferably, apply an alkaline cleaner with a foam unit. If foam is unavailable, use a high-pressure, low-volume unit. Never spray more containers than can be rinsed before the cleaning compound dries.

3. Use a high-pressure spray of 55 °C (130 °F) water as a rinse.

4. Inspect all rinsed containers and reclean as needed.

Trolley Wash

Frequency: Depends on the physical appearance.

Procedure

1. Skim off excess waste material from the cleaning solution.
2. Check the cleaning solution strength with a test kit. If it registers under the recommended strength, add the appropriate compound and retest.
3. Open the main steam valve. Maintain a solution temperature of 82–88 °C (180–190 °F).
4. Lower the trolleys into the tank.
5. After the trolleys have soaked for 25–30 min, remove them, and rinse thoroughly.
6. Inspect the clean trolleys. Place the unsatisfactory ones on a rack for recleaning.
7. Place the clean trolleys in an oil bath while another rack is being cleaned.
8. Place the oiled trolleys over a drip pan or allow sufficient drip time while suspended over the oil tank.

Offices, Locker Rooms, and Rest Rooms

Frequency: Offices, daily; locker rooms and restrooms, at least every other day.

Procedure

1. Cover electrical connections with plastic sheeting.
2. Clean areas with a foam or high-pressure unit (or scrub brush and/or mop).
3. Within 20 min after application with the cleaning compound, rinse with 55 °C (130 °F) water.
4. If the cleanser and rinse do not clean dirty areas or if drains are not present, hand scrub with scouring pads.

Garments

Frequency: Daily.

Procedure

1. Place dirty garments into the washer extractor. Do not load the washer beyond its rated capacity.
2. Place the programmer dial at the start of the cycle and push the "On" and "Run" buttons. The drum programmer will automatically select the wash time and water temperature. An example would be a mixture of 1 kg (2.2 lb) of a laundry compound, and 0.25 kg (0.55 lb) of chlorine bleach should not be used when washing gloves.
3. After the wash extract cycle, remove the garments and place them in the dryer. Set aside garments that were not thoroughly cleaned for rewashing. Do not load the dryer beyond its rated capacity.
4. Set the temperature at 121 °C (250 °F) for 30 min. Dry gloves for only 20 min.
5. Place dried garments in a clean wire crib or equivalent container. They do not need to be folded.

Troubleshooting Tips

- *Discoloration of floors*: To restore the original color of darkened concrete floors, spread a bleach solution on them, and allow it to stand for at least 30 min prior to using a mechanical scrubber.
- *White film buildup on equipment*: This condition is caused when too much cleaning compound is used, when the equipment is not being properly rinsed, or when hard water is used to clean.
- *Conveyor wheel freezing*: The cleaning water temperature is probably too high. Wheels lose lubricant at about 90 °C (194 °F). The cleaning temperature should not exceed 55 °C (130 °F).
- *Sewer lines plugged*: Sediment bowls are probably not being cleaned daily and/or floor sweepings are being flushed into sewer pipes.
- *Yellow protein buildup on equipment*: This condition may be caused by using too high of a water temperature during cleaning. Brushing

away all organic material will remove daily buildup. If heated soil is allowed to remain on equipment for long periods of time, rubbing with steel wool will remove it.

- *To avoid trouble, do not spray*: Liver slicers, cube steak machines, electronic scales, patty machines, any electrical outlet, motor or equipment with open connections (cover all possible outlets with polyethylene bags), wrapping film or containers, or wrapping units.

Preoperation Flood Sanitizing Considerations

Flood sanitizing is applying a sanitizer at a high flow rate. This allows a flow rate capable of flushing off soils and penetrating cracks and crevices with sanitizer solution without taxing the water supply.

Either sanitizer compounds can be injected at the hose station or sanitizer solution can be pumped through a central piping system. From a cost and durability standpoint, wall-mounted sanitizer stations with dual orifice inlets for sanitizer selection provide the best results. Central sanitizing system concentrations (PPM) are difficult to change and require pumps, control panels, and a separate piping layout.

According to Carling-Kelly (2003), most modern production areas can be physically cleaned relatively soil free during the sanitation process. But, recontamination issues can become apparent during the pre-op or start-up phase of production. This contamination is caused by factors such as:

1. Poor consistency during final rinse inspection by sanitation operators prior to pre-op. This problem may be caused by a short sanitation window or lack of trained sanitors to perform the final inspection.
2. Area or equipment recontamination during the actual pre-op inspection and setup process before production begins. This complication is caused during the setup process by bringing in supplies, preparing equipment for opera-

tion, and the influx of personnel getting the area ready for production.

Whatever the recontamination cause, two-stage flood sanitizing will provide a more effective method for controlling area results during this critical start-up time. The sanitizer solution should be applied at tap water temperature to reduce the potential for condensation in refrigerated areas. The basic concept is:

First Step
Use wall-mounted sanitizer injectors (or a central sanitize system) to flood all surfaces in the production room with 600–800 ppm of sanitizer solution as part of the sanitation final inspection process.

- Training the sanitors to inspect their area as they flood sanitize will result in a more thorough application of sanitizer at a disinfecting rate. Walls, equipment, framework, and floors should all be flood sanitized.

Second Step
This step should occur after pre-op and area setup, but immediately before production actually starts. Flood all product contact surfaces with sanitizer solution at the allowable no-rinse limit.

- This sanitizing step will remove any soils deposited on product contact surfaces during the setup phase and bring these surfaces into no-rinse compliance to avoid any contamination issues. Leaving the walls, framework, and floors with the higher sanitizer concentrations will provide additional bacteria control as the day progresses.

The benefits of two-stage flood sanitizing become readily apparent as pre-op inspections find less visible soils and bacteria growth is reduced throughout production areas. In effect, two-step sanitizing adds additional antimicrobial controlling rinses without increasing overall sanitation time.

Study Questions

1. How do microorganisms affect meat color and/or flavor?
2. What is the function of air curtains?
3. What are limited uses of CIP equipment in a meat or poultry plant?
4. Why is chlorine dioxide an effective sanitizer in meat and poultry plants?
5. How is *Salmonella* and *Campylobacter* currently controlled in broiler carcass chill tanks?
6. Why does the meat and poultry sanitarian need to know something about HACCP?
7. How should an a thermal processing plant be designed to minimize the spread of *Listeria* through the plant?
8. How can the discoloration of darkened concrete floors be removed?
9. What causes a white film buildup on equipment in a meat and poultry plant?
10. What causes a yellow protein buildup on equipment in a meat and poultry plant?
11. Where is foam cleaning in a meat or poultry plant especially beneficial?
12. How much reduction in labor costs may be obtained through an efficient cleaning system for meat and poultry plants?
13. What is the significance of activated lactoferrin to the meat processor?
14. What are the three alternative levels of *Listeria* control in a meat or poultry plant?
15. Provide an example of an antimicrobial that can be used as an ingredient in RTE meat products to help control *Listeria*.

References

Abbar S, Schilling MW, Phillips TW (2017). Efficacy of selected pesticides against Tyrophagus putrescentiae (Schrank): Influence of applied concentration, application substrate, and residual activity over time. *J Pest Sci* 90: 379.

Allen D (2004). It's a wash. *Meat Mark Technol* November: 62.

Barboza Y, Martinez D, Ferrer F, Salas EM (2002) Combined effects of lactic acid and nisin solution in reducing levels of microbiological contamination in red meat carcasses. *J Food Prot* 65: 1780.

Bashor MP, Curtis PA, Keener KM, Sheldon, BW, Kathariou S, Osborne JA (2004). Effects of carcass washers on campylobacter contamination in large broiler processing plants. *Poult Sci* 83: 1232.

Butts J (2003). Seek and destroy: Identifying and controlling Listeria monocytogenes growth niches. *Food Saf* 9(2): 24.

Carling-Kelly T (2003). Personal communication. Saratoga Specialties: Elmhurst, IL.

Castillo A, McKenzie KS, Lucia LM, Acuff GR (2003) Ozone treatment for reduction of Escherichia coli O157:H7 and salmonella serotype typhimurium on beef carcass surfaces. *J Food Prot* 66: 775.

CDC. Centers for Disease Control (2016). Foodborne Outbreak Online Database (FOOD Tool). https://www.cdc.gov/foodsafety/fdoss/data/food.html. Accessed October, 2016.

Clark JP (2004). Ozone-cure for some sanitation problems. *Food Technol* 58(4): 75.

Cooper CT (2016). Personal communication. Ecolab: Minneapolis, MN.

Doyle MP (1987). Low-temperature bacterial pathogens. *Proc Meat Ind Res Conf*, 51. American Meat Institute: Washington, DC.

FDA. Food and Drug Administration (2008). Guidance for industry: Control of Listeria monocytogenes in refrigerated or frozen ready-to-eat foods. http://www.fda.gov/Food/GuidanceRegulation/GuidanceDocumentsRegulatoryInformation/FoodProcessingHACCP/ucm073110.htm. Accessed October, 2016.

Frank JF, Ehlers J, Wicker L (2003). Removal of Listeria monocytogenes and poultry soil-containing biofilms using chemical cleaning and sanitizing agents under static conditions. *Food Prot Trends* 23(8): 654.

Goetz G (2013). 11 years of data show poultry, fish, beef have remained leading sources of food-related outbreaks. Food Safety News. http://www.foodsafetynews.com/2013/06/20-years-of-food-borne-illness-data-show-poultry-fish-beef-continue-to-be-leading-sources-of-outbreaks/. Accessed October, 2016.

Gombas DE, Chen Y, Clavero RS, Scott VN (2003). Survey of *Listeria monocytogenes* in ready-to-eat foods. *J Food Prot* 66: 559.

Koohmaraie M, Arthur TM, Bosilevac JM, Brichta-Harhay DM, Kalchayanand N, Shackleford SD, Wheeler TL (2007). Interventions to reduce/eliminate *Escherichia coli* O157:H7 in ground beef. *Meat Sci* 77: 90.

Lazar V (2004). Reaching alternative level one. *Meat Process* 43(7): 28.

McEntire JC, Montville TJ, Chikindas ML (2003). Synergy between nisin and select lactates against *Listeria monocytogenes* is due to metal cations. *J Food Prot* 66: 1631.

MMWR (2013). Incidence and trends of infection with pathogens transmitted commonly through food — *Foodborne diseases active surveillance network, 10 U.S. sites, 1996–2012*. https://www.cdc.gov/mmwr/preview/mmwrhtml/mm6215a2.htm. Accessed October, 2016.

Mukhopadhyay S, Ramaswamy R (2012). Application of emerging technologies to control *Salmonella* in foods: A review. *Food Res Int* 45: 666.

Oyarzabal OA, Hawk C, Bilgili SF, Warf C, Kemp GK (2004). Effects of postchill application of acidified sodium chlorite to control Campylobacter spp. and Escherichia coli on commercial broiler carcasses. *J Food Prot* 67: 2288.

Petrak L (2003). Ingredients for success. *The National Provisioner* 217(9): 88.

Russell J (2003a). Swiping pathogens. *The National Provisioner* 217(4): 63.

Russell SM (2003b). Advances in automated rapid methods for enumerating E. coli. *Food Saf* 9(1): 16.

Sabah JR, Thippareddi H, Marsden JL, Fung DYC (2003). Use of organic acids for the control of Clostridium perfringens in cooked vacuum-packaged restructured roast beef during an alternative cooling procedure. *J Food Prot* 66: 1408.

Sebranek J (2003). Managing listeria. *Meat Process* 42(5): 66.

Shapton DA, Shapton NF, eds. (1991). Buildings. In *Principles and practices for the safe processing of foods*, 37. Butterworth-Heinemann: Oxford.

Smith BS (2016). Use of buffered vinegar in processed meat products. Personal communication.

Sofos J (2009). Biofilms: Our constant enemies. *Food Saf Mag* February/March: 1.

Sommers C, Fan X (2003). Gamma irradiation of fine-emulsion sausage containing sodium diacetate. *J Food Prot* 66: 819.

Sommers C, Fan X, Handle AP, Sokorai KB (2003a). Effect of citric acid on the radiation resistance of Listeria monocytogenes and frankfurter quality factors. *Meat Sci* 63: 407.

Sommers C, Fan X, Niemira BA, Sokorai K (2003b). Radiation (gamma) resistance and post irradiation growth of Listeria monocytogenes suspended in beef bologna containing sodium diacetate and potassium lactate. *J Food Prot* 66: 2051.

Stahl NZ (2004). 99.99 percent sure. *Meat Process* 43(7): 36.

Stier R (2002). The story of O_3. *Meat Poultry* September, p. 36.

Stopforth JD, Samelis J, Sofos JN, Kendall PA, Smith GC (2003). Influence of extended acid stressing in fresh beef decontamination runoff fluids on sanitizer resistance of acid-adapted Escherichia coli O157:H7 in biofilms. *J Food Prot* 66: 2258.

Sunen E, Aristimuno C, Fernandes-Galian B (2003). Activity of smoke wood condensates against *Aeromonas hydrophila* and *Listeria monocytogenes* in vacuum-packaged, cold-smoked rainbow trout stored at 4 °C. *Food Res Int* 36(2): 111.

The Poultry Site (2015). Challenges of campylobacter control during poultry processing. July 2015. http://www.thepoultrysite.com/articles/3518/challenges-of-campylobacter-control-during-poultry-processing/. Accessed October, 2015.

USDA-FSIS (2014). FSIS Compliance Guideline: Controlling *Listeria monocytogenes* in post-lethality exposed ready-to-eat meat and poultry products. https://www.fsis.usda.gov/wps/wcm/connect/d3373299-50e6-47d6-a577-e74a1e549fde/Controlling-Lm-RTE-Guideline.pdf?MOD=AJPERES. Accessed October, 2016.

USDA-FSIS (2015a). Microbiological testing program and other verification activities for *Escherichia coli* O157:H7 in raw ground beef products and raw ground beef components and beef patty components. https://www.fsis.usda.gov/wps/wcm/connect/c100dd64-e2e7-408a-8b27-ebb378959071/10010.1Rev3.pdf?MOD=AJPERES E coli directive. Accessed December, 2016.

USDA-FSIS (2015b). Draft FSIS compliance guideline for controlling *salmonella* and *campylobacter* in raw poultry December 2015. Guidance. https://www.fsis.usda.gov/wps/wcm/connect/6732c082-af40-415e-9b57-90533ea4c252/Controlling-Salmonella-Campylobacter-Poultry-2015.pdf?MOD=AJPERES.. Accessed October, 2016.

Velazco J (2003). Searching for solutions. *Meat Mark Technol* September, p. 121.

Yovich DJ (2003). Cargill sharpens its edge on E. coli O157:H7. *Meat Mark & Technol* October, p. 119.

Seafood and Aquaculture Plant Sanitation

18

Abstract

A hygienically designed plant can improve the wholesomeness of seafood and/or aquaculture and the sanitation program. The location of the seafood plant can contribute to the sanitation of the facility, and the design and construction materials used in the plant and equipment are also critical to an effective sanitation program.

Personnel allocation and an organized cleaning schedule with required cleaning steps are essential in maintaining a hygienic operation. This portion of the sanitation program should be matched with the most effective cleaning compounds, cleaning equipment, and sanitizers. The sanitation operation can be enhanced by the recovery of by-products, adoption of recommendations provided by regulatory agencies, and participation in voluntary inspection programs. Siluriformes fish include catfish, basa, swai, and other species. These fish are now under USDA-FSIS inspection, which means that their processing plants have an inspector present during all hours of operation. This does not change sanitation requirements but does add another layer of accountability to ensure that a cleaning and sanitation plan is carried out effectively.

Keywords

Seafood • Aquaculture • Siluriformes • Contamination Source • *Vibrio vulnificus* • *Vibrio parahaemolyticus* • Traceability

Introduction

Sanitation programs in the seafood industry are essential to provide the processor with guidelines that will give the consumer a high-quality, wholesome food. Because these guidelines relate to the facility and work practices, proper planning of new, expanded, and renovated plants should be considered. Every production phase of the distribution chain, from harvest to the consumer, must ensure that only wholesome products are provided to the ultimate consumer. Effective sanitation contributes to the maintenance of desired seafood quality.

© Springer International Publishing AG, part of Springer Nature 2018
N.G. Marriott et al., *Principles of Food Sanitation*, Food Science Text Series,
https://doi.org/10.1007/978-3-319-67166-6_18

Seafood and aquaculture processors should be familiar with microorganisms that cause spoilage and foodborne illness. Also, they need to know about characteristics of various types of soil, effective cleaning compounds and sanitizers, available cleaning equipment, and effective cleaning procedures.

Each processor should be equally familiar with existing federal, state, and local public health regulations. Regulatory requirements are by no means the only reason that the seafood processor should practice strict sanitary procedures. Another important factor is the consumer's increased awareness of nutritional value, wholesomeness, and processing conditions of all foods, including seafood. Seafood processing is regulated by the US Food and Drug Administration (FDA) and is required to utilize Hazard Analysis Control Points (HACCP) plans. Seafood is therefore minimally impacted by the Food Safety Modernization Act. However, Siluriformes including catfish, balsa, swai, and other species now fall under the jurisdiction of the US Department of Agriculture Food Safety Inspection Service (USDA-FSIS). The transitional period for domestically produced fish is from March 1, 2016, to September 1, 2017. As of September 1, 2017, foreign countries seeking to import products to the United States must provide documentation showing that their inspection program for Siluriformes is equivalent to that of the United Sates. If this documentation is not provided, imports will not be allowed. In addition, it will be a requirement to present imported shipments of Siluriformes to USDA-FSIS for reinspection.

Sanitary Construction Considerations

A hygienically designed plant can enhance the wholesomeness of all foods and dramatically improve the effectiveness and efficiency of the sanitation program. Even a well-designed plant is not a safeguard against microbial infection or other contamination unless it is accomplished by sound maintenance and sanitation. In a hygienic operation, the employer or management team should ensure good housekeeping and should be constantly vigilant against ineffective sanitary practices for all physical facilities, unit operations, employees, and materials. Chapter 14 contains design and construction considerations to supplement those to be discussed here.

Site Requirements

A clean and attractive site is necessary. Clean premises should be maintained for a satisfactory public image, to promote the individual firm and the industry. First impressions of a site are important to regulatory personnel and to the public, who are favorably impressed by a clean, neat, and orderly plant. The condition of the plant premises frequently reflects the caliber of the plant hygienic practices. According to the US Food and Drug Administration (FDA), areas that are inadequately drained "may contribute to contamination of food products through seepage or foodborne filth and by providing an environment conducive to the proliferation of microorganisms and insects." Excessively dusty roads, yards, or parking lots constitute a contamination source in areas where food is exposed. Improperly stored refuse, litter, equipment, and uncut weeds or grass within the immediate vicinity of the plant buildings or structures may provide a breeding place or harborage for rodents, insects, and other pests.

The site should be equipped with the capability to dispose of the seafood plant wastes. Solids, liquids, vapors, and odors emanating from a plant present a poor image and can result in legal action by either regulatory groups and/or concerned citizens. Waste disposal facilities must be designed to meet federal, state, and local requirements.

The site must also supply an ample amount of potable water for plant operations. If water is drawn from wells, analysis for mineral content and microbial load should be conducted, and the water must meet the standards established by the appropriate regulatory agency. After water use, adequate provisions should be made for wastewater discharge.

Construction Requirements

Although construction requirements are addressed in Chap. 14, this information relates to considerations for seafood processing plants. Materials that do not absorb water and are easily cleaned with resistance to corrosion and other deterioration should be incorporated. Openings should be equipped with air or mesh screens to prevent entry of insects, rodents, birds, and other pests. A brief discussion of sanitary features of various construction phases will be covered to provide guidelines for establishing a hygienic facility that is designed for effective cleaning.

Floors

Floors should be constructed of an impervious material, such as waterproof concrete or tile. The material should be durable with a surface that is even enough to prevent accumulation of debris but not smooth enough to cause slipping and falling. A rough finish or use of embedded abrasive particles can reduce accidents. A frequently used surface is a water-based acrylic epoxy resin that provides a durable, nonabsorbent, easy-to-clean surface that can double the life of the concrete floor. This finish should contain an abrasive material to provide a skid-resistant surface. Although the cost is nearly prohibitive, acid brick floors are known to be satisfactory and durable.

Floor Drains

A drainage outlet should be provided in the processing area for each 37 m² (400 ft²) of floor space. As with other processing plants, floors in the processing areas should have a slope to a drainage outlet of 2%. It is imperative that this slope be uniform, with no dead spots to trap water and debris. All drains should contain traps. Drainage lines should have an inside diameter of at least 10 cm (4 in²) and should be constructed of cast iron, steel, or polyvinyl chloride tubing. State and local codes should be checked to verify that these materials are permitted. Drainage lines should be vented to the outside air to reduce odors and contamination. All vents should be screened to prevent entrance of pests into the plant. It is also recommended that contamination

be further reduced by connection of drain lines from toilets directly into the sewage system instead of into other drainage lines.

Ceilings

Ceilings should be constructed at least 3 m (10 ft) high in work areas with a material impervious to moisture. One acceptable material is Portland-cement plaster, with joints sealed by flexible sealing compound. A false ceiling reduces debris from overhead pipes, machinery, and beams from falling onto exposed products.

Walls and Windows

Walls should be smooth and flat with a nonabsorbent material such as glazed tile, glazed brick, smooth-surface Portland-cement plaster, or other nonabsorbent, nontoxic material. Concrete walls are satisfactory if they contain a smooth finish. Although painting is discouraged, a nontoxic paint that is not lead based can be applied. Window sills, if present, should be slanted at a 45° angle to reduce debris accumulation.

Entrances

Entrances should be constructed of rust-resistant materials with tightly soldered or welded seams. Double-entry screened doors should be provided for outside entrances, as well as air curtains (or equivalent) over outside doorways in the processing areas.

Processing Equipment

Processing equipment should have a durable, smooth finish that is easily cleaned. Surfaces should be free of pits, cracks, and scale. The equipment should be designed to prevent contamination of products from lubricants, dust, and other debris. In addition to hygienic design for cleaning ease, equipment should be installed and maintained to facilitate cleaning of equipment surfaces and surrounding areas.

Where metal construction is essential, stainless steel should be used to protect seafood or other edible products. Galvanized metal is discouraged because it is not sufficiently resistant to the corrosive action of seafood products, cleaning compounds, or salt water. However, galvanized construction can be economically used for handling of waste materials. If galvanized material is used, it should be smooth and have a high-quality dip.

Cutting boards should be fabricated of a hard, nonporous, moisture-resistant material. They should be easy to remove for cleaning and should be kept smooth. This material should be abrasion and heat resistant, shatterproof, and nontoxic. Cutting boards should not contain material that will contaminate products.

Conveyor belts should be constructed of moisture-resistant material (such as nylon or stainless steel) that is easy to clean. Conveyors should be designed to eliminate debris-catching corners and inaccessible areas. This equipment, like other processing equipment, should be easily broken down for cleaning. Cleaning is facilitated through use of sealed or closed steel tubing, instead of angle or channel iron. Drive belts and pulleys should be protected with guard shields that are easily removed during cleaning. Motor mounts should be elevated enough to permit effective cleaning. Motors and oiled bearings should be located so that oil and grease will not come in contact with the product.

As with other food plants, stationary equipment should not be located within 0.3 m (1 ft) of walls and ceilings, so that access for cleaning is available. Equipment should be mounted at the same distance above the floor or have a water-tight seal with the floor. All wastewater should be discharged through flumes or tanks, so that it is delivered with an uninterrupted connection to the drainage system without flowing over the floor.

Contamination Sources

The environment at a seafood plant location can contribute to contamination within the plant, as well as contamination to the products. The processing equipment, containers, and work surfaces are other contamination sources. An effective sanitation program is necessary to reduce contamination and to monitor program effectiveness. Raw fish and processing environments are potential sources for *Listeria monocytogenes* contamination. Although this pathogen is destroyed through

pasteurization and thermal processing, it often enters cooked, ready-to-eat (RTE) products as a post-processing contaminant. *Listeria monocytogenes* can grow at temperatures between 45 °C (113 °F) and 2.7 °C (37 °F) and pH values of 5.0–9.6 and is commonly isolated from soil and water.

Because seafood involves so many varieties of flesh foods, the amount of contamination varies among species. The initial contamination source can be the raw product, especially if the product is improperly harvested and subjected to unsanitary practices on a vessel or truck. Delayed refrigeration after harvest and other improper handling between harvesting and processing can result in product decomposition and increase the microbial load.

Seafood quality, including microbial load, should be satisfactory for processing the day after harvesting if:

- Chilling begins immediately after harvesting
- Chilling reduces product temperature to 10 °C (50 °F) within 4 h
- Chilling continues to approximately 1 °C (34 °F)

Storing fish at 1 °C (33.8 °F) or higher for 4 h, with subsequent chilling to 27 °C (80.6 °F), will provide an acceptable product for only 12 h.

Workers contribute to contamination, especially through unsanitary practices. Other sources of contamination are processing equipment, boxes, belts, tools, walls, floors, utensils, supplies, and pests. Contaminants of greatest concern are those that come in direct contact with RTE products. Therefore, effective cleaning and sanitizing of equipment are vital. Scombroid contamination is associated with some of the dark-fleshed, fast-swimming fish. This contamination could be properly called *histamine poisoning* and causes an allergic reaction. Nardi (1992) indicated that scombrotoxin is always associated with temperature abuse and resultant decomposition, so it is entirely avoidable. Undercooked shellfish can be contaminated with *Vibrio vulnificus* and can contain viral infections from hepatitis A. The National Shellfish Sanitation Program (NSSP) (NSSP 2015) provides details on time/temperature requirements for the harvest of molluscan shellfish. Effective 2012, all states must have a *Vibrio vulnificus* control plan that uses at least one of the following five triggers: (1) water temperatures, (2) air temperatures, (3) water salinity, (4) harvesting techniques, and (5) other risk factors. When the average water temperature for the month is over 21 °C (70 °F), control measures such as labeling, modeling and sampling, and/or post-harvest processing must be taken. All states must also have control plans for *Vibrio parahaemolyticus* when waters bordering the Pacific or Atlantic Oceans are 15.6 °C (60 °F) or greater or waters bordering the Gulf of Mexico are 27.2 °C (81 °F) or greater. The NSSP Guide contains time/temperature controls for shellfish which are important components of a HACCP plan as well as important elements of sanitation during harvesting and processing that are necessary for seafood and aquaculture products. There are generally four critical control points in post-harvest processing: (1) reception of shellfish based on temperature; (2) temperature of cold storage; (3) post-harvest processing including hydrostatic high pressure, irradiation, low temperature pasteurization, and freezing; and (4) finished product storage temperature.

Studies of fishery products serving as foodborne vehicles for listeriosis have been less focused than for some other foods in the past. However, samples found positive for *L. monocytogenes* include raw and cooked shrimp, lobster tails, crab meat, squid, finfish, and surimi analogs.

Condensation on the ceiling, walls, equipment, and floor is a common problem in seafood and aquaculture plants. Humidity and temperature need to be controlled such that condensation is minimized since it can be a vehicle for cross-contamination.

Biofilm Prevention

Biofilms are a community of bacterial cells that (1) adhere to each other and surfaces such as steel, glass, and plastic, (2) are held together and protected by polysaccharides that act as a glue-like material, and (3) differ in gene expression when compared to normal planktonic cells (Sofos 2009). Open-air ponds are an area of concern for biofilm

formation. Some control methods include using plastic liners on the bottom of the ponds and treating pond water with low levels of chlorine. During harvesting, metal and plastic buckets can contain bacteria and biofilms and need to be cleaned and sanitized using good aquaculture practices. Water quality and water disinfestation are very important in recirculating aquaculture systems. Sand filtration and ultraviolet light are common methods to control biofilms in these systems. During wild seafood catches, cleaning and sanitizing the harvesting vessel are crucial. Deheading and removing guts can lead to bacterial contamination of the ship's deck, which must be cleaned and sanitized to prevent biofilm formation. Temperature control on the boats is also crucial since fish and seafood must be cooled quickly to ensure that they are safe, wholesome food products.

When seafood arrives at processing plants, it has a biofilm on their outer skin which can contaminate surfaces. Heating the fish will inactivate the bacteria but does not physically remove the biofilm. The only effective methodology to control biofilms in seafood plants is similar to other food plants and includes proper cleaning, sanitizing and temperature control.

Sanitation Principles

A seafood sanitation program must encompass proper handling of the sanitation tasks as well as personnel allocation. Other information on sanitation control procedures can be found at CFR 21 123.11.

Sanitation Inspection Critical Factors

Stanfield (2003) suggested the following critical factors to remember when a sanitation inspection of a processing plant for fresh or frozen fish is conducted:

1. Look for evidence of rodents, insects, birds, or pets within the plant.
2. Observe employee practices including hygienic practices, clothing cleanliness, and use of proper strengths of hand-dip solutions.

3. Check to determine if fish are inspected upon receipt and during processing for decomposition, off-odors, and parasites.
4. Determine if equipment is washed and sanitized during the day and at the beginning and end of the daily production cycle.
5. Check to determine if the fish are washed with a spray after evisceration and periodically throughout the process prior to packaging.
6. Determine the method and speed of freezing for frozen fish and fish products.
7. Check the use of rodenticides and insecticides to assure that no contamination occurs.
8. Observe handling from boats to the finished package and observe any significant objectionable conditions.

Manufacturing Inspection

The following manufacturing inspection suggestions were adapted from those provided by Stanfield (2003):

1. The flow plan and manufacturing procedure should be evaluated.
2. Processing equipment should be evaluated for construction, materials, and ease of cleaning.
3. Equipment cleaning and sanitizing procedures should be observed and evaluated to determine their adequacy.
4. All harvesting procedures should be observed and evaluated.
5. Water source should be determined and evaluated to confirm that only potable water from an approved source should be utilized.
6. If a long production delay occurs during processing fish at room temperature, the product should be checked for decomposition.
7. All handling steps and intermediate steps in processing that may cause contamination should be examined.
8. Holding times and temperatures during processing should be determined.
9. If battering and/or breading of fish is involved, the process should be reviewed carefully, including temperature and possible contamination sources.
10. Compliance with good manufacturing practices should be evaluated.

Personnel Allocations

In addition to the need for adequate cleaning methods and seafood facilities, a well-qualified sanitarian is required. Although the seafood plant manager is ultimately responsible for an effective sanitation program and the production of wholesome products, sanitation employees who are trained to maintain a clean plant must be provided. Employees should be adequately instructed in seafood product knowledge and in proper sanitary techniques, so that they are informed of the importance of the effect of proper sanitation on product wholesomeness. Any employee with a contagious illness should not work around processing areas, even during cleanup (see Chap. 6 for further discussion related to employee health requirements).

The typical seafood processing plant should have one or more employees responsible for daily inspection of all equipment and processing areas for hygienic conditions. Any sanitation deficiencies should be corrected before production operations are initiated.

Cleaning Schedule

A cleaning schedule with sequential cleaning steps is essential. The schedule should be adopted for each area of the plant and should be followed. Continuous-use equipment, such as conveyors, flumes, filleting machines, batter and breading machines, cookers, and tunnel freezers, should be cleaned at the end of each production shift. If there are no refrigerated areas, batter machines and other equipment in contact with milk or egg products should be cleaned at 4 h intervals by draining the batter, flushing the batter reservoir with clean water, and subsequently applying a sanitizer. At the end of the production shift, this equipment should be disassembled, and all parts should be cleaned and sanitized. These parts, as well as portable equipment, should be stored off of the floor in a clean environment to protect against splash water, dust, and other contamination sources.

The following steps are applicable when cleaning seafood and aquaculture plants:

1. Cover electrical equipment with polyethylene or equivalent film.

2. Remove large debris and place it in receptacles.

3. Manually or mechanically remove soil deposits from the walls and floors by scraping, brushing, or by the action of a hose from mechanized cleaning equipment. Proceed from the top to the bottom of equipment and walls, toward the floor drains or exit.

4. Disassemble equipment as required.

5. Conduct a prerinse for wetting action and removal of large and water-soluble debris, with water at 40 °C (104 °F) or lower. This temperature is important. A higher temperature can cause denaturation of seafood residues and other proteins, with subsequent baking onto the contact surface.

6. Apply a cleaning compound that is effective against organic soil (usually an alkaline cleaner) by portable or centralized high-pressure, low-volume, or foam equipment. The temperature of the cleaning solution should not exceed 55 °C (131 °F). Cleaning compounds such as sodium tripolyphosphate, tetrasodium pyrophosphate (a general-purpose cleaner), or a chlorinated alkaline detergent are usually considered satisfactory. More than one cleaner should be incorporated because of the nature of the soiled equipment material characteristics. (Chapter 9 discusses appropriate cleaning compounds for various cleaning applications. Chapter 11 provides a detailed discussion of the optimal cleaning equipment for various cleaning applications.)

7. After the cleaning compound has been applied and given approximately 15 min to aid in soil removal, rinse the equipment and area with water that is 55 °C (131 °F) –60 °C (140 °F). Hotter water is more effective in removing fats, oils, and inorganic materials, but the cleaning compound aids in emulsification of these solids. Also, a higher water temperature contributes to higher energy costs and more condensation on the equipment, walls, and ceilings.

8. Inspect equipment and the facility for effective cleaning, and correct deficiencies.

9. Ensure plant sanitation through application of a sanitizer. Although chlorine compounds are the most economical and widely used, other methods (as discussed in Chap. 10) are available. Table 18.1 provides the recommended concentrations for various sanitizing operations. Washing raw salmon with an acidified sodium chlorite (ASC) solution reduces the microbial load on the skin of whole salmon and in fillets as well as $I_{\cdot}$ monocytogenes in the fillets. The antimicrobial activity of ASC is enhanced when salmon is washed with an ASC solution and stored in ASC ice (Su and Morrissey 2003). Sanitizers are most effectively applied by use of a portable sprayer in small applications or with a centralized spraying or fogging system in large-volume operations. Chapters 9, 10, and 11 discuss available cleaning compounds, sanitizers, and sanitation equipment.

10. Avoid contamination during maintenance and equipment setup by requiring maintenance workers to carry a sanitizer and to use it where they have worked.

Table 18.1 Recommended sanitizing concentrations for various applications

Application	Available chlorine (ppm)	Available iodine (ppm)	Quaternary ammonium compounds (ppm)
Wash water	2–10	Not recommended	Not recommended
Hand dip	Not recommended	8–12	150
Clean, smooth surfaces (rest rooms and glassware)	50–100	10–35	Not recommended
Equipment and utensils	300	12–20	200
Rough surfaces (worn tables, concrete floors, and walls)	1,000–5,000	125–200	500–800

The following sanitation checks should be conducted:

1. Compliance with good manufacturing practices (GMPs) should be confirmed.
2. Effectiveness of equipment cleaning and sanitizing should be inspected.
3. Handwashing and sanitizing facilities and the appropriate solution strength should be checked.
4. The correct usage and storage of pesticides should be verified.
5. The proper processing and storage temperature should be verified to ensure reduced microbial growth.

High Hydrostatic Pressure Treatment

High hydrostatic pressure (HHP) processing is a viable treatment technique for use in reducing pathogenic microorganisms associated with food and in extending shelf life. HHP has been applied to a variety of foods, including seafoods, fruit juices, sauces, and meats. Dong et al. (2003) found that HHP was effective in killing microorganisms in raw fish fillets, but its significant effect on the color and overall appearance of the product limits its application to the processing of fish for raw fish markets.

Flick (2003) indicated that HHP offers seafood processor advantages such as reduced process time; retention of freshness, flavor, texture, appearance, and color; and reduced functionality alterations compared to traditional thermal processing. HHP of 250–300 Mpa (36,300–43,500 psi) for 120 s curtails many of the disease risks (such as from V*ibrio parahaemolyticus, V. cholera*, and *V. vulnificus*) associated with the consumption of raw oysters (Cook 2003), but some consumers find the product changes that occur due to HHP unacceptable. Campus (2010) stated that pressures of 205–275 MPa (29,700–39,900 psi) can be used at temperatures of 10–30 °C (50–86 °F) for 1–3 min are commonly used for live oysters. A pressure greater than 350 MPa (50,800 psi) at temperatures between 1–35 °C (2 °F and 95 °F) for 2 min is necessary for a 5-log

reduction of *Vibrio parahaemolyticus* on oysters (Kural et al. 2008).

Ozone Generation

Although sanitizing principles that are discussed in Chaps. 10 and 17 apply here, ozone has utility in aquaculture to disinfect water, assist in filtration, and cool tower water. Production units are available that concentrate oxygen from the air using pressure swing absorption (PSA), use air directly, or feed pure oxygen from another source (Clark 2004). The most common is PSA, because the feed gas must be dried away (to prevent formation of undesirable by-products from ozone formation) and the drying process is similar to the concentration process. In addition, many fish processors use ozonated water in direct contact with fish to knock down spoilage bacteria prior to freezing fish (Higgins 2014). Some plants have installed ozone-resistant piping that delivers treated water to processing rooms. Use of ozonated water prior to freezing can inhibit lipid oxidation that occurs in frozen fish and seafood.

Recovery of By-Products

Waste management, including the recycling of seafood waste products, has become increasingly important. In addition to the economic considerations, an effective recovery system can contribute to a more hygienic operation. Today, many food processors are recycling and/or reducing their liquid discharges.

Innovations in water conservation are:

- Wastewaters used for non-contaminating purposes in one area of a food processing operation are now being redirected to other areas that do not require potable water.
- Closed water system food processing operations in which all process waters are continuously filtered to remove solid materials have been established.
- Dry conveying equipment has been utilized to replace water transport of solids.

Hazard Analysis Critical Control Point Models

Seafood processing regulations, which became effective on December 18, 1997, require that a seafood processing plant (domestic and exporting foreign countries) represent a preventive system of food safety controls known as HACCP. The basic concept of HACCP is to (1) identify food safety hazards that, in the absence of controls, are likely to occur in products and (2) establishment of controls at those operations in the process that will eliminate or minimize the possibility that an identified hazard will occur. HACCP provides a systematic approach for taking those measures that demonstrate to the Food and Drug Administration (FDA), customers, and consumers that food safety and design are being practiced.

Four raw fish workshops conducted by the National Marine Fisheries Service developed HACCP models for each region that identified between 23 and 26 steps with 5–11 critical control points. The HACCP model for breaded shrimp production identified 30 process steps, with 9 identified as critical. Similar evaluations were made through analysis of cooked and raw shrimp processing. This surveillance model is designed to develop a seafood product inspection program to protect consumers, based on the HACCP concept. More information about HACCP is provided in Chap. 7. HACCP forms and plans are available at the Seafood Network Information Center (SNIC), a site maintained by Oregon State University (SNIC 2016).

Biodefense

In 2016, a final rule was instituted under the Food Safety Modernization Act (FSMA) that requires all food plants registered with FDA to develop a food defense plan. This includes all seafood processing plants, with the exception of Siluriformes. Biosecurity and food defense plans are discussed in Chap. 2 and will be required for all plants over the next 3–5 years, based on plant size. Siluriformes fish plants fall under the jurisdiction of USDA and are therefore encouraged but not required to have a food defense plan. However, the USDA (2015) has provided food defense guidelines for Siluriformes fish plants. Key areas in a defense plan include farms, ponds, hatcheries, processing facilities, general facility security (outside and inside), processing security, shipping/receiving security, storage security, water/ice security, and personnel security.

Traceability

Seafood is the most highly traded food commodity in the world. Market demand and weak legislation have contributed to the selling of seafood products with a misleading label or description (Anderson 2016). Therefore, the United States, Japan, and EU have introduced traceability to their import regulations. A recent analysis indicated that approximately 30% of the world's seafood supply is mislabeled or misdescribed (Pardo et al. 2016), with higher mislabeling rates for more valuable fish such as chinook salmon and bluefish tuna. In addition, fish are often misrepresented in restaurants where labels are not needed. Substituting fish that should be presented with health warnings can lead to consumer health issues. Some consumers also choose fish based on sustainability, and some consumers prefer wild-capture fish over farm-raised fish products. Misrepresenting and mislabeling can lead to consumers eating fish that are not sustainable and/or do not meet their ethical criteria.

The color, texture and flavor of fish can be altered during processing such that it is not possible to accurately identify species or source. However, there are voluntary supply chain traceability programs that ensure the integrity of supply chains (Anderson 2016). For example, the Marine Stewardship Council (MSC) has a Chain of Custody Standard that includes over 3,000 members of the seafood industry worldwide. Consumers can be sure that MSC-labeled seafood is sourced legally and that the product matches the label.

DNA testing of seafood products can be used to verify the authenticity of many seafood products, which has been the greatest development in seafood traceability over the last 10–15 years. The genetic code differs between seafood species and therefore can be determined for different products. This is achieved by comparing the genetic code to the database of known species to identify and verify species. The MSC uses biannual DNA testing to monitor the effectiveness of their traceability program. Current results of these analyses indicate that 99.6% of MSC products are correctly labeled. In addition, there is an ongoing effort to improve methods for DNA testing and increase the number and accuracy of specific genetic codes for different species. The FDA covers DNA testing protocols and information for seafood at www.fda.gov/food/foodscienceresearch/dnaseafoodidentification/default.htm (FDA 2016).

Study Questions

1. How much floor slope should exist in seafood processing plants?
2. How much chlorine sanitizer should be applied to equipment and utensils in seafood plants?
3. How much quaternary ammonium sanitizer should be applied to equipment and utensils in seafood plants?
4. How much iodine sanitizer should be incorporated in a hand dip for seafood plants?
5. What is the maximum cleaning solution temperature for a seafood plant?
6. What is the maximum rinse temperature for a seafood plant?
7. What kind of paint should be applied in seafood plants?
8. What measure can conserve water in a seafood plant?
9. How can entrances into seafood plants be designed to provide a more hygienic operation?
10. How can drainage lines from seafood plants be designed to reduce contamination?
11. How can fish species be determined?
12. What is the role of the Marine Stewardship Council?
13. What is a Siluriformes?
14. What products pertaining to this chapter are under FDA regulations? Which are under USDA-FSIS regulations?
15. How does FSMA affect the seafood and aquaculture industries?
16. What is ozonated water used for in seafood and aquaculture plants?

References

Anderson L (2016). From ocean to plate: Ensuring traceable supply chain in the seafood industry. *Food Saf Mag Digest* 1.

Campus M (2010). High pressure processing of meat, meat products and seafood. *Food Eng Rev* 2: 256.

Clark JP (2004). Ozone-cure for some sanitation problems. *Food Technol* 58(4): 75.

Cook DW (2003). Sensitivity of vibrio species and phosphate-buffered saline and in oysters to high-pressure processing. *J Food Prot* 66: 2276.

Dong FM, Cook AR, Herwig RP (2003). High hydrostatic pressure treatment of finfish to inactivate Anisakis simplex. *J Food Prot* 66: 1924.

FDA United States Food and Drug Administration (2016). DNA testing protocols for seafood. www.fda.gov/food/foodscienceresearch/dnaseafoodidentification/default.htm. Accessed August 19, 2016.

Flick GJ (2003). High pressure processing-improve safety and extend freshness without sacrificing quality. Unpublished data. Virginia Polytechnic Institute & State University.

Higgins K T (2014). *Is ozone the next sanitation superstar?* Food Proc http://wwwfoodprocessing.com/articles/2014is-ozone-the-next-sanitation-superstar/. Accessed August 19, 2016.

Kural AG, Shearer AEH, Kingsley DH, Chen H (2008). Conditions for high pressure inactivation of *vibrio parahaemolyticus* in oysters. *Int J Food Microbiol* 127: 1.

Nardi GC (1992). Seafood safety and consumer confidence. *Food Protection Inside Report* 8: 2A.

NSSP (2015). The 2015 National Shellfish Sanitation Program Guide. http://www.fda.gov/Food/GuidanceRegulation/FederalStateFoodPrograms/ucm2006754.htm. Accessed August 21, 2016.

Pardo MA, Jimenez E, Perez-Villarreal B (2016). Misdescription incidents in the seafood sector. *Food Control* 62: 277.

SNIC. Seafood Network Information Center (2016). http://seafood.oregonstate.edu/Seafood-HACCP.html. Accessed August 16, 2016.

Sofos J (2009). Biofilms: Our constant enemies. *Food Saf Mag* February/March: 1.

Stanfield P (2003). Seafood processing: Basic sanitation practices. In *Food plant sanitation*, eds. Hui YH, et al. Marcel Dekker, Inc.: New York. p. 543.

Su Y-C, Morrissey MT (2003). Reducing levels of Listeria monocytogenes contamination on raw salmon with acidified sodium chlorite. *J Food Prot* 66: 812.

USDA. United States Department of Agriculture (2015). Food defense guidelines for siluriformes fish production and processing. http://www.fsis.usda.gov/wps/wcm/connect/fc8c5e29-acf2-4636-8963-5f0011075f08/Food-Defense-Guidelines-Catfish.pdf?MOD=AJPERES. Accessed August 6, 2016.

Fruit and Vegetable Processing Plant Sanitation

19

Abstract

An effective sanitation program for fruit and vegetable processing facilities requires a sanitary design of facilities and equipment, training of sanitation personnel, use of appropriate cleaning compounds and sanitizers, adoption of effective cleaning procedures, and effective administration of the sanitation program—including evaluation of the program through visual inspection and laboratory tests. Effective sanitation starts with reduced contamination of raw materials, water, air, and supplies. If the facility and equipment are hygienically designed, cleaning is easier and contamination is reduced.

Cleaning labor can be reduced through the use of portable or centralized high-pressure or foam cleaning systems, and cleaning-in-place (CIP) systems can be used in large operations. Many facilities, if designed of durable material, can be cleaned effectively with acid cleaning compounds and sanitized most adequately and economically by using paints and other protective coatings as additional sanitary precautions. The effectiveness of a sanitation program can be evaluated through the establishment of standards as guidelines, visual inspection, and laboratory tests.

Keywords

Contamination • Food Safety Modernization Act • Cleaners • Sanitizers • Cleaning procedures • Wash water • Disinfestation

Introduction

Effective preservation of fruits and vegetables depends on the prevention of contamination by spoilage-causing and pathogenic microorganisms during production, processing, storage, and distribution. It is important to consider raw materials as a potential source for food spoilage microorganisms and as a contributor to bacterial pools within a processing plant.

Federal laws mandate that processed foods shipped interstate be free of pathogenic microorganisms. The normal sterilization process for

352

19 Fruit and Vegetable Processing Plant Sanitation

commercially canned foods is sufficient to destroy pathogenic bacteria that may exist in the container at the time of sterilization. Also, washing and peeling operations contribute to the physical removal of organisms. Therefore, if the canning and freezing processes are properly conducted, the finished product should be wholesome. Chapter 5 provides more information on the contamination.

Raw materials are exposed to many unclean sources and can provide additional contamination in the receiving, raw material storage, and processing areas. Raw materials may possess biological hazards such as certain fruits and vegetables contaminated with microorganisms. Furthermore, sucrose may be contaminated with bacterial spores and yeasts, and water can be contaminated with pathogenic microorganisms. The incoming materials may contain hazardous chemicals. Fruit may contain pesticide residues, and water could be contaminated with heavy metals and chemical residues, whereas packaging materials may contain harmful chemical residues that could leach into the product. Furthermore, the intermediate products may become contaminated in the processing steps from cleaning compound residues due to improper rinsing. Incoming materials may be contaminated with hazardous extraneous material such as metal, plastic, glass fragments, and wood slivers.

Washing fresh produce with water cannot be relied upon to completely remove pathogenic bacteria. Washing with water can also result in cross-contamination. Traditionally, chlorinated water has been the most frequently used sanitizer for the washing of fresh produce. However, this treatment has minimal effect and results in less than a 2 log CFU/g reduction of pathogens on fresh produce (Beuchat et al. 1998). Other sanitizers such as chlorine dioxide, hydrogen peroxide, organic acid, calcinated calcium solution, ozone, and acidic electrolyzed water have antimicrobial effects that are similar to chlorinated water (Bari et al. 1999; Han et al. 2000; Kim et al. 1999; Lin et al. 2002; Koseki et al. 2003). Acidic electrolyzed water has effectively inactivated pathogens such as E. coli O157:H7, Listeria monocytogenes, Salmonella, and Bacillus cereus (Kim et al. 2000; Koseki et al. 2001; Park et al.

2001; Hung et al. 2010). High-pressure ultrasound can be used in combination with chlorinated water to achieve an additional log reduction in comparison to chlorine alone since ultrasound removes bacteria that are entrapped on the surface (Seymore et al. 2002). Combining ultrasound with other technologies such as heat, high pressure, pulsed electrical field, or cleaning and sanitizing solutions is a more viable option than using ultrasound alone (Sao Jose et al. 2014). A three-step process can be used in which acidic electrolyzed water is used first as a rinse, peroxyacetic acid is used second, and a neutral electrolyzed water rinse is the final treatment that is used to reduce pathogen growth on cabbage (Lee et al. 2014). Human norovirus is a contamination concern of fresh produce. Use of chlorine wash water alone can reduce counts by approximately 1 log. However, surfactants can be incorporated into chlorine washes to increase efficacy such that there is a 3 log reduction (Predmore and Li 2011).

Produce plants currently use peroxyacetic acid, chlorine, or chlorine dioxide to disinfect wash water. Peroxyacetic acid with a residual concentration of up to 80 parts per million (PPM) peroxyacetic acid can be used to reduce pathogens such as Salmonella enterica, Escherichia coli O157:H7, and Listeria monocytogenes by 3 log and control spoilage microbes, including bacteria, yeast, fungi, and molds.

The increase in bagged salads and cut vegetables that are available to consumers has contributed to a number of outbreaks on products such as lettuce, spinach, and tomatoes (Jung et al. 2014). Hsu et al. (2006) reported that Salmonella and E. coli O157:H7 populations declined rapidly when the products were stored under 4 °C (40 °F) but still survived for up to 24 days. Fruit and vegetables are susceptible to pathogen contamination at many points in the supply chain. Matthews (2013) reported that 23% of total foodborne illness outbreaks from 1996 to 2010 were produce related. The predominant pathogens responsible for these outbreaks were Salmonella, Listeria monocytogenes, and E. coli O157:H7.

The Food Safety Modernization Act (FSMA) of 2013 requires preventative controls for allergens, sanitation, and supply chain. These items were previously covered under prerequisite programs such as sanitation previously being covered by processing plants under standard sanitation operation procedures (SSOPs). Three areas of focus for sanitation preventive control processing plants include (1) pathogen contamination, (2) cross-contamination due to water, and (3) cross-contamination of allergens. The first two of these areas are important to fruit and vegetable processing plants. Environmental samples should be taken on both food contact and non-food contact surfaces for *Listeria* and potentially other pathogens. For water cross-contamination, wash water should be evaluated for the presence of pathogens, disinfectant concentration, and pH. Water solutions with chlorine or peroxyacetic acid should be maintained at 6.5–7.5. Wash water disinfestation is also optimized through spray washing as compared to dip washing in chill tanks. Spray washing occurs in rotating buckets and prevents soil buildup in the wash water, which enhances effectiveness.

FSMA has also developed standards for growing, harvesting, packing, and holding of produce for human consumption that indirectly impact sanitation effectiveness and vegetable and fruit processing plants. These standards apply at the farm level and require generic *E. coli* testing as an indicator of water quality. In addition, stabilized compost must meet microbial standards for *Listeria monocytogenes*, *Salmonella* spp., fecal coliforms, and *E. coli* O157:H7. These standards include requirements to prevent the contamination of sprouts, including pathogen testing of sprout irrigation water; environmental testing of the growing, harvesting, packing, and holding environments for the presence of *Listeria* in the environment; and corrective actions when positive samples exist. Other sections of this final rule include worker training and health and hygiene and sanitation of equipment tools and buildings.

Soil Contamination

Heat-resistant bacteria are present in the ground and can cause "flat sour" and other spoilage of canned vegetables if washing is not thorough. Microbial population is affected by the degree of wind, humidity, sunlight, and temperature, as well as by domestic and wild animals, irrigation water, bird droppings, harvesting equipment, and workers. Most pathogens are introduced to fruits and vegetables via irrigation shortly before harvesting and before the sun dehydrates and destroys pathogens.

Air Contamination

Contaminated air contributes to less sanitary raw products. Besides normal microorganisms and pollutants found in the air, this medium serves as a transport of pathogens. Infiltration of unclean air into the processing plant can be improved by the use of air filters.

Pest Contamination

Certain pests can invade fruits and vegetables during the process of forming on the tree or vine. Contamination by pests can be expressed through the spread of viruses, spoilage bacteria, and pathogens, as well as by physical damage. Infesting microorganisms frequently remain inactive because of the protective skin layer of fruits and vegetables and because of the low availability of moisture (measured as minimum water activity [A_W]) on the surface. As these products reach maturity or shortly thereafter, profound changes in the medium can cause spoilage. The action of pests, such as the pollinating fig wasp (*Blastophaga psenes*), introduces microbes that persist and develop in quantity throughout the ripening period until the fruit is mature. Although a portion of the microorganisms that are introduced do not cause spoilage, these microbes attract other organisms, such as *Drosophila*, which carries spoilage yeasts and bacteria. When the protective covering of fruits and vegetables is broken by bruises, mechanical injury, or attack of insects, microorganisms can enter readily.

The presence of coliforms on processing grade fruit as it arrives at the processing plant is not truly indicative of the amount of these microorganisms in the manufactured juice or of positive evidence of unsanitary conditions in the processing plant. However, the presence of lactic acid bacteria constitutes an accurate index of processing sanitation for high-quality frozen citrus products. Lactic acid bacteria are a more accurate indicator of unsanitary conditions caused by inadequate cleaning because these microorganisms are the most likely to accumulate in the bacterial pools that can exist when proper sanitation practices are not followed.

Although several mycotoxins occur in nature, few are regularly found in fruits. The formation of mycotoxins depends more on endogenous and environmental factors than fungal growth. Mycotoxins may remain in fruits even when the fungal mycelium has been removed. The diffusion of mycotoxins into the sound issues of fruits may occur, depending on the food and mycotoxin. Proper selection, watching, and sorting of fruits is the most important factor in the reduction of mycotoxin contamination during the production of fruit juices. However, the processing of foods does not result in the complete removal of mycotoxins (Drusch and Ragab 2003).

Harvesting Contamination

Contamination may occur during harvesting fruits and vegetables out of the field through contact with contaminated boxes, buckets, knives, gloves, etc. (Matthews 2013). One example is coring of lettuce in the field which leads to tissue damage and additional human contact, which can contribute to product contamination (Jung et al. 2014). Rapid cooling to 4 °C (40 °F) is needed after harvesting to decrease respiration and the growth of any bacteria (pathogenic and spoilage) that are present on the fruit or vegetables.

Processing Contamination

Cross-contamination has the potential to occur during processing since cutting, washing, sanitizing, packaging and storing are involved. Cutting of produce releases moisture and nutrients that pathogenic bacteria can use to grow (Matthews 2013). *E. coli* O157:H7 has been associated with

cutting lettuce, and viruses such as norovirus and hepatitis A can be transferred to produce during cutting or grating (Wang et al. 2013). Bacterial contamination should be decreased through removal of soil through washing and sanitizing. However, if the sanitizer concentration is not sufficient and continuously monitored, cross-contamination of produce can occur through the dispersion of the microbes in the wash water (Holvoet et al. 2012). Temperature control is the most important factor during packaging and storage. Maintaining the temperature below 4 °C (40 °F) will help prevent bacterial growth in the environment, thus decreasing bacterial growth and the chance of cross-contamination.

Prior to 2005, the use of recirculated water was not recommended for washing fruits and vegetables because of the contamination caused through a rapid buildup of microorganisms in the wash water. Chlorination effectiveness of the wash water is minimal because bacterial spores exhibit resistance to chlorine. The benefit of chlorinated water for recirculation is further reduced through absorption of free chlorine and subsequent neutralization by the accumulated organic content of the water. However, the rinsing of lettuce with common household sanitizers such as distilled water, apple cider vinegar (5%), lemon juice (13%), bleach (4%), and white vinegar (35%) can reduce aerobic bacterial populations by averages of 0.6, 1.2, 1.8, and 2.3 log/g (log/0.035 oz), respectively, without severely affecting sensory attributes (Vijayakumar and Wolf-Hall 2002). Use of a mixture of peroxyacetic acid, hydrochloric acid, and potassium hydroxide as a buffer makes it possible to use the oxidation mechanism to control microbial growth. The concentration of peroxyacetic acid must be continually monitored to ensure that the concentration is effective at controlling microbial growth.

Sanitary Construction Considerations

A well-designed processing plant does not eliminate microbial infiltration unless the design incorporates hygienic features, such as easy-to-clean areas and equipment with optimal cleaning features and instructions. If the processing plant is newly constructed, expanded, or renovated, functional layouts, mechanical and plumbing layouts, and equipment and construction specifications should be reviewed by all professional personnel associated with the processing organization. This includes mechanical engineers, industrial engineers, food chemists, microbiologists, sanitarians, and operations personnel. This approach permits integration of operating procedures and process control (frequently called *quality control*).

Construction of new and expanded fruit and vegetable processing plants must reflect hygienic design because most of today's plants are volume oriented. High-volume plants operate under the principle that greater capacity is attained through pushing more materials through a larger-capacity production pipeline. With increased mechanization, there has been less emphasis on manual cleaning and visual inspection and more reliance on a CIP system. However, there is still limited use of CIP equipment in fruit and vegetable processing plants, except in the manufacture of juices. This concept also incorporates more emphasis on mechanized start-up and shutdown of production equipment and cleaning and sanitizing equipment. This approach provides less opportunity for human error but also reduces the possibility of spotting a performance error in cleaning.

High-volume processing plants, by design, operate with longer production periods and much greater product volume flow than do lower-volume plants. There is much more microbial buildup in the plant because of the longer dwell time and larger volume output. To reduce the microbial buildup, safe levels should be set by a saturation device that senses the buildup, stops production, and triggers an automatic cleaning procedure. It is suggested that this device would be activated only under excessive buildup, such as 150% of normal conditions.

Sanitary design features are necessary to minimize downtime for cleaning and sterilizing. The need for maximum utilization of equipment and facilities and for minimum discharge of sewage has mandated that the minimum effective cleaning approach to a process cycle is a short cleaning time and less effluent discharge from cleaning.

More mechanization and automation have been developed for cleaning tasks to equipment previously done by hand. Prior to CIP cleaning, machines and storage equipment were disassembled every production day and hand cleaned. After CIP cleaning was made available, control was initially conducted through a control panel with push buttons. Increased automation has incorporated use of an automatic panel with computer-controlled timers to provide automatic start-up and cutoff of cleaning, rinsing, and sanitizing. (Additional features of the CIP cleaning system have been previously discussed in Chap. 11.)

One of the most important features of hygienic design is the absence of crevices (narrow and deep cracks or openings) and pockets (large cracks and openings) in the construction of buildings and equipment. Crevices frequently present greater cleaning obstacles than do pockets because penetration and access are more of a challenge.

Principles of Hygienic Design

Minimum standards should be adopted when constructing or remodeling a fruit or vegetable processing plant. Effective hygienic design should incorporate the following principles:

- Equipment should be designed so that all surfaces in contact with the product can be readily disassembled for manual cleaning or CIP.
- Exterior surfaces should be constructed to prevent harboring of soil, pests, and microorganisms on the equipment, as well as on other parts of the production area, including walls, floors, ceilings, and hanging supports.
- Equipment should be designed to protect food from external contamination.
- All surfaces in contact with food should be inert to reaction with food and under conditions of use and must not migrate to or be absorbed by the food.
- All surfaces in contact with food should be smooth and nonporous to prevent accumulation of tiny particles of food, insect eggs, or microorganisms in microscopic surface crevices.

- Equipment should be designed internally, with a minimum number of crevices and pockets where soil particles may collect.

The interior and exterior of the plant should have the following sanitary features:

- Ledges and dirt traps should be avoided.
- Projecting bolts, screws, and rivets should be avoided to reduce the accumulation area for debris.
- Recessed corners and uneven surfaces and hollows should be avoided to reduce accumulation areas for debris.
- Sharp and unfilled edges should be avoided to reduce debris accumulation and microbial contamination.
- Proofing against pest entry through double-door construction, heavy-duty strips, and self-closing mechanisms is essential.

Certain pitfalls should be avoided when a processing plant is being built, expanded, or renovated to minimize contamination from external sources. Requirements may change as technology advances. Thus, the layout should reflect maximum flexibility and accommodate existing systems that are compatible with the proposed plant. The following points should be considered as a means of reducing contamination:

- Adequate storage space should be provided for raw materials and supplies. With inadequate storage space, contamination from the packaging material of supplies can occur. Sufficient space is also needed for thorough screening of raw material because foreign bodies may accommodate these products. Segregated materials that are contaminated should be salvaged and cleaned to prevent the spread of contaminants. Tainting can occur when raw materials share the same storage area as cleaning and maintenance materials.
- Separate storage space should be provided for finished products. Insufficient space may dictate use of the production area for this function. This practice can cause cross-contamination of raw materials.

- Congestion in areas of open food production should be eliminated. Insufficient space complicates cleaning and maintenance and increases contamination and risks of personnel injury and equipment damage.
- Short and direct routes for waste removal are necessary so that waste is not transported through open production areas. This design is especially critical because of the unsanitary condition of equipment used for waste collection.
- Location of the returned goods area is important. These foodstuffs are frequently infested and may be partially decomposed. It is essential to isolate these products from all raw material and production areas.
- Control of the environment should be exercised to reduce pests and to provide cleaner air through location of the waste collecting, waste treatment, and incineration areas as far as possible from the plant. This control also includes adequate surface drainage to prevent accumulation of water, outside surfaces that are easily cleaned, control of weed and grass growth, and control of stocks of surplus supplies and equipment.
- Employee personal hygiene is essential (discussed in detail in Chap. 6).

Cleaning Considerations

As with other food plants, management has the legal and moral responsibility to provide the consumer with a wholesome product. An effective sanitation program is needed to provide a clean environment for processing.

Housekeeping

Housekeeping relates to orderliness and tidiness. Careful arrangement of supplies, materials, and clothing contributes to a tidier operation, reduces contamination, and makes cleaning easier. Attention to neatness and orderliness contributes to the performance of responsibilities. Although the responsibility for housekeeping should be assigned to the sanitarian, the maintenance of good housekeeping depends on the cooperation of all employees—production, maintenance, and sanitation. Cooperation is needed to ensure that trash containers, tools, supplies, and personal belongings of employees are kept in the proper location. Convenient location of trash receptacles is necessary to encourage that anything not likely to be used further be discarded immediately.

Insects, rodents, and birds increase contamination. Knowledge of their biological characteristics and habits is necessary for their control. Sanitary practices can eliminate nutrition and protection for pests and, thus, can provide an important means of control. Hygienic design—air and mesh screens and filling of holes, cracks, and crevices—will discourage pests from entering the plant. Periodic inspection for the presence of pests is another prevention technique. (Methods of detection and other discussion related to pests are included in Chap. 13.)

Waste Disposal

Wastes can be handled more effectively and salvaged more efficiently as by-products if solid and liquid wastes are separated. Solid wastes are frequently separated through some method of pickup and/or transfer of solid materials before being flushed into drains or gutters. The liquid waste that is flushed away is usually handled as liquid waste and is treated as effluent, according to methods discussed in Chap. 12. Some food processing plants are processing waste by-products. The citrus industry incorporates more than 99% of the raw material for juices, concentrates, or dried cattle feed. Salvage efficiency has increased with reduced the cost of waste disposal.

Water Supply

As with other cleaning applications, an abundant, high-quality water supply is necessary to produce a wholesome product and to effectively clean the plant. In addition to being used as a cleaning medium, water is important as a heat transfer medium, and it is incorporated in the processed products. The sanitary condition of water should be monitored daily for two criteria: bacterial content and organic or inorganic impurities. Bacterial content serves as a guide for acceptability for use in contact with the food or any surface responsible for indirect contamination. The effectiveness of water in washing the product or equipment is dependent on organic and inorganic impurities.

Role of HACCP

The juice industry now requires Hazard Analysis Critical Control Points (HACCP) plans. As with meat and poultry firms, those classified as retail operations are exempted from coverage under the juice HACCP regulation. Contributions by the industry, academic, and government communities have been instrumental in advancing juice safety through the application of HACCP. The FDA places the highest inspection priority to firms that produce non-pasteurized juice because of the possibility of production through novel processing methods which merit closer regulatory monitoring when implemented in HACCP (Kashtock 2004).

Fresh-Cut Produce Washing and Sanitizing

Fresh-cut produce washing should involve the following:

1. Pre-shower washing to remove dirt and soil from the cut surfaces.
2. Placement into a washing tank with a sanitizing agent. Buffered peroxyacetic acid is the most efficacious and common sanitizer.
3. Rinsing with water is sometimes done depending on the product and the process.
4. An efficient water disinfection and recirculation system should be used with water flow and product flow going in opposite directions.

Biofilm Prevention

Inadequacy of cleaning/sanitation can lead to biofilm formation. Biofilms are a community of bacterial cells that (1) adhere to each other and surfaces such as steel, glass, and plastic, (2) are held together and protected by polysaccharides that act as a glue-like material, and (3) differ in gene expression when compared to normal planktonic cells (Sofos 2009). Bacteria are able to attach to surfaces when there are nutrient and soils on a surface due to inadequate cleaning and form biofilms. The bacteria in biofilms include *Listeria*, *Salmonella*, *Escherichia coli*, and spoilage bacteria. They can contain many bacteria, but are often dominated by a single species. Proper cleaning and sanitation are needed to prevent biofilm formation. Environmental testing and swabbing of equipment after cleaning and sanitation can be used as an indicator of the presence of biofilms. Sporadic spikes in microbial counts may indicate that biofilms exist in the plant (Mejias-Sarceno 2011). ATP bioluminescence can be used to detect immature biofilms but are not effective against mature biofilms. Other indicators of biofilms are rainbow-like appearance on stainless steel and use of touch to detect a slimy feel on the surface of surfaces that appear clean. Biofilms are very difficult to remove but can be inactivated and removed by combining proper cleaning and sanitizing agents, adequate exposure time, proper temperature, and mechanical action (Sofos 2009). This allows the biofilm and soil to dissolve and allow the sanitizing agent to kill the cells that were part of the biofilm. Cramer (2012) stated that the following method can be used to control biofilms in a food plant:

1. Dry clean: remove as much visible soil as possible.
2. Rinse with potable water at 49–55 °C (120–130 °F).
3. Apply detergent such as a chlorinated alkali or a combination of oxidative agents and acids such as peroxyacetic acid since they break the chemical bonds of food and soil. Mechanical action, such as scrubbing of surfaces, is the most effective way to remove biofilms.
4. A 60 °C (140 °F) water rinse should be used to remove all cleaner and soils. It is crucial that all soil and cleaner are removed for the sanitizer in step 5 to be effective.
5. Apply a sanitizer such as quaternary ammonium compounds or an acid-based sanitizer at a high concentration to prevent and control biofilms.

Cleaning of Processing Plants

A hygienic product results from rigid sanitation and effective destruction of microbes during processing. Conventional fruit and vegetable canning operations may be characterized as pouring food into containers (i.e., metal, glass, or plastic), followed by sealing and heat treatment. This heat treatment is referred to as *terminal sterilization* and is designed to eliminate extremely large numbers of *Clostridium botulinum* spores and to reduce the chance of survival of the much more heat-resistant spores of spoilage organisms. This condition is called *commercial sterility*. The process of aseptic packaging is sometimes called *aseptic canning*. In the aseptic process, the food and containers are commercially sterilized separately. The food is cooled to an acceptable filling temperature with subsequent filling and sealing of the containers under aseptic conditions. The microbial destruction (kill step) during terminal sterilization is accomplished for sealed containers, and, because of the excellent control that is technically possible over container integrity, conventional canning is safe technology. This technology is also suitable for the HACCP approach.

Aseptic packaging is a relatively new technology; thus, development of test methods is important. Active areas of development and concern are package integrity and maintenance of sterility, package performance in distribution, package sterilization techniques, and package residual. An online continuous monitoring method is needed. Several methods are available for measurement of concentration levels of H_2O_2 solutions (Shapton and Shapton 1991). An efficient layout of cleaning equipment is essential to reduce cleaning labor. It is much easier to install cleaning equipment when the processing equipment is put in place. The type of soil found in fruit and vegetable processing plants is most easily cleaned by portable cleaning systems in small plants and by a combination of CIP and centralized foam cleaning in large plants.

Hot Water Wash

Water provides transport of cleaning compounds and suspended soil. Sugars, other carbohydrates, and other compounds that are relatively soluble in water can be cleaned rather effectively with water. The main advantage of a hot water 60–80 °C (140–176 °F) wash for fruit and vegetable processing plants is minimal investment of cleaning equipment. Limitations of this cleaning method include labor requirements, energy costs, and water condensation on equipment and surroundings. This cleaning technique is not effective for the removal of heavy soil deposits.

High-Pressure, Low-Volume Cleaning

High-pressure spray cleaning has utility in the fruit and vegetable processing industry because of the effectiveness with which heavy soils can be removed. Difficult-to-reach areas can be cleaned more effectively with less labor, and there is increased effectiveness of the cleaning compounds below 60 °C (140 °F). Water temperature should not exceed 60 °C (140 °F) because high-temperature sprays tend to bake the soil to the surface being cleaned and to increase microbial growth. More discussion on this cleaning method is provided in Chap. 11.

Foam Cleaning

Portable foam cleaning is widely used because of the ease and speed of foam application in cleaning ceilings, walls, piping, belts, and storage containers in fruit and vegetable processing plants. Equipment size and cost is similar to that of portable high-pressure units. Centralized foam cleaning applies cleaning compounds by the

same technique used in portable foam equipment. The equipment is installed at strategic locations throughout the plant. The cleaning compound is automatically mixed with water and air to form foam, which is applied at various stations installed throughout the plant.

Gel Cleaning

Here, the cleaning compound is applied as a gel rather than as a high-pressure spray or foam. Gel is an especially effective medium for cleaning canning and packaging equipment because it clings for subsequent soil removal.

Slurry Cleaning

This method is identical to foam cleaning, except that less air is mixed with the cleaning compound. A slurry is more fluid than foam and penetrates uneven surfaces in a canning plant more effectively, but it lacks the clinging ability of foam.

Combination Centralized High-Pressure and Foam Cleaning

This system is the same as a centralized high-pressure system, except that foam can also be applied through the equipment. This method is more flexible because foam can be used on large surface areas, and high pressure can be applied to belts, stainless steel conveyors, and hard-to-reach areas in a canning plant.

Cleaning-in-Place

With this closed system, a recirculating cleaning solution is applied by nozzles that automatically clean, rinse, and sanitize equipment. However, this equipment is expensive and ineffective for heavily soiled areas. Nevertheless, CIP cleaning has application for vacuum chambers, pumping and circulation lines, and large storage tanks.

Since most fruits contain sugar and are low in fat content, water will flush most of the materials away. An acid cleaner or rinse should be incorporated to reduce scale buildup. Higher-volume operations are better adapted to CIP cleaning because labor savings provide a quicker payout of the equipment. Additional information about cleaning equipment is provided in Chap. 11.

Sanitizers

Soil remaining on equipment or at any location in the plant after cleaning is contaminated with microorganisms. Thorough physical cleaning of all equipment and rooms is necessary to prevent microorganisms from contacting chemical sanitizers. (Readers are referred to Chap. 9 for additional information on cleaning compounds.) Residual soil can also reduce the strength of chemical sanitizing solutions. Combination cleaners (detergent sanitizers) are used most frequently with smaller operations that perform manual cleaning at a temperature below 60 °C (140 °F). If the cleaning medium temperature exceeds 80 °C (176 °F), the solution will destroy spoilage microorganisms and most pathogenic bacteria without application of a chemical sanitizer.

Halogen Compounds

Chlorine and its compounds are the most effective sanitizers of the halogens for sanitizing food processing equipment and containers and for disinfecting water supplies. Calcium hypochlorite and sodium hypochlorite are two of the most frequently used sanitizers in fruit and vegetable processing plants. Although elemental chlorine is less expensive on an available chlorine basis, calcium hypochlorite and sodium hypochlorite are easier to apply in low concentrations. Hypochlorite solutions are sensitive to changes in temperature, residual organic matter, and pH. These compounds are quick acting and less expensive than other halogens but tend to be more corrosive and irritat-

ing to the skin. Additional information about chlorine and iodine sanitizers is provided in Chap. 10.

Chlorine Dioxide

Chlorine dioxide is approved as a flume water treatment for fruits and vegetables (that are not raw agricultural commodities) at a concentration of up to 3 parts per million (PPM) and to control microorganisms in process waters. Also, it is incorporated in wastewater treatment and for slime control in cooling towers. The typical use concentration of this sanitizer is 1–10 PPM (Anon 2003).

Quaternary Ammonium Compounds

Quaternary ammonium compounds ("quats") are effective against most bacteria and molds. These compounds are stable as a dry powder, a concentrated paste, or a solution at room temperature. They are heat stable, water-soluble, colorless, odorless, noncorrosive to common metals, and nonirritating to the skin in normal concentrations. These compounds are more active if soil is present than are other sanitizers, and they express the greatest antimicrobial activity in the pH range of 6.0 and above. The quats have limited bacterial effectiveness when combined with cleaning compounds or when dissolved in hard water.

Acid Sanitizers

Peroxyacetic acid-based sanitizers provide microbial control for use in fresh-cut, further processed, and post-harvest fruit and vegetable flume and wash water systems. They reduce the population of spoilage microorganisms including yeasts, molds, and bacteria on processed fruit and vegetables and pathogenic bacteria on processed fruit and vegetable surfaces. This sanitizer is EPA registered for use in fresh-cut, further processed, and post-harvest processing facilities. Also, it is cleared for all other process applications after a processing step has occurred. Wright et al. (2000) reported that 5% acetic acid and peroxyacetic acid solutions are effective in the reduction of *Escherichia coli* O157:H7 on apples relegated to cider manufacture. An acidified sodium chlorite rinse can provide pathogen reduction and offers a possible alternative sanitizer for fresh-cut produce (Gonzalez et al. 2004).

Ozone Sanitizing

Ozone effectively sanitizes raw materials, packaging materials, and the processing environment. It has gained acceptance by many industries, such as fresh-cut produce processing, produce storage facilities, and fruit and vegetable processing. Ozone applied as potatoes are transferred in a covered conveyor to storage can reduce the incidence of pathogens (Clark 2004). Williams et al. (2004) concluded that ozone treatment of apple cider and orange juice may provide an alternative to thermal pasteurization for the reduction of *E. coli* O157:H7 and *Salmonella*.

Ozone systems are generally mounted or fixed in place, to simplify management of ozone monitoring for safety and efficacy. Ozone is an unstable gas and readily reacts with organic substances. It sanitizes by interacting with microbial membranes and denaturing metabolic enzymes. It does not leave a chemical residue, and under ambient conditions, it has a half-life of 10–20 min. Ozone must be electrically generated on demand and cannot be stored for later use. An advantage of ozone is its ability to readily oxidized microbes in solution. Once a surface is spray washed, the microorganisms physically lifted from the surface will be destroyed as they are conveyed to a drain. Because ozone requires no storage or special handling or mixing considerations, it may be viewed as advantageous over other chemical sanitizers.

Phenolic Compounds

These compounds are used most frequently in the formulation of antifungal paints and antifungal protective coatings, instead of as sanitizers applied after cleaning. Phenolic compounds have limited utility in fruit and vegetable plants because of their low solubility in water.

Ultraviolet (UV) Light

This sanitizing technique has limited utility for equipment and processing and storage areas, but has been incorporated to reduce microbial growth on fresh fruits and vegetables. Accumulation of ethylene gas during storage is a potential detriment to fruit and vegetable quality after harvest. Potential solutions to this problem are development of a titanium dioxide photocatalytic reaction technology to decompose ethylene gas in the storage environment, and UV irradiation has an energy source for the titanium oxide photocatalytic reaction. Maneerat et al. (2003) found that UV doses improve appearance and do not adversely affect fruits stored in a dark environment.

Cleaning Procedures

A rigid set of procedures cannot be adopted for use in every fruit and vegetable processing plant. Procedures depend on plant construction, size, operations, age, and condition. Those discussed here are used only as guidelines and should be adapted to the actual cleaning application.

Facilitating Effective Cleaning

The following practices are recommended to aid in cleaning:

1. Reduce burn-on through careful, controlled heating of vessels.
2. Promptly rinse and wash equipment after use to reduce drying of soil.
3. Replace facility gaskets and seals to reduce leakage and splatter.
4. Handle food products and ingredients carefully to reduce spillage.
5. Work in an orderly manner to keep areas tidy throughout the operating period.

6. During a breakdown, rinse equipment and cool to 35 °C (94 °F) to arrest microbial growth.
7. During brief shutdowns, keep washers, dewatering screens, blanchers, and similar equipment running and cooled to 35 °C (94 °F) or below.

Preparation Steps for Effective Cleaning

To facilitate effective cleaning, it is necessary to prepare equipment and the area for cleaning:

1. Remove all large debris in the area to be cleaned.
2. Dismantle equipment to be cleaned as much as possible.
3. Cover all electrical connections with a plastic film.
4. Disconnect lines or open cutouts to avoid washing debris onto other equipment that has been cleaned.
5. Remove large waste particles from equipment by use of an air hose, broom, shovel, or other appropriate tool.

Processing Areas

FREQUENCY daily.

Procedure

1. Prerinse all soiled surfaces with 55 °C (131 °F) water to remove extraneous matter from the ceilings and walls to the floor drains. Avoid direct hosing of motors, outlets, and electrical cables.
2. Apply a strongly acid cleaner through portable or centralized foam cleaning equipment. A centralized system is more appropriate for large plants. Portable equipment is more practical for smaller plants. For heavily soiled areas, cleaning compounds are more effective if applied by portable or centralized high-pressure, low-volume cleaning equipment. If

metal other than stainless steel is present, the acid cleaning compound should be replaced with a heavy-duty alkaline cleaning compound. Hand brushing may be necessary to remove tenacious soil deposits left from foam cleaning. The cleaning compound should reach all framework, table undersides, and other difficult-to-reach areas. Soak time for the cleaning compound should be 10–20 min.
3. Rinse surfaces within 20 min after application of the cleaner to remove residues. The same rinse pattern as with prerinse and cleaning compound application should be followed by the application of 50–55 °C (122–131 °F) water.
4. Thoroughly inspect all surfaces and conduct any necessary touch-ups.
5. Apply a chlorine compound sanitizer to clean equipment with centralized or portable sanitizing equipment. The sanitizer should be sprayed as a 100 PPM of chlorine solution. Water pipes used for recirculating wash water and for pumping peas, corn, and other vegetables, as well as brines and syrup, should be sanitized by the same method. Frequently drain, clean, and sanitize water storage tanks to reduce microbial buildup.
6. Thoroughly backwash and sanitize water filters and water softeners.
7. Eliminate scale (as needed) from the surfaces of pipeline blanchers, water pipes, and equipment to reduce the chance of thermophiles and other microorganisms being harbored.
8. Remove, clean, and replace drain covers.
9. Apply white edible oil only to surfaces subject to rust or corrosion. Further use of oil is discouraged because the protective film harbors microorganisms.
10. Avoid contamination during maintenance by requiring maintenance workers to carry a sanitizer and to use it where they have worked.

Large processing plants can effectively utilize a CIP system for cleaning piping, large storage tanks, and cookers. The CIP system can be used as an alternative to steps 1, 2, 3, and 5 above.

Packaged Storage Areas

FREQUENCY at least once per week where processed products are stored, more frequently in a high-volume operation, and daily in areas where raw products are stored.

Procedure

1. Pick up large debris and place in receptacles.
2. Sweep and/or scrub with a mechanical sweeper or scrubber, if one is available. Use cleaning compounds provided for mechanical scrubbers, according to directions provided by the vendor.
3. Use a portable or centralized foam or slurry cleaning system with 50 °C (122 °F) water to clean areas heavily soiled, unpackaged products, or other debris. Rinse as described for the processing areas.
4. Remove, clean, and replace drain covers.
5. Replace hoses and other equipment.
6. Wash and sanitize vegetable boxes after each trip. Replace wooden husker and cutter bins with metal containers, which should be cleaned and sanitized.

Evaluation of Sanitation Effectiveness

A sanitation program must be evaluated to determine the effectiveness of cleaning and sanitation. Performance data not only measure sanitation effectiveness but also provide documentation of the program being conducted. Sanitation goals and checks are vital in the determination of sanitation effectiveness.

Sanitation Standards

To evaluate sanitation procedures, a yardstick measuring the current performance against past performance and desired goals should be incorporated to determine progress. Sanitation standards, derived through visual inspections and microbial counts, can be established. This approach has limitations due to variations, especially in microbial counts. Visible contamination and microbial load are not always highly correlated. However, the sanitarian can compensate for variables and still effectively evaluate the program.

Inspections can be conducted by the sanitarian or by a sanitation committee consisting of the sanitarian, production superintendent, and maintenance supervisor. Evaluations should be made in writing. A form that uses a numerical rating system is considered the most appropriate. The report should be divided into areas with specific sanitary aspects itemized in each area, as shown in Exhibit 19.1. The completed report should be provided to each supervisor associated with the inspected areas.

Laboratory Tests

The sanitarian must know the genera, characteristics, and sources of microorganisms found in the plant before laboratory tests have applicable value. With this knowledge, laboratory tests can be a monitoring device to evaluate the effectiveness of a sanitation program. The sanitarian should strive to reduce the total count of microorganisms found on clean equipment and among processed products but should also recognize that total plate count is not always highly correlated with spoilage potential or with the presence of microorganisms of public health concern. It is important to identify microorganisms, such as coliforms, as indicators of contamination or thermophiles and certain mesophiles as potential spoilage microbes. Large numbers of sporeformers can also be significant because these bacteria can reduce shelf life, and certain microorganisms can cause foodborne illness.

Spot checks for microbial load can verify opinions formed through visual inspections. Microbial sampling of products and equipment from various stages of manufacturing can identify trouble spots in the processing control cycle. Use of laboratory tests further utilizes the "think sanitation-practice sanitation" concept.

Exhibit 19.1 Sanitation evaluation sheet for food processing plants

Name of plant:	Location:		Date:
Scoring system: *1* unsatisfactory, *2* poor, *3* fair, *4* satisfactory			
	Location	**Score**	**Comments**
1.	Premises		
	Property outside of building		
	Waste disposal facilities		
	Other		
2.	Receiving		
	Dock		
	Containers		
	Conveyors		
	Floors, walls, ceilings, and gutters (or drains)		
	Other		
3.	Preparation		
	Washers and flumes		
	Conveyors		
	Graders and snippers		
	Blanchers, hoppers, and dewaterers		
	Pulpers and finishers		
	Floor, walls, ceilings, and gutters (or drains)		
	Other		
4.	Canning		
	Conveyors		
	Packaging or filling equipment		
	Floors, walls, ceilings, and gutters (or drains)		
	Other		
5.	Cooking		
	Exhaust box		
	Syrupers		
	Steamers		
	Floors, walls, ceilings, and gutters (or drains)		
	Other		
6.	Storage		
	Tanks and pipes		
	Other containers		
	Floors, walls, ceilings, and gutters (or drains)		
	Other		
7.	Welfare facilities		
	Lockers		
	Wash basins		
	Toilets and urinals		
	Floors, walls, ceilings		
	Other		
8.	Personnel		
	Cleanliness		
	Head covering		
	Health records		
	Other		

Study Questions

1. Where is CIP cleaning used most in fruit and vegetable processing plants?
2. What percentages of raw materials from the citrus juice industry are normally handled as waste products?
3. What is the maximum water temperature that should be used for cleaning fruit and vegetable processing plants?
4. Which sanitizer that can be applied in fruit and vegetable plants is the most stable and acts the longest amount of time?
5. What is the most likely cause of "flat sour" in canned vegetables?
6. Why do infesting microorganisms frequently remain inactive on fruits and vegetables?
7. Why is the use of recirculated water not recommended for washing fruits and vegetables?
8. Why is the chlorination of wash water ineffective?
9. Which pest introduces microorganisms that persist and multiply throughout the ripening period until fruit matures?
10. How can microbial buildup in a fruit or vegetable processing plant be reduced?
11. What are biofilms and how are they best controlled?
12. What changes have the Food Safety Modernization Act led to with respect to sanitation of fruit and vegetable processing plants?

References

Anon (2003). *Making the right choice-sanitizers*. Ecolab Inc.: St. Paul, MN

Bari M, Kusunoki H, Furukawa H, Ikeda, H, Isshiki K, Uemura T (1999). Inhibition of growth of *Escherichia coli* O157:H7 in fresh radish (Raphanus sativus L.) sprout production by calcinated calcium. *J Food Prot* 62: 128.

Beuchat LR, Nail NV, Adler BB, Clavero MRS (1998). Efficacy of spray application of chlorinated water in killing pathogenic bacteria on raw apples, tomatoes, and lettuce. *J Food Prot* 62: 845.

Clark JP (2004). Ozone-cure for some sanitation problems. *Food Technol* 58(4): 75.

Cramer C (2012). Biofilms: Impact on the food industry. *Food Saf Mag* June/July: 1.

Drusch S, Ragab W (2003). Mycotoxins in fruits, fruit juices, and dried fruits. *J Food Prot* 66: 1514.

Gonzalez RJ, Yaguang L, Ruiz-Cruz S, McEvoy JL (2004). Efficacy of sanitizers to inactivate Escherichia coli O157:H7 on fresh-cut carrot shreds under simulated process water conditions. *J Food Prot* 67: 2375.

Han Y, Sherman DM, Linton RH, Nielsen SS, Nelson PE (2000). The effects of washing and chlorine dioxide gas on survival and attachment of *Escherichia coli* O157:H7 to green pepper surfaces. *Food Microbiol* 17: 521.

Holvoet K, Jacxsens L, Sampers I, Uyttendaele M (2012). Insight to the prevalence and distribution of microbial contamination to evaluate water management in the fresh produce processing industry. *J Food Prot* 75: 671.

Hsu WY, Simonne A, Jitareerat P. (2006). Fates of seeded *Escherichia coli* O157:H7 and *Salmonella* on selected fresh culinary herbs during refrigerated storage. *J Food Prot* 69: 1997.

Hung YC, Tilly P, Kim C (2010). Efficacy of electrolyzed oxidizing (EO) water and chlorinated water for inactivation of *Escherichia coli* O157:H7 on strawberries and broccoli. *J Food Qual* 33: 559

Jung Y, Jang H, Matthews KR. (2014). Effect of the food production chain from farm practices to vegetable processing on outbreak incidence. *Microb Biotech* 7(6): 517.

Kashtock ME (2004). Juice HACCP approaches a milestone. *Food Saf Mag* 9(6): 11.

Kim JG, Yousef AE, Chism GW (1999). Use of ozone to inactivate microorganisms on lettuce. *J Food Saf* 19: 17.

Kim C, Hung YC, Brackett RE (2000). Efficacy of electrolyzed oxidizing (EO) and chemically modified water on different types of foodborne pathogens. *Int J Food Microbiol* 61: 199.

Koseki S, Yoshida K, Isobe S, Itoh K (2001). Decontamination of lettuce using acidic electrolyzed water. *J Food Prot* 64: 652.

Koseki S, Yoshida K, Kamitani Y, Itoh K (2003). Influence of inoculation method, spot inoculation site, and inoculation size on the efficacy of acidic electrolyzed water against pathogens on lettuce. *J Food Prot* 66: 2010.

Lee HH, Hong SI, Kim D (2014). Microbial reduction efficacy of various disinfection treatments on fresh-cut cabbage. *Food Sci Nutr* 2(5): 585.

Lin CM, Moon SS, Doyle MP, McWatters KH (2002). Inactivation of *Escherichia coli* O157:H7, *Salmonella enterica*, serotype Enteritidis and *Listeria monocytogenes* on lettuce by hydrogen peroxide and lactic acid and by hydrogen peroxide with mild heat. *J Food Prot* 65: 1215.

Maneerat C, Hayata Y, Muto N, Kuroyanagi M (2003). Investigation of UV-A light irradiation on tomato fruit injury during storage. *J Food Prot* 66: 2168.

Matthews KR (2013). Sources of enteric pathogen contamination if fruits and vegetables: Future directions
of research. *Stewart Postharv Rev* 9: 1.

Mejias-Sarceno G (2011). Inadequate sanitation results in
biofilm formation. *Food Saf Mag* April/May: 1.

Park CM, Hung YC, Doyle MP, Ezeike GOI, Kim C
(2001). Pathogen reduction and quality of lettuce
treated with electrolyzed oxidizing and acidified chlorinated water. *J Food Sci* 66: 1368.

Predmore A, Li J (2011). Enhanced removal of human
norovirus from fresh vegetables and fruits by a combination of surfactants and sanitizers. *Appl Environ
Microbiol* 77(1): 4829.

Sao Jose JFB, Andrade NJ, Ramos, AM, Vanetti MCD,
Stringheta PC, Chaves JBP (2014). Decontamination
by ultrasound application in fresh fruits. *Food Control*
45: 36.

Seymore IJ, Burfoot D, Smith RL, Cox LA, Underwood
AU (2002). Ultrasound decontamination of minimally
processed fruits and vegetables. *Int J Food Sci Technol*
37(1): 547.

Shapton DA, Shapton NF, eds. (1991). Aspects, microbiology safety in food preservation technologies. In
*Principles and practices for the safe processing of
foods*. Butterworth-Heinemann: Oxford. p. 305.

Sofos J (2009). Biofilms: Our constant enemies. *Food Saf
Mag* February/March: 1.

Vijayakumar C, Wolf-Hall CE (2002). Evaluation of
house sanitizers for reducing levels of *Escherichia coli*
on iceberg lettuce. *J Food Prot* 65: 1646.

Wang Q, Erickson M, Ortega YR, Cannon JL (2013). The
fate of murine norovirus and hepatitis A virus during
preparation of fresh produce by cutting and grating.
Food Environ Virol 5: 52.

Williams RC, Sumner SS, Golden DA (2004). Survival
of Escherichia coli O157:H7 and salmonella in apple
cider and orange juice as affected by ozone and treatment temperature. *J Food Prot* 67: 2381.

Wright JR, Sumner SS, Hackney CR, Pierson MD,
Zoecklein, BW (2000). Reduction of Escherichia coli
O157:H7 on apples using wash and chemical sanitizer
treatments. *Dairy Food Environ Sanit* February: 120.

Beverage Plant Sanitation

20

Abstract

Most soils found in beverage plants are high in sugar content, water soluble, and relatively easy to remove. However, control of raw materials is essential to ensure a method of detoxifying a finished product that is contaminated. Bacteria of greatest significance in breweries are non-sporeformers. Spray cleaning is most effective, through the incorporation of a properly blended, low foaming cleaning compound with specific cleaning properties for the soil that exists. Sanitizers such as chlorine, iodine, or an acid anionic surfactant are recommended for the final rinse in fermenters, cold wort lines, and coolers.

Rigid sanitation increases during the winemaking process and peaks at bottling time. A combination of wet and dry cleaning is usually most appropriate. Wine manufacturing equipment should be dismantled as much as possible, thoroughly washed with water and a phosphate or carbonate cleaner for nonmetallic surfaces and caustic soda or equivalent for cleaning metal equipment, and then sanitized with hypochlorite or an iodophor. Installation of a circular spray head inside a tank will help remove tartrates, as will soaking with soda ash and caustic soda. Fillers, bottling lines, and other packaging equipment can be cleaned with a cleaning-in-place (CIP) system. Prompt processing of grapes after picking will reduce fly infestation.

The control of raw materials is essential for distilled spirits. Yield and product acceptability are compromised if sanitary conditions are not maintained.

Keywords

Cleaning • Cleaning compounds • Construction • Contamination • Pasteurization • Sanitizing

Introduction

Increased emphasis is being placed on contamination prevention (instead of after-the-fact reaction) because of developing industry requirements such as the Food Safety Modernization Act rules. Because the soils found in most beverage plants are primarily high in sugar content and are water soluble, they are less difficult to remove than those described in some plants. Soil removal and microbial control present more of a problem in breweries, wineries, and nonalcoholic beverage plants. Therefore, a large percentage of the discussion in this chapter will concentrate on these areas.

Mycology of Beverage Manufacture

Because beverage plants such as breweries must maintain a pure yeast culture, it is important to retain the desirable microbes and to remove those that cause spoilage and unsanitary conditions. Ineffective sanitation can cause product acceptability problems because contaminating microorganisms (although kept under control) are not eliminated from the environment. Three categories of microorganisms found in beverage producing and filling plants are (1) product spoilage microorganisms, (2) hygiene indicators, and (3) pathogenic microorganisms.

Breweries differ from most plants in that commonly recognized pathogenic microorganisms are normally of minimal concern, primarily because of the nature of the raw materials, processing techniques, and limiting environmental characteristics of the final product (low pH, high alcohol concentration, and carbon dioxide tension). An exception to this is the unlikely possibility that significant levels of toxic metabolic products from certain fungi may pass from infected raw materials into finished products. Rigid control of raw materials is essential to ensure an acceptable product because there is no satisfactory method to detoxify a finished product that is contaminated.

Sanitation Principles

An adequate supply of urinals should be provided, kept in a sanitary condition, and located near the bottling area and other production areas. Employees must be required to wash their hands after using the toilet facilities. Drinking fountains should contain guards to prevent contact of the mouth or nose with the metal of the water outlet.

Employee Practices

As with other food operations, sanitation is a team job. It is important in beverage plants that employees clean as they go. Periodic cleaning increases tidiness, reduces contamination, and minimizes cleanup time at the end of the production shift or during a production change from the manufacture of one product to another. Furthermore, one or more employees who operate equipment that fills bottles or cans frequently have time to pick up debris or to hose down spills or other extraneous matter.

Effective housekeeping in a beverage plant depends on training and standards for the development of appropriate employee working habits. Rigid sanitation practices and work habits should be cultivated through effective communication, training programs, educational material, and continuous supervision and instruction. Employees should be instructed how, when, and where to clean to immediately remove soil and debris that can provide nutrition for pests and microorganisms. Leaking equipment should be corrected immediately. If rodents, birds, insects, or molds are detected, employees should either perform the necessary corrective steps or report the problems. Employees should be instructed regarding proper storage practices so that pest harborages are not created and proper cleaning can be accomplished. Further instruction should be related to closing doors and windows, removal of infested and extraneous matter, and proper storage of tools and cleaning and sanitizing equipment.

The following sanitation rules apply for beverage plants:

1. All employees visiting a lavatory must wash their hands before returning to work.
2. Any spilled materials or products must not be returned to the production area.
3. Waste materials must be placed in containers (with tight-fitting covers) suitable for disposal.
4. Each employee is required to keep the immediate work area clean and tidy.
5. Tobacco use is forbidden, except in designated areas.
6. Spitting is prohibited anywhere in the plant.
7. A periodic inspection of clothing, lunchrooms, and lockers by management should be conducted to ensure proper cleanliness.
8. Headgear should be worn at all times.

Cleaning Practices

There are six standard steps for cleaning (except CIP) a beverage plant:

1. Prerinse to remove large debris and nonadherent soil, to wet the area to be cleaned, and to increase the effectiveness of the cleaning compound.
2. Apply a cleaning compound (usually via foam) to provide intimate contact of water with the soil for removal through effective wetting and penetrating properties.
3. Hand detail and inspect for cleanliness.
4. Postrinse for removal of the dispersed soil and the cleaning compound to increase the effectiveness of the sanitizer.
5. Sanitize with quaternary ammonium compounds (with or without acid), acid anionic sanitizers, peracetic acid, chlorine compounds, or iodophors to destroy residual microorganisms.
6. Rinse quaternary ammonium sanitizers (if present in more than 200 parts per million (PPM) before exposing the cleaned area to any beverage materials.

Hygienic Lubrication

An important component of beverage sanitation is the adopted lubrication program and the various tasks and methods used for proper implementation. The objective for a lubrication program is to implement it as a successful prerequisite program. According to Briseño (2015), machinery lubrication practices are considered prerequisite programs within the Hazard Analysis and Critical Control Points (HACCP) system. These programs are defined by the World Health Organization as practices and conditions needed prior to and during HACCP implementation, which are essential for food safety. Machinery lubrication addresses "good housekeeping" concerns and reduces the occurrence of hazards.

To maintain hygienic conditions, it is important to identify where potential contamination related to lubrication may occur. It is essential to discern if lubricants will come in contact with food during production. An example is a developed machine leak. Lubricants for beverage or other food processing facilities are identified according to their intended use through industry classification systems. Some lubricants are intended for direct contact uses such as release agents, pan oils, and divider oils. Other lubricants are intended for incidental contact applications such as hydraulic systems, gear drives, and conveyors. Proper use identification and potential food contamination locations are essential.

A detailed lubrication survey is essential in constructing the required prerequisite program documentation. Work records should identify important details including where food-grade lubricants are needed, required use classification, correct lubrication intervals, responsibility assignments for correct application, and optimal practices for storage and handling. Furthermore, the survey should inventory all lubricated equipment, lubricant products, and applications and identify critical machines and operations.

As with any sanitation operation, employees responsible for machinery lubrication should receive proper training to ensure successful food-grade lubricant use. Information provided for employees should include knowledge of product classification differences; application methods, amounts, and intervals; and handling and storage procedures. Employees should be cautioned about potential seal rupture through over-lubrication, overflow, and/or foaming due to the addition of excessive oil. Furthermore, under-lubrication can cause foaming and reduce equipment life. Improper pump handling techniques and/or use of dirty containers will cause lubricant contamination. Also, employees should be advised that improper storage practices, such as storage outdoors, can cause contamination from dirt, water, and other lubricants. Temperature fluctuation can cause containers to breathe and introduce moisture from the environment even if they are sealed. Lubricant cross-contamination should be minimized by the conversion of as many lubricant applications as possible, prevention of mixing of lubricants, and separation of food-grade lubricants from those that are non-food grade. Employees should also clearly label lubricant filling containers and dispensers and store these and all lubricants in a locked area.

Inspection of Ingredients and Raw Materials

Because foreign objects and microbial contamination occur in both raw materials and the finished product, they should be inspected, including rodent and insect inspection, for foreign matter. Letters of compliance should be required of suppliers stating that the material was processed under the HACCP system.

Bottled Water Establishment Sanitation

As with other beverage plants, bottled water must be produced under hygienic conditions. Both tap water and bottled water are federally regulated. Bottled water is comprehensively regulated by the US Food and Drug Administration (FDA). By federal law, FDA regulations governing the safety and quality of bottled water must be at least as stringent as US Environmental Protection Agency (EPA) standards for tap water.

Water bottling should be conducted in a separate room from other operations to protect against potential contamination. Water contact surfaces and other locations in the bottling area must be kept physically clean and sanitized. Hui (2015) indicated that the sanitizing operation should involve steam or 77 °C (170 °F) water for 15 min or 93 °C (200 °F) for 5 min. Chemical sanitizing with an equivalent bactericidal action is accomplished with 50 PPM available chlorine for 2 min at 57 °C (135 °F) when used as an immersion or circulating solution or 100 PPM available chlorine at the same temperature when applied as a spray or fog. Ozone water solutions must be 0.1 PPM for 5 min in an enclosed system (Hui 2015).

Nonalcoholic Beverage Plant Sanitation

It is beyond the scope of this text to discuss sanitation principles for all nonalcoholic beverage plants. Although ultrahigh temperature as a technique for aseptic packaging is becoming more common, the ramifications of this technology are too extensive and specific for this general discussion. If further information about sanitation in these specialized operations is desired, a technical publication about aseptic technology should be reviewed.

A viable sanitizer for the skin of fruits for juice manufacture is chlorine dioxide. An alternative to thermal pasteurization for the reduction of *E. coli* and *Salmonella* in apple cider and orange juice is the incorporation of ozone treatment. If sanitizing is not practiced, pathogens, such as *E. coli* O157:H7 in apple juice (or cider), can become incorporated into the product.

Proper hygiene in a beverage processing facility includes the use of sanitary water, steam, and air. High-quality liquids and gases are required when they are incorporated into finished products or included in the packaging material that contacts the product. The desire to manufacture acceptable products and to meet safety standards has resulted in several beverage processors incorporating various types of filtration to remove microorganisms and other particulate or suspended materials. Filtration for the clarification or microbial control of water, air, and steam is accomplished by absolute filtration to prevent contaminants larger than the filter pore size to pass through and into the filtrate.

Because beverages such as soft drinks, bottled water, beer, and distilled spirits should be manufactured from microbial- and particulate-free water, some form of treatment is necessary. Various treatments include flocculation, filtration (i.e., through a sand bed), chlorination, sterile filtration, reverse osmosis, activated carbon, and deionization.

Conditioning of water for use in beverage plants is accomplished primarily through particulate removal and microbial control. Particulate contaminants that may be present in water are most frequently removed by flocculation and sand filtration. The installation of an absolute-rated depth filter behind the sand filter will remove all of the contaminants, larger than the rated pore size, prior to chlorination and activated carbon treatment.

Activated carbon is incorporated to remove excess chlorine, trihalomethanes, and other compounds associated with chlorine disinfection. However, activated carbon sheds carbon fines and provides sites for microbial growth. Carbon beds are potential microbial contamination sources and are difficult to disinfect. Thus, the use of filtration before and after carbon beds will reduce the loading of microorganisms and particles.

Resin beds for deionization of water are potential sites for microbial growth and can unload or shed resin beads into the treated or conditioned water. An absolute-rated filter will ensure that particles or microorganisms larger than the removal rating of the filter do not enter the treated water. As a final treatment, the incorporation of a sterilizing nylon 6.6–9.2-μm (0.00026–0.00036 in.) filter will remove microbes present in the water if the unit has been presterilized. Sterile filtration requires no chemicals and is beneficial because of its ease of use and low

energy input. A microbially stable product may be produced through use of a combination of flocculation and filtration steps followed by an absolute-rated filter.

Although steam is frequently incorporated in a production operation, it can be a viable contamination source. Steam is normally generated in carbon steel boilers, which are highly susceptible to rusting. A fine impervious film of rust, which acts as a protective barrier against further corrosion, normally deposits as a result of a continual operation of a boiler. Intermittent use allows a continual supply of fresh air containing oxygen into a boiler and promotes the oxidation of iron-to-iron oxides, or rust. The continual generation of rust causes flaking and steam contamination. Use of steam contributes to contamination and the particles of rust from the boiler transfer lines will damage equipment surfaces, block steam valves, fill orifices and filter pores, and stain equipment surfaces. Processing efficiency is reduced through the alteration of the heat transfer characteristics of heat exchangers. This problem is reduced by the injection of culinary steam with an uninterrupted supply provided through the installation of porous stainless steel filters in parallel to permit the cleaning of one set while the other is in use. Non-culinary-grade steam will add contaminants.

During the past, bottlers have been installing CIP equipment to clean tanks, processing lines, and filters. Most bottlers that manufacture multiple flavors prefer CIP as a tool to prevent flavor "carryover" (especially of root beer). Remus (1991) advocates the TACT (time, action, concentration, and temperature) approach to cleaning beverage plants (see Chap. 9). Parameters can be varied; for example, a 1% cleaning compound concentration at 43.5 °C (110 °F) can be equivalent to a 0.5% concentration at 60 °C (140 °F).

Increased efficiency and superior lubricity may be attained through an automated solid-lubricant dispensing system. This equipment saves labor and lubricant costs and reduces contamination during lubrication.

The following discussion relates how common soils found in beverage plants can be removed.

Although this discussion addresses carbonated beverage plants, these cleaning principles apply to other beverages. Principles of cleaning of floors, walls, and the bottling area, as discussed under winery sanitation later in this chapter, should also be considered for carbonated beverage plants. Cleaning applications and principles not discussed here are normally similar to those discussed for dairy processing plants (see Chap. 16).

Automated Cleaning Equipment

A portion of the carbonated beverage industry has turned to mechanized equipment to facilitate cleaning. A variety of automated solutions exists such as automated chemical formulation and an allocation and control system to streamline the operation.

A microprocessor-controlled system can be assessed by keying in an identification number or by using custom magnetic swipe cards. The controller contains a detailed application list that indexes sanitation procedures and equipment types with the proper chemicals and usage rates. Then the system dispenses the product into a reusable chemical container for use in plant sanitation. A smaller auxiliary dispensing station may allocate acids and other specialty chemicals to cleaning stations. This equipment can maintain detailed records to help monitor regulatory compliance, perform cost analyses, and create custom reports. Data reporting includes which chemicals have been incorporated into each application, when and in what quantity, and times and dates. Chemical barrel labels may be color-coded so that workers need to replace only the empty drums that correspond to colored spots on the floor. A computer-controlled CIP unit directs water and solutions to the appropriate location and automatically maintains operating conditions. The four basic parameters to be controlled are time, temperature, chemical concentration, and impingement which relate to flow velocity through a pipeline. Rinse water may be recycled one time for reuse and cleaning compounds several times. The initial prerinse may use recycled water from the previous final rinse.

Tire Track Soil Accumulation

Tire tracks are a difficult soil to remove. The most effective cleaning compound for this application is one that is solvated and alkaline. To facilitate cleaning ease and effectiveness, a mechanical scrubber should be considered. Floor soils should be removed daily to enhance cleaning ease and to avoid soil being further ground into the floor surface.

Conveyor Track Soil Accumulation

This accumulation is most likely to be spilled product, bearing grease, container and track filings, and precipitated soap. Incorporation of a track lubricant containing a detergent with a blend of fatty acids will reduce contamination. An effective way of removing this soil is through foam cleaning with a high-pressure rinse.

Film Deposits

Film deposits most frequently occur inside storage tanks, transfer lines, and filters. Thin films cause a dull surface, but as buildup increases, a bluish hue develops. As the film becomes thicker, a white appearance may occur. Although residues from sugars are relatively easy to remove, films from aspartame and certain gums are difficult to eliminate. Tanks may be cleaned manually, but circulation cleaning is frequently practiced. To remove surface films, a chlorinated cleaning compound (or one specially formulated with surfactants for food soils) should be applied.

Biofilms

Residual beverages or their ingredients provide nutrients for microbial growth and their biofilms. Biofilms can occur inside cooling towers, in and outside of warmers and pasteurizers, and inside carbocoolers. As with film accumulation, biofilm removal is enhanced by use of a chlorinated alkaline cleaning compound. A quaternary ammonium sanitizer or another biocide should be applied to reduce biofilm deposition because this formation can occur within 24 h after use.

Hot Sanitizing

Sanitation of beverage plants differs from that of other food facilities. During the past few years, a trend toward hot sanitizing has occurred. Hot sanitizing can be incorporated when cleaning contact surfaces of production equipment, such as batch tanks, low-mix units and fillers, and carbocoolers. Although this sanitizing method is not economical because of the required energy costs and ineffectiveness in bacteria removal, it has some merit because of its penetrating ability. Heat can effectively penetrate equipment and destroy microorganisms behind gaskets or in tiny crevices.

Hot sanitizing is not sterilization. It involves raising the surface temperature to 85 °C (185 °F) for 15 min. Sterilization requires 116 °C (240 °F) for 20 min. Sanitizing only reduces the microbial population to an acceptable level. A few of the more resistant microorganisms (yeasts and spores) remain viable. Chemical sanitizers can accomplish the same microbial kill as hot sanitizing, with a much quicker action.

Specially formulated cleaning compounds can be incorporated in a hot sanitizing procedure to loosen and remove soils and biofilms. These compounds are specifically formulated to handle the soil, to condition the water, and to be free rinsing in the hot sanitizing procedure. Soil and biofilm removal are essential for effective sanitation. A nonviable but intact biofilm is an easy attachment site and nutrient base for other films to develop.

Membrane Technology

Membrane technology applied to water treatment for the beverage industry includes a wide range of polymeric and ceramic impurity removal techniques, including treatments such as microfilters to remove granular activated carbon fines and reverse osmosis. Particulate filters remove relatively large suspended matter and are incorporated at the end of the water treatment chain as a "polishing filter" to remove small floc particles, oxidized iron, carbon, or precipitated calcium carbonate that may have occurred from the primary treatment process. Microfilters are incorporated for their controlled pore size distribution that facilitates mechanical removal of bacteria from water. Frequently, this application is a stepped removal approach that includes filters of decreasing pore size oriented in series, to minimize the plugging potential of the smallest pores. This is an important tool for the removal of particulates, large organic matter, and many microorganisms, including viruses, bacteria, and protozoa. The major contribution of membranes used in water treatment applications for the beverage industry is pressure, which is applied across a membrane to force the filtered or purified water through the membrane, leaving the impurities behind.

Container Handling

Bottles, cans, jugs, and other containers used for nonalcoholic beverages are a viable contamination source from foreign objects such as metal shavings, wood, and other materials. Product containers should be checked before use according to a standard sampling plan. Single-use containers should be rinsed with water immediately before filling. Returnable containers such as bottles and kegs should be washed with a cleaning compound that is effective against organic soils and rinsed thoroughly to remove cleaning compound residue.

Bottle Filler

Bottling equipment for nonalcoholic and alcoholic beverages may break glass bottles, creating a physical hazard. Plant personnel should keep a constant watch for broken glass that may fall into product containers when bottles become stuck on approach to the filler and the conveyor remains in motion smashing bottles against each other.

Brewery Sanitation

Rigid sanitation is of paramount importance to successful brewing. Because breweries have been traditionally production oriented, prophylaxis has normally superseded detailed taxonomic interest in microorganisms associated with these operations. The environment typical of a brewery can inhibit pathogen activity and impose limitations on the array of spoilage microbes. In fact, no pathogens are known to survive in beer with normal alcohol content, pH, bitterness, and carbonation (Donald 2014).

Bacteria of greatest significance in this environment are non-sporeformers. However, spore-forming bacteria, such as *Clostridium* species, may be involved in the spoilage of brewery by-products, such as spent grain. Non-spore-forming bacteria that are found in breweries may contribute to a wide variety of problems in wort, including pH elevation, acidification, acetification, incomplete fermentation, ropiness, and slow runoff time. Such infection may also be directly or indirectly responsible for various off-odors and biological hazes in finished beer (Kinney 2013).

Lactobacillus brevis is usually regarded as the most common and troublesome genus of bacteria in the brewery because its species represents a potential spoilage hazard at the various stages of production, including finished beer. This microbe is classified as an obligate beer spoiler since it deteriorates it by haze, acid formation, and off-flavors.

Acetobacter lovaniensis is an acidic bacterium that occurs in early stages of biofilm formation. It is commonly found on fruits and flowers and in vinegars and fermented foods and drinks. This microbe can proliferate and cause an acidic off-flavor in wort and beer with high oxygen content.

Pseudomonas aeruginosa is usually introduced to the brewing environment through the water supply. It is slime forming and contributes to biofilm formation.

Wickerhamomyces anomalus is associated with spoilage of grain products and can be present among various production steps of the production of alcoholic beverages such as beer. It is involved in biofilm formation.

Other genera are less versatile under brewery conditions; therefore, their spoilage potential is more limiting. However, enterobacteria may have an impact on the fermentation, flavor, and aroma of beer. The most common techniques used for detecting and differentiating the various brew contaminants are selective and differential culture media (either alone or in combination with centrifugation) or millipore filtration, depending on the expected cell density and various serologic techniques and impedance measurement.

Construction Considerations

Sanitation is enhanced through the design and construction of insect and rodent proof construction materials such as concrete, brick, and tile. The floors should be dense, impervious, readily cleanable, and durable. Preferred flooring materials are acid-resistant concrete or epoxies. A sufficient number of grains should be provided to convey liquid from all rooms. Rounded gutters are preferred to right-angle corners and should contain grid covers of corrosion resistant materials. They should be screened to exclude rodent entry.

Double or hollow walls and ceilings should be avoided or all areas should be tightly sealed. Insulation should be completely sealed into walls or ceilings. Unnecessary recesses or ledges should be avoided because they trap dirt and debris. If ledges are necessary, they should be beveled so that dirt and wash water will slide off easily. Covered or shatterproof lights are needed to reduce physical hazards.

Equipment must be designed to protect the product from contamination. Outside fermentation tanks must be constructed to protect against insect entry, damp, and dusty weather. It is acceptable to use food-grade plastic or stainless steel tanks, but the use of threaded fittings is discouraged since they can cause a cleaning problem.

Control of Microbial Infection

Contamination may be controlled through removing excess soil and microorganisms that cause off-flavors. Although beer will self-sterilize in

5–7 days, undesirable bacteria, yeasts, and molds grow rapidly in freshly cooled wort that is contaminated through poor sanitation. Therefore, it is necessary to clean and sanitize the brewery equipment that processes wort. Clean kettles and coolers are known to transfer heat faster because 1 mm (0.04 in.) of soil on the inner cooling surface is equivalent to 150 mm (6 in.) of steel. It is hypothesized that 1 mm (0.04 in.) of soil could have a similar insulating effect. Furthermore, high-speed equipment, such as fillers, cappers, casers, and keggers, performs more effectively if kept clean. Used kegs should be washed upon return to the brewery and cleaned internally with a caustic and an acid and steam before being rinsed and refilled. A keg line monitor system to validate the cleaning process can be incorporated.

The most effective means of preventing spoilage of beverage products is to control infection by developing and maintaining a comprehensive cleaning and sanitizing program. A program can be developed by sanitation personnel or with the help of a reliable sanitation consulting group or a dependable cleaning compound and sanitizer supply firm. Discussions in other chapters relating to equipment and facility design, cleaning equipment, cleaning compounds, and sanitizers should be reviewed to determine guidelines for the implementation of a sanitation program. It is especially important to review discussion related to CIP equipment (Chap. 11). These systems are quite adaptable to cleaning beverage equipment, and the trend in the industry is toward automation through this concept.

Fermentation facilities such as breweries require sterile air for the production of starter cultures or the maintenance of sterile conditions within a storage tank. The optimum practice has been recognized as coarse-filtering air with a coarse depth or pleated filter to remove the bulk of contaminants, followed by filtering with a 0.2 μm (0.00008 in.) membrane or sterile filter. Thus, the sterile air can blanket the stored product by creating a positive pressure within the storage vessel. An inert gas can be substituted for air to reduce oxidation. Blanketing a storage tank is an easy way to create a sterile environment, especially with large storage tanks.

The control of microorganisms may be enhanced through ultraviolet (UV) light to reduce the airborne microbes, eliminate pests, and treat water. Several breweries have implemented UV light in water treatment as it is the main ingredient of the final product and allows for residue-free water that will not affect the chemistry of beverage manufacture, as do most sanitizer residues. This treatment does not have a detrimental effect on water since UV light is a nonionizing and nonresidual disinfectant.

This disinfectant functions through irreparably damaging microbial DNA, which absorbs these high-energy wavelengths. The disruption of DNA prevents the microorganism from repair and replication. The violet-colored light of the nearby visible wavelength region can be generated by the UV lamps, which are beneficial in alerting personnel to the presence of UV light but can ultimately diminish its effectiveness. In some applications, UV light is cost effective and can be easily incorporated into an existing sanitation program. The nonselective nature of UV light permits the nonresidual cleaning of air, water, packages, and some foods.

Different microorganisms can contaminate (from maturation to storage stage) barley designated for malting. Fungi that cause a serious plant disease are known as *Fusarium* (FHB). Mycotoxins may occur in infected grain, and the consumption of these mycotoxins may lead to health complications for humans and animals.

The use of FHB-infected grain in the malting and brewing industry has posed a challenge and compromised product acceptability. The growth of *Fusarium* during the malting process results in mycotoxin production and impaired malting characteristics of barley. FHB-infected grain possesses reduced kernel plumpness with increased wort soluble nitrogen and free amino nitrogen with less acceptable wort color.

The incorporation of *Fusarium*-infected barley, for malting, may cause mycotoxin production and decreased product acceptability. Physical methods for the treatment of this condition may prevent safety and quality defects and permit the use of otherwise acceptable barley. Through previous evaluation of hot water and electronic beam irradiation for the reduction of *Fusarium* infection in malting barley, it was found that at higher water temperatures, *Fusarium* was nearly eliminated, but germination was also reduced severely. Electron beam irradiation of *Fusarium*-infected barley reduced FHB infection at doses of >4 kGy. Thus, it appears that physical methods may have potential for the treatment of FHB-infected malting barley.

Cleaning Compounds

Efficient cleaning can be attained only if the proper cleaning compound is incorporated. Spray cleaning is most effective with the incorporation of a properly blended cleaner having specific cleaning properties for the soil that exists. The cleaning compound should be low foaming because foam reduces velocity during circulation and tends to prevent contact of the solution with part of the surface. The appropriate cleaning compound will prevent "Beerstone" formations. It should also be formulated to prevent metal attack and must be easily rinsed to avoid the uptake of objectionable flavors by the beer. (Other information on cleaning compound selection, application, and safety during use is discussed in Chap. 9.)

Sanitizers

Sanitizers such as chlorine, iodine, or an acid anionic surfactant should be incorporated with the final rinse in fermenters, cold wort lines, and coolers. Because water can contain viable microorganisms exceeding 100/mL (0.21 pints), it is possible to have a sterile surface after cleaning but bacteria or yeasts deposited on the equipment surface after the final rinse.

According to Kinney (2013), silver has become recognized as an effective antimicrobial agent for beverage (including breweries) plants. The "smart" release of silver ions through manufacturing innovations and new developments with polymers that accommodate this technology for silver have made it possible to include this technique in the delivery and dispensing systems of beverage (also foods) products. Tubing and fittings are being manufactured with a polyvinyl chloride (PVC)-free polymer that aids in the elimination of potentially harmful health and environmental effects of PVC coextruded with a silver lining constructed with this "smart" technology. Silver lined tubing is effective against the main beverage spoiling bacteria. (Additional information related to sanitizers and their application is given in Chap. 10.)

Heat Pasteurization

Heat pasteurization is still the most common method for microbial control in beverage plants, such as those producing packaged beer. Although the energy costs are high, it is, nevertheless, a convenient method. Alternative procedures have been investigated because of energy costs and the adverse effect of heat on the flavor of drinks such as beer. Such alternative procedures, frequently called cold pasteurization, include the use of chemical compounds, such as propyl gallate, as well as millipore filtration, either followed by aseptic packaging or used in conjunction with other chemical treatments. It appears that the practice of cold pasteurization will increase in the future.

Official approval of chemical compounds is subject to change as new technology and information related to safety become available. The bacterial count of pitching yeast may be reduced by treatment with dilute acids such as phosphoric, sulfuric, and tartaric acid. Acid treatment can reduce bacterial infection, but it has an adverse effect on the yeast culture, and retarded fermentations can occur in the first few cycles after treatment. Sulfur dioxide (SO_2) has been used in the past for control of wort bacteria.

Aseptic Filling

Aseptic filling is considered to be a non-pasteurization process that involves ultrafiltration techniques to remove the spoilage organisms from beer before packaging. Because ultrafiltration occurs before packaging, spoilage microorganisms can enter the product. The comments that follow are provided to ensure deliverance of a high level of sanitation in aseptic packaging.

Hygiene Practices

It is important to have closed filling rooms with a positive pressure of filtered air. The workers' apparel should always be clean, and before workers enter a room, their hands should be washed with a sanitizing soap. A conveyor lubricant system that reduces microorganisms should also be utilized.

The interior of the filler should be cleaned and sanitized daily, utilizing recirculating CIP equipment. The exterior of the filler, conveyors, associated equipment, floors, and walls should be foamed or gelled, then sanitized daily. This process should provide a residual antimicrobial activity because a detergent or sanitizer remains on a surface after its application and subsequent drying, preventing recontamination of the sanitized surfaces.

There should be a regular program of surface and air monitoring for bacteria, yeasts, and molds in the filling area. HACCP utilizes chemical and microbial monitoring to guarantee safe food production. These monitorings are always compared against reference standards. Microbial monitoring in aseptic beer filling needs to be developed within that aspect of the beer industry. A baseline of data should be gathered and statistically evaluated against finished product quality.

Bottle Cleaning

When returned, the bottler should examine the empty bottles. New bottles should also be inspected at the bottling plant to detect any obvious contamination. All new and used bottles should be mechanically washed immediately before filling with a washer that applies a heavy spray of caustic solution, both internally and externally, with subsequent rinsing.

A critical phase of bottle washing involves the transition from the caustic to rinsing phase. Several bottle washers contain a prerinse module to reduce the alkaline carryover. With this arrangement, the prerinse accumulates alkalinity and is either discharged every day or diluted with freshwater or water from the first rinse. Also, alkalinity may be reduced through direct addition of an acid such as sulfuric acid to the second rinse to neutralize part of the existing alkalinity. Sulfuric acid provides the best alternative because of its cost, activity, and environmental compatibility. Phosphoric acid forms a precipitate with calcium and hydrochloric acid and is corrosive (Stanga 2010). The spray and rinse temperature should be 60–70 °C (140–159 °F). Casein and starch from glue act as generic dispersants and help sequestrants to prevent deposits. In this application, sequestrants keep the inert matter and salts dispersed to enhance the detergent activity on molds, insect eggs, and larvae and to detach labels. Sodium gluconate and gluconic acid are commonly utilized in bottle washing. Also, sorbitol is considered an effective cleaner for the removal of labels in bottle washing. The second rinse should be accompanied by the addition of a disinfectant to achieve a bacteriostatic effect.

Chlorination of the final rinse with up to 0.5 PPM concentration can be incorporated without affecting flavor. This operation is not necessary unless the water characteristics dictate this purification technique.

Bottle cleaning and disinfection for alcoholic and nonalcoholic beverages involve (1) label detachment, (2) interior soil removal, (3) scale removal (i.e., calcium carbonate and aluminate) through rinsing and caustic baths, and (4) sanitizing. Centralized high-pressure, low-volume cleaning equipment has improved the efficacy of bottle cleaning of beer and other beverages. (Chapter 11 describes the principle and capabilities of this equipment.) Tenacious soil must be removed from very difficult-to-reach areas such as conveyors, bottle fillers, cappers, and casers. Temperature is an important factor in label removal since detachment is not accomplished below 60 °C (140 °F) unless the label material is made of weak paper.

Label removal effectiveness is affected by:

1. Wettability: dependent upon label finish and time required for penetration by a 1% solution.
2. Inorganic coating: coating integrity is determined by sulfate determination after stirring in a 1% caustic soda solution at 60 °C (140 °F) for 5 min.
3. Dyes, wax, and varnish: affect wettability because of their detachment resistance.
4. Resistance of paper structure: dependent upon paper thickness and cellulose properties.

Sanitation in Storage Areas

In addition to suggestions provided for storage areas of other food facilities discussed in this book, it is appropriate to recognize the need for proper storage of materials such as grain, sugar, and other edible dry products. Screw conveyors should be cleaned on a schedule basis. This is especially true for the dead ends of conveyors where dormant residues can accumulate. The ends and junctions of conveyors should be cleaned at least once a week. The free-flowing section of a conveyor should be equipped with hinged covers for easy cleaning and inspection. After conveyors have been cleaned thoroughly, they should be fumigated with a nonresidual fumigant. Empty bins should be thoroughly swept (and preferably vacuumed) prior to fumigation. Regular checks should be made of material cleaned out for possible infestation. A description of cleaning equipment is discussed in Chap. 11. Chapter 13 provides a detailed discussion of recommended pest control measures that may apply to storage areas at beverage plants.

Brewing Area Sanitation

Spray cleaning is faster and more dependable than manual cleaning and can reduce downtime. Although unheated water can be used, a water temperature of up to 45 °C (116 °F) can increase the chemical reaction of the cleaning compound with the soil. If glass-lined tanks are used, the maximum water temperature should be 28.5 °C (80 °F) to reduce damage due to sudden temperature fluctuation. Temperatures above 45 °C (116 °F) should be discouraged because of condensation problems and increased refrigeration requirements. In fact, it is advisable to lock in specific temperature or high-temperature cutoff switches to control water temperature. Caustic soda cleaning compounds should not be used because they attack soldered ends. Scale formation in aluminum vessels can be removed with 10% nitric acid, applied as a paste mixed with kaolin.

Initial and maintenance costs of hoses and fittings suggest the viability of stainless steel lines (even though stainless steel is expensive). Circulation cleaning of product-in and product-out lines can be accomplished by the use of U-type fittings to connect the tank valve to both lines. Industrial spray nozzles for equipment cleaning can be positioned to clean areas such as vapor stacks on kettles and strainer troughs in hop strainers and to provide continuous cleaning for conveyor belts. The brewing area should be cleaned at least once per week, and debris and other soils should be removed daily.

Beerstone (a primarily organic matter in a matrix of calcium oxalate) is one of the most difficult beverage soils to remove. This deposit is most effectively removed with extensive scrubbing and use of a strong chelating agent and alkaline cleaning compound.

Cleaning of beer residue is relatively easy in bright beer tanks (BBT); however, cleaning of the fermenting beer ring is difficult because of hop resins, proteins, polyphenols, calcium, oxalate, and polysaccharide linked and polymerized in the course of wort fermentation. Kettle and fermenting tanks need caustic cleaning complemented with hydrogen peroxide. However, soil in BBTs is lighter and can be removed with an acid detergent.

Beer Pasteurization

Most brewers pasteurize their beer to maintain a stable condition, flavor, and smoothness. Certain brewers have incorporated sterile filtration as a substitute for sterilization. If filtration is incorporated, the filters should be replaced every other week to reduce the risk of microbes penetrating the series of filters. In a sanitary operation, sterile filtration can be effective.

Pasteurization during containerizing is practiced by much of the brewing industry because it can protect the beer against contamination after packaging. Overheating during pasteurization, however, can have an adverse effect on flavor and can cause haze. Therefore, it is essential to subject the beer to the minimum time and temperature for effective microbial destruction. Most of the brewing industry now has conveyor systems for a pasteurization cycle of approximately 45 min. During pasteurization, the beer temperature is gradually raised from 1° or 2 °C (34° or 35 °F) to 61° (144 °F) up to 63 °C (146 °F), with subsequent cooling to an ambient temperature at the end of the cycle. The moving belt speed can determine the length of exposure time in the pasteurization environment. The total air content of packaged beer should not exceed 1 mL/220 mL (0.002 pints/0.46 pints) of beer.

Haze may develop in beer. A nonbiological haze may form from the slow precipitation of products with unstable solubility—a condition caused or accelerated by oxidation. A biological haze may be caused by the growth of bacteria or yeasts. A sufficient period in the cold conditioning tank and fine filtration will minimize the chances of nonbiological haze. The exclusion of air in the beer container, as well as the selection of suitable container materials, will also minimize the chances of nonbiological haze occurrence. Other hazes have been traced to metallic influences, especially that of tin. Beer haze in brightly colored bottles due to bacteria or yeast growth suggests either imperfect filtration or subsequent infection. A bacterial or yeast haze can be attributable to lack of proper sanitation in the plant or unclean storage containers or filters.

Cold pasteurization provides another production technique. The bacterial count of yeast added to wort (pitching yeast) can be reduced by treatment with diluted acids such as phosphoric, sulfuric, and tartaric. However, acid treatment can affect yeast cultures through retarded fermentations.

Cleaning of Air-Conditioning Units

The following procedure for cleaning air-conditioning units is suggested:

1. Clean air-conditioning units every 6 months. Insert a ball spray through a special opening above the coils on top of the air-conditioning units.
2. Run water for 10 min to flush the unit.
3. Run a hypochlorite solution (200 PPM) at 40 °C (102 °F) for 5 min.
4. Let the unit soak for 5 min.
5. Rinse the unit with warm water for 10 min.
6. Check the unit and clean the pan bottom.

Water Conservation in Brewery Sanitation

Water usage during cleaning can be reduced with a wash-rinse cycle sometimes called the *slop cycle*. This cycle includes a prerinse, in which a cleaning solution is pumped through a spray device for 20 s, with 1 min permitted for chemical action and a subsequent burst rinse with water—the same procedure as used in most home dishwashers. Reuse of cleaning solutions is practical and economical. The length of reuse can be increased if the solution tank has a top overflow to skim off floating soil and a drain valve to permit bottom draining of the heavy soil. Furthermore, the final rinse water can be salvaged for the prerinse on the next tank to be cleaned. This technique can reduce water and sewage treatment costs in areas where both water supply and sewage are metered.

Winery Sanitation

It is essential for a winery to be maintained in a sanitary condition. Dirty storage conditions can cause off-odors and off-flavors since wine absorbs various odors.

Soil contaminants that affect the taste, appearance, and perishability of wine should be removed. Included among the contaminants are the reddish tartrate deposits that form or build up on tank interiors as a result of fermentation. Other tenacious soils should be cleaned from the surfaces of processing equipment to reduce microbial growth throughout the winery. In general, the more sanitary a winery is, the smaller the quantities of SO_2 that must be added to the wine at the end of the winemaking process. Although SO_2 has been used to control microbial growth, use of this compound has been discouraged and may be discontinued in the future. As a complement to SO_2, sorbic acid is effective in the prevention of fermentation of sweet wines if there is a low initial yeast count and free SO_2 is still active to prevent bacteria from destroying the sorbic acid. Rigid sanitation is a viable alternative strategy for microbial destruction.

Because rigid sanitation will not destroy all microorganisms, as sterilization, the reduction of viable cell number to an acceptable level may be attained. Effective sanitation accomplishes another important goal—the elimination of hospitable environments for growth.

Although the requirements for sanitation increase during the winemaking process and peak at bottling time, it is important to recognize that the vineyard tools and harvesting equipment must be washed to remove dirt, pomace, soil, and leaves. Destemmers, crushers, and grape processing and bulk storage areas require a brush, detergent, and water. Harvester heads, pipes, hoses, pumps, faucets, spigots, and anything else coming in direct contact with the juice or wine will require the five cleaning steps discussed early in this chapter. The same steps apply to the bottling line, but additional control and checking are necessary to reduce the microbial load of the wine.

Basic Sanitation Principles

Basic sanitation principles supported by the Wine Institute include:

1. A winery should be kept free of refuse both inside and outside.
2. Equipment should be arranged in an orderly way and the work areas kept free of clutter.
3. The entire winery should be cleaned on a regularly scheduled basis.

4. The winery should be protected against harmful bacteria, yeasts, molds, insects, and rodents with necessary measures to prevent a recurrence of these pests in the future.
5. The winery premises, equipment, and cooperage should be inspected at least once each month.

Cleaning Compounds

Several cleaning compounds are available for use in wineries. The sanitation operation is more successful if an appropriate cleaning compound is utilized. The selected cleaning compound should be easy to rinse away. Cleaning compounds with artificial odors should be avoided to decrease the possibility of adverse effects on product quality. Sodium phosphate is an effective winery cleaning compound in addition to other phosphate-based compounds. Chlorinated trisodium phosphate is considered a "workhorse cleaning agent" that provides an appropriate defense against contamination. Sodium hypochlorite is inexpensive and can serve as a potent disinfectant but lacks utility because it is a powerful oxidizer and does not rinse away easily in cold water.

Cleaning Aids

An adjustable nozzle attached to a hose is a primary piece of cleaning equipment in most small wineries. The nozzle should provide several spray patterns including a strong, high-velocity stream. Long-handled brushes are inexpensive and convenient for scrubbing small tanks, containers, and most winemaking equipment.

Water Quality

Water used in a winery must have certain chemical and microbial properties. A low pH is inimical to steel and other surfaces, and a high pH will favor calcium precipitation. The biochemical oxygen demand (BOD) should be less than 3 mg/L (0.0001 oz. /1.05 quarts). Because water can be a potential carrier of molds, yeasts, and acetic or lactic acid bacteria, pure water should be used.

Winery design and layout should incorporate hygienic principles. Floors must be easy to clean and have nonslip, sloped surfaces. Walls and ceilings should be impervious and easily cleaned. Sanitation in a winery can be enhanced through proper location of equipment to reduce the creation of corners and crevices that are difficult to clean. As with other food manufacturing facilities, equipment should be constructed with sanitary features that enhance effective cleaning.

Cleaning Floors and Walls

Although a winery may be somewhat seasonal in operation, year-round sanitation is necessary. A combination of wet and dry cleaning is usually most appropriate. A heavy-duty, wet-dry vacuum cleaner can be effectively incorporated in cleaning. Floors should be cleaned at least once a week by dry or wet methods, depending on the nature of the soil. To facilitate cleaning, floors should be constructed of concrete and sloped and should contain trench drains. Spilled wine, especially any that has spoiled, should be washed away immediately.

It is necessary to remove as much of the visible debris as possible before the use of cleaners. This task is accomplished manually through mechanized cleaning (spray balls, etc.). Spray applications should be directed at an angle to the surface being cleaned.

The area should be mechanically scrubbed, washed with lime or a strong hypochlorite solution, and rinsed with water. Floors and the outside of the wood cooperage (if applicable) should be periodically washed and disinfected with a dilute hypochlorite solution. When dry cleaning is possible, the humidity of the winery can be kept lower than when wet cleaning is practiced, with resultant reduced mold growth in areas where wood is present. Tank tops, overhead platforms, and ramps can be vacuumed, cleaned, or washed, taking precautions that no water gets into the wine. The walls should be washed with a warm alkaline solution, such as a strong solution of a mixture of soda ash and caustic soda, followed by rinsing with water and spraying with a hypochlorite solution containing 500 mg/L (0.017 oz./1.05 quarts) of available chlorine. All free chlorine should be removed through washing.

Equipment Cleaning and Sanitizing

Improper equipment cleaning is one of the most viable sources of contamination. Crushers, must pumps and lines, presses, filters, hoses, pipes, and tank cars are all difficult to clean completely. Less complicated equipment, such as wine thieves, hydrometer cylinders, buckets, and shovels, can also be difficult to clean. This equipment should be dismantled as much as possible, thoroughly washed with water and a phosphate or carbonate cleaner for nonmetallic surfaces and caustic soda or equivalent for cleaning metal equipment, and sanitized with *hypochlorite* or an *iodophor* if the material being cleaned is adversely affected by chlorine.

Ozone has gained popularity as a sanitizer for winery equipment. It is an unstable gas and readily reacts with organic substances and does not leave a chemical residue. Under ambient conditions, it has a half-life of 10–20 min. This sanitizer has the ability to readily oxidize microorganisms in solution. Ozone has an advantage over other sanitizers because it requires no storage or special handling or mixing considerations. When ozone is generated, it is important that the concentration and flow rates be verified and checked periodically.

Enzymes are useful as cleaning agents because they can hydrolyze proteins, fats, and pectins. They are currently used in enology because their maximum efficiency is at nearly a neutral pH. Where possible, circulating the cleaning solutions is recommended. Hoses, after cleaning and rinsing, should be placed in sloping racks, instead of on the floor, to facilitate drainage and drying. Thorough cleaning and sanitizing are essential for equipment that has been in contact with spoiled or contaminated wine.

During the harvest season, conveyors, crushers, and must lines should be kept clean. They should not be permitted to stand with must in them for more than 2 h. After use for 2 days, they should be washed, drained, and thoroughly flushed with water before reuse.

Bottling Area Cleaning

Effective cleaning of the bottling area is essential to reduce bacterial or metallic contamination. This area is usually observed most closely by public health agencies. To facilitate effective sanitation in this area, the room should be well lighted and ventilated and should have glazed tile walls and epoxy finished floors. Ample space between equipment is essential to facilitate easy cleaning, and the equipment should be easy to disassemble. All pumps, pipes, and pasteurizers should be constructed of stainless steel. Although bottles should be sterile through ethylene dioxide treatment prior to arrival at a winery, they should be flushed with nitrogen and rinsed before filling. Uncleaned bottles should be cleaned and sterilized before use by soaking for 2 h in low pH 1% SO_2 and a little glycerol, then rinsed with water. Corks can be gamma radiation sterilized to prevent off-odors occurring because of mold growth.

Pomace Disposal

It is essential to dispose of the pomace as rapidly as possible after pressing. This material must not stand in or close to the fermentation room because it rapidly acetifies, and the fruit flies carry acetic acid bacteria from the pomace pile to clean fermenting vats. Pomace should be further processed or scattered as a thin layer on fields, where it dries quickly and does not become a serious breeding ground for fruit flies.

Cleaning of Used Cooperage

Alkaline solutions (soaking with 1% sodium carbonate) are most effective in removing tannins from new barrels. If further treatment is necessary, steam and several rinsings should be applied.

Other viable cleaning compounds are sodium ortho- and metasilicates (Na_2SiO_3) that are less caustic and less corrosive than sodium hydroxide (NaOH) with superior detergent properties. A lighter organic load permits the application of milder alkalies, such as sodium carbonate (soda ash) or trisodium phosphate. Sodium carbonate (Na_2CO_3) is an inexpensive, frequently used, cleaning compound. However, it contributes to precipitate formation in hard water.

Polyphosphates included in cleaning compound formulations chelate calcium and magnesium and prevent precipitation. Examples are sodium tetraphosphate (Quadrofos) and sodium hexametaphosphate (Calgon). The amount to be included in the formulation depends on water hardness. Acid cleaners are formulated in specialized detergent formulations (approximately 0.5%) to reduce mineral deposits and to soften water. Phosphoric acid is preferred because of its low corrosiveness and compatibility with non-ionic wetting agents.

Past practices have involved washing empty containers with water and spraying with a hot (50 °C or 121 °F) 20% solution of a mixture of 90% soda ash and 10% caustic soda or caustic potash (KOH). Both NaOH and KOH have effective detergent properties and are strongly antimicrobial against viable cells, spores, and bacteriophage. After subsequent washing with hot (50 °C or 122 °F) water, containers should be sprayed with a chlorine sanitizer solution containing 400 PPM of available chlorine. A cold-water rinse should follow, with subsequent drainage and drying using a dry-wet vacuum. If mold is present, it should be scraped off because it cannot be removed by washing. Further precautions include washing with a quaternary ammonium compound. Paints containing copper-8-quinolate can also control mold growth. Burning a sulfur wick in the tank or adding SO_2 from a cylinder of gas is also effective. Before use, the tanks should be rewashed, and the cooperage should be inspected visually and by smelling before being filled. A warm 5% soda ash concentration is too high, and if exposed too long, the wood can deteriorate. The outside surface of wooden containers should be washed with a dilute solution once a year.

Propylene glycol can be applied to discourage mold growth on the surface of the tanks. Stainless steel tanks should be cleaned with a 400 PPM or less concentrated solution to prevent mold growth.

Removal of Tartrate Deposits

It is necessary to remove tartrate deposits to smooth the inner surface, which becomes very rough. Scraping is labor intensive and may injure the wood. Installation of a circular spray head inside the tank can help remove tartrates. Soaking with 1 kg (2.2 lbs.) of soda ash and caustic soda in 100 L (26 gallons) of water will also aid in the removal of tartrates.

Storing Empty Containers

Concrete tanks should be left open and kept dry when not in use. Before reuse, they should be inspected and cleaned. An example of fermentation tanks is shown in Fig. 20.1. Open wooden fermenters are sometimes painted with a lime

paste when not in use, but this surface is difficult to remove. A better approach is to clean the fermenters thoroughly with an alkaline solution, followed by a chlorine solution. They can then be filled with water and approximately 1.6 kg (3.6 lbs.) of unslaked lime per 1000 L (260 gallons) of water added.

A caustic soda solution dissolves potassium tartrate and bitartrate which are the major components of the deposit on storage tank walls. Stored empty barrels can be sulfured by a sulfur wick or by introducing SO_2. However, the use of sulfur wicks can be disadvantageous because sulfur may sublime into the walls of the container, and pieces of elemental sulfur from the wick may fall to the bottom of the cask. If containers with elemental sulfur are used, hydrogen sulfide might be reduced.

Other Cleaning and Sanitary Practices

Fillers, bottling lines, and other packaging equipment can be cleaned with CIP systems. A chlorinated alkaline cleaning compound can clean,

Fig. 20.1 Red wine fermentation tanks (Courtesy of Bruce Zoecklein, Virginia Polytechnic Institute & State University, Blacksburg, Virginia)

sanitize, and deodorize in one operation if the soil is light. However, the presence of organic matter can negate the effect of the chlorine sanitizer because chlorine will react with the organic matter in the soil. The addition of approximately 7 g (0.28 oz.) of a sanitizer with 4.25% available chlorine per liter (1.05 quarts) of water should provide approximately 200 PPM of available chlorine for rapid destruction of microorganisms.

Heat is the safest sterilization process available, but it does not yield high quality premium wines. Also, it can potentially transform some taste elements. Bottle sterilization is the best method for sterilizing sparkling wine. Bottling line sterilization can be accomplished (albeit expensively) with steam or hot water. Where hot water is employed to sanitize lines, a minimum temperature of 82 °C (180 °F) for at least 20 min should be attained. The temperature should be monitored at the farthest point from the steam source (i.e., fill spouts, end of the line, etc.) Ultraviolet light is effective against microbes, but it has low penetrative capabilities, and a thin film provides a barrier between radiation and the microbes. Ozone can be used to sanitize in cold-water recirculation.

Sterile Filtration

Sterilization by filtration is attained through use of sterile filter pads, or better, with membranes. Diatomaceous earth filtration reduces yeasts but will not eliminate bacteria. This step is followed by membrane filtration.

Reinfection

Any efforts in sterile filtration are ruined by post-filtration infections if the entire bottling line is not sterile. The filter, as well as the bottling line, should be sterilized before admitting the wine. The most effective method of sterilization is to use a steam generator hooked to the filter, which is hooked to the bottling line. A slow flow of low-pressure steam is run for 30 min through the entire system. The steaming is followed by cold water to cool off the machinery before allowing

the wine to enter. Steam may not be available or may injure some equipment, such as plastic filter plates. Some parts of the bottling line, such as the corker, are more difficult to sterilize. The corker jaws or diaphragm should be sterilized with alcohol. Membrane filters may be sterilized with water at 90 °C (194 °F).

Corks

Modern cork suppliers provide sterile corks. In case of doubt, corks should be dipped before use in a SO_2 solution.

Bottles

Bottles can be recontaminated from dust and cardboard. A rinsing and sterilization station must be provided for a solution of SO_2 at 500 PPM. Sterile water (obtained after cooling the sterile filter) is used to wash off excess SO_2 solution from the bottle. A dispenser for SO_2 can be set in line off of the main water supply using a medicator or other similar device. Iodophors are used frequently for bottling sanitation, followed by a cold-water rinse.

Barrels

Spent empty barrels are difficult to maintain. Approximately 12 liters (3.1 gallons) of wine soak into the wood surfaces of a new barrel the first time is filled. Wood is a living substance which releases aroma, oxygen, and tannins from its spongy structure. Organic deposits and yeasts can accumulate during aging in wooden barrels. *Brettanomyces spp.* is among the most difficult microbes to control when wine is aged in wooden barrels. Four by-products from the growth of *Brettanomyces* are esterase, volatile esters, volatile phenols, and tetrahydropyridines.

The spongy organic structure of wood is sensitive to strong chemicals. Caustic cleaners hydrolyze cellulose causing a rotting effect, extract aromatic compounds, and convert wood into an inert container with no beneficial flavor contribution to wine. Therefore, cleaning should involve noncaustic salts such as sodium tripolyphosphate and the monophosphates which are effective in cleaning and buffering pH.

When barrels are stored empty, the wine soaked into the wood acidifies, turns to vinegar, and becomes contaminated with acidophilic bacteria. Although soda ash can be used for cleaning contaminated barrels, sterilization is practically impossible. Thus, contaminated barrels with a vinegar smell should be discarded.

Empty barrels may be safely stored for several weeks if they are gassed with sulfur dioxide and kept tightly sealed. Barrels gassed with sulfur dioxide should be washed with clean water before they are filled. Another limitation of using spent barrels is that the wood in an empty barrel becomes dry and the staves shrink. As the wood shrinks, the hoops become loose and barrels lose their shape and leak.

Control

Good bottling practices require checking sanitation standards. Special kits are available to evaluate the level of sanitation through a count of the number of receivable yeasts or spoilage bacteria left in the wine after filling.

Pest Control

Fruit flies are especially attracted to fermenting musts. A large proportion of the fly population is brought to the winery from the vineyard. The most effective control measures are the prompt crushing of grapes after picking, removing of all dropped and culled fruit from the winery, disposing of all organic wastes, using repellent insecticides around the vineyard, washing of all containers and trucks after handling grapes, and using attractant insecticides on dumps. Maximum fly activity occurs in the range of 23.5–27.0 °C (74–80 °F) in low light intensity and low wind velocity. Fans blow air out of winery entrances. Mesh screens and air curtains are also helpful.

Insecticides that kill fruit flies are available, but the heavy fly population in adjacent unsprayed areas makes the effectiveness of this practice questionable. If insecticides are used, FDA tolerances must be observed. (Chapter 13 provides additional information related to fly, rodent, and bird control.)

Sanitation Monitoring

The most common method of sanitation evaluation is sensory. Visual appearance and smell are assessed and, sometimes, touch to determine whether the surface feels clean. A slippery surface suggests inadequate cleaning and/or rinsing. In some instances, microbial sampling should be conducted as a means of verification.

Each microbial technique has limitations, such as surface characteristics, definition of area to be sampled, amount of pressure applied to the surface, and time of application. Furthermore, cotton swabs will not recover all microbes. Thus, standardization of sampling procedures will improve the success of sanitation monitoring.

Distillery Sanitation

As with breweries and wineries, the commonly recognized microorganisms are normally of minimal concern in distilleries because of the nature of the raw materials, processing techniques, and high alcohol concentration. A possible safety exception is the potential for contamination by significant levels of toxic metabolic products. Control of raw materials is essential because a contaminated finished product cannot be effectively detoxified. Yield and product quality are compromised when sanitary conditions are not maintained.

Reduction of Physical Contamination

To practice effective sanitation, corn and other grains are inspected upon arrival at the plant. Insects are the major concern at this stage, because a contaminated grain shipment can infect the storage silos, as well as the entire plant. The most common insect pests for grain are flour beetles and weevils. Off-odors are also important to detect at this point, because many will persist through the fermentation and become detectable in the final product. Grain storage silos are routinely emptied 2–4 times a year, sprayed with high-pressure hoses, and allowed to air-dry. The area surrounding the outside of the silos should be kept clean from grain dust by washing the area with water and by periodic insecticide spraying.

When the grain enters the plant by auger conveyors, it should be sifted on shaking screens to remove any corncobs, debris, or insects that may have found their way into the silos. The mill room should be washed down with water to reduce grain dust, and approximately every 2–6 months, it should be heated to up to 55 °C (131 °F) for 30 min to kill any insects that may be present.

Reduction of Microbial Contamination

Bacterial and wild yeast contamination of the fermentation is the most important sanitation aspect to control in whiskey production. The source of most of the contamination is the malted barley. Malt routinely has bacterial counts of 2×10^5 to 5×10^8. The malt is added at 60–63 °C (140–145 °F). Thus, many of these microbes survive to propagate during the fermentation. Bacterial levels of the corn and other grains added before the cooking process or at temperatures greater than 88 °C (190 °F) are not given much attention because they are killed at these high temperatures.

The most common bacterial contaminants include *Lactobacilli*, *Bacilli*, *Pediococci*, *Leuconostoc*, and *Acetobacter*. These microbes propagate at the expense of yeasts. Microbial contamination will cause plant yield to decrease because these bacteria utilize sugar substrates to produce compounds other than alcohol. Many of these bacteria produce acids, mainly lactic and acetic acid, which can alter fermentation conditions, as well as lower product quality. Other compounds that can alter the consistency of the

whiskey, such as esters and aldehydes, are also produced.

The fermentation process is a rather hostile environment for many microbes. Initially, sugar concentrations may exceed 16%, which provides for high osmotic tension. Initial pH is between 5.0 and 5.4 for a sour mash whiskey and, by completion, will be between 4.0 and 4.5. Final alcohol concentration is approximately 9%, and little, if any, oxygen will be present in this high carbon dioxide environment. These conditions severely restrict the types of contaminating organisms that will proliferate.

Contamination of the fermentations should be minimized. Because many bacteria are airborne, dust is kept to a minimum by washing all plant surfaces (walls, floors, etc.). Incoming shipments of malt are probed for the determination of bacterial counts.

Equipment Cleaning

Large fermentation vessels (120,000–180,000 L or 31,000–49,000 gallons) should be cleaned through filling with hot water and detergent, while steam is spread through a CIP sparger in the center of the tank. This process should continue for 30 min at which time the tank is emptied, rinsed with water, and steamed for 2–3 h for sterilization purposes before new mash is pumped into the vessel.

The cooling coils should keep fermentation temperatures below 32 °C (90 °F). Higher temperatures promote yeast cell death and off-flavor production. These coils tend to sustain a buildup of beerstone, which is a hard, rocklike composition of calcium carbonates, phosphates, and, sometimes, sulfates. As this material builds on the coils, heat transfer efficiency is reduced. To combat this problem, every 6 months, the vessels should be filled with a 1% caustic solution (NaOH) and water and allowed to soak for 3 days to remove the buildup.

The cookers, where the mash is prepared, and the beer still, where the finished beer is pumped, tend to get residual grain buildup through continuous operation. To remedy this, a 1% caustic solution should be prepared weekly to wash the cookers, the beer still, and all connecting lines. Some distillers use a 1% solution of acetic acid instead. The caustic solution can be prepared in a large tank and pumped to the beer still, then through the connecting lines, and finally into each of the cookers. These areas should then be rinsed with water to remove caustic residue.

The lines and stainless steel tanks that accommodate alcohol that is distilled from the beer and pumped to receiving tanks and into barrels for maturation should be periodically rinsed. The nature of the product (crystal clear alcohol at 140 proof) alleviates the need for more stringent sanitation.

Water quality for distillery products is important to ensure an acceptable end product. The blending of water for distilled spirits is typically from a chlorinated and carbon-treated well or city water supply that is visually clarified using a depth filter prior to blending with the high-proof spirits. The microbial safety/acceptability is ensured through water chlorination, and polishing only prior to blending is required for visual clarity.

Study Questions

1. What is the significance of the machinery lubrication practices prerequisite program(s)?
2. What temperature is used in hot sanitizing a beverage plant?
3. What is the maximum water temperature for cleaning glass-lined tanks in a brewery?
4. What spray and rinse temperature should be used for bottle washing in a brewery?
5. What are the two major methods of pasteurizing beer?
6. What cleaning solution is recommended for washing empty wine storage containers?
7. How can tartrates be removed in a winery?
8. How are wine fermenters cleaned most effectively?
9. How should the mill room of a distillery be cleaned?
10. How should large fermentation vessels in a distillery be cleaned?
11. Why are soils in beverage plants less difficult to remove than those from most other food plants?
12. Why is rigid control of raw materials so important for beverage plants?
13. What concentration of ozone is required to sanitize a bottled water operation?
14. What is the most troublesome microorganism in breweries?

References

Briseño BA (2015). Moving your lubrication program in the right direction. *Food Qual Saf* 22 (3): 42.
Donald T (2014). The art behind quality craft beer. *Food Qual Saf* 21(4):16.
Hui YH (2015). *Plant sanitation for food processing and food service*, 514. CRC Press: Boca Raton, FL.
Kinney G (2013). Alternative tubing eliminates bacteria in beverage delivery systems. *Food Qual* 20(2): 34.
Remus CA (1991). When a high level of sanitation counts. *Beverage World* March: 63.
Stanga M (2010). *Sanitation cleaning and disinfection the food industry*, 221. Wiley-VCH: Weinheim, Germany.

Foodservice Sanitation

21

Abstract

Increased handling of food is responsible for a more complicated and critical challenge of protecting food from contamination. Microbial contamination causes food spoilage and foodborne illness. Sanitary practices in the receiving, storage, preparation, and serving areas safeguard food. Hygienic handling of food involves equipment and utensils and in a physical facility that has been thoroughly cleaned and sanitized.

Hygienic design for cleanability of the facility and equipment improves sanitation in foodservice establishments. If managed properly, mechanized cleaning can effectively remove contamination from utensils and equipment. A written, supervised, evaluated, and subsequently documented cleaning and sanitizing program enhances the effective management of a foodservice facility.

Keywords

Contamination · Equipment · Foodservice · Illness · Microorganisms · Sanitation

Introduction

Foodservice establishments range from mobile restaurant stands to large industrial cafeterias and multiunit fast-food chains to plush restaurants. As the foodservice industry has grown, methods of food production, processing, distribution, and preparation have changed. The major changes have included increased prepackaged food as partially or fully prepared bulk or preportioned servings and centralized food production.

As food production, handling, and preparation techniques and eating habits change, food is a source for microorganisms that can cause illness. Food handlers can act as vectors of disease and cross-contamination. Handling and modern processing methods increase the journey from the production area to the table; contamination with microorganisms becomes a public health concern. Over 90% of the illnesses are attributable to bacteria that cause foodborne illness. Food is the source of over 50% of the foodborne illnesses,

© Springer International Publishing AG, part of Springer Nature 2018
N.G. Marriott et al., *Principles of Food Sanitation*, Food Science Text Series,
https://doi.org/10.1007/978-3-319-67166-6_21

and over 50% of the foodborne illness outbreaks are from food served in restaurants. The estimated relative risk of foodservice-related illness is 1 in 9,000. A growing number of centralized kitchens compound this problem. Large feeding operations increase the number of people affected by any contamination. Thus, the challenge of protecting food from contamination is more complicated and critical.

The primary goal of a foodservice sanitation program is to protect the consumer from contamination and to reduce its effects. It is difficult to protect food from all contamination because pathogenic microorganisms are in so many locations and on approximately 50% of the people who handle food.

Sanitary Design

Maintaining proper sanitation standards in an improperly designed foodservice facility is difficult. High-touch surfaces require frequent cleaning.

Cleanability

The primary requirement for sanitary design of foodservice facilities and equipment is cleanability. Cleanability of an item or surface suggests exposure for inspection or cleaning without difficulty and with construction for soil removal effectively by normal cleaning methods. Minimal inaccessible locations for soil, pests, and microorganisms to collect will enhance the maintenance of a clean establishment. A facility that is easier to clean is more hygienic.

Design Features

Design for sanitary features should begin during planning of the facility. Although most managers inherit an established foodservice facility, they can improve the environment every time remodeling or renovation takes place or with new equipment purchase.

In most areas, government regulations influence sanitary design of foodservice establishments. Public health, building, and zoning departments may all have the power to regulate construction of a facility. Regulatory agencies frequently provide checklists of features considered desirable or necessary for good sanitation.

Floors, *walls*, and *ceilings* require construction materials that are cleaned and maintained easily and are attractive. The materials used should be inert, durable, resistant to soil absorption, and smoothly surfaced. Evaluation includes absorbency or porosity of floor material. When liquids are absorbed, damaged flooring occurs and microbial growth increases. The incorporation of nonabsorbent floor covering materials in all food preparation and food storage areas merits consideration including carpeting, rugs, or similar materials.

Although flooring material is a critical aspect of sanitation, the way the floor is constructed is also important. Covering at a floor-wall joint facilitates cleaning by preventing accumulation of bits of food that attract insects and rodents. Sealed concrete and terrazzo floors make the floors nonabsorbent and reduce possible health hazards from cement dust.

Many of the same factors apply to the selection of wall and ceiling materials. Ceramic is a popular and satisfactory wall covering for application in most areas. Grouting should be smooth, waterproof, and continuous, without holes to collect soil. Stainless steel, although expensive, is a satisfactory finish because it is resistant to moisture and most soil and is durable. Walls of plaster painted with nontoxic paint or cinder block walls are satisfactory for relatively dry areas if sealed with soil-resistant and glossy paint, epoxy, acrylic enamel, or similar materials. Toxic paints, such as those with a lead base, do not belong in a foodservice facility because flaking and chipping can result in food contamination. Ceilings need smooth, nonabsorbent, and easily cleanable materials. Smoothly sealed plaster, plastic panels, or panels of other materials coated in plastic are all good choices.

When purchasing equipment, the foodservice manager should specify that all acquisitions comply with generally acceptable standards. The following characteristics are examples of sanitary features needed in foodservice equipment:

- Minimal number of parts necessary to perform effectively
- Easy disassembly features for cleaning
- Smooth surfaces free of pits, crevices, ledges, bolts, and rivet heads
- Rounded edges and internal covers with finished smooth surfaces
- Coating materials resistant to cracking and chipping
- Nontoxic and nonabsorbent materials that impart no significant color, odor, or taste to food

Cutting boards are a means of cross-contamination. Wooden cutting boards absorb juices, and plastic boards can harbor microbes in crevices. Knife cuts on plastic surfaces do not heal and offer crevices in which bacteria can evade removal during manual cleaning and contaminate surface sampling. Used foamed polypropylene cutting boards retain high numbers of bacteria, even after a thorough wash. Continued use of polyethylene boards results in numerous knife marks, holes, cracks, and a furry and shaggy appearance, which contribute to bacterial entrapment.

Equipment Arrangement and Installation

Proper equipment design reduces food contamination and makes all areas accessible and cleanable. For example, the soiled-dish table should not be located next to the vegetable preparation sink. Waste processing and the food preparation areas should be located as far apart as possible, and food preparation equipment not placed under an open stairway.

When feasible, mobile equipment permits easy cleaning of walls and floors. Sealed immobile equipment to the wall or adjoining equipment reduces debris accumulation. If sealing is not practical, equipment should be located approximately 0.5 m (20 in) from the wall or adjoining

equipment to permit easy cleaning. Immobile equipment mounted approximately 0.25 m (10 in) from the floor or sealed to a masonry base reduces debris accumulation. If the latter approach is used, a 3–12-cm (1.2–4.4 in) toe space is essential.

A nontoxic sealant is required for sealing equipment to the floor or wall. Wide gaps caused by faulty construction should not be covered with a sealant because such buried mistakes will be exposed ultimately, opening new cracks to soil, insects, and rodents.

Handwashing Facilities

Past observations have revealed that 8.6 hand washes by restaurant staff per employee hour were required to comply with the Food and Drug Administration's (FDA's) Model Food Code. To comply with this requirement, hand hygiene requires a participation of all involved-from entry-level food workers to managers and owners.

Hands are the most viable source of microbial contamination. Therefore, management should provide handwashing facilities in areas such as food preparation areas, locker or dressing rooms, and areas adjacent to toilet rooms. Because employees may be reluctant to walk very far to wash their hands, these facilities should be conveniently located. Handwashing facilities should consist of foot-operated or mechanized handwashing equipment or a bowl equipped with hot and cold water, liquid or powdered soap, and individual towels or other hand-drying devices, such as air dryers. More discussion about handwashing is included in Chap. 6.

Welfare Facilities

Employees need dressing rooms or locker rooms. Street clothes are a viable source of microbial contamination; therefore, employees need uniforms to wear during the production shift. Dressing rooms should be located outside of the area where food is prepared, stored, and served and separated physically from the other areas by a wall or other barrier. The best location for hand-washing facilities is next to the dressing rooms and toilet rooms, with mirrors hung away from the handwashing equipment. The washing facilities and toilet rooms require daily scrubbing. Receptacles for waste materials and should be emptied daily.

Waste Disposal

Disposal of garbage and trash is important in foodservice sanitation because waste products attract pests that can contaminate food, equipment, and utensils. Waste containers should be leak-proof, pest-proof, easily cleaned, durable, and lined with plastic or wet-strength paper bags.

Waste removal from food preparation areas as soon as possible with disposal will prevent the formation of odor and the attraction of pests. Accumulation of waste materials should occur only in waste containers. Waste storage areas should be easily cleaned and pest-proof. If a long holding time is required, refrigerated indoor storage is necessary with effective cleaning and protection against pests. Large waste containers, such as dumpsters and compactors located on the outside, should be stored on or above a smooth surface or nonabsorbent material, such as concrete or machine-laid asphalt. An area equipped with hot and cold water and a drain is essential for waste containers. They should be located and washed in an area that avoids food contamination.

Pulpers and mechanical compactors reduce trash volume in a foodservice facility. This equipment grinds waste material into components small enough to be flushed away with water. Mechanical compacting of dry, bulk waste materials is beneficial for establishments with limited storage space because the process can reduce volume to 20% of its original bulk.

Incineration of burnable trash and garbage is another alternative, provided the area and incinerator construction meets all federal and local clean-air standards. Most waste from a foodservice establishment is high in moisture content and does not burn well. Incinerators are only temporary collection containers for waste.

Contamination Reduction

Although Hazard Analysis and Critical Control Points (HACCP) is incorporated in the operation of many processors, those in foodservice frequently may not be familiar with this prevention program because the US Food Code makes it only a voluntary exercise for most. However, the FDA has encouraged participation by issuing a HACCP manual for foodservice and retail establishments. Chapter 7 contains a discussion of HACCP for the food industry.

Sanitary practices in the kitchen and storage areas safeguard the wholesomeness of prepared food. The appearance of clean surfaces is not an effective solution for assessing the existence of harmful bacteria. Microbial swab testing is the fail-safe tool to test for the presence of microorganisms. The use of ATP technology reduces the risk of spreading contamination through expanding the ability to ensure that surfaces are sanitary. Another simple and low-cost measurement technique is to mark target surfaces with an invisible UV-sensitive "ink" and track its removal with a pocket UV disclosure lamp. This system can provide a base by reporting the percentage of targets eliminated (Mann 2012).

Preparation Area

The prevention of contamination with food poisoning and food spoilage microorganisms, as well as with filth, is especially important in the food preparation area after preparation and during service. Contamination during service can permit transfer of disease-causing microorganisms directly to the consumer.

Utensils

Contamination reduction and maintenance of effective hygiene result from thorough washing and disinfecting of utensils. Subjecting utensils to a 77 °C (170 °F) environment for at least 30 s after cleaning provides disinfection. The application of chemical germicides at room temperature for 10 min or longer is required. Contamination reduction occurs through the disposal of cracked, chipped, creviced, or dented dishes or utensils. Food particles and microorganisms can collect in the damaged parts and are more difficult to reach during cleaning and sanitizing.

Requiring server contact with any surface that will come in direct contact with the mouth or with food reduces contamination further. Dishes or utensils where surfaces touch the counter or tabletop should not contact foods. Microorganisms from contact with hands and surfaces that touch dishes, utensils, or food transfer to consumers.

Factors that affect safe preparation and handling of food to reduce the risk of foodborne disease are:

- *Multiple step preparation*: Increased handling leads to more exposure to contamination.
- *Temperature changes*: Heating and cooling places foods in the "danger zone" (3–60 °C or 38–140 °F).
- *Large volume*: Products in large volumes require multistep handling and longer times to heat and cool, giving microorganisms more time for growth.
- *Naturally contaminated foods*: Field dirt or pesticides contaminate raw produce, whereas harvesting contaminates raw red meats and poultry, and raw seafood can carry a variety of viruses, bacteria, or parasites.

Another consideration is to survey the progress of items through the establishment—from delivery at the receiving area to service at the table. Each handling step requires a recording of temperatures and times. Time/temperature curves determine whether existing procedures are adequate to retard microbial growth. Although several control points exist, only a few will be critical control points (CCPs) as follows:

- Prevent microbial growth by holding foods below 2 °C (34 °F) or above 60 °C (140 °F).
- Ensure microbial destruction by cooking foods above 74 °C (165 °F).

Cooling

Excessive time for the cooling of potentially hazardous foods is one of the factors contributing to foodborne illness. Foods that have been cooked and held at improper temperatures provide an environment conducive to the growth of disease-causing microorganisms that may have survived the cooking process (i.e., sporeformers). Furthermore, recontamination of a cooked food item through poor employee practices of cross-contamination from other food products, utensils, and equipment is possible. Large food items, such as roasts, turkeys, thick soups, chili, stews, and large containers of rice or refried beans, require a long time to cool because of their mass and volume. A tightly covered hot food container decreases further the cooling rate. Through reducing the volume of food in individual containers and providing an opening for heat to escape, increases the cooling rate. Avoiding a large mass through the preparation of smaller batches closer to the time of service, stirring hot food while a food container is within an ice bath, and recipes redesigned for the preparation and cooking of smaller or concentrated bases with subsequent addition of cold water or ice to make up the volume needed enhances cooling. A record-keeping system provides scheduled product temperature checks to ensure that the process is working.

Reheating

If food is at an improper temperature for too long, pathogens can multiply and subsequently cause foodborne illness. Proper reheating provides an opportunity to destroy these microorganisms. Heating is especially effective in reducing contamination from bacterial sporeformers that survive the cooking process and remain viable during storage at an improper temperature.

Holding

To avoid pathogen growth (especially sporeformers), food should not be held between 5 °C and 60 °C (40 °F and 140 °F).

Serving

Employees that work with food and food contact surfaces can easily spread bacteria, viruses, and parasites. Personal hygiene management is essential to control these hazards. An effective management program includes proper handwashing, personal cleanliness, control of personal contact with food and food surfaces, proper food temperature maintenance (minimal exposure to the danger zone), and control of customer contamination. Suggestions for protection during prepared food storage and serving include (1) the use of packaging, (2) salad bar food guards, (3) appropriate utensils for dispensing, (4) separation of freshly prepared and other products, and (5) employee monitoring of self-serve stations.

Floors

A newer generation of steam vapor machines hygienically cleans floors. It combines pressurized water heated to 116 °C (240 °F). The steam vapor cleans floors, surfaces, and other areas. However, the limitations of this device are the slow process, potential for injury from the hot water, potential condensation, and possible need for mopping after use to remove residue from the floor. Another option is a spray-and-vac or no-touch cleaning system. These systems inject cleaning compounds onto the floor or surfaces. After a few minutes of dwell time, soils are loosened and suspended with subsequent vacuuming of the cleaned area with the equipment's built-in vacuum system. This cleaning approach can be as much as two-thirds faster than traditional cleaning methods.

Use of Gloves

The major contribution of gloves appears to lie in the perception of a safer food and reduced consumer anxiety. The use of gloves for the protection of the food chain is controversial and appears to be more elusive and dependent on numerous factors such as frequency of replacement and overall hygienic practices.

Sanitary Procedures for Food Preparation

Three sanitary procedures can reduce contamination:

1. *Wash food.* Processed foods do not necessarily require washing; however, washing and draining are essential for all fruits and vegetables eaten raw and cooked, dried fruits and raisins, raw poultry, fish, and variety meats. Washing poultry will reduce contamination of the body cavity of *Salmonella* and other microorganisms. Refrigeration until time for cooking is necessary for items not cooked immediately. If insect infestation occurs, fresh foods should be soaked in salted water for 20 min; any insects will rise to the top of the water.
2. *Protect food from contamination.* Protection from contamination of all foods with poisonous substances and bacteria responsible for foodborne illness is part of a sanitation program. Cleaning compounds, polishes, insect powders, and other compounds used in a foodservice operation can get into food inadvertently. To prevent contamination, all chemicals should be stored separately from food and never in the food preparation area or other locations where food is stored and handled.
3. *Heat thoroughly questionable foods.* When feasible, heat thoroughly all foods that harbor illness-producing microorganisms such as raw meat, poultry, and any foods recontaminated after processing. Heating to 77 °C (170 °F) is

necessary to destroy nonspore-forming bacteria, such as staphylococci, streptococci, and salmonellae. Time-temperature exposure for the destruction of spore-forming bacteria depends on the genus and species.

Sanitation Principles

The foodservice operator has an arsenal of cleaning and sanitizing procedures and products available for selection. The challenge is to determine the most appropriate procedures and products and to apply them properly.

Cleaning Principles

A clean and sanitary establishment is the result of a planned program properly supervised and followed according to schedule. Rushed workers who are trying to meet the needs of customers frequently neglect correct practices. A knowledgeable, alert, and strong manager prevents a breakdown in sanitation discipline. He or she must be able to recognize and institute proper sanitary conditions. Cleaning and sanitizing are the basis of good housekeeping and are essential for all food contact surfaces after every use, service interruption, or at regularly scheduled intervals.

Cleaning is "a practical application of chemistry." Selection of a specific cleaning compound depends on its special cleaning properties. A compound that is effective for one application may be ineffective for other uses. In addition to being effective and compatible with its intended use, a cleaning compound should fit the needs of the establishment. Chapter 9 discusses the important characteristics of cleaning compounds. Some cleaning agents are more effective, and a smaller quantity is required to achieve desired results, which necessitates cost comparisons.

Alkaline cleaners do not affect some soils, for example, lime encrustations on dishwashing machines, rust stains in washrooms, and tarnish that darkens copper and brass. Acid cleaners, usually in a formula that contains a detergent, are more viable for these purposes. The kind and strength of the acid depend upon the purpose of the cleaner.

A soil attached so firmly to a surface that alkaline or acidic cleaners will not be effective requires a cleaner containing a scouring agent, usually finely ground feldspar or silica, to attack the soil. Abrasives clean effectively worn and pitted porcelain, rusty metals, or seriously soiled floors. Abrasives in a foodservice facility receive limited use, especially on food contact surfaces, because they can mar a smooth surface.

Sanitary Principles

It may appear to be unnecessary to sanitize cooking utensils subjected to heat during cooking. However, heat from cooking is not always uniform enough to raise the temperature of all parts of the item high enough for a long enough time to ensure effective sanitizing.

Heat or chemicals sanitize. Heat sanitizing occurs through a high enough temperature to kill microorganisms. Chemical sanitizing acts primarily through interference of metabolism of the bacterial cell. Regardless of the method used, it is necessary to clean and rinse thoroughly the area and equipment. Soil not removed by cleaning may protect microorganisms from the sanitizer. (Chapter 10 discusses sanitizing methods and compounds.)

For foodservice establishments, chemical sanitizing is accomplished by immersing the object in the correct concentration of sanitizer for approximately 1 min or by rinsing, swagging, or spraying twice the normally recommended concentration on the surface to be sanitized. The strength of the sanitizing solution needs testing frequently, because the bacterial-killing process depletes the sanitizing agent. Loss of the effectiveness indicates need for a sanitizer change. Sanitizer manufacturing firms normally provide free test kits for monitoring sanitizer strength. Apply sanitizing agents that are toxic to humans only on nonfood contact surfaces.

Sanitizers blended with cleaning compounds create detergent sanitizers. These products can sanitize, but sanitizing should be a separate step from cleaning. A separate step is necessary because of the sanitizing power destruction during cleaning. The chemical sanitizer can react with organic matter in the soil. Generally, detergent sanitizers are more expensive than regular cleaning compounds and are more limited in their applications than are detergents.

Cleaning and sanitizing most portable food contact items require a washing area away from the food preparation area. Work areas need three or more sinks, separate drain boards for clean and soiled items, and an area for scraping and rinsing food wastes into a garbage container or disposal. If hot water is used to sanitize, the third compartment of the sink must be equipped with a heating unit to maintain water near 77 °C (170 °F) and with a thermometer. Requirements for cleaning and sanitizing equipment vary among areas necessitating a check of regulations that apply to the area.

Cleaning Steps

There are eight basic steps for manual cleaning and sanitizing of a typical foodservice facility:

1. Clean sinks and work surfaces before each use.
2. Scrape heavy soil deposits and presoak to reduce gross deposits that contribute to deactivation of the cleaning compound. Sort items to be cleaned, and presoak silverware and other utensils in a solution designated for that purpose.
3. Wash items in the first sink in a clean detergent solution at approximately 50 °C (122 °F), using a brush or dish mop to remove any residual soil.
4. Rinse items in a second sink. It should contain clear, potable water that is approximately 50 °C (122 °F) for the removal of all traces of soil and cleaning compound that may interfere with the activity of the sanitizing agent.

5. Sanitize utensils in a third sink by immersing the items in hot water (82 °C or 180 °F) for 30 s or in a chemical sanitizing solution at 40–50 °C (104–122 °F) for 1 min. For items immersed in water, mix the sanitizing solution to twice the recommended strength. Therefore, water carried from the rinse sink will not dilute the sanitizing solution below the minimum concentration required to be effective. Avoid air bubbles that shield the interior from the sanitizer.
6. Air-dry sanitized utensils and equipment. Wiping can recontaminate sanitized utensils and equipment.
7. Store clean utensils and equipment in a clean area more than 20 cm (8″) off the floor for protection from splash, dust, and contact with food.
8. Cover the food contact surfaces of fixed equipment when not in use.

Stationary Equipment

Clean stationary food preparation equipment according to the manufacturer's instructions for disassembly and cleaning. Generally, these procedures are:

1. Unplug all electrically powered equipment.
2. Disassemble, wash, and sanitize all equipment.
3. Wash and rinse the balance of the food contact surfaces with a sanitizing solution mixed to twice the strength required for sanitizing by immersion.
4. Wipe all nonfood contact surfaces. Periodically wring out cloths used for wiping down stationary equipment and surfaces in a sanitizing solution. Keep them separate from other wiping cloths.
5. Air-dry all cleaned parts before reassembling.
6. Clean stationary items designed to have a detergent and sanitizing solution pumped throughout according to the manufacturer's instructions. High-pressure, low-volume cleaning equipment (as discussed in Chap. 11) assists in cleaning, and spray devices aid in sanitizing. For sanitizing, spray for 2–3 min with double-strength solution of the sanitizer.
7. Scrub wooden cutting boards with a nontoxic detergent solution and stiff-bristled nylon

brush (or a high-pressure, low-volume cleaning wand). Apply a sanitizing solution after every use. Replace wooden cutting boards that reflect wear from cuts and scars with polyethylene boards. Submerge wooden cutting boards in a sanitizing solution.

Floor Drains

Clean floor drains daily at the end of the cleaning operation. Sanitation workers should wear heavy-duty rubber gloves to remove the drain cover and the debris with a drain brush. After cover replacement, it should be flushed with a hose through the drain. Water should not splatter. Pour a heavy-duty alkaline cleaner down the drain, following the manufacturer's directions for solution preparation. Wash the drain with a hose or drain brush and rinse it. If a quaternary ammonium compound (quat) plug is not used, pour a chlorine or quat sanitizing solution down the drain.

Light Fixtures

Clean light fixtures at least monthly and when a light bulb is changed. Installed light bulbs above food require more frequent cleaning. Turn the electricity off and subsequently remove and wash the fixtures with a warm, low foaming cleaning compound.

Cleaning Tools

Cleaning tools should be stored separately from those used to sanitize equipment and other areas. Rinse, sanitize, and air-dry clothes, scrubbing pads, brushes, mops, and sponges after use. Launder clothing daily. Empty, wash, rinse, and sanitize all buckets and mop pails daily.

Mechanized Cleaning and Sanitizing

Through proper operation and maintenance, mechanized cleaning can more effectively remove contamination from utensils and equipment than

hand cleaning. A trend toward more emphasis on sanitation, combined with increased volume, has been responsible for extensive use of dishwashing machines. In addition, portable high-pressure, low-volume cleaning can be effectively adapted to larger foodservice establishments. The two basic types of dishwashing machines are high-temperature washers and chemical sanitizing machines.

High-Temperature Washers
Discussion follows of the major high-temperature washers according to model. The sanitizing temperature for these washers should be a minimum of 82 °C (180 °F) and a maximum of 90 °C (195 °F).

1. *Single-tank, stationary-rack type with doors.* This washer contains racks that do not move. A compound and water at 62–65 °C (145–150 °F) introduced from beneath washes utensils, with headers installed above the rack. A hot water final rinse follows the wash cycle.
2. *Conveyor washer.* This equipment features a moving conveyor that takes utensils through the washing (70–72 °C or 158–162 °F), rinsing, and sanitizing (82-90 °C or 180-195 °F) cycles. Conveyor washers may contain a single tank or multiple tanks.
3. *Flight-type washer.* This washer is a high capacity, multiple-tank unit with a peg-type conveyor. It may have a build-on dryer. Large foodservice facilities commonly install this washer.
4. *Carousel or circular conveyor washer.* This multiple-tank washer moves a rack of dishes on a peg-type conveyor or in racks. Some models have an automatic stop after the final rinse.

Chemical Sanitizing Washers
A brief description of the major chemical sanitizing dishwashers will follow. Glassware washers are also chemical sanitizing machines:

1. *Batch-type dump washer.* The water temperatures for chemical sanitizing should be 49–55 °C (120–130 °F). This washer combines the wash and rinse cycle in a single tank. Each timed cycle dispenses the cleaning compound and sanitizer automatically.
2. *Re-circulatory door-type, non-dump washer.* This washer incompletely drains water between

Table 21.1 Dishwashing difficulties and solutions

Symptom	Possible cause	Suggested solution
Soiled dishes	Insufficient detergent	Use enough detergent in wash water to ensure complete soil removal and suspension
	Low wash-water temperature	Keep water temperature within recommended ranges to dissolve food residues and to facilitate heat accumulation (for sanitation)
	Inadequate wash and rinse times	Allow sufficient time for wash and rinse operation to be effective (time should be automatically controlled by a timer or by conveyor speed)
	Improper cleaning	Unclog rinse and wash nozzles to maintain proper pressure-spray pattern and flow conditions; overflow should be open; keep wash water as clean as possible by pre-scraping gross soil from dishes, etc.; change water in the tanks at proper intervals
	Improper racking	Verify that racking or placement is done according to size and type; silverware should always be presoaked and placed in silver holders without sorting or shielding
Films	Water hardness	Use an external softening process; use the proper detergent to provide internal conditioning; check temperate of wash and rinse water (water maintained above recommended temperature ranges may cause a precipitate film)
	Detergent carryover	Maintain adequate pressure and volume of rinse water; worn wash jets of improper angle of spray may cause wash solution to splash over into final rinse spray
	Improperly cleaned or rinsed equipment	Prevent scale buildup in equipment by adopting frequent and adequate cleaning practices; maintain adequate pressures and volume of water
Greasy films	Low pH; insufficient detergent; low water temperature; improperly cleaned equipment	Maintain adequate alkalinity to saponify greases; check cleaning compounds and water temperature; unclog all wash and rinse nozzles to provide proper spray action (clogged rinse nozzles may also interfere with wash tank overflow); change water in tanks at proper intervals
Foaming	Detergent dissolved or suspended solids in water	Change to a low-sudsing product and reduce the solid content of the water
Streaking	Alkalinity in the water; highly dissolved solids in water	Use an external treatment method to reduce alkalinity; selection of a proper rinse additive will eliminate streaking; above this range, external treatment is required to reduce solids
	Improperly cleaned or rinsed equipment	Maintain an adequate pressure and volume of rinse water; alkaline cleaners used for washing must be thoroughly rinsed from dishes
Spotting	Rinse-water hardness	Provide external or internal softening; use additional rinse additives
	Rinse-water temperature too high or too low	Check rinse-water temperature; dishes may be flash-drying, or water may be drying on dishes rather than draining off
	Inadequate time between rinsing and storage	Change to a low-sudsing product; adopt an appropriate treatment method to reduce the solid content of the water
	Food soil	Adequately remove gross soil before washing; the decomposition of carbohydrates, proteins, or fats may cause foam during the wash cycle; change water in the tanks at proper intervals
Coffee, tea, metal staining	Improper detergent	Food dye or metal stains, particularly where plastic dishware is used, normally require a chlorinated detergent for proper destaining
	Improperly cleaned equipment	Keep all wash sprays and rinse nozzles open; keep equipment free from deposits of films or materials, which cause foam buildup in future wash cycles

cycles. Fresh water dilutes the wash reused during the next cycle.

3. *Conveyor-type washer with or without a power prerinse.* The name of this equipment defines its function.

The following considerations are important for the procurement and operation of dishwashing equipment:

1. Provide optimal capacity.

2. Include a booster heater with sufficient capacity to supply the dishwashing equipment with 82 °C (180 °F) water for a sanitizing rinse in hot water.
3. Proper installation, maintenance, and operation are necessary to ensure that the equipment adequately cleans and sanitizes.
4. Incorporate dishwashing equipment in an efficient layout for optimal utilization of the unit and personnel.
5. Require accurate thermometers to ensure that appropriate water temperature is used.
6. Include a prewash cycle to omit scraping and soaking of soil utensils.
7. If machines have compartments, protect rinse water tanks through a device to prevent washwater flow into the rinse water.
8. Clean larger dishwashers at least daily, according to the manufacturer's instructions.

Table 21.1 provides further information about symptoms, causes, and cures related to dishwashing problems.

Cleaning-in-Place (CIP) Equipment

Cleaning of soft-serve ice cream and frozen yogurt dispensers and ice machines occurs by passing a detergent solution, hot water rinse, and sanitizing solution through the unit. These machines require design and construction for cleaning, and the chemical sanitizing solution remains within a fixed system of tubes and pipes for a predetermined amount of time. The cleaning water and solution cannot leak into the remainder of the machine. Cleaning and sanitizing solutions reach all contact surfaces. The CIP equipment must be self-draining, and the units designed for inspection through exposure of the cleaned area.

Cleaning Recommendations for Specific Areas and Equipment

Area: Floors
Frequency: Daily and weekly
Supplies and equipment: Broom, dustpan, cleaning compound, water, mop, bucket, and powered scrubber (optional)

Daily
1. Stack chairs on table surfaces or remove from the area
2. Sweep and remove all trash from floor
3 Clean all table surfaces

4. Post signs warning of wetness
5. Mop floors or use a mechanical scrubber for larger operations

Other requirements
1. Self-explanatory
2. Use push broom
3. Wipe food particles into a container. Wash table with warm soapy water. Rinse with clean water
4. Self-explanatory
5. Mix 15 g (0.505 oz.) of detergent per liter (2.1 pints) of clean water. Rinse with clean water

Weekly
1. Apply steps 1–4 for daily cleaning
2. Scrub floors

Other requirements
1. See above
2. Use powered scrubber and/or buffer on floor. Rinse with clean 40–55 °C (104–130 °F) water. Squeegee the floor and dry mop

Area: Walls
Frequency: Daily and weekly
Supplies and equipment: Hand brush, sponge, cleaning compound, bucket, water, and scouring power

Daily
1. Spot clean as necessary

Other requirements
1. Use of 15 g (0.505 oz.) of cleaning compound per liter of water. Hand-wipe all dirty areas. Rinse with clean water. Wipe dry. Mop floor areas to remove spillage

Weekly	Other requirements
1. Remove all debris from walls	1. Self-explanatory
2. Assemble cleaning equipment	2. Mix 15 g (0.505 oz.) of cleaning compound per liter (2.1 pints) of water
3. Scrape walls	3. Use hand brush. Scrub tiles and grout
4. Rinse wall surfaces	4. Use clean, warm water
5. Wipe dry	5. Use clean cloths, or paper towels
6. Scrub floor area to remove any spillage	6. Self-explanatory

Area: Shelves
Frequency: Weekly
Supplies and equipment: Hand brush, detergent, sponge, water, and bucket

Weekly	Other requirements
1. Remove items from the shelves	1. Store items on a pallet or other shelves
2. Brush off all debris	2. Brush debris into a pan or container
3. Clean shelves in sections	3. Mix detergent and warm water and scrub shelves
4. Replace items on the shelves	4. Check for damaged cans and discard as appropriate
5. Mop floor to remove soil	5. Use a clean, damp mop

Equipment: Stack oven
Frequency: Clean once a week thoroughly; wipe daily
Supplies and equipment: Salt, metal scraper with long handle, metal sponges, cleaning compound in warm water, 4-L (4.2 quarts) bucket, sponges, stainless steel polish, ammonia, vinegar, or oven cleaner, as appropriate

Weekly	Other requirements
1. Turn off the heat and scrape the interior	1. Sprinkle salt on hardened spillage of oven floor. Turn thermostat to 260 °C (500 °F). When the spillage has carbonized completely, turn off the oven. Cool thoroughly. Scrape the floor with a long-handled metal scraper. Use a metal sponge or hand scraper on the inside of doors, including handles and edges
2. Brush out scraped carbon and other debris	2. Begin with top deck of stack oven. Brush out with stiff-bristle brush and use dustpan to collect
3. Wash doors	3. Use a hot detergent solution on enameled surfaces only; rinse; wipe dry
4. Brush interior chamber	4. Use a small broom or brush for daily cleaning
5. Clean and polish exterior	5. Wash the top, back, hinges, and feet with warm cleaning compound solutions; rinse; wipe dry. Polish all stainless steel

Note: Do not squeegee, drip, or pour water inside oven to clean

Equipment: Hoods
Frequency: Once a week (minimum)
Supplies and equipment: Rags, warm soapy water, stainless steel polish, degreaser for filters

Weekly	Other requirements
1. Remove filter	1. After removal, carry filter outside and rinse it with a degreaser, and run it through the dishwasher after the cleaning of all dishes and eating utensils
2. Wash hood inside and outside	2. Use warm, soapy water and a rag to wash hoods completely on the inside and outside to remove grease. Clean drip trough in an area below filters

Weekly	Other requirements
3. Shine hood with polish	3. Spray polish on the hood and wipe it off. Use a clean rag on the inside and outside
4. Replace filters	4. Put filters back into proper place after they have completely drained

Equipment: Range surface unit
Frequency: Thoroughly once a week
Supplies and equipment: Putty knife, wire brush, damp cloth, hot detergent water, 4-L (4.2 quarts) container, vinegar or ammonia, as appropriate

Weekly	Other requirements
1. Clean back apron and warming oven (or shelf)	1. Let surfaces cool before cleaning. Use a hot, damp cloth wrung almost dry. Wipe back apron and warming oven. Remove hardened substances with a putty knife; scrape edge of plates. Scrape burned material from top, flat surfaces with a wire brush
2. Remove top sections and scrape edges and flat surfaces	2. Lift plates. Remove burned particles with a putty knife; scrape edge of plates. Scrape burned material from top, flat surfaces with a wire brush
3. Wipe the heating element	3. Wipe the heating elements with a damp cloth
4. Clean base and exterior	4. Wipe with a cloth and hot detergent water
5. Clean grease receptacles and drip pans	5. Soak grease receptacles and drip pans in a detergent for 20–30 min; scrub, rinse, and dry

Note: Do not immerse heating elements in water

Equipment: Griddles
Frequency: Daily
Supplies and equipment: Spatula, pumice stone, paper towels, hot detergent solution

Daily	Other requirements
1. Turn off heat. Remove grease (after each use)	1. Scrape surface with a spatula or pancake turner after surface has cooled. Wipe clean with dry paper towels. Use pumice stone block to clean hard-to-remove burned areas on plates after use. Avoid daily use of pumice stone where possible
2. Clean grease and/or drain troughs	2. Pour a hot detergent solution into a small drain and brush. Rinse with hot water
3. Empty grease receptacles	3. Remove grease from scrapings and supporting pans with hot detergent solution. Rinse and dry
4. Scrub guards, front, and sides of the griddle	4. Using a hot detergent solution, wash off grease, splatter, and film. Rinse and dry

Equipment: Rotary toaster
Frequency: Daily
Supplies and equipment: Warm detergent, brush, rags, stainless steel polish, nonabrasive cleaner

Daily	Other requirements
1. Disconnect and disassemble	1. After cooling, remove pan, slide, and baskets. Move basket midway up front. Press to left carrier chain to permit pins to slip out of holes in the basket
2. Clean surface and underneath	2. Use a soft brush to remove crumbs from the front surface and behind break racks

Daily	Other requirements
3. Clean frame and interior as far as is accessible	3. Wipe with a warm detergent solution. Rinse and dry. Polish if necessary with a nonabrasive cleaning power. The exterior casing should not collect excessive grease or dirt. Prevent water and cleansing compounds from touching the conveyor chains. Polish if the frame is stainless steel

Equipment: Coffee urns
Frequency: Daily
Supplies and equipment: Outside cleaning compound (stainless steel polish), inside cleaning compound (baking soda), urn brush, faucet, and glass brush

Daily	Other requirements
1. Rinse urns	1. Flush with cold water after use
2. Heat water and half fill the urn tank	2. Be certain that outer jacket is three-quarters full of water. Turn on heat. Open water inlet valve and fill coffee tank with hot water to the coffee line. Add recommended quantity of cleaning compound (15 g/L or 0.505 oz./2.1 pints). Allow solution to remain in the liner for approximately 30 min, with the heat on full
3. Brush the liners, faucet, gauge glass, and draw-off pipe	3. Scrub inside of tank, top rim, and lid. Draw-off pipe 2 L (2.1 quarts) of solution, and pour it back to the fill valve and sight gauge. Insert the brush in the gauge glass and coffee draw-off pipe, and brush briskly
4. Drain	4. Open the coffee faucet and completely drain the solution. Close the faucet
5. Rinse	5. Open the water inlet valve into the coffee tank. Use 4 L (4.2 quarts) of hot water. Open the faucet for 1 min to allow water to flow and sterilize the dispensing route
6. Disassemble the faucet and thoroughly clean it	6. Scrub with brush. Rinse spigot thoroughly. Clean
7. Refill (twice weekly)	7. Make a solution (1 cup of baking soda in 4 L (4.2 quarts) of hot water), and hold in the urn for approximately 15 min. Drain. Flush thoroughly with hot water before use

Note: Place a tag on the faucet, while the urn is soaking with the cleaning compound

Biweekly	Other requirements
1. Fill urn with a destaining compound solution	1. Fill the urn with 80 °C (176 °F) water. Add destaining compound in the ratio of 2 tablespoons to 20 L (5.25 gallons) of water (or as directed by manufacturer)
2. Draw off mixture and re-pour	2. Open spigot and draw off 4 L (1.05 gallons); thoroughly remix to allow the mixture to come into faucet. Allow the solution to stand for 1 h at 75–80 °C (168–176 °F)
3. Scrub liner, gauge glass	3. Use a long-handled brush to loosen scale
4. Clean faucet	4. Take a faucet valve apart and clean all components. Soak in hot water until reassembled

Biweekly	Other requirements
5. Rinse and reassemble faucet	5. Rinse urn liner three or four times with hot water. Repeat until complete removal of all traces of the compound
6. Refill urn	6. Fill urn with hot water until next use. Drain and replace fresh water when ready to make coffee

Note: To de-stain vacuum-type coffee makers, use a solution of 1 teaspoon of compound per liter (2.1 pints) of warm water. Fill the lower bowl up to within 5 cm (2 in) of the top and assemble the unit

Equipment: Iced tea dispensers
Frequency: Daily
Supplies and equipment: Rags and warm, soapy water

Daily	Other requirements
1. Clean the exterior	1. Wipe exterior parts with a damp cloth
2. Wash the drip pan	2. Empty the drip pan and wash it and the grill with a mild detergent and warm water
3. Wash the trough	3. Open the front jacket, remove, mix trough, and wash in a mild detergent and warm water
4. Inspect all parts	4. When inspecting parts, remember their order of removal, to ensure proper replacement
5. Wash the plastic parts	5. Do not soak plastic parts in hot water or wash them in dishwashing machines

Equipment: Steam tables
Frequency: Daily
Supplies and equipment: Dishwashing detergent, spatula, scrub brush, and rags

Daily	Other requirements
1. Turn off heating unit	1. Turn steam valve counterclockwise (steam heated). Turn dial to OFF position (electrically heated)
2. Remove insert pans, and transport them to the dishwashing area	2. Lift one end up until clear; then pull forward, grasping the other with the free hand, and remove. Clean inserts thoroughly after each use by hand cleaning and sanitizing processes. Air-dry. Store in a clean area until needed
3. Drain water from the steam	3. Remove the overflow pipe, using a cloth to prevent injury
4. Prepare the cleaning solution; assemble the supplies	4. Dissolve 30 mL (0.0525 pints) of dishwashing
5. Scrape out food particles from the steam table	5. Use a spatula or dough scraper
6. Scrub the interior and clean the exterior	6. Use a scrub brush and cleaning solution
7. Rinse exterior	7. Use enough clear water to remove all traces of the detergent

Note: Hot-food tables, electric, mobile clean corrosion-resistant steel after each use. Remove ordinary deposits of grease and dirt with a mild detergent and water. Whenever possible, thoroughly rinse and dry after washing

Equipment: Refrigerated salad bars (with ice beds or electrically refrigerated)
Frequency: After each use
Supplies and equipment: Detergent, plastic brush, and sanitizing agent

Daily	Other requirements
1. Transfer shallow pans or trays to preparation areas following meal service	1. Run insert pans and/or trays through a dishwasher

Daily	Other requirements
2. Clean and sanitize the table counter	2. Wash and/or scrub table surfaces with detergent and plastic brush. Rinse. Sanitize by swabbing with a solution containing a sanitizer
3. Periodically descale to prevent rust, lime, or hard-water scale formation (non-refrigerated types)	3. Fill the table bed with boiling water. Add a descaling compound in proportions recommended by the manufacturer. Allow to stand for several hours. Scrub with a plastic brush. Drain. Rinse thoroughly. Sanitize by spraying on solution
4. Defrost electrically refrigerated units	4. Turn off electric current and defrost ice formation from the coils as often as required. Follow up with a cleaning procedure, as described above

Equipment: Milk dispenser
Frequency: Daily
Supplies and equipment: Sanitized cloth or sponge, mild detergent, sanitizing agent

Daily	Other requirements
1. Remove empty cans from the dispenser	1. Place a container under the valve. Open the valve and tip can forward in dispenser to drain out the remaining milk. Extract tube. Lift out the oar
2. Wipe up spillage as it occurs	2. Use a sanitized cloth or sponge to prevent possible contamination
3. Clean interior when units are empty	3. Wash entire inner surface with a milk cleaning solution. Rinse
4. Clean exterior	4. Follow procedures for cleaning stainless steel. If steel shows discoloration or stains, swab with a standard chemical to stand 15–20 min before rinsing with clean water and polishing with a soft cloth
5. Disassemble and clean valves daily or as frequently as empty cans can be removed to keep valves clean and sanitary	5. To remove life valves, swing valve upward and slide pins free of recesses to disengage from the plastic well upward to remove. Wash in detergent water. Rinse and sanitize
6. Place full cans in the dispenser	6. Wipe the bottom of milk cans with a sanitizing solution. Thoroughly clean and sanitize clamp-type dispensing valves before reuse

Equipment: Deep fat fryer
Frequency: Daily
Supplies and equipment: Knife, spatula, wire brush, detergent, long-handled brush, vinegar, nylon brush, dishwashing compound

Daily	Other requirements
1. Turn off the heating element	1. Allow fat to cool to 65 °C (148 °F)
2. Drain and filter the fat (after each use)	2. Open drain valve and catch drained fat in a container. Drain entire kettle contents and filter into a container. Place a clean fat container into the well or wash and replace the original container
3. Remove baskets	3. Scrape off the oxidized fat with a knife. Remove loose food particles from the heating units with a spatula or a wire brush. Flush down sides of the kettle with a scoop of hot fat. Soak basket and cover in a deep sink with a hot detergent
4. Remove strained container or cup as often as necessary for cleaning	4. Clean off sediment and place container in the kettle. Stir hot fat and whirl sediment to permit settling in the sediment container

Daily	Other requirements
5. Close the drain. Fill the tank with water	5. Add water up to fat level; then add 60 mL (2 oz.) of dishwashing compound
6. Turn on the heating element	6. Set control at 121 °C (250 °F) and boil 10–20 min, depending on need
7. Turn off heat	7. Open drain. Draw off cleaning solution
8. Scrub interior	8. Using a long-handled brush, scrub the interior. Flush out with water. Clean the basket with a nylon brush and place it back in the kettle
9. Rinse and sanitize	9. Fill the kettle with water. Add one-half cup of vinegar to neutralize the remaining detergent. Turn on the heating element. Boil 5 min and *turn off heat.* Drain. Rinse with clear water
10. Air-dry parts	10. Expose baskets and strainer to air and dry
11. Clean exterior	11. When kettle is cool, wipe off exterior with a grease solvent or a detergent solution. Rinse

Weekly

1. Fill the kettle to fat level with water. Heat to at least 80 °C (176 °F) or boil for 5–10 min.
2. Add one-half tablespoon of destaining compound (stain remover, tableware) per liter of water. Agitate solution and loosen particles remaining on sides of the kettle.
3. Place screens and strainers in 80 °C (176 °F) water containing one-half tablespoon of destaining compound per liter (2.1 pints). Allow to stand overnight. Rinse thoroughly and air-dry.
4. Drain kettle and rinse thoroughly before replacing cleaned screen and strainer.

Equipment: Vegetable chopper
Frequency: Daily
Supplies and equipment: Brush, sponge, cloth, bucket, detergent, sanitizer solution

Daily	Other requirements
1. Disassemble parts after each use	1. Turn off power. Wait until knives stop revolving
2. Clean knives, bowl guard, and bowl	2. Remove blades from the motor shaft and clean them. Wash with a hand detergent solution. Rinse and air-dry. Remove all food particles from the bowl guard. If the bowl is removable, wash it with other parts; if the bowl is fixed, wipe out food particles from table or base. Clean with a hand detergent solution; rinse and air-dry
3. Clean parts and under chopper surface	3. Immerse small parts in a hot hand detergent solution; wash, rinse, and air-dry
4. Reassemble detachable parts	4. Replace comb in guard. Attach the bowl to the base and knife blades to the shaft. Drop guard into position

Note: Choppers vary considerably in mechanical operational details

Equipment: Meat slicer
Frequency: Daily
Supplies and equipment: Bucket, sponge, cloth, brush, detergent, sanitizer solution

Daily	Other requirements
1. Prepare equipment for cleaning	1. Disconnect. Remove meat holder and chute by loosening screw. Remove scrap tray by pulling it away from the knife. Remove the knife guard. Loosen the bolt at the top of knife guard in front of the sharpening device. Remove the bolt at the bottom of the knife guard behind chute. Remove the guard
2. Clean the slicer parts	2. Scrub parts in a sink filled with hot detergent solution. Rinse with hot water. Immerse in a sanitizer solution. Air-dry
3. Clean the knife blades	3. Use a hot detergent solution to wipe off knife blade. Wipe from center to edge. Air-dry
4. Clean the receiving tray and underneath tray	4. Wipe the receiving tray with a hot detergent solution. Rinse in hot clear water. Air-dry

Note: Do not pour water on or immerse this equipment in water

Area: Welfare facilities (see Chap. 17)

Foodservice Sanitation Requirements

Sanitation managers must fully ensure that available sanitation tasks are not omitted and must plan ahead to maximize the use of resources, familiarize new employees with cleaning routines, establish a logical basis for such supervisory tasks as inspections, and save employees time that might be spent in deciding which tasks to perform. Table 21.2 provides a partial cleaning schedule. A full schedule can incorporate the same format. The schedule adopted should constitute a detailed and comprehensive list arranged logically.

The ideal schedule for major cleanup is when minimal contamination of foods is least likely to occur and with minimal service interference. Vacuuming and mopping should not occur during preparation and serving of food. However, prompt cleaning after these operations prevents soil from drying and hardening and reduces bacterial multiplication. Optimal cleaning operations involve even spacing of periodic cleaning and arrangement of jobs in the proper order.

Although mops serve certain applications for cleaning up spills and moisture buildup, they do spread soils and make floors slippery and unhealthy. The use of a disinfectant does reduce the spread of contamination. Repeated use of a mop increases contamination, reduces the efficacy of the disinfectant, and increases the spread of soils.

It is important to discuss a new cleaning program with employees at a meeting, which also can serve as an opportunity to demonstrate the use of new equipment and procedures that relate to the program. It is essential to explain the need for the program and its anticipated benefits and to emphasize the importance of following the procedures exactly as written. Communication with employees can reduce deviation from specified procedures.

Effective evaluation of the sanitation program during continuous supervision and self-inspection increases its effectiveness. Monitoring is necessary to verify procedure compliance. Documented evaluations in the form of periodic inspection reports verify that the program compliance and that expected results have occurred.

Employee Practices

Three areas of employee behavior responsible for the majority of foodborne illness outbreaks are:

1. Poor personal hygiene such as improper hand-washing, personal cleanliness, and dirty clothes. The FDA Food Code requires that foodservice workers wash their hands with soap and water with subsequent application of an alcohol-based sanitizer.
2. Working while ill with a potential foodborne illness.
3. Inadequate cleaning and sanitizing of food equipment.

Table 21.2 Sample cleaning schedule (partial), food preparation area

Item	When	What	Use	Who
Floors	As soon as possible	Wipe up spills	Broom, bucket, mop, and dustpan	— — —
	Once per shift between rushes	Damp mop	Mop, bucket, or scrubber	— — —
	Weekly, Thursday evening	Scrub	Brushes, bucket detergent (brand)	— — —
	January, June	Strip, reseal	See procedure	— — —
Walls and ceilings	As soon as possible	Wipe up splashes	Cloth; portable high-pressure, low-volume cleaner; or portable foam cleaner	— — —
	February, August	Wash walls	Same as above	— — —
Work tables	Between uses and at end of day	Empty, clean, and sanitize drawers; clean frame, shelf	See cleaning procedure for each table	— — —
	Weekly Saturday p.m.		See cleaning procedure for each table	— — —
Hoods and filters	When necessary	Empty grease traps	Container for grease	— — —
	Daily, closing	Clean inside and out	See cleaning procedure	— — —
	Every Wednesday evening	Clean filters	Dishwashing machine	— — —
Broiler	When necessary	Empty drip pan, wipe down	Container for grease; clean cloth	— — —
	After each use	Clean gird tray, inside, outside, top	See cleaning procedure for each broiler	— — —

Source: Adapted from *Applied Foodservice Sanitation*, 4th Edition. Copyright 1992 by the Educational Foundation of the National Restaurant Association

When using, cleaning, and sanitizing cutting boards, the following practices should be adapted:

1. It is unacceptable to cut or chop ready-to-eat (RTE) foods, such as salad or vegetables, on an unwashed cutting board used to trim or slice raw meats, poultry, or seafood.
2. Use a clean, separate, color-coded cutting board for RTE foods to avoid cross-contamination.
3. Use white pads for difficult to remove materials instead of stainless steel pads or wire brushes that will damage the finish. Dishwashers are effective for thicker cutting boards but may damage thin plastic boards.
4. To enhance cleaning effectiveness, clean and sanitize cutting boards within 4 h after use through the incorporation of one teaspoon of chlorine bleach to 1 L (2.1 pints) of water or an approved sanitizing solution.

Rags and sponges are potential sources for disease-causing microorganisms. The continually moist cellulose sponge provides multiple surfaces for harmful microbes to adhere. A safer and more hygienic tool is a sanitary wipe rag.

Proper utilization of a sanitized wipe rag destroys microorganisms present on a soiled rag.

Effective use of wipe rags involves starting to clean from the top and work downward to avoid recontaminating cleaned surfaces. A putty knife or a white scrubbing pad loosens adhered soils. Rinsing should occur prior to surface drying. A cleaning solution applied and allowed to soak for at least 10 min before scrubbing, rinsing, and sanitizing cleans heavily deposited soils. A viable solution is to replace pails and rags with a spray bottle of cleaner-sanitizer and single-use paper towels or a disposable wipe.

Custodial workers should not clean welfare room facilities with a cleaning cloth subsequently used for high-touch areas such as door handles, light switches, ledges, railings, and other frequently and commonly touched locations. Kravitz (2015) suggested that a color-coding system prevents this poor practice. He suggested the following color-coded program for cleaning cloths:

- Red: welfare facilities and fixtures
- Blue: kitchen area surfaces, counters, etc.

- Yellow: high-touch areas
- Green: office desks, equipment, chairs, counters, etc.

Employee Training

Training requires time away from the job for both workers and management and training specialists. Printed material, posters, demonstration, slides, and films serve as training devices.

It is difficult to measure the return on the investment in sanitation training. Benefits are not always measurable, but a potential realized savings exists through the prevention of a foodborne illness outbreak or of closed establishments until meeting local health standards. It is difficult to measure the improved image attained through a sanitary operation, even though increased sales will result.

Employee training is important because it is difficult to recruit competent and motivated workers. Periodic training is essential because the industry has a higher rate of employee turnover than do most organizations. On-the-job training can be effective for certain tasks but is not comprehensive enough for sanitation training. Each employee involved in foodservice sanitation should become familiar with the sanitation concept and sanitary practices required for job performance.

An ideal method for training employees of a large firm is to set up a training department and hire a training director. This adopted approach occurs in many large and medium-sized food service operations. In fact, foodservice trainers have established their own professional association, the Council of Hotel and Restaurant Trainees. Furthermore, the Association of Food and Drug Officials develops and publishes food sanitation codes and encourages food protection through the adoption of uniform legislation and enforcement procedures.

In most foodservice operations, the supervisors, rather than professional trainers, normally conduct the sanitation training. Therefore, a previously trained employee or one certified in foodservice sanitation should personally conduct the training.

The effectiveness of a training program depends on the ability of employees to perform their assigned tasks. If standards of achievement have been set before training, progress can be deter-

mined by measuring individual achievement against those standards. Employee turnover data, absenteeism and tardiness reports, and performance data determine the value of the training program. The quality of training reflects in the amount of guest complaint reports and customer return rates.

Two methods exist to evaluate the effectiveness of training. An objective method involves the use of tests or quizzes to determine employee comprehension. The other method is job performance by employees, as evaluated by management. Praise of employees, wall charts that recognize superior performance, pins, and certificates, enhances training effectiveness. Organizations and some regulatory branches provide certification courses that provide both training and recognition.

Study Questions

1. What are the best construction materials for (a) floors, (b) walls, and (c) ceilings of foodservice facilities?
2. What kind of faucets are the best for handwashing in foodservice facilities?
3. What temperature disinfects utensils?
4. What endpoint cooking temperature ensures microbial destruction?
5. What is the required water temperature for the third compartment sink?
6. What is the required water temperature for the first and second compartment sinks?
7. What are three areas of employee behavior responsible for foodborne illness?
8. What is the required water temperature for a dishwasher?
9. How can serving practices reduce food contamination?

References

Kravitz R (2015). Floor cleaning eye-opener. *Food Qual and Safety* 22 (1): 44.
Mann J (2012). All hands on deck. *Food Qual* 18 (6): 31.
National Restaurant Association Educational Foundation (1992). *Applied foodservice sanitation*. 4th ed. Education Foundation of the National Restaurant Association: Chicago, Illinois.

Management's Role in Food Sanitation

22

Abstract

A major challenge of management in the food industry is to recruit, train, and retain employees for an effective sanitation operation. The success or failure of a sanitation program depends on the extent to which company leadership is committed to and supports the program.

An effective sanitation program includes provisions for constant education and training of employees. Educational information can be disseminated through sanitation training manuals, websites, booklets, job aids, and tip sheets as well as through company training programs and those offered by trade associations, professional organizations, universities, consultants, or regulatory agencies.

The major functions of sanitation management are to delegate responsibilities and to train and supervise employees. Self-supervision and self-inspection are two tools that contribute to a more effective sanitation program.

Keywords

Employee selection • Management • Management selection • Sanitation education • Sanitation management • Supervision • Training

Introduction

Since many entry-level jobs in food processing, food retailing, and foodservice (including sanitation) do not require previous formal training or education, many unskilled workers select the food industry as the area of their first employment. High school and college students frequently work part time in entry-level jobs in retail food stores, restaurants, and fast-food operations. The age and multiple interests of these employees, as well as the modest salaries of these jobs, and the repetitive tasks often associated with them, have been blamed for absenteeism, poor job performance, and high employee turnover in these sectors of the food industry.

Most food company executives will agree that the rapid turnover rate of employees can be attributable to a lack of education and training and,

© Springer International Publishing AG, part of Springer Nature 2018
N.G. Marriott et al., *Principles of Food Sanitation*, Food Science Text Series,
https://doi.org/10.1007/978-3-319-67166-6_22

sometimes, the nature of the job. These conditions have apparently contributed to a lower salary scale, especially among retail food and food service employees. Therefore, company management has a challenge in recruiting, training, and retaining employees for sanitation activities, which provide the foundation of a food safety management system. Another challenge is the need to give sanitation a more professional and exciting image so that employees will proudly and enthusiastically accept their responsibilities related to the maintenance of a hygienic operation. It is clear that an effective sanitation program will reduce microbial contamination, promote cleanliness in the facility, improve product stability, reduce cleaning expenses through increased efficiencies, and, ultimately, save money. Sanitation can also serve as a source of pride and morale to all employees who enjoy working in a clean facility (Cramer 2008).

Company leaders play a vital role in developing, implementing, maintaining, and sustaining the effectiveness of a company sanitation program and are held accountable for its results. In recent years, several high-profile foodborne illness outbreaks, including one from Salmonella-contaminated peanut butter and peanut butter paste (CDC 2009) and another from Listeria-contaminated cantaloupes (CDC 2012), resulted in multiple illnesses, hospitalizations, and fatalities, as well as nationwide product recalls. Executives from the company that produced the peanut butter were convicted of introducing adulterated food into interstate commerce and sentenced to lengthy prison terms (Flynn 2016). The owners of the cantaloupe farm pleaded guilty to introducing contaminated food into interstate commerce and were given probation, home detention, and community service and had to make restitution to the victims for their role in this outbreak (Food Safety News 2015). The consequences of poor sanitary conditions that lead to foodborne outbreaks are substantial. These unfortunate incidents have sent a key message to company executives that sanitation and food safety needs to be a very high priority in their businesses.

Management Requirements

The top management of a food company certainly affects the success of the sanitation program in their operation, and the success or failure of it depends on the level of commitment they have to the program. The discussion that follows will outline the key role that company leadership plays in the organization and implementation of an effective sanitation program.

Management Philosophy and Commitment

Company executives in the food industry need to be *totally committed* to food sanitation and food safety in their companies. Through their words, actions, decisions, and commitment of financial resources and personnel, they set the tone for the importance that sanitation and, consequently, the role of food safety in their organization. Through this deep level of commitment, they can create a culture of food sanitation and food safety within their companies that is passed down to all employees and becomes part of fabric of the organization. The mission and vision statements of the company should also reflect this philosophy and commitment and should be widely communicated to all employees, suppliers, customers, and consumers. It is vital that company leadership "model the way" and continuously advocate for food sanitation and food safety. Through effectively designed and delivered food sanitation/food safety education and training programs, all employees, from the day they are hired and onboarded, throughout their entire employment with the company should be taught about their important role in sanitation. They also need to understand the impact that doing a consistently good job has on the cleanliness of the facility and how it influences the health and welfare of all who consume the products that they make. This concept, as well as new developments in the science and technology of sanitation, should be continuously reinforced with all employees.

In the past, some managers did not always support sanitation, since it reflects a cost where the dividends of increased sales and profits could not always be accurately measured. Frequently, lower and middle management had difficulty selling the sanitation concept when top management did not fully support it. This is not usually the case today, as most company leaders understand the importance of sanitation and the consequences that can result if it is neglected which can lead to product contamination, a foodborne outbreak, product recall, and regulatory and legal actions.

In describing the challenges of food sanitation, Keener (2009) shared a wonderful analogy. He said that "A poor sanitation program is similar to a ship taking on water. If the hole is not plugged, eventually the ship will sink, regardless of how fast the boat moves." An effective sanitation program provides a solid foundation for prerequisite programs (Good Manufacturing Practices) on which a strong food safety management system (HACCP/Preventive Controls for Human Food) is built (FSPCA 2015).

Sanitation Preventive Controls

In the Food Safety Modernization Act (FSMA), the Preventive Controls for Human Food regulation requires that if the hazard analysis identifies hazards related to sanitation, then sanitation preventive controls need to be implemented (FSPCA 2015). These preventive controls should be appropriate to the facility and the food being produced and should be designed to significantly minimize or reduce hazards including environmental pathogens, biological hazards related to employee handling, and food allergen cross-contact (FSPCA 2015). Sanitation preventive control focuses on procedures, practices, and processes for the cleanliness of food contact surfaces and prevention of allergen cross-contact and cross-contamination. The sanitation preventive controls describe the monitoring activities and frequency, corrective actions that apply to environmental pathogens and allergens that need to be made when requirements are not met, as well as verification activities appropriate to the facility (FSPCA 2015).

Management Knowledge of Sanitation

Company management should be knowledgeable about the principles of food sanitation and food safety, food microbiology, and sanitary design and keep up-to-date on the latest innovations in the field. With FSMA and the Preventive Controls for Human Food regulation (FSPCA 2015), companies have to demonstrate and document that the sanitation procedures and preventive controls contained in a food safety plan are efficacious. Executives must fully understand these regulations and implement those that pertain to their products and the processes by which they are produced. Management must support and promote sanitation because of its direct impact on corporate planning, marketing, product safety and quality, and compliance with laws and regulations. Sanitation programs have a direct impact on the industry-regulatory interface and relationships with regulatory agencies. The US Food and Drug Administration (USFDA) has a detailed definition of adulterated food in the Federal Food, Drug, and Cosmetic Act (FD&C Act) and also in the Good Manufacturing Practice regulation. One section of that definition states that food is adulterated "if it has been prepared, packed or held under insanitary conditions whereby it may have become contaminated with filth, or whereby it may have been rendered injurious to health" (USFDA 2017a, b). This provision is very interesting because instead of having to prove that the food is adulterated, the agency considers insanitary conditions sufficient to show that the food might have become adulterated (USFDA 2017a). The FDA has used this law to enforce violations of the Act and seek prosecutions of company executives in several high-profile cases, so it is vital that companies practice due diligence when processing food. Company management must be aware of the latest food laws and regulations as well as scientific and technological innovations in food sanitation and food safety.

Program Development

A complete and detailed sanitation program along with Good Manufacturing Practices (GMPs) and other prerequisite programs provide the foundation of a Hazard Analysis Critical Control Point (HACCP) plan. GMPs (devised by the FDA) are a driving force for sanitary program design and hygienic operations because the primary objective of these practices is the prevention of adulteration (contamination) of foods. There are many methods and procedures that can be used to design, develop, implement, and maintain an effective and successful sanitation program. One method that has been proposed by Keener (2009) and adapted from Marriott and Gravani (2006) includes seven steps and four areas of emphasis. The steps involved in the development of a sanitation program are discussed below (Keener 2009) and include:

1. *Develop a sanitation philosophy and policy*
   A total commitment of company executives to sanitation, food safety, and the hygienic production of safe and wholesome food should be the overarching philosophy of the company. This commitment should be incorporated into the mission and vision statements, as well as in a policy statement of the company that is known to all employees, suppliers, customers, consumers, and other interested parties. This makes sanitation a priority and helps to create a culture of sanitation and food safety within the company.

2. *Gather information*
   Gather all the local, state, and federal regulations and other pertinent information that pertain to the product(s) being produced. It is critical to know the regulatory requirements for the facility including who will inspect the operation, frequency of inspection, record-keeping requirements, verification procedures, employee training programs, etc.

3. *Establish a sanitation team*
   Establish a sanitation team, made up of representatives from departments throughout the company (quality control/quality assurance, microbiology, engineering, maintenance, pro-

duction, operations, purchasing, management, and other departments as appropriate), to:
- Conduct a thorough and detailed survey to evaluate current conditions throughout the facility (from incoming raw materials to the storage and shipment of finished product, including buildings as well as the roof and grounds, equipment, utilities, air handling systems, processing environment (including floors, walls, drains, drainage, ceilings, hoses, overhead structures, etc.), personnel and product flow through the operation, sanitary design of equipment and facilities, etc.).
- Assess sanitation processes and procedures as well as GMPs.
- Determine all potential sources of contamination and document areas in the plant that need special attention and improvement.
- Strengthen sanitation and prerequisite programs and GMPs.
   It is important that all aspects of the operation be surveyed so that a complete and thorough evaluation of the current conditions can be made.

4. *Create written procedures for preventive sanitation*
   From the compiled survey results, written procedures should be developed to address prerequisite programs (GMPs): a HACCP plan, Sanitation Standard Operating Procedures (SSOPs), an environmental monitoring program, sanitation preventive controls if needed, and a maintenance plan. Cleaning and sanitizing procedures identified in the Preventive Controls for Human Food regulation (FSPCA 2015) include the following items:
- The *purpose* of performing the task so the worker understands the importance of it.
- The *frequency* of when the procedure needs to be performed to be effective.
- The *person responsible* for performing the procedure or task.
- The *procedural instructions* to accomplish the task and these should include the tools, supplies, personal safety equipment, and instructions and the specific steps necessary to successfully complete the procedure.

These are known as Sanitation Standard Operating Procedures or job aids and also include how long it takes to complete the task and the skill level required.

- *Monitoring* to provide a record that the task was completed.
- *Corrections* or what to do when inspection determines that the procedure was not adequate to produce a sanitary situation.
- *Verification* procedures to "double check" that the process was completed.
- *Records* that include the name of the form used to record monitoring activities.
- *Other special considerations*.

5. *Establish a record-keeping system*

Develop a daily, effective record-keeping system for the sanitation program so that inspections, monitoring of processes and procedures, deviations, and corrections are documented and verified. The old adage "if you didn't write it down, you didn't do it" certainly addresses the importance of record keeping. Today, many companies are moving away from cumbersome paper-based recording keeping systems and using sanitation management software programs for recordkeeping. With more regulatory requirements and the increasing complexity of sanitation, electronic systems can be used to document HACCP plans, corrective actions for Critical Control Points (CCPs), track third-party audits, cleaning supplies, inventory, as well as other functions of the sanitation process (Anon 2004c).

6. *Develop an effective education and training program*

Using best practices, the principles of adult learning and current motivational techniques develop an effective education and training program that will provide practical information to new employees, refresh and renew the concepts for longtime workers, and establish a sanitation incentive program for all. Well-trained employees who perform their job task regularly and routinely are an invaluable asset to the company.

7. *Conduct regular assessments of the sanitation program*

Regularly scheduled reviews can be used to provide an accurate assessment of the effectiveness of the ongoing sanitation program, with the goal of continuous improvement of the entire program. Close scrutiny and attention to detail are needed to keep the sanitation program operating at a high level and maintain sanitary conditions and cleanliness throughout the facility. Milestones in the program and successes achieved should be celebrated with all the sanitation team and incentives should be provided for those who consistently perform outstanding work.

For the sanitation program to be successful, several other important items must be considered, and they include the personnel who do the work, the sanitary design of the facility and equipment, and the preventive maintenance (Keener 2009). Since sanitary design has already been addressed in Chap. 14, it will not be discussed here.

Program Follow-Through

The sanitation department of a company needs to be well designed and have a good organizational structure. The sanitation manager should be experienced in the field and possess scientific and technical skills, leadership characteristics, and good problem-solving ability (Cramer 2007, 2008). An effective manager assures that everyone involved with sanitation works as a team and communicates effectively to share problems, solutions, knowledge, and skills. A successful sanitation program that has been developed and implemented must be regularly checked through monitoring, verification, and record review. Another effective check is through periodic outside, third-party sanitation audits. Trained and experienced independent auditors can be hired to conduct a thorough and detailed inspection of the facility to provide a fresh perspective and new ideas and innovations. A company internal audit by the sanitation manager or plant manager should also be conducted on a regular basis. Any deficiencies or areas needing improvement observed during these inspections should be captured on a detailed list, and prioritized action should be taken to correct them.

Sanitation is more than just cleaning and includes the documentation of scheduled tasks, employee training, inspection, and corrective actions (Daniel-Sewell 2017). Comprehensive sanitation scheduling is used to identify tasks and how to perform them, as well how frequently they are to be performed, personnel requirements, tools and supplies needed, and follow-up to ensure that they have been done correctly. Sanitation tasks usually fall in three categories including the master sanitation schedule, shift sheets, and housekeeping tasks (Daniel-Sewell 2017). The master schedule normally includes the SSOPs for cleaning that typically occurs before and after production and longer-term tasks such as cleaning overhead fixtures, walls, ceilings, etc. Shift sheets outline tasks assigned to specific workers to perform during a shift. Housekeeping tasks are those not directly associated with production and may include offices and welfare facilities (Daniel-Sewell 2017). Housekeeping also involves the general cleanliness of the facility including the pickup of paper, packaging materials, product spills, etc.

Employee Selection

Employees who work with food in any food facility should be carefully selected to be free of infectious diseases. They should have a personal hygienic level above the average of the population and be conscientious and motivated to work in a food facility. The current FDA Good Manufacturing Practices regulations summarize this concept very well by stating that "All persons working in direct contact with food, food contact surfaces and food packaging materials shall conform to hygienic practices while on duty to the extent necessary to protect against allergen cross-contact and against contamination of food" (FDA 2017b). In addition, prospective employees should be physically able to do the work, be available to work the hours needed (often on second or third shift), and be willing to work in a challenging job in wet, humid environments with cleaning and sanitizing chemicals (Cramer 2007, 2008). When hiring new members of the sanitation team, the level of education, ability, prior knowledge, and motivation of potential candidates should also be considered.

The level of expertise of the sanitation team is changing rapidly. In the past, it was a standard practice to hire inexperienced employees and assign them to the sanitation team on the second or third shift in a food processing plant without any training and with minimal supervision. Today, many companies start new sanitation workers with an orientation to the company, including information on the importance of sanitation and food safety to the company, the nature and specific details of the job, and its importance to the health and safety of consumers. This type of onboarding program is often mandatory for new sanitation workers, and they must successfully complete it after being hired and before beginning work. Employees who are involved in the cleaning and sanitizing process are paramount to food safety, so their turnover should be minimized. To improve the retention of sanitation workers, an incentive (or job enrichment) program should be developed to let people know that they are valued and are an important part of the team. Successful incentive programs use rewards such as a pay bonus, time off, coupons for lunch, movie passes, discount coupons to area attractions, etc. These incentives acknowledge excellent performance and help workers feel valued and important (Keener 2009). It is important to celebrate sanitation team successes and milestones including reduced microbial levels as determined by swab tests, improved product shelf life, etc. (Cramer 2008).

Employee Training

In any food facility, whether it is a food processing plant, a retail food store, or a foodservice operation, it's the people that are the most important part of successful sanitation and food safety programs. When thinking about education and training programs for newly hired workers and refresher training for current and longtime employees, it is important to consider the results of a worldwide survey about why people don't accomplish the work that they were assigned. The survey of almost 25,000 managers identified a number of reasons why workers don't accomplish assigned tasks (Fournies 2007). Some of the reasons for not accomplishing tasks that were identified in the manager survey can be applied to the performance of sanitation and food safety tasks and include the following (Fournies 2007):

They don't know why they should do it

It is important to stress the benefits of performing sanitation tasks properly and the harm of doing them incorrectly. It is essential that job tasks be linked to the cleanliness and safety of foods that customers purchase and consume. The successful completion of sanitation tasks should be strongly associated with public health, food safety, and the prevention of foodborne illnesses. It is important to stress what is in it for them (the workers) to perform the job correctly each time that they do it. The task and its goals should be explained in detail, as well as the benefits of success of doing it correctly (increased knowledge, prestige, opportunity, financial rewards, comfort, and security) and the consequences for not performing the task correctly.

They don't know what they are supposed to do

Provide workers with accurate job descriptions that contain the job behaviors that they should perform. This way, they know exactly how and when to perform the job that is required of them.

They don't know how to do it

This can be easily remedied through well-designed, well-developed, and well-delivered education and training programs, job aids, tip sheets, and other instructional methods to provide the "how to" information to employees. Employees need to practice the skills that they learn to perform a new job, so enough time should be provided for this practice until the employee becomes proficient at the tasks and can perform them regularly and routinely, with "unconscious competence."

They think that they are doing it

This occurs when employees don't receive enough feedback about the quality of their work. Feedback is the vital information that lets people know how they are doing. If it is not provided, then employees think that their performance is fine, and they think they are doing the right things.

There is no positive consequence to them for doing it

Workers need to be recognized for a job well done, and positive reinforcements (verbal compliments) and incentives provided immediately following a good job have an impact on future performance. Research has shown that these types of rewards are more effective than larger rewards given long after the performance has taken place (Fournies 2007).

This information provides some important solutions to why people don't do what they're supposed to do and should be incorporated into employee education and training programs, group meetings, "team huddles," and coaching sessions.

The importance of adequate education and training of employees has been suggested several times throughout this book. It is especially important to train sanitation employees in the basics of sanitation because *nothing happens in a food establishment until the facility is clean*. Sanitation employees should be hardworking, dedicated people who clearly understand the company's sanitation and food safety policies and their role in achieving them. A finely tuned sanitation program consists of effective interaction between a QA department and a research and development laboratory—within the organization or in a private laboratory—for an accurate assessment of sanitary practices.

A group of food safety experts who participated in an elicitation study to evaluate the frequency and severity of food safety risk in five food processing industry sectors and in three different plant sizes identified the top 10 food safety problems in the US food processing industry (Sertkaya et al. 2006). The experts collectively ranked "deficient employee training" as the top problem facing processors. They also identified a number of other problems in their top 10 that have been addressed in this book, including poor plant and equipment sanitation, poor plant design and construction, no preventive maintenance, difficult to clean equipment, and poor personal hygiene. The results of this study reinforce how important effective education and training of employees is to a knowledgeable workforce.

Management must ensure that sanitation workers are well qualified and should have a good working knowledge in the operations of the food facility, food microbiology, the role of cleaning compounds and sanitizers, and the fundamentals of sanitary design. Additionally, sanitation managers should have knowledge of specific surface design and hardness, porosity, vulnerability to oxidation, and corrosion of surfaces to be cleaned, so that the appropriate cleaning equipment, cleaning compounds, and sanitizers may be determined.

An effective management team should ensure that sanitation workers are educated in the safety and efficacy of cleaning compounds, the functions of detergent auxiliaries and sanitizers, and the most effective cleaning equipment. A person well versed in these areas can reduce waste and personal injury and simultaneously optimize cleaning efficiency. Additional benefits include reduced water consumption, sewage load, and sanitation labor.

Management must provide practical information to sanitation workers in an easy-to-understand form. It should be presented in a clear, easily accessible instruction manual or booklet that provides facts related to cleaning all areas and equipment, including the selection and application of cleaning compounds and sanitizers for all cleaning applications. The instruction manual should also include a sanitation plan and material on operational methods, pest control, hygienic practices, and preventive maintenance. The adoption and application of these principles will affect operational appearance, practices, and performance and will positively reflect on the company image.

The sanitation team is a valuable asset, and through its sustained efforts and attention to detail, future food safety issues can be prevented. A stable and well-trained food sanitation team should cross-train new members of their group to attain maximal efficiency and reduce facility downtime.

One important consideration when thinking about training programs is that some workers may be from different cultures, do not use English as their primary language, and may have trouble understanding sanitary practices and principles (Keener 2009; Neal et al. 2015). If this is the case, then instructions should be given in their native language, as well as through the use of pictures and photographs of the correct procedures being performed.

Another important area that needs to be addressed during training is the use of personal protective equipment (PPE) such as protective eyewear (goggles or face shields), rain gear, aprons, boots, gloves, and respirators (Keener 2009; Cramer 2007). New sanitation team members need to know how and when to use PPE and why it is important to do so to insure their personal safety.

Some companies conduct intensive, formal in-house training programs for sanitation employees. These firms can provide sanitation technology based on their QA program needs. Those needs that are most frequently addressed include determination of required manpower and effective communication to nontechnical personnel.

Most company leaders and responsible managers recognize that consumers desire and deserve wholesome, safe products and acknowledge the importance of well-trained employees in producing these foods. The delivery of carefully designed and sound employee training programs is an integral part of their processing or foodservice operations. Therefore, sanitarian managers and supervisors should attend training courses, professional society meetings, seminars, and workshops that provide the latest scientific information on the science of sanitation. In addition, sanitation professionals should seek the assistance of regulatory agency personnel for discussion of current sanitation standards and public health needs and incorporate this information in employee training programs. As a necessary follow-up to education and training programs, company leaders should provide materials, facilities, resources, and opportunities necessary for employees to practice what they have learned. During these practice sessions, it is important to provide employees with feedback on their performance of newly learned tasks. This is an important part of the education process and will ensure that workers perform their tasks accurately and reliably. Managers and supervisors also benefit from such employee training activities and that the success of the activity depends on their own leadership, proper and thorough preparation, and

Fig. 22.1 An annotated five-step model for structuring an effective employee training program (Stolovitch et al. 2011) (Courtesy of the American Society of Training and Development)

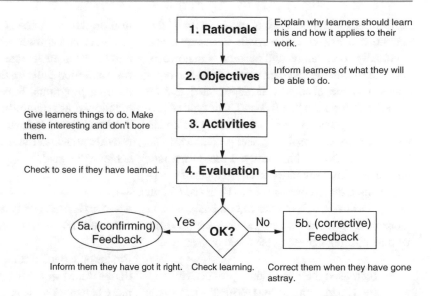

attention to detail. An annotated five-step model for structuring an effective employee training program (Stolovitch et al. 2011) is shown in Fig. 22.1.

The training model is quite simple but, if properly designed and delivered, is very effective in changing behavior and results in employees truly learning a specific task (Stolovitch et al. 2011). Instructors should provide the rationale for why employees should learn what is being shared with them and tell them how it applies to their work. Providing information about why the material is important and what's in it for them should also be highlighted. The learning (and performance) objectives of the training program should then be shared with participants in a meaningful way, and they should be told what they will be able to do (what tasks they will be able to perform) at the end of the training program. Activities that provide challenges, stimulate curiosity and are fun should be created to enhance learning. Stimulating and interesting instruction and hands-on activities, including exercises, games, problem-solving scenarios, and small group cases or projects, should be included in the training program to allow participants to apply and practice the skills they have learned. After the training program concludes, instructors need to determine, based on the program objectives, what participants have learned. By evaluating their knowledge and skills, feedback can be provided to participants on whether they learned the tasks or if they need more practice to perform them properly.

Another important component to consider in a training program is motivation. Motivation is the desire to achieve something, and because it comes from within an individual, it cannot be taught. Motivation to obtain new knowledge and skills is affected by three major factors including the value placed on it (the more we value something, the more we are interested in it), the confidence that people have in learning it (if confidence is high, motivation increases), and the mood that they are in while learning it (a positive working/learning environment improves mood and motivation) (Stolovitch et al. 2011). This information is important to keep in mind when developing training activities.

A well-trained sanitation team will reduce production downtime, reduce product recalls, and improve employee morale, since a clean plant is a more productive plant. It has been estimated (Keener 2009) that a well-trained sanitation worker can save a company between $5000 and $10,000 per year in retraining costs and in waste and water consumption.

Additionally, if the sanitation team is not perceived as a key component of the operation, is not recognized for their efforts and accomplishments in performing their jobs effectively and efficiently, and is considered a necessary expense, the team will not

22   Management's Role in Food Sanitation

perform at its highest level of productivity. If this occurs, sanitary conditions will deteriorate, and eventually food safety will become a major concern. One of the needs of all people within an organization is the need to feel appreciated and be recognized for a job "well done." This type of recognition inspires greater interest in performing tasks properly and inspires greater productivity. The Gallup organization has surveyed more than four million employers worldwide on the topic of recognition and praise in organizations (Rath and Clifton 2009). Their research of over 10,000 businesses, in more than 30 industries, indicated that individuals who receive regular recognition and praise:

- Increased their individual productivity
- Increased the engagement of their colleagues
- Are more likely to stay with a company
- Have better safety records and fewer accidents on the job

The sanitation team needs to know that through its diligence and hard work, company management recognizes the important role that it plays in assuring that food processed, prepared, or merchandised in their operation will be clean and safe.

Other Sources for Sanitation Education and Training

In addition to in-house courses that can provide specific information about company policies and procedures, equipment and schedules, etc., there are also a wide variety of educational opportunities (seminars, webinars, short courses, workshops, and online and in-person training programs) offered by professionals societies (International Association for Food Protection (IAFP), Institute of Food Technologists (IFT), and the National Environmental Health Association (NEHA)), food trade associations (the Grocery Manufacturers Association, the North American Meat Institute, and the National Restaurant Association Educational Foundation), university extension programs, and consultants that are designed to provide up-to-date informa-

tion on the latest science, technology, sanitation products, and equipment in the field. In addition, regulatory agencies are also excellent resources for sanitation information that can be used in training programs. Several of the organizations mentioned are committed to improving the professional status and image of food sanitation and provide education and training opportunities to achieve this goal.

Management of a Sanitation Operation

Business experts define *management* as "getting things done through people." Sanitation management has three basic responsibilities:

1. Training employees by providing knowledge, skills, and motivation and then showing them how job responsibilities should be executed, while providing opportunities for practice and then offering feedback
2. Delegation of responsibilities or telling employees what jobs must be done, why they are important, and when they must be completed
3. Supervision to ensure that all responsibilities are properly executed

Managers should continue to make certain that job assignments are being properly performed by conducting regular inspections. Although employees are properly trained, they must be supervised to ensure that they are regularly and routinely carrying out their job responsibilities.

The technical aspects of sanitation merit serious consideration because of the complexity of products being manufactured, prepared, and merchandised. Technical competence of management should include the understanding of employees, including the principles of adult learning and how they should be educated, motivated, and supervised. Employees achieve more if they are clearly informed of their expectations and why they are essential to the attainment of safe food.

Technical knowledge should include knowledge of organic residues and how they should be removed. Furthermore, knowledge of microorganisms will enhance their control as well as knowledge of appropriate environmental microbial sampling and testing.

The competence of food sanitation personnel and the effectiveness of the program administration are major factors in achieving the objectives of a food sanitation program, regardless of the type of enforcement methods employed. Managers cannot afford to be mistaken in their judgment or unreasonable in their decisions, because such actions affect the health and well-being of consumers. Success in food sanitation and consumer protection programs also depends on understanding, interest, and support within the top levels of the regulatory authority and other branches of government.

Major problems that are the most detrimental to effective sanitation (Anon 2004b) include:

1. Lack of support from company managers
2. Improper training of sanitation supervisors and other workers
3. Lack of effective written procedures
4. Improper equipment disassembly procedures
5. Improper selection of cleaning compounds and sanitizers
6. Lack of sanitizer concentration checks
7. Ineffective preoperative inspection procedures
8. Ineffective microbial monitoring

All of these problems can be remedied through a well-designed company sanitation program that includes top management commitment, effective training programs, excellent supervision, thorough inspection of facilities and equipment, and a detailed microbial monitoring program.

In summary, there are six areas (Anon 2004a) that are critical to effective sanitation:

1. Employee training. Continuous training should focus on sanitation fundamentals and the role of employees in maintaining the safety and hygiene of foods. Chemical and equipment suppliers frequently provide training programs on the use of the products that they sell and can also suggest information about other training programs.

2. Personal hygiene. Employees cannot create hygienic conditions unless they exhibit appropriate personal hygiene and have the facilities to carry it out. A detailed discussion about personal hygiene is included in Chap. 6.

3. Sanitation product labeling. Employee headgear, cleaners, sanitizers, and equipment (such as brushes, etc.) should be color coded to reduce their misuse. Containers of mixed product used throughout the facility should identify the product by name with any hazardous warning, so that it can be traced back to the Material Safety Data Sheets (MSDS). Furthermore, it is prudent to include other information such as the name, address, telephone number, and website of the manufacturer to provide quick access to desired information.

4. Personal protective equipment (PPE). Without appropriate training and supervision, employees may take unnecessary risks based on their habits or insufficient information about the potential hazards associated with the improper use of cleaning compounds, sanitizers, and equipment. Employees should have proper training in the importance of PPE and be required to use this equipment during their work shift.

5. Chemical selection. Information from chemical suppliers, as well as the investigation of other available data, should be utilized to ensure that appropriate cleaning compounds and sanitizers are used. A detailed discussion about cleaning compounds and sanitizers is included in Chaps. 9 and 10, respectively.

6. The use of chemical dispensers. This equipment requires consistent use and dependable performance to ensure sanitation effectiveness, worker safety, and economical use of cleaning compounds and sanitizers. The correct concentration of cleaners and sanitizers is essential to the sanitization process.

Management and Supervision

The key to success of any sanitation program is not only the people who perform the tasks but those who manage and supervise the function. The role of management in supervision involves the audit of the sanitation program to ensure that the work is being done regularly and routinely by conscientious workers. Program requirements may be considered the cement that holds the building blocks of sanitation together. Supervisors should always be on the alert to identify "shortcuts" in procedures that lead to unsafe practices in an operation. Thorough supervision should be reinforced, with a continuous training program to keep employees informed of their responsibilities.

A major challenge of the supervisor is to set a good example or "model the way" for other employees. A supervisor who does not follow the rules and doesn't do what is asked of others does not "walk the talk" and will not be effective. These managers clearly don't practice what they preach and because of this, don't have the trust and support of their workers. The supervisor is usually the most experienced employee in the sanitation operation and must lead by example and set the tone for the department. They must be able to quickly identify and correct situations that could lead to unclean equipment and product contamination. Supervisors should clearly recognize that their overarching goal is to provide their customers with a safe and wholesome product.

Monitoring a food production facility involves an organized supervision routine. Supervising food workers should involve the same health standards that are used in screening prospective employees, including daily checks of employees for infections and illnesses that can be transmitted through food. In fact, many local health ordinances require that the facility manager who knows or suspects that an employee has a contagious disease or is a carrier must notify health authorities immediately.

When employees are motivated to do a good job, the burden of managing the department is reduced, and supervision is made easier. Effective training programs and the professional treatment of workers can be a positive motivating force that will improve morale. As mentioned previously, the efforts of workers who do an excellent job should be recognized, not ignored. Instead of ignoring their efforts and criticizing failures, they should be commended for maintaining a hygienic environment and for their contribution to safe food products. This approach provides positive reinforcement and motivates employees to perform to a higher level. Management should convey to the sanitation team that its work is highly valued and critical to food safety.

Total Systems Approach

A company sanitation program should involve an all-encompassing, holistic, total systems approach, where company leaders take a very broad view of the role of the program and its impact on departments throughout the organization and stakeholders external to it. Proactive planning is essential, and this is accomplished by keeping abreast of new technologies in the field, including new regulations, scientific updates in microbiology, biofilms, sanitary design and environmental testing, equipment, cleaning and sanitizing compounds, automation and computer technologies, etc.

Effective Communication

Sanitation managers should have a good understanding of communication and public relation skills and apply them in their daily activities. It is important to be an effective communicator and be able to share thoughts, concepts, and ideas with workers and supervisors, as well as with colleagues in other departments in a clear, concise, and understandable manner, both orally and in writing. They must interpret the program's needs and objectives and motivate people to cooperate and give their best efforts at work. Whenever a food sanitation manager makes recommendations to company executives, the operating costs of the department are frequently increased. Providing a good rationale that includes the scientific reasons for the new piece of equipment or chemical and then effectively communicating the need for and benefits of such recommendations are essential.

It is vital that the sanitation manager has good working relationships and communication with the managers of other departments (including the communication or public relations department) and executives within the company. It is highly desirable to share information about improvements, achievements, new programs, appointments, promotions, and similar developments between and among all interested parties. These practices promote openness, better understanding, and an appreciation of the program by everyone within the company and help to promote a culture of food sanitation and food safety in the organization.

The duties and responsibilities of sanitation managers are much broader than just making inspections of facilities and equipment. These professionals should have a practical understanding and working knowledge of the fundamentals of human motivation.

Cooperation with Other Agencies

Joint regulatory/industry advisory committees have frequently provided valuable assistance in the evaluation of new developments, techniques, and procedures. Consumers can also be represented on these advisory committees, which may be helpful in gaining insights and advice on broad policy matters and in establishing and maintaining strong industry and regulatory agency relationships. Through cooperative efforts like these, benefits accrue to everyone.

Job Incentive (Enrichment) Program

Many employees, including managers and supervisors, consider the sanitation operation to be a second-rate job. Yet, sanitation workers should be made aware of the importance and vital nature of their responsibilities. Sanitation can be highlighted and made more exciting. An effective job incentive or enrichment program can create more interesting and rewarding work for employees. This program can also make them feel more a part of the operation and can actually be more demanding of employees by assigning them more responsibilities and emphasizing self-inspection.

Self-inspection

Self-inspection should be considered a regular task performed by trained personnel who are familiar with the establishment's operation. Inspections should be conducted through the owner/operator or managers, supervisors, or sanitation consultants. These inspections are more beneficial if they are conducted with the aid of a detailed checklist.

Contract Sanitation

Many food processors hire an outside firm to accept the responsibility of cleaning the plant. This arrangement, called contract sanitation, offers the processor benefits of a firm that specializes in cleaning, reduces the responsibility of the plant management team, and provides a more consistent and predictable cleaning budget (Dawson 2010). There are advantages and disadvantages to performing sanitation both in-house or through contract sanitation, so company management needs to make that decision based on a variety of factors. In-house sanitation can save the processor cleaning costs and provides more flexibility, because the contract sanitation team is present during only one shift. In-plant personnel offer additional flexibility through the use of employees where they are needed such as production, maintenance, or cleaning. In-plant sanitation offers the processor more control since the training of employees can be controlled and protects the processor against contract firms that may lack effectiveness.

Study Questions

1. What is management?
2. What health requirements should be considered when selecting employees?
3. What sources exist for sanitation training and education?
4. What are three basic responsibilities of sanitation management?

5. What is the major key to success in a sanitation program?
6. What are some important steps in developing a company sanitation program?
7. What is contract sanitation?
8. What are the advantages of contract sanitation?
9. What are the advantages of in-house cleaning?
10. What is motivation and what factors affect it?

References

Anon (2004a). 6 common food sanitation mistakes-and how to fix them. *Food Saf Mag* 10(1): 40.

Anon (2004b). Top 10 sanitation problems. *Meat Poul* 49(5): 58.

Anon (2004c). Top reasons sanitarians fear sanitation software-and why these fears are unfounded. *Food Saf Mag* 10(1): 48.

Centers for Disease Control and Disease (2009). Multistate outbreak of *Salmonella typhimurium* infections linked to peanut butter, 2008–2009. (Final Update). https://www.cdc.gov/salmonella/2009/peanut-butter-2008-2009.html

Centers for Disease Control and Disease (2012). Multistate outbreak of listeriosis linked to who cantaloupes from jensen farms, Colorado. (Final Update). https://www.cdc.gov/listeria/outbreaks/cantaloupes-jensen-farms/

Cramer MM (2007). Sanitation best practices. *Food Saf Mag* 13(1): 20. http://www.foodsafetymagazine.com/magazine-archive1/februarymarch-2007/sanitation-best-practices/

Cramer MM (2008). Food plant sanitation: Have you found your niche? *Food Saf Mag* 14(5): 12. http://www.foodsafetymagazine.com/magazine-archive1/octobernovember-2008/food-plant-sanitation-have-you-found-your-niche/

Daniel-Sewell S (2017). Sanitation scheduling. In Nextor Technologies white paper. http://www.nexcortech.com/Sanitation%20Scheduling%20Article.pdf

Dawson MG (2010). Outsourcing sanitation a smart alternative. *Food Quality and Safety* August. http://www.foodqualityandsafety.com/article/outsourcing-sanitation-a-smart-alternative/?singlepage=1

Flynn D (2016). Parnell brothers finally in prison for deadly peanut butter outbreak. *Food Safety News* February 17. http://www.foodsafetynews.com/2016/02/123674/#.WSL9R4eGPIU

Food Safety News (2015). Settlement reached for 66 victims of 2011 cantaloupe listeria outbreak. *Food Safety News* February 11. http://www.foodsafetynews.com/2015/02/settlement-reached-in-2011-cantaloupe-linked-listeria-outbreak/#.WSL-j4eGPIU

Food Safety Preventive Controls Alliance (FSPCA) (2015). FSPCA preventive controls for human food training curriculum. 1st ed.

Fournies FF (2007). *Why employees don't do what they're supposed to do and what to do about it.* 2nd ed. McGraw Hill Publishing: New York, NY.

Keener K (2009). Cleaning and sanitizing operations. In *Microbiologically safe foods*, eds. Heredia N, Wesley I, Garcia S. John Wiley and Sons: Hoboken, NJ. p. 415.

Marriott NG, Gravani RB (2006). *Principles of food sanitation.* 5th ed. Springer Scientific: New York, NY.

Neal J, Crandall PG, Dawson M, Madera J (2015). Food safety and language barriers on the food processing line. *Food Saf Mag* 21(1): 22. http://www.foodsafetymagazine.com/magazine-archive1/februarymarch-2015/food-safety-and-language-barriers-on-the-food-processing-line/

Rath T, Clifton DO (2009). *How full is your bucket?* Gallup Press: New York, NY.

Sertkaya A, Berlind A, Lange R, Zink DL (2006). Top ten food safety problems in the United States food processing industry. *Food Prot Trends* 26(5): 310.

Stolovitch HD, Keeps EJ, Rosenberg MJ (2011). *Telling ain't training.* 2nd ed. Association for Talent Development: Alexandria, VA.

U.S. Food and Drug Administration (2017a). Federal Food, Drug and Cosmetic Act. Code of Federal Regulations. Title 21. Chapter 9, Section 402 (342) a Definition of Adulterated Food. https://www.law.cornell.edu/uscode/text/21/342

U.S. Food and Drug Administration (2017b). Current good manufacturing practice in the manufacturing, packing or holding human food. Title 21, Code of Federal Regulations, Part 117.10 (21 CFR 117). Personnel. https://www.gpo.gov/fdsys/pkg/CFR-2016-title21-vol2/pdf/CFR-2016-title21-svol2-part117.pdf

Index

A

Acarus siro, 250
Acetobacter lovaniensis, 374
Acid anionic sanitizers, 189
Acid cleaner hazards
  acetic acid, 170
  citric acid, 170
  hydrochloric acid (muriatic acid), 170
  hydrofluoric acid, 171
  phosphoric acid, 170
  sodium acid sulfate and phosphate, 170
  sulfamic acid, 170
Acid sanitizers, 187–188, 360–361
Acidified sodium chlorite (ASC), 183, 324,
      325, 347
Acid-quat sanitizers, 189–190
Activated carbon adsorption, 239
Activated lactoferrin (ALF), 195
Activated sludge system, 236–237
Adenosine triphosphate (ATP), 135
Adulteration, 412
Advanced wastewater treatment, 239
Agricultural production, 2
Agroterrorism, 20
Airborne contamination, 273
Air contamination, 353
Air filtration system, 273
Air-handling system, 273
Allergen-containing product contamination, 75
Allergen control plan (ACP)
  cleaning, 76
  color-coding, 77
  employees education, 76
  identification, 75
  label review, 75, 77
  layout, 77
  product changeover, 77
  program effectiveness, 78
  sanitation, 75
  segregation, 77
  storage, 77
  supplier monitoring, 76
  systematic method, 76
Allergenic reactions, 73–75

Allergen(s)
  airborne, 74
  allergen-related recalls, 74
  antigen and antibody, 75
  avoidance, 73
  contamination, 73
  food industry, 74
  foodservice, 74
  hostile invader, 74
  human, 74
  immune system, 74
  labeling, 80
  management, 80–81
  mast cells, 75
  prevalence, 74
  product contamination, 75
  symptoms, 74
  undeclared, 73
American Cockroach (*Periplaneta americana*), 245
American Meat Institute Foundation (AMIF), 274
American National Standards Institute
      (ANSI), 130
Amorphous phosphates, 165
Anaerobic lagoons, 234–235
Angoumois grain moth, 286
Anionic wetting agents, 162
Anthium dioxcide, 183
Antimicrobial agents, 176, 241
Aquaculture, *see* Seafood
Aquaculture production operation, 238
Arcobacter butzleri, 54
Aseptic canning, 358
Association of Official Analytical Chemists
      (AOAC), 226
Attribute control charts, 144–146
Auditing, 122
  assessments, 136
  class I, 137
  class II, 137
  class III, 137
  preparation, 137
  sampling, 138, 139
  SQF certification, 136
Azacosterol, 261

© Springer International Publishing AG, part of Springer Nature 2018
N.G. Marriott et al., *Principles of Food Sanitation*, Food Science Text Series,
https://doi.org/10.1007/978-3-319-67166-6

Printed in the United States
By Bookmasters